# Modern Pavement Management

# Modern Pavement Management

RALPH HAAS
The Norman W. McLeod Engineering Professor
Department of Civil Engineering
University of Waterloo
Waterloo, Ontario

W. RONALD HUDSON
The Dewitt C. Greer Centennial Professor
Department of Civil Engineering
The University of Texas at Austin

JOHN ZANIEWSKI
Associate Professor
Department of Civil Engineering
Arizona State University
Tempe, Arizona

KRIEGER PUBLISHING COMPANY
MALABAR, FLORIDA
1994

Original Edition 1994
(Based on Pavement Management Systems)

Printed and Published by
**KRIEGER PUBLISHING COMPANY**
**KRIEGER DRIVE**
**MALABAR, FLORIDA 32950**

Copyright © 1994 by Krieger Publishing Company

All rights reserved. No part of this book may be reproduced in any form or by any means, electronic or mechanical, including information storage and retrieval systems without permission in writing from the publisher.
*No liability is assumed with respect to the use of the information contained herein.*
Printed in the United States of America.

> FROM A DECLARATION OF PRINCIPLES JOINTLY ADOPTED BY A COMMITTEE OF THE AMERICAN BAR ASSOCIATION AND A COMMITTEE OF PUBLISHERS:
>
> This publication is designed to provide accurate and authoritative information in regard to the subject matter covered. It is sold with the understanding that the publisher is not engaged in rendering legal, accounting, or other professional service. If legal advice or other expert assistance is required, the services of a competent professional person should be sought.

**Library of Congress Cataloging-In-Publication Data**

Haas, Ralph, 1933–
  Modern pavement management / Ralph Haas, W. Ronald Hudson, John Zaniewski.
    p.  cm.
  Includes bibliographical references and index.
  ISBN 0-89464-588-9
  1. Pavements—Design and construction—Management.  2. Pavements—Maintenance and repair—Management.   I. Hudson, W. Ronald.
II. Zaniewski, John P.   III. Title.
TE251.H27   1994
625.8—dc20                                                        92-30178
                                                                     CIP

10  9  8  7  6  5  4  3  2

# Contents

Preface    xvii

**PART ONE    THE PAVEMENT MANAGEMENT PROCESS**

**Chapter 1    Introduction to Pavement Management    3**

    1.1 Focus of the Book    3
    1.2 Historical Background    5
    1.3 Role of Pavements in Today's Transport System    7
    1.4 Types of Pavements    8
    1.5 Concept of Pavement Management    10
    1.6 Essential Features of Pavement Management    13
    Review Questions    15

**Chapter 2    General Nature and Applicability of Systems Methodology    16**

    2.1 Nature of the Systems Method    16
    2.2 Structure of the Systems Method    17
    2.3 Some Basic Terminology    17

2.4 Some Precautions in Application   19
2.5 Some Analytical Tools for Systems Engineering   20
Review Question   22

**Chapter 3   Application of Systems Concepts to Pavement Management   23**

3.1 Introduction   23
3.2 The General Structure of Systematic Pavement Management   24
3.3 The Process of Pavement Management: Levels, Decisions, and Issues   27
3.4 The PMS Does Not Make Decisions—The People Who Use It Do   28
Review Questions   30

**Chapter 4   Pavement Management Levels and Functions   31**

4.1 The Ideal Pavement Management System (PMS)   31
4.2 The Network and Project Levels of Pavement Management   32
4.3 Influence Levels of PMS Components   32
4.4 Pavement Management at Three Levels   34
4.5 PMS Function   37
4.6 Key Considerations in Application of a Total Pavement Management System Concept   43
4.7 The Function of Pavement Evaluation   44
4.8 Summary   48
Review Questions   48

**Chapter 5   Using PMS as a Research Planning and Technology Improvement Tool   50**

5.1 Identifying Research Needs   50
5.2 System Parameters and the State of the Art   51
5.3 Future Advances in Pavement Management   52
5.4 Toward Realizing Future Prospects for Pavement Management   54
5.5 Establishing Priorities   54
5.6 Implementing Research Results   55
Review Questions   55

**References to Part One   56**

Contents vii

## PART TWO  DATA REQUIREMENTS

### Chapter 6  Review of Pavement Management Data Needs  61

6.1 Classes of Data Required  61
6.2 The Importance of Construction and Maintenance History Data  63
6.3 The Importance of Performance Related Pavement Evaluation  66
6.4 Objectivity and Consistency in Pavement Evaluation  67
6.5 Combining Pavement Evaluation Measures  68
Review Questions  68

### Chapter 7  Inventory Data Needs  69

7.1 Inventory Data Needs  69
7.2 Types of Inventory Data  70
7.3 Collecting and Processing Inventory Data  75
Review Questions  75

### Chapter 8  Pavement Performance  76

8.1 The Serviceability-Performance Concept  76
8.2 Characterization of Pavement Roughness  79
8.3 Equipment for Evaluating Roughness  82
8.4 A Universal Roughness Standard  91
8.5 Relating Roughness to Serviceability  101
8.6 Applications of Roughness Data  105
Review Questions  107

### Chapter 9  Evaluation of Pavement Structural Capacity  108

9.1 Basic Considerations  108
9.2 Nondestructive Measurement and Analysis  110
9.3 Deflection Devices  112
9.4 Destructive Structural Evaluation  124
9.5 Structural Capacity Index Concepts  125
9.6 Network Versus Project Level Applications of Structural Capacity Evaluation  129
Review Questions  130

### Chapter 10  Evaluation of Pavement Distress: Condition Surveys  131

10.1 Principles of Surface Distress Surveys  131
10.2 Survey Methodology  133
10.3 Types of Distress  134
10.4 Examples of Distress Survey Procedures  134

10.5 Equipment for Distress Evaluation   135
10.6 Pavement Distress Indexes   144
10.7 Applications of Distress Data   150
Review Questions   153

## Chapter 11  Evaluation of Pavement Safety   154

11.1 Major Safety Components   154
11.2 Skid-Resistance Evaluation   155
11.3 Basic Concepts of Skid Resistance and Uses of Measured Values   155
11.4 Methods of Measuring Skid Resistance   156
11.5 Change of Skid Resistance with Time, Traffic, and Climate   158
Review Questions   160

## Chapter 12  Combined Measures of Pavement Quality   161

12.1 Concept of Combined Measures of Pavement Quality   161
12.2 Reasons for Uses of a Combined Index   161
12.3 Methods of Developing a Combined Index   162
12.4 Precautions to the Use of Combined Indexes   165
Review Question   165

## Chapter 13  Data Base Management   166

13.1 Data Base Considerations   166
13.2 Types of Data Bases   168
13.3 Data Base Issues   168
13.4 Computer Issues   169
13.5 Examples of Data Base Management   171
Review Questions   172

## Chapter 14  Describing the Present Status of Pavement Networks   173

14.1 Uses of Present Status Information   173
14.2 Examples of Pavement Current Status and Trend Reports   176
Review Question   176

**References to Part Two   177**

Contents ix

## PART THREE   DETERMINING PRESENT AND FUTURE NEEDS, AND PRIORITY PROGRAMMING OF REHABILITATION AND MAINTENANCE

### Chapter 15   Establishing Criteria   187

15.1 Reasons for Establishing Criteria   187
15.2 Measures or Characteristics to Which Criteria Can Be Applied   188
15.3 Factors Affecting Limits, and Some Examples   189
15.4 Effects of Changing Criteria   189
Review Questions   191

### Chapter 16   Prediction Models for Pavement Deterioration   192

16.1 Clarification of Performance and Deterioration Prediction   192
16.2 Parameters or Measures to Be Predicted, and the Requirements   192
16.3 Basic Types of Prediction Models and Examples   193
Review Questions   201

### Chapter 17   Determining Needs   202

17.1 Needs Years and Action Years   202
17.2 Effect of Prediction Model Errors   203
17.3 A Need Versus Type of Action That Should Be Taken   203
17.4 Graphical and Tabular Portrayal of Needs   204
Review Questions   205

### Chapter 18   Rehabilitation and Maintenance Strategies   206

18.1 Identification of Alternatives   206
18.2 Decision Processes and Expert Systems Approaches to Identifying Feasible Alternatives   207
18.3 Deterioration Modeling of Rehabilitation and Maintenance Alternatives   212
18.4 Costs, Benefits, and Cost-Effectiveness Calculations   214
Review Questions   223

### Chapter 19   Priority Programming of Rehabilitation and Maintenance   224

19.1 Basic Approaches to Establishing Alternatives and Policies   224
19.2 Selecting A Length of Program Period   226
19.3 Basic Functions of Priority Programming   226
19.4 Priority Programming Methods   228
19.5 Mathematical Programming (Optimization Method)   229
19.6 Examples and Comparisons   231

19.7 Budget Level Evaluation    236
19.8 Funding Level Requirements for Specified Standards    236
19.9 Final Program Selection    238
Review Questions    239

**Chapter 20  Developing Combined Programs of Maintenance and Rehabilitation    240**

20.1 Desirability of Combined Programs and Barriers    240
20.2 Pre-Rehabilitation Maintenance Strategies    241
20.3 Post-Rehabilitation Maintenance Strategies    241
20.4 Example Result of a Combined Program    241
Review Questions    245

**References to Part Three    248**

**PART FOUR  PROJECT LEVEL DESIGN: STRUCTURAL AND ECONOMIC ANALYSIS**

**Chapter 21  A Framework for Pavement Design    255**

21.1 Introduction    255
21.2 Evolution of Pavement Design Technology    255
21.3 Pavement Design Framework and Components    258
21.4 Design Objectives and Constraints    264
Review Questions    266

**Chapter 22  Characterization of Physical Design Inputs    268**

22.1 Introduction    268
22.2 Material Properties    268
22.3 Traffic Characteristics    275
22.4 Environmental Effects    278
22.5 Interaction Effects    279
Review Questions    279

**Chapter 23  Basic Structural Response Models    280**

23.1 Introduction    280
23.2 Elastic Layer Theory    281
23.3 Viscoelastic Layer Analysis    283
23.4 Thin Plate Theory    284
23.5 Numberical Methods    285
23.6 Environmental Models    286
23.7 Cumulative Damage Theory    292
Review Questions    293

Contents                                                                  xi

**Chapter 24  Variability, Reliability, and Risk in Pavement Management  294**

    24.1 Definitions and Basic Concepts   294
    24.2 Formulation of Pavement Reliability   296
    24.3 Influence of Variability on Pavement Management   298
    Review Questions   299

**Chapter 25  Generating Alternative Design Strategies   300**

    25.1 Introduction   300
    25.2 New Pavement Alternatives   302
    25.3 Rehabilitation and Reconstruction Alternatives   310
    Review Questions   312

**Chapter 26  Structural Analysis and Design of Asphalt Concrete Pavements   313**

    26.1 Introduction   313
    26.2 Predicting Fatigue Cracking   314
    26.3 Predicting Permanent Deformation   318
    26.4 Low-Temperature Shrinkage Cracking   320
    26.5 Performance Prediction   321
    26.6 Use of Structural Models for Project Level Pavement Management of Asphalt Pavements   327
    Review Questions   327

**Chapter 27  Structural Analysis and Design of Portland Cement Concrete Pavements   328**

    27.1 Introduction   328
    27.2 Structural Design of the Pavement Slab   329
    27.3 Joint Design   333
    27.4 Reinforcement Design   334
    27.5 Other Concrete Pavement Design Considerations   335
    Review Question   335

**Chapter 28  Rehabilitation Design Procedures   336**

    28.1 Types of Rehabilitation   336
    28.2 Design Inputs and Field Data Required   338
    28.3 Performance or Deterioration Models   342
    28.4 Summary of Rehabilitation Design   344
    Review Questions   344

## Chapter 29 Economic Evaluation of Alternative Pavement Design Strategies  345

29.1 Introduction  345
29.2 Basic Principles  345
29.3 Pavement Cost and Benefit Factors  346
29.4 Methods of Economic Evaluation  352
29.5 Economic Analysis Example  360
29.6 Limitations of Economic Analysis  363
Review Questions  364

## Chapter 30 Selection of an Optimal Design Strategy  368

30.1 Role of the Decision Maker  368
30.2 Basis for Optimal Strategy Selection  369
30.3 Communicating Results  371
Review Questions  372

**References to Part Four  373**

## PART FIVE  IMPLEMENTATION

## Chapter 31 Implementing a Pavement Management System  385

31.1 Introduction  385
31.2 Major Steps in Implementing a Pavement Management System  386
Review Questions  394

## Chapter 32 Construction  395

32.1 Introduction  395
32.2 Construction as Related to Other Phases of Pavement Management  397
32.3 Pavement Construction Management  400
32.4 Pavement Construction and the Environment  403
32.5 Construction Control  404
32.6 Documentation of Construction Data  410
32.7 Summary  415
Review Questions  415

## Chapter 33 Maintenance  416

33.1 Introduction  416
33.2 Maintenance Management Systems  417
33.3 Maintenance Policies  421
33.4 Effects of Policy Variations  424

33.5 Costs, Economics, and Decision Criteria   425
33.6 Maintenance Information Needs   427
33.7 Summary   431
Review Questions   434

### Chapter 34   Research Management   435

34.1 Introduction   435
34.2 Levels of Research Management   436
34.3 Major Elements in Generating and Carrying Out a Research Program   436
34.4 Elements of Good Research Management   441
Review Questions   442

**References to Part Five   443**

## PART SIX   EXAMPLES OF WORKING SYSTEMS

### Chapter 35   Basic Features of Working Systems   447

35.1 Introduction   447
35.2 Structure of Available Project Level Systems   448
35.3 Network Level Systems Examples   464
Review Questions   466

### Chapter 36   Network Level Examples of PMS   467

36.1 Minnesota DOT Network Level PMS   467
36.2 Metropolitan Transportation Commission PMS   472
36.3 Conclusion   479
Review Questions   479

### Chapter 37   The SAMP System   480

37.1 Introduction   480
37.2 Input for SAMP-5   481
37.3 SAMP Program Operation   483
37.4 Performance Model   488
37.5 Interpretation and Use of Results   490
37.6 Improvement of SAMP System   490
37.7 Implementation Study   492
Review Questions   494

### Chapter 38   The Highway Design and Maintenance Standards Model   495

38.1 Background   495
38.2 Description of the Model   495

38.3 Conclusion  502
Review Questions  503

**Chapter 39 Municipal Pavement Design System  504**

39.1 Introduction  504
39.2 Characteristics of MPDS-1  505
39.3 MPDS-1 Example  505
39.4 Summary  509
Review Questions  509

**References to Part Six  510**

## PART SEVEN  LOOKING AHEAD

**Chapter 40  Analyzing Special Problems  515**

40.1 Introduction  515
40.2 Historical Problems  515
40.3 Energy Issues  516
40.4 Alternate Sources of Materials  517
40.5 New Types of Pavements and New Materials  517
40.6 Changes in Load Conditions  518
40.7 Summary  520
Review Questions  520

**Chapter 41  Applications of Expert Systems Technology  521**

41.1 Introduction  521
41.2 Usefulness of Expert Systems Technology  521
41.3 Expert Systems Versus Conventional Softwares  522
41.4 Expert Systems Architecture  523
41.5 Expert System Development Tools  524
41.6 Examples of Existing Expert Systems for Pavement Engineering  525
41.7 Expert Systems Technology Benefits and Limitations  529
Review Questions  530

**Chapter 42  New and Emerging Technologies  531**

42.1 Introduction  531
42.2 Geographic Information Systems  531
42.3 New Software, Hardware, Data Bases, and Personal Computers  533
42.4 New Measurement Technology  533
42.5 Interface with Other Systems  535

42.6 Summary   535
Review Questions   536

**Chapter 43  Institutional Issues and Barriers Related to Pavement Management Implementation   537**

43.1 Introduction   537
43.2 Skepticism of Pavement Management   538
43.3 Managerial and Organizational Issues   539
43.4 Legal and Regulatory Requirements   545
43.5 Agencywide PMS   546
Review Questions   547

**Chapter 44  Cost and Benefits of Pavement Management   548**

44.1 Introduction   548
44.2 Pavement Management Costs   548
44.3 Benefits and Costs Associated with A Pavement Management System   549
44.4 Evaluation Methodologies   552
44.5 Concluding Remarks   556
Review Questions   556

**Chapter 45  Future Directions and Need for Innovation in Pavement Management   557**

45.1 Introduction   557
45.2 Changing Nature of Pavement Research and the Issues and Needs   560
45.3 Standardized (Generic) Structure for Pavement Management   561
45.4 Major Types of Research and Benefits of a Coordinated Plan   562
45.5 Elements of Successful Research   564
45.6 Opportunities for Innovation   566
45.7 The Future of Pavement Management   571
45.8 Closure   573
Review Questions   573

**References to Part Seven   574**

**Index   579**

# Preface

This book is a successor to the *Pavement Management Systems* text originally published in 1978 by McGraw-Hill, Inc. and then reissued in 1982 by Krieger Publishing Company. While it continues the concept of the original book, and uses some of the material, it does represent a full rewrite.

Both books have been based on the premise that the provision of pavements for highways, airports, and other uses involves a large variety of interrelated activities, and when these activities are carried out in a coordinated and integrated way they can constitute a *pavement management system*. When the first book was published, a subset of these activities, concerned with design, was occupying much of the attention of practitioners and researchers. Consequently, a major effort was made to focus on the broader concept of pavement management. During the ensuing years and leading up to the publication of this book, a vast expansion of pavement management research and implementation has occurred to the point where numerous federal, state, and local agencies around the world now have working systems in place.

This book is written to provide a basic understanding of the principles of evaluating, planning and programming, designing, constructing, maintaining, and carrying out research on pavements. It also addresses the implementation of pavement management systems. Although the book is directed mainly to senior and graduate students, practicing engineers and researchers, others, such as those in an administrative function or those desiring some general knowledge of the pavement field, should find many useful sections.

The book attempts to organize existing pavement management information in an efficient, understandable format. In doing so, it makes use of much of the material that we have presented to various classes, seminars, and other forums in recent years. It also relies heavily on the research efforts and practices of a large number of individuals and agencies. The availability of this other material, both from those acknowledged in the references and from those unintentionally overlooked, has been most valuable to the preparation of the book. It is impossible to consider the various subject areas of the book in depth, because there are literally hundreds of documents and papers forming the basis for each topic. We have tried to incorporate a representative set of these references, knowing that many other good ones will unfortunately have been omitted.

There are some differences in both opinion and results between the work of various people in the pavement field. We have attempted to recognize these in the book, but by doing so, we may have unintentionally misinterpreted some people, for which we ask indulgence. The fact that a variety of contradictions or differences exist in various areas of the pavement field should not be considered unusual. Wherever the state of knowledge is imperfect, as in nearly all areas of human endeavor, such differences of opinion will be found.

Any book such as this can really present only the existing state of knowledge to a given time. Because the state of technology in the pavement field continues to advance, some of the material in the book concerning current practices will undoubtedly be superseded by better practices in a very short time. We have attempted to recognize this inevitability by providing a basic framework within which such new knowledge can be efficiently structured as it becomes available. Also, we have attempted to foresee a few future developments that we expect to occur within the next decade or two.

There are a number of things that the book does not attempt to do. It does not attempt to be all things to all people. For example, it is not a book on pavement materials per se. Materials types and materials characterization are discussed in the book, but not in the depth required for a full understanding of this very important subject area. It is not a book on the methodology of constructing pavements. Construction practices and control and their effects on the behavior and performance of pavements are considered in the book, but, again, not in the depth required for a full understanding of this subject. There are a number of other such subject areas in the book that certain readers might wish to have treated in more depth. We have made arbitrary judgments on the extent of material to be included in each subject area. This has been necessary to provide adequate space for treatment of the total system as the central theme.

The book is divided into seven major parts, starting with a relatively extensive introduction to the pavement management process. We considered it important to provide general background, definitions, and a brief description of systems principles to structure this process. By doing so, we hope that the student, the researcher, or the practitioner can understand our basic philosophy and approach in the subsequent parts of the book. Also, those people desiring only a general appreciation of the pavement field may satisfy some of their requirements by reading Part One.

Part Two, Data Requirements, provides a major component of the "foundation" for pavement management, because good data is essential to the success of any pavement management system. Included are discussions of why such data is needed, as well as the actual methodology for acquiring and evaluating or interpreting the various types of data relating to performance, structural capacity, surface distress, and safety. The concept of pavement performance measurement and evaluation is introduced in this part, largely because it is important to fully understand the purposes for which a pavement is provided before carrying out programming, design, construction, or maintenance activities.

Part Three, Determining Present and Future Needs, and Priority Programming of Rehabilitation and Maintenance, builds on the data base concepts and methodology described in Part Two. It emphasizes that criteria must be established, and deterioration or performance prediction models must be available in order to identify current and future needs. As well, it considers how maintenance and rehabilitation strategies can be analyzed to develop network level priority programs under budget restraint conditions.

Part Four, on project level design, comprises a major portion of the book for a variety of reasons. These include the fact that much of the technology relating to pavements is concentrated in design. It is in this part that we first provide a general framework for pavement design and then attempt to combine an in-depth consideration of the basic response of pavements to load, climate, and other input variables with their use in actual, practical design applications. We have not attempted to summarize the numerous existing methods of design practiced by a variety of agencies in North America, Europe, and elsewhere. Rather, we have attempted only to choose a few representative examples. More extensive and detailed examples of working systems are contained in Part Six.

Part Five deals with implementation, including the implementation of a pavement management system. It then considers the implementation of construction and maintenance, mainly as related to their effects on the behavior and performance of pavements. In other words, it is the end result of implementation and the information required to assess this, rather than the techniques or practice of construction and maintenance per se, on which we have concentrated. This is not intended to minimize the extreme importance of these areas; rather, we consider that they are both large enough within themselves for a separate book or books and subject to wide variation between regions and contractors.

In Part Six, several examples of working pavement management systems, both at the network and project levels, and involving states and local agencies, are presented. These examples illustrate that systems engineering principles and the comprehensive, efficient use of up-to-date technology are well represented in the pavement field.

Part Seven concludes the book by looking ahead at some special problems, some new and emerging technologies, and how institutional barriers to more effective pavement management implementation may be overcome in the future. It also includes a chapter on the costs and benefits of pavement management because this is an issue which will have to be fully addressed in the future. Finally, it considers future directions and the need for innovation in pavement management.

For those wishing to use this book for senior undergraduate instruction, Parts One through Four might comprise a reasonable one-semester load, whereas the addition of the remaining parts might comprise a reasonable one-semester load for graduate students.

Because the intent of the book is largely to present pavement management concepts and principles, some chapters do not lend themselves to the usual short, specific questions found in many textbooks. Moreover, where specific technology is presented, it should be recognized that this can and probably will be very quickly dated. There is a danger of losing sight of the concept and principles by concentrating on the importance of specific methods, procedures, equipment, computer programs, etc. Consequently, we have included review questions with most of the chapters. Also, we have developed an Instructor's Manual that should be useful to those wishing to develop more extensive problem sets and exercises.

We have a mixture of SI and Imperial units in the book, although the latter dominate. It would of course have been ideal to have every graph, table, equation, etc. in both units. But for reasons of practicality, the fact that much of the research and development work on which the book is based was done in Imperial units and a variety of difficulties associated with conversions, we felt that it was appropriate to use these units. Hopefully, for the next edition, the United States will have converted to SI units, as for most of the rest of the world, and it will cease to be an issue.

Finally, we would like to acknowledge with considerable gratitude the contributions made by many colleagues in governmental and in private agencies, and in our own and other universities. Their efforts, and our association with them, have provided an invaluable source of information and knowledge for this book. As well, the original authors (Haas and Hudson) take pleasure in welcoming Dr. John Zaniewski to the "team" for this book.

Ralph Haas<br>
W. Ronald Hudson<br>
John Zaniewski

Part One

# The Pavement Management Process

Chapter 1

# Introduction To Pavement Management

## 1.1 FOCUS OF THE BOOK

This book concentrates on pavement management as a necessary part of the process of providing adequate pavements. It attempts to carefully show that pavement management in no way replaces or interferes with good pavement design, materials, maintenance, rehabilitation, or other activities. Rather, pavement management provides a methodology for synthesizing these activities to maximize pavement life and benefits.

The book starts with a historical background and perspective. It provides a framework for pavement management and concentrates on developments in the last decade. Major sections related to these developments include:

- A clearer definition of project versus network level pavement management.
- A review of significant progress in implementation of pavement management at the small to medium city and county level, but not a city-by-city report.
- A review of state level pavement management system (PMS) implementation in the United States, Canada, and other countries but not on an individual basis.
- An understanding of PMS as a sound basis for planning and coordinating

pavement research, including a foundation for long-term pavement performance monitoring.
- Projections and predictions of the expected future of pavement management worldwide; ideas are presented on future directions in pavement management and the benefits of more complete implementation.

### 1.1.1 The Intent of Good Pavement Management

Good pavement management is not business as usual. It requires an organized and systematic approach to the way we think and in the way we do day-to-day business. Pavement management, in its broadest sense, includes all the activities involved in the planning and programming, design, construction, maintenance, and rehabilitation of the pavement portion of a public works program. A pavement management system (PMS) is a set of tools or methods that assist decision makers in finding optimum strategies for providing and maintaining pavements in a serviceable condition over a given period of time. The function of a PMS is to improve the efficiency of decision making, expand the scope, provide feedback on the consequences of decisions, facilitate the coordination of activities within the agency, and ensure the consistency of decisions made at different management levels within the same organization.

The details of a PMS depend on the organization of the particular agency within which it is implemented. Nevertheless, an overall, generally applicable framework can be established without regard to such particular departmental organization. This book describes a rather complete, long-term concept of pavement management and also provides guidelines for more immediate application based on existing technology.

Pavement management systems can provide several benefits for highway, airport, and other agencies at both the network and project levels. Foremost among these is the selection and implementation of cost-effective alternatives. Whether new construction, rehabilitation, or maintenance is involved, a total PMS can help management achieve the best possible value for the public dollar.

At the network level, agencywide programs of new construction, maintenance, or rehabilitation are developed which will have the least total cost, or greatest benefit, over the selected analysis period.

At the project level, detailed consideration is given to alternative design, construction, maintenance, or rehabilitation activities for a particular section or project within the overall program which will provide the desired benefits or service levels at the least total cost over the analysis period.

In order to realize the full benefits of such a management system, proper information for each management level must be collected and periodically updated; decision criteria and constraints must be established and quantified, where possible; alternative strategies must be identified; predictions of the performance and costs of alternative strategies must be made; and optimization procedures that consider the entire pavement life cycle must be developed. Moreover, the proper implementation of all of these management activities and the use of the optimum strategies selected are essential to the full realization of benefits.

**Introduction To Pavement Management** 5

### 1.1.2 Basis of the Book

This book addresses the status of pavement management system (PMS) development and implementation. It is based largely on the work and experience of the authors, existing literature, and interactions with various people actively involved in pavement management. The intent is to describe a total, long-term concept of pavement management and to provide a framework for application. It is also intended to recognize the various functions and activities of people involved in pavement management.

Two major organization management levels and the people who function at these levels are considered:

- Administrative levels, where decisions are made regarding a program, or set of projects, and the budgets and priorities appropriate to the program
- Technical management levels, where decisions are made on "best" design, monitoring of in-service pavements, maintenance procedure, etc., for an individual project

## 1.2 HISTORICAL BACKGROUND

We often think of paved highways as beginning with the era of the automobile in the late 1800s. This of course is erroneous, because overland travel has been second only to water travel in the history of the development of the world.

The first real road builders moved southwestward from Asia toward Egypt [AASHO 52] soon after the discovery of the wheel, about 3500 B.C. It is not surprising that the cradle of civilization was also the cradle of early road building, because roads and population have always gone together.

The Romans were the first scientific road builders, with the *Via Appia*, or the "Appian Way," being initiated in 312 B.C. The oldest, most famous long-distance highway, approximately 1,755 miles long, was named the Royal Road by the Persians. It was constructed over a 4,000-year period, ending in 323 B.C., across Southwest Asia and Asia Minor. Travel time, according to Heroditus (457 B.C.), was 3 months and 3 days, or an average of 19 miles per day.

The Appian Way was generally 3 to 5 ft thick, consisting of three layers. All the work was hand-placed stone, and this type of construction became standard practice for over 2,000 years until it was superseded by MacAdam's light-wearing course surface in the nineteenth century. These early roads had to withstand the wear of hoofed animals, and great attention was given to the wearing surface. Speeds were slow, and therefore overall smoothness of the roads was of less importance until after the introduction of the automobile.

### 1.2.1 Pioneer Road Builders

Road building became recognized as a profession requiring application of scientific principles in the latter part of the eighteenth century. The main trouble with roads of that day, as well as with many roads of today, seemed to be the lack of adequate drainage and the lack of a hard-wearing surface. Perhaps the real founder of pavement management systems was Pierre Marie Jerome Tresaguet, the first modern highway engineer, who was named French Inspector General of Roads in 1775 by King Louis XVI. He introduced the innovation of relatively light road surfaces designed on the principle that the subsurface of the pavement must be well drained and support the load as opposed to the massive pavements designed by the Romans. More important to pavement management, however, was the fact that Tresaguet recognized the need for continuous maintenance.

Thomas Telford was responsible for the construction, in 1816, of the Carlisle-Glasgow Road, said to be the finest road ever built up to that time. It placed emphasis on flat grades and, because of Telford's early training as a stone mason, involved surfacing the road with stones capable of carrying the heaviest expected traffic of that day.

The most famous of early road builders was probably John MacAdam (1756–1836). He is known as the father of modern pavement construction. His road cross-section design was based on the principle that a drained and compacted base should support the load applied to a pavement, whereas the stone surfacing should act only as a wearing course. The construction techniques involved compaction of the materials by normal traffic and probably would not have been at all satisfactory for modern highways. The era of modern road construction actually started in 1869 when the original steam road roller was used for the first time in New York City. It made compaction of macadam roads better, quicker, and easier.

In the post-1900 time period, the rapid growth of the automobile and the decline of horse-drawn vehicles and bicycles brought about a major change in pavement construction. The faster automobile caused serious dust problems on roads, and the use of oils and other agents to cut down on dust began. This led to experiments in 1905 with coal tars and crude oil in Jackson, Tennessee, to determine their benefit in pavement construction. In 1906, bituminous macadam roads were built in Rhode Island. The conclusion drawn from the experiments was that highways used heavily by high-speed motor cars should be built with bituminous macadam surfaces and that existing roads subjected to similar high-speed traffic should be resurfaced using bituminous materials.

These first bituminous roads were followed closely in 1909 by the first rural portland cement concrete roads built in Wayne County, Michigan. The pavement was 17.8 ft wide with natural earth shoulders and expansion joints every 25 ft.

In 1920, the Highway Research Board was organized and major research efforts in the pavement field began with the objective of improving pavement design and construction methods. This research was highlighted by a variety of theoretical and empirical studies, including the well-known AASHO Road Test in Ottawa, Illinois, in 1958–1961. The five-year Strategic Highway Research Program (SHRP), started in 1987–88, has continued this type of Road Test approach, but incorporating many different test site locations.

### 1.2.2 Development of Pavement Systems Methodology

In 1966, the American Association of State Highway Officials, through the National Cooperative Highway Research Program, initiated a study to make new breakthroughs

**Introduction To Pavement Management** 7

in the pavement field. The intent was to provide a theoretical basis for extending the results of the AASHO Road Test. As a result, researchers at the University of Texas [Hudson 68] in 1968 began a basic new look at pavement design using a systems approach.

Similar independent efforts were being conducted at the same time in Canada [Hutchinson 68, Wilkins 68] to structure the overall pavement design and management problem and several of its subsystems.

A third concurrent keystone effort in this area was that of Scrivner and others at the Texas Transportation Institute of Texas A&M University as a part of their work for the Texas Highway Department [Scrivner 68].

The work of these three groups provides the overall historic perspective for pavement management systems.

In the late 1960s and early 1970s the term pavement management system began to be used by these groups of researchers to describe the entire range of activities involved in providing pavements [Haas 70]. At the same time, the initial operational or "working" systems were developed in two major projects. The largest of these was Project 123, conducted by the Texas Highway Department, Texas A&M University, and the University of Texas. A series of reports and manuals have resulted from this research, beginning with Report 123-1 in 1970 [Hudson 70]. The project has produced many of the modern innovations in pavement analysis.

The other major continuous research effort in this field was that carried out in NCHRP Project 1-10, initiated in 1968 [Hudson 68]. A second phase was carried out by Hudson and McCullough to develop an actual working system for implementation at the national level [Hudson 73]. A third phase on implementation was carried out by Lytton et al. at Texas A&M University [Lytton 74], whereas a fourth phase was continued at Materials Research and Development, Inc., in California under the direction of Finn.

Much of the early pavement management work was summarized in two books on the subject in 1977 and 1978 [Haas 77, 78]. Subsequently, a large amount of effort has gone into developing the component technology of pavement management, and into implementation of systems at the state/provincial and local levels. Much of this is reported in the two North American Pavement Management Conferences of 1985 and 1987 [NA Conf. 85, 87]. Research and further development of the process itself has continued, as summarized in a 1991 ASTM Symposium [Hudson 92].

## 1.3 ROLE OF PAVEMENTS IN TODAY'S TRANSPORT SYSTEM

Today's transport system includes marine, highway, rail, air, and pipeline transportation. Of these, only marine and pipeline transportation do not make use of pavements. Certainly the major structural load-carrying elements of the highway system are the pavements. For air travel, pavements are required in the form of runways, taxiways, and parking aprons. Likewise, the railroads operate on a form of pavement historically made up of rails, ties, and ballast, not dissimilar to a highway pavement design. In fact, modern design principles show that rails can easily be mounted on a properly designed continuous pavement.

It is difficult to define precisely the dollar value of the expenditures in each of these modes of transportation in the United States, in Canada, or in the world. However, it is safe to say that the expenditures in the highway sector in the United States represent the largest amount in U.S. transportation and exceed $20 billion annually [FHWA 87]. Including maintenance as well as new construction, pavements represent approximately one-half of this total highway expenditure. In effect, pavements, along with bridges and other structures, represent the major investment in fixed facilities of highway transport. It is also important to point out that after the initial development of a highway system, expenditures for right-of-way and other initial costs cease but expenditures on pavements continue to grow as maintenance and rehabilitation are required.

Although the function of the pavement varies with the specific user, in modern highway, airport, and rail facilities the purpose of the pavement is to serve traffic safely, comfortably, and efficiently, at minimum or "reasonable" cost.

With the relatively large investments involved in pavements, even marginal improvements in managing this investment, and in the technology involved, may effect very large absolute dollar savings. In addition to the direct savings in capital costs and maintenance, the indirect benefits to the road user can be equally significant, although much more difficult to ascertain. Pavement construction in itself will probably not continue to develop as fast in the future as it has since World War II, but the investment we now have in pavements must be protected through periodic rehabilitation and maintenance. Otherwise, this investment can be lost if pavements are allowed to deteriorate too much.

## 1.4 TYPES OF PAVEMENTS

Many definitions are applied to the term *pavement*. In this book, the pavement is considered as the upper portion of the road, airport, or parking area structure and includes all the layers resting on the subgrade. Additionally, the pavement is considered to have a bound surface and includes the load-carrying capacity of the subgrade.

Many so-called types of pavement are discussed in modern technology, such as *rigid pavement, flexible pavement, composite pavement, asphalt pavement, concrete pavement*, and others. Each of these terms has been developed for a particular reason and each has some useful connotation. Perhaps the most straightforward terminology is the definition of pavement by its structural function or response. Two basic types can be considered: (1) flexible pavements and (2) rigid pavements. These definitions provide a framework to which all others can be related. Rigid pavements normally use portland cement concrete as the principal structural layer. Flexible pavements normally use asphaltic concrete for the surface, and sometimes for the underlying layers.

Pavements can also be defined in terms of the mechanical theory normally used to describe their behavior. In this context, slab analysis is commonly used to define the behavior of rigid pavements, which usually carry their load in bending. On the other hand, layered system analysis is commonly used to analyze the behavior of flexible or asphaltic concrete pavements, which carry their load in shear deformation.

# Introduction To Pavement Management

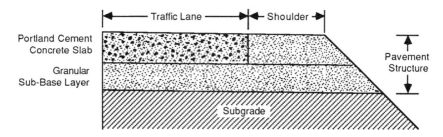

**Figure 1.1** A typical cross section of any of the four basic types of rigid pavements.

The term *composite pavements* has been introduced from time to time but normally is not a very effective definition. It is intended to describe a pavement that combines rigid and flexible elements, such as a portland cement concrete surface on an asphalt concrete base, or an asphalt concrete surface (usually an overlay) over an old portland cement concrete (rigid) base or over a portland cement-treated (rigid) base. A more useful definition is to assign this type of pavement to one of the other two types, depending on the basic load-carrying element and not the visible surface type.

There are many definitions involved in pavement technology and they are provided in various published glossaries. Perhaps one of the best sources is the Transportation Research Board (see their most up-to-date catalogue of publications).

## 1.4.1 Rigid Pavements

There are four basic types of rigid pavements. These derive from the combination of reinforcement and load-transfer devices within the concrete. Rigid pavements may be unreinforced, lightly reinforced, continuously reinforced, or prestressed. A typical cross section for any of these types is shown in Figure 1.1. The basic tenant of the diagram is to show that the pavement structure, per se, consists of those layers above the subgrade and not just the slab.

Unreinforced concrete pavements can be placed without joints, but this practice is seldom followed. Therefore, unreinforced concrete pavements are normally jointed and are placed with or without load-transfer devices across the joints. Jointed pavements are also placed with light steel reinforcement between the joints. Spacing of the joints in such lightly reinforced pavements is substantially greater than for unreinforced pavements. The purpose of the steel is to hold together the cracks so that the concrete can fulfill its function of carrying traffic loads. These pavements nearly always are placed with load-transfer devices across the joints.

Continuously reinforced pavements are placed without regular spacing of joints and contain adequate steel reinforcement to carry the load in the cracked concrete sections.

Prestressed concrete pavements consist of slabs that are placed with adequate steel to allow it to be stressed. This provides a prestress on the concrete (so that the slab is in compression) and increases its tensile load-carrying capacity in a way similar to other structural applications of prestressed concrete. Such pavements have found only very limited use to date in North America, largely because of the high labor

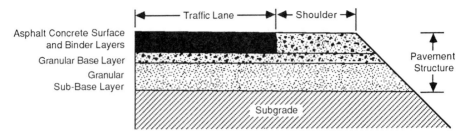

**Figure 1.2** A conventional asphalt concrete pavement section.

costs involved, but they may have potential for future or certain specialized applications.

### 1.4.2 Flexible Pavements

Flexible pavements have recently been developed into a wider variety of types than formerly used. A conventional pavement section, as shown in Figure 1.2, involves the use of an asphalt concrete surface layer, an asphalt concrete binder or leveling course layer (often of the same material as the surface layer but sometimes with a larger maximum particle size), and one or more base and subbase layers. The granular base and subbase layers may or may not be bound (i.e., with asphalt or some other treatment). Also, the shoulder may or may not be surfaced with asphalt concrete or a surface treatment. Often, the shoulder simply consists of the same material as the base.

Recent variations of flexible pavement types are Full Depth$^R$ and Deep Strength$^R$. (These are terms registered by The Asphalt Institute with the U.S. Patent Office.) The former refers to an asphalt pavement in which asphalt mixtures are employed for all courses above the subgrade, whereas the latter term refers to an asphalt pavement in which the base layer, in addition to the surface, is constructed of an asphalt mixture.

## 1.5 CONCEPT OF PAVEMENT MANAGEMENT

Pavement management is not a new concept; management decisions are made as a part of normal operations every day in highway agencies. The idea behind a pavement management system is to improve the efficiency of this decision making, expand its scope, provide feedback as to the consequences of decisions, and ensure the consistency of decisions made at different levels within the same organization.

Good pavement management is not business as usual. It requires people to think, respond to information, and make decisions in a logical, effective, and coordinated way. Often this means change, which is difficult. Many people are reluctant to change and thus pavement management is sometimes difficult to get started.

### 1.5.1 Pavement Management Primer

The performance of a pavement certainly depends in part, but not exclusively, on the design concepts that were used. The success of any design is also largely dependent on subsequent construction, maintenance, and rehabilitation. Historical studies by

## Introduction To Pavement Management

many agencies show that the concept of a 20-year new pavement design is generally fictitious. Often pavements provide adequate service for only up to 10 or 12 years, and sometimes less, without major maintenance or rehabilitation. If such action occurs, sometimes more than once, it is quite feasible to provide a total of 20 or 25 years performance or total service life. Consequently, many agencies have recognized the need to link together explicitly the activities of planning, designing, constructing, and maintaining pavements. In other words, they have recognized the need to "manage" the technology of providing pavements on a comprehensive basis.

After development of the basic pavement systems concepts in the mid 1960s, it became more and more obvious that these concepts should in fact be applied not only to design but to the entire process of providing pavements, from planning and design to construction to periodic maintenance. Nowhere was this more clearly shown than in December 1970, when the Federal Highway Administration-University of Texas-HRB Conference on Structural Design of Asphalt Pavement Systems was held [HRB 71]. As a result of this conference, it became clear that pavement systems technology was here to stay and that the concept of pavement management systems was the basis on which technology could move forward in an efficient manner.

Unfortunately, however, it has also become common to use the term *system* in a rather loose way in the pavement field. For example, *joint system* might mean all the joints and associated physical details in a rigid pavement, or the method used to design the joints, etc.

Design technology for pavements has traditionally been both prescriptive and deterministic. It has been prescriptive in the sense that designers have set limits on such factors as deflection, stability, or other parameters in an attempt to avoid premature failure, rather than to predict the type and degree of damage that might occur and the time at which it might occur under a specified set of conditions. It has been deterministic in the sense that the equations or models predict a single answer and do not account for statistical variation or reliability factors.

The concepts of pavement performance developed in the late 1950s and early 1960s in association with the AASHO Road Test [HRB 62] and studies in Canada [CGRA 65] provided the necessary system output function, in terms of performance concept, for pavement management systems. Without this performance concept, it is relatively difficult to understand and relate the differences between pavement design alternatives.

It has been only recently that "design" itself has been elevated from the concept of specifying an initial structural section to that of a "strategy" where the strategy is an optimized design involving not only the best initial construction and structural section but also the best combination of materials, construction policies, maintenance policies, and overlays. This concept will be fully explored in subsequent chapters, but basically design must involve more than just the initial construction to be carried out.

The technology of designing, constructing, and maintaining pavements has been considerably enhanced in recent years by research into the fundamental characteristics of pavement materials. We are now better able to understand how materials behave when subjected to a variety of loads under various climatic or environmental conditions and the effects of time and loading.

Another thing that has markedly aided the development of pavement technology is the use of computers. Computers have enabled us to process large volumes of data and to perform complex and extensive calculations on the response of pavements as simulated, for example, by layered systems or slab analysis.

The disparate atmosphere in which pavements are financed, constructed, and maintained has tended to fragment the development of codified pavement management policies over the years. Many administrative levels are involved, with different reporting functions, and they provide a wide variety of input into the process. This does, however, inhibit the codification of data into a single body of knowledge.

To summarize, the reader should keep in mind the variety of considerations and detailed inputs that have affected the development of pavement management systems throughout history. A great deal of work has been done to overcome some of the resulting problems, but there is also scope for considerable improvement in all areas of pavement technology.

### 1.5.2 Terminology

There is a great deal of difference of opinion as to the terminology, key factors, and related items associated with pavement management systems. Some people think the words *management* and *systems* are nothing more than "buzz words" coined to gain attention. At the other extreme, some people feel that a PMS is a highly sophisticated computer based technology that is a panacea for all pavement problems. Both of these views are, of course, false.

> The word *system* has been appropriated for many purposes, such as circulatory system, drainage system, sprinkler system, and highway system. The dictionary says that a system is a regularly interacting or interdependent group of items forming a unified whole.
>
> The word *management* means many things to many people. To some it means "to administer," to others it is "to control," and to still others it means "to coordinate the various elements of." The dictionary definition of management is "the act or art of managing," or, less circularly, "the judicious use of means to accomplish an end."
>
> *Pavement management* in its broad sense encompasses all the activities involved in providing and managing the pavement portion of a public or private works program, large or small. The objective of the management system is to use reliable information and decision criteria in an organized framework to produce a cost-effective pavement program.

### 1.5.3 Definitions

The following definitions are intended to provide a common and consistent basis for the use of certain fundamental terminology in the pavement management field. This list is not all-inclusive and those terms considered to enjoy reasonable agreement among most agencies and people in the field are not repeated.

An obvious omission from this list is *pavement*. The authors define *pavement* to include shoulders as well as all structural elements of the roadway (i.e., all layers). Also, the load-carrying capacity of the subgrade is implicitly included in this defi-

**Introduction To Pavement Management** 13

nition. More restrictive definitions may be preferred by some highway agencies. The framework presented in this book is flexible enough to incorporate a wide range of definitions.

> *Pavement management* involves the identification of optimum strategies at various management levels as well as the implementation of these strategies. It is an all-encompassing process that covers all those activities involved in providing and maintaining pavements at an adequate level of service. These range from initial information acquisition to the planning, programming, and execution of new construction, maintenance, and rehabilitation, to the details of individual project design and construction; to periodic monitoring of pavements in-service.
>
> A *pavement management system* (PMS) provides decision makers at all management levels with optimum strategies derived through clearly established rational procedures. A PMS evaluates alternative strategies over a specified analysis period on the basis of predicted values of quantifiable pavement attributes, subject to predetermined criteria and constraints. It involves an integrated, coordinated treatment of all areas of pavement management, and it is a dynamic process that incorporates feedback regarding the various attributes, criteria, and constraints involved in the optimization procedure.
>
> A *total pavement management system* consists of a coordinated set of activities, all directed toward achieving the best value possible for the available public funds in providing and operating smooth, safe, and economical pavements. This is an all-inclusive set of activities, which may be characterized in terms of major components or subsystems. A pavement management system must serve different management needs or levels and it must interface with the broader highway, airport, and/or transportation management system involved (Figure 1.3).

## 1.6 ESSENTIAL FEATURES OF PAVEMENT MANAGEMENT

Pavement management must be capable of being used in whole or in part by various technical and administrative levels of management in making decisions regarding both individual projects and an entire highway network. All types of decisions should be incorporated in the process, including those related to information needs, projected deficiencies, or improvement needs for the network as a whole, budgeting, programming, project design, construction and maintenance, resource requirements, in-service monitoring, and research.

All functions involved in providing pavements are essential to comprehensive pavement management, but not all functions need to be active at the same time. In planning future construction, for example, it is necessary to consider individual project design in only a very approximate way. Thus, a pavement management system can be viewed as a set of connected modules or "building blocks." In this sense, a pavement management system may be likened to a kaleidoscope: the whole thing exists at all times, but what part of it one sees depends on how one looks at it.

In addition to defining pavement management, it is useful to list some of the essential requirements:

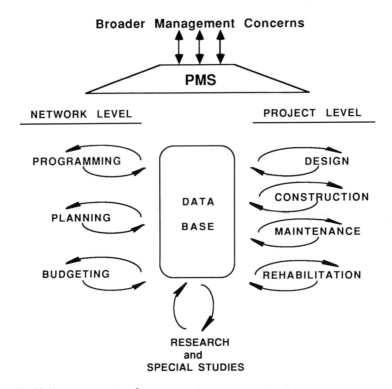

**Figure 1.3** Major components of a pavement management system.

- Capability of being easily updated and/or modified as new information and better models become available
- Capability of considering alternative strategies
- Capability of identifying the optimum alternative or strategy
- Capability to base decisions on rational procedures with quantified attributes, criteria, and constraints
- Capability to use feedback information regarding the consequences of decisions

Pavements are complex structures involving many variables, e.g., combinations of load, environment, performance, construction, maintenance, materials, and economics. In order to design, build, and maintain better pavements, it is important that the various technical and economic factors be well understood. Pavements are not inexpensive parts of the transportation infrastructure. An investment of approximately $30 billion has been made in pavements for the U.S. Interstate Highway System alone, and billions more will be spent annually on maintenance and upgrading. Thus, even marginal improvements in the component technologies of pavement management, and in the process itself, can result in large absolute savings.

**Introduction To Pavement Management**

## REVIEW QUESTIONS

1. Describe why pavements are such an important component of a nation's transportation infrastructure.
2. What is the difference in the way flexible and rigid pavements carry load?
3. What is the difference between traditional pavement design and a pavement design strategy?

Chapter 2

# General Nature and Applicability of Systems Methodology

## 2.1 NATURE OF THE SYSTEMS METHOD

Systems methodology comprises a body of knowledge that has been developed for the efficient planning, design, and implementation of new systems, and for structuring the state of knowledge on an existing system or modeling its operation. It is a comprehensive problem-solving process, and the framework that characterizes it has been formally developed in the postwar decade from observations of a large number of efficiently and systematically conducted projects [Hall 62].

There are two main, interrelated uses of systems methodology:

1. The framing or structuring of a problem, or body of knowledge
2. The use of analytical tools for actually modeling and solving the problem

These uses are complementary and interrelated; one is insufficient without the other. The framing of a problem is usually too generalized by itself to achieve a useful operational solution, whereas the application of analytical techniques to an inadequately structured problem may result in an inappropriate solution [Stark 72].

## 2.2 STRUCTURE OF THE SYSTEMS METHOD

The structure or framework of any problem-solving process should provide for systematic incorporation of all the technical, economic, social, and political factors of interest. Moreover, it should be a logical simulation of the progression of activities involved in efficiently solving a problem.

Figure 2.1 presents the major phases and components of such a process. In this general form, it is applicable to a wide variety of engineering and other problems. The diagram illustrates that the recognition of a problem comes from some perceived inadequacy or need in the environment. It leads to a definition of the problem that involves a more in-depth understanding. This provides the basis for proposing alternative solutions. These alternatives are then analyzed in order to predict their probable outputs or consequences. Evaluation of the outputs is the next step in order that an optimal solution may be chosen. Implementation involves putting this solution into service, and its operation. Feedback for improving future solutions, or checking on how well the system is fulfilling its function, is provided by periodic performance measurements.

The process of Figure 2.1 is continuous and iterative. It is applicable to both the overall problem being considered and to its many component subproblems, basically at three levels:

1. The systems approach
2. Systems analysis
3. Systems engineering

Generally, these levels increase in complexity and utility. The systems approach often means nothing more than broad consideration of a problem, or as many aspects of the problem as convenient. In terms of Figure 2.1, we might say that the systems approach involves only the problem-recognition phase, and the problem-definition phase in an initial manner, with perhaps a cursory look at the generation of alternative strategies.

Systems analysis encompasses the systems approach and extends it to a more complete consideration of alternate strategies. More important, it provides a methodology for analyzing and optimizing these alternatives.

Finally, systems engineering is a more complete manifestation of the systems method, with design, implementation, and performance evaluation aspects given strong attention.

## 2.3 SOME BASIC TERMINOLOGY

The systems terminology most often confused is that associated with the problem-definition phase. Inputs can be thought of as those factors that place some demand on the system (i.e., loads, stresses, etc.). They, together with the constraints, usually represent information that must be acquired by problem solvers. Objectives also represent necessary information, but they must usually be developed or specified by

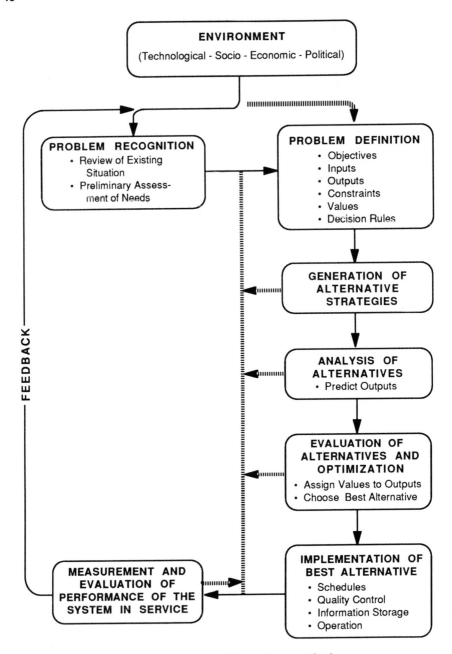

Figure 2.1 Major phases and components of the systems method.

problem solvers. Similarly, they must decide as a part of this problem definition what outputs will be of concern when they subsequently analyze alternative solutions, and what output prediction models they will use. They should additionally specify what types of values they will place on these outputs, what sort of function they will use to combine them, and what decision rule they will eventually use to choose the best solution.

It is important that the "system" under consideration be clearly recognized and identified; otherwise there can be confusion in determining the inputs and in specifying the applicable objectives, constraints, and so on. For example, consider the frequently used term *pavement system*. It is sometimes unclear whether the actual physical structure, the design method, the construction or maintenance policies, or some combination of the foregoing are being considered.

Chapter 3 extends the foregoing generalizations to some more specifically defined concepts for the pavement management field.

## 2.4 SOME PRECAUTIONS IN APPLICATION

The general systems method of problem solving, as shown in Figure 2.1, models the logical, systematic pattern that is used by efficient problem solvers. It must, however, be used with full recognition that there are certain limitations.

First, successful application of the systems method inherently depends on the capabilities of the people involved. The method is no substitute for poor engineering, and it does not represent a framework for only classifying all the factors pertinent to a problem.

Second, the point of view of the individual or agency involved must be clearly recognized and identified. Otherwise, confusion and contradiction can be the result. For example, a materials processing problem for a public works project might well be viewed differently by the contractor than by the government agency involved. They could have competing objectives, and they would undoubtedly have different constraints.

Third, the components or extent of the system under consideration should be clearly identified. For example, the term *parking system* might mean the actual physical parts of the parking lot, such as pavement, curbs, gates, and so forth, to one person; to another it might mean the method used to operate the facility; and to still another person it might mean a combination of the two. Without such clear identification of the extent of the system, inconsistency among objectives, constraints, inputs, and outputs can results.

A fourth point concerns the oversimplification imposed on many problems by considering the system as a black box, with an arrow on the left labeled inputs and one on the right labeled outputs. This is a convenient conceptualization for certain hardware problems, but it is quite inadequate for most other problems.

A fifth point concerns the inherent danger of generating precise solutions to an imperfectly understood problem. That is, the problem has been recognized but not yet rigorously defined. It is, of course, common to perceive some general solutions in the problem-recognition phase of Figure 2.1. However, these may be inadequate or incomplete if the problem solver does not go on to define the problem.

## 2.5  SOME ANALYTICAL TOOLS FOR SYSTEMS ENGINEERING

The structuring of a problem is usually too general to be used alone to find a useful, operational solution, and the application of analytical techniques to an inadequately structured problem may result in an inappropriate solution. In other words, the analytical techniques that are used as part of the systems method for solving problems have maximum usefulness when the problems are well formulated or structured; otherwise, they can be an exercise in mismanagement.

There is a large variety of available techniques (or tools, or models) that can be used in applying the systems method to the solution of a problem. It might be noted that these techniques are also applicable to what is commonly known as the operations method. Basically, the operations method and the systems method differ only in the scope and nature of the problems solved, rather than in the form of the methodology. This section of the book provides only a "catalogue" of some of the more widely used systems and operations methods. These methods have varying degrees of potential applicability to the pavement field. The references cited provide a means for further exploration on the part of the reader.

The use of systems or operations models or techniques should facilitate reaching a decision on as objective a basis as possible. The type of "objective function" used depends largely on the available knowledge of the outputs of the system, which can be classified in terms of the three following problem types:

1. Certainty, where definite outputs are assumed for each alternative (i.e., deterministic type of problem)
2. Risk, where any one of several outputs, each of known probability, can occur for each alternative
3. Uncertainty, where the outputs are not known for the alternative courses of action; thus probabilities cannot be assigned

A major amount of engineering practice has treated problems in terms of decisions under certainty (type 1), because of convenience and because of the available information. However, there is considerable current effort being directed toward incorporating probabilistic concepts (type 2) into practice.

Where practical problems are too complex for symbolic representation, they may be modeled on an analogue or a scale basis. Alternatively, it is possible to "force" a solution by experimentation, gaming, or simulation for some types of problems [Stark 72].

One of the most widely applied and useful classes of systems models involves linear programming. These techniques have been used in everything from construction to petroleum refinery operations because they are well suited to allocation-type problems [Gass 64]. A typical problem for linear programming application might involve the determination of how much of each type of material a contractor should produce,

given production capacity, the number and capacities of trucks, available materials and their costs, the delivery distance, profits for each type of material, and so forth. There are several variations of linear programming models and several methods of solution, including parametric linear programming, integer linear programming, and piecewise linear programming. The latter is used to reduce a nonlinear problem to approximate linear form.

Nonlinear methods can range from the so-called classical use of differential calculus, Lagrange multipliers (and their extension to nonnegativity conditions and inequality constraints) and geometric programming, to the iterative search techniques [Künzi 66]. These latter techniques start from an initial solution and seek improvements until an acceptable tolerance is reached. They are often applicable where more rigorous methods are impractical.

There are some types of nonlinear problems not easily solved by analytical techniques that may lend themselves quite well to graphical solution. A variety of methods may be used to obtain optimum values. Their applicability and use, which have received comparatively little attention, are directly dependent on the nature of the problem and the way in which it is formulated by the problem solver. For example, it has been illustrated that a simple graphical solution can be applied to a construction problem involving a discontinuous cost function [Haas 73].

Problems involving multistage decisions can be represented as a sequence of single-stage problems. These can be successively solved by a method known as dynamic programming [Dreyfus 65]. Each single variable or single-stage problem that is involved can be handled by the particular optimization technique that is applicable to that problem. These techniques are not dependent on each other from stage to stage and can range from, say, differential calculus to linear programming. Combinatorial-type problems are often well suited to dynamic programming. A typical example might be that of an aggregate producer with several mobile crushers and several sources of raw materials who wants to determine how many crushers should be assigned to each site for a given profit matrix.

Random and queuing models can have a wide range of applicability to systems problems, and there is a large amount of literature available [Wiley 66]. For example, one class of models involves the Monte Carlo methods, which are quite useful when adequate analytical models are not available. These methods require distribution functions for the variables. They are, however, somewhat inefficient and are applied mainly to complex problems that are otherwise unmanageable. There are also a large range of problems to which reliability, random walk, and Markov chain techniques can be applied. The latter can be used to extend stochastic and chance-constrained programming models. Queuing models have been used very extensively in engineering, including various air terminal operations, traffic facility operations, rail operations, and canal operations.

Many systems problems involve the allocation and scheduling of personnel, equipment, money, and materials. Several techniques have found widespread use for these types of problems, including sequencing, routing, and scheduling. Sequencing involves the ordering of various tasks in sequential manner to minimize total time or effort [Conway 67]. Routing involves the identification of a path through a network to minimize time, cost, or distance. There are graphical methods and matrix methods available for the minimum path type of routing problems.

Scheduling involves the allocation of time or resources to various tasks whose sequence is fixed but whose cost is time-dependent, to minimize total completion time or cost. There are two well-known types of scheduling methods, the critical path method (CPM) and the program evaluation and review technique (PERT). They have received especially wide application in the construction field [Moder 64]. Single-value time estimates are used for each activity in CPM, whereas PERT uses a range of possible completion times (i.e., including stochastic aspects). Thus PERT has tended to be used more for research and development purposes whereas CPM has found more routine application in construction. Critical path scheduling problems can often be formulated as linear and dynamic programming situations.

This section has only noted the analytical tools that have potential applicability to various aspects of the pavement field. Those desiring more in-depth information may consult some of the many references available in libraries in these areas.

## REVIEW QUESTION

1. Describe how the systems method, as represented by Figure 2.1 could be applied to the selection of the number and types of trees for a landscaping project.

Chapter 3

# Application of Systems Concepts to Pavement Management

## 3.1 INTRODUCTION

A 1967 NCHRP project was one of the first efforts in applying systems engineering to pavement design [Hudson 68]. In a similar but independent effort [Hutchinson 68], a systems approach was applied to structuring the overall problem and several of the subsystem design problems. Simultaneously, the Texas Transportation Institute developed a working design model in connection with a cooperative research project with the Texas Highway Department. As a result of these studies, the Texas Highway Department, recognizing the need for a system for organizing and coordinating their pavement research program and updating their design system, initiated a project in cooperation with The University of Texas Center for Highway Research and the Texas Transportation Institute of Texas A&M University [Hudson 70].

The overall purpose of these efforts was to develop the basic framework required to adapt to local environments information such as that obtained on the AASHO Road Test. This was to be accomplished by carrying out the following specific objectives:

1. Development of descriptions of significant basic properties of materials used in roadway structures

2. Development of procedures for measuring these properties in a manner applicable to pavement design and evaluation
3. Development of procedures for pavement design utilizing the measured values of the basic properties which would be applicable to all locations, environments, and traffic loads

In essence, the goal was to "formulate the overall pavement problem in broad conceptual and theoretical terms," which would enable the solution of a variety of pavement problems which have long plagued engineers.

## 3.2 THE GENERAL STRUCTURE OF SYSTEMATIC PAVEMENT MANAGEMENT

As a part of pavement management system development at the project level, the design process was structured and its components were identified more specifically in 1965–67 as shown in Figure 3.1 [Hudson 68]. This detailed version was further simplified into Figure 3.2 as a summary representation. The components of Figure 3.1 are further discussed as follows:

1. *Inputs*. Inputs, including a number of different variables, plus objectives, must be established.
2. *Models*. The need for analysis of alternatives was identified. A simple model of the pavement response expressed, for example, as a "simple design chart" does not adequately treat improved pavement materials.
3. *Behavior—Distress*. Most pavement models calculate or predict pavement response such as stress or strain. Given the prediction of response, if carried to its limit the result will be distress. Prediction models for pavement response, and thus for cracking and other pavement distress, are essential.
4. *Performance—Output Function*. Accumulated distress reduces pavement serviceability, and serviceability history defines pavement performance.
5. *Safety*. Skid resistance and other safety response are important.
6. *Costs*. Life cycle economic analysis is a vital part of the pavement management process.
7. *Decision Criteria*. Closely tied to economics are decisions on allowable costs versus the resulting benefits related to a particular choice. These factors must be explicitly defined and considered in the analyses.
8. *Compare—Optimize*. Selecting the optimal alternatives or strategy is an important step in decision making.
9. *Implementation*. Construction of the selected alternatives or strategy, and periodic maintenance plus rehabilitation when required constitutes implementation. Another view of implementation, as subsequently discussed in the book, is that of implementing a pavement management system itself, as opposed to implementing a decision made within the PMS.

### 3.2.1 Applying Systems Engineering to Pavement Behavior

A pavement is a complex structure which is subjected to many diverse combinations of loading and environmental conditions. Adding to this complexity is materials char-

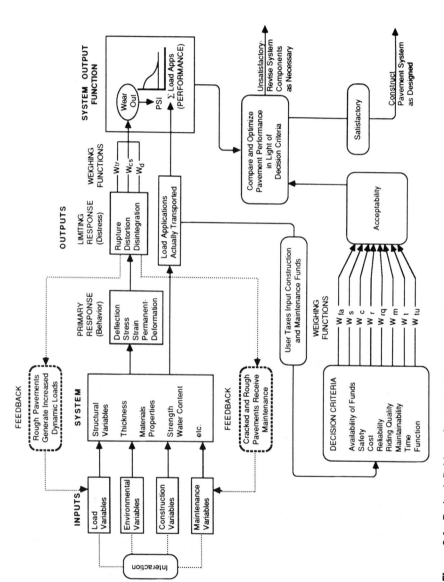

**Figure 3.1** Early definition of a project-level pavement design system and its major components [Hudson 68].

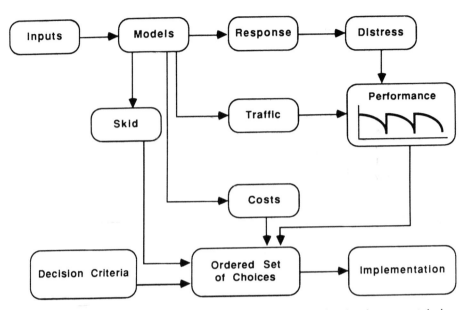

**Figure 3.2** Simplified diagram of the major components of a project-level pavement design system [Hudson 68].

acterization, varying pavement performance and their interrelationships. Because of this, a coordinated framework applicable to the overall problem of pavement design is needed. Examination of available techniques for analyzing such complex relationships revealed that the concepts of systems engineering (which evolved in the electronics, communications, and aerospace industries) are quite appropriate to pavement structures. The use of systems engineering does not, per se, lead to a new and dramatic solution of the pavement design problem, but it does provide a means of organizing the various segments of the total problem into an understandable framework.

The scientific and engineering aspects of a systems problem usually span a broad spectrum of activities:

1. Use of physical observations or measurements to characterize behavior
2. Statement of mathematical models that describe or approximate the physical phenomena
3. Development of a system for prescribed behavior using the mathematical models
4. Physical realization of the design

Thus, it is essential that systems engineers be able to formulate a system in terms of a mathematical or physical model, or, failing this, the system must be simulated in some realistic way to observe the necessary outputs.

## 3.2.2 Systems Engineering Applied to Pavement Design

The design process involves several distinct features:

1. Appropriate input and response variables must be identified and described quantitatively.
2. Methods of selecting both construction materials and construction techniques must be adopted.
3. Response of the system to all classes of input expected to occur in service must be measured, whether directly in the system itself or in some type of simulated way.
4. Quality of the response or measure of the performance of the system must be judged by appropriate criteria.
5. Modification of the system must be permitted in order to attain a condition as nearly optimum as possible.

In order to treat quantitatively the foregoing features, it is necessary to define terms and operations more precisely. The input to the system consists of traffic, environment, and maintenance. The effect of traffic is to impart, through wheel loads, certain strains, stresses, and deflections on the pavement surface. Spatial distribution and time variations (both dynamic and cyclic) should be represented. The environmental input consists of, among other things, diffusion of heat and moisture into the system. These inputs should also be represented as functions of space and time. In certain instances, a chemical input may occur, e.g., the use of deicing salts. The response consists of the generation of a mechanical state identified by deformation and internal stress. The mechanical state is most readily described in terms of stress and strain.

## 3.2.3 Pavement Behavior and Performance

Examination of the pavement design system diagrammed in Figure 3.1 illustrates the complex interrelationships which necessarily exist between (1) materials comprising the structure, (2) manifestations of pavement behavior, and (3) pavement performance. A good deal of effort has been directed to trying to develop improved models for relating behavior to the serviceability-age history of the pavement (its performance). The problem is difficult but is well worth pursuing. It should be realized that good materials evaluation and good mechanistic modeling of pavement structures are essential to pavement management.

## 3.3 THE PROCESS OF PAVEMENT MANAGEMENT: LEVELS, DECISIONS, AND ISSUES

Pavement management is a process for carrying out in a coordinated, systematic way all those activities that go into providing pavements [Haas 77]. It can be viewed in terms of two basic working levels: network and project. The major activities occurring at each level are identified in the next chapter.

Network level management has as its primary purpose the development of a priority program and schedule of rehabilitation, maintenance, or new pavement construction work, within overall budget constraints. Project level work comes "on stream" at the appropriate time in the schedule.

### 3.3.1 Network and Project Levels Versus Organizational (User) Issues

Pavement questions and issues can be viewed from three basic organizational or user levels: legislative, administrative, and technical, whereas pavement management actually operates at two major levels: network and project. The network level is the primary responsibility of administrators who also work, of course, with input of a technical nature while the project level essentially involves technical considerations and decisions. The legislative level also requires answers from pavement management and these can range from general network impacts of budgeting to plans for specific projects.

Network level pavement management, where rehabilitation and maintenance work, due to limited budgets, is handled through a priority analysis, involves a "from the top down" flow. In other words, individual projects come on stream from the priority program established at the network level. However, project level activities are equally important to the overall process. Moreover, "from the bottom up" data is vital to the updating of network level estimates. The network level works on much more approximate data and technical analyses, and, therefore, considerable "fine tuning" may be required when individual, detailed project testing, design, and construction are done.

Thus, the structure of pavement management incorporates a project level and a network level that are both directed to providing answers to legislative, administrative, and technical level users within an organization. The major issues and questions that require these answers are identified in the next section.

## 3.4 THE PMS DOES NOT MAKE DECISIONS—THE PEOPLE WHO USE IT DO

It will not likely ever be possible to rely entirely on quantitative criteria and decision rules for engineering projects. Qualitative judgment is an important element in all human endeavors; thus, the system or the computer used in the system does not make decisions; rather, it processes the information for use by decision makers. Thus, it is the people who use a PMS who assume the responsibility for making decisions.

In order to make these decisions, the legislative, administrative, and technical level users of a pavement management system face certain issues and questions. These may differ in focus and scope, depending on the agency (i.e., federal, state/provincial, city, county), and the management level involved. The following paragraphs provide examples under these three categories.

### 3.4.1 Legislative Level Users

The issues and questions at the legislative or elected level are fairly broad in scope but have to be recognized by the administrative and technical levels. They include the following:

1. Justification of Budget Requests: Legislators are faced with a variety of competing demands and those that "make the case" in a clear, properly supported manner are likely to receive more favorable consideration.
2. Effects of Less Capital and/or Maintenance Funding: Legislators may well ask what the short- and long-term effects are of less funds, perhaps even a zero capital budget, on the deterioration of serviceability, extra maintenance costs, eventual replacement costs, effects on users, etc.
3. Effects of Deferring Work or Lowering Standards: The related question to lower funding is that of deferring maintenance and rehabilitation, and/or lowering the standards.
4. Effects of Budget Requests on Future Status of the Network: If a funding level matching the budget request is approved, a key question relates to the effect this will have on the status of the network. Will the average serviceability or condition decrease, improve, or stay the same? Alternatively, the question may be asked as to what level of funding is required to keep the network in its present state.
5. Effects of Increased Load Limits: This is an example of the type of issue facing many legislative bodies. Obviously a good pavement management system should be able to supply the technical and economic answers.

### 3.4.2 Administrative Level Users

The administrative and planning people responsible for developing capital spending and maintenance programs (i.e., state/provincial highway administrators, division heads for planning, maintenance, city engineers) need to explicitly recognize and respond to legislative level issues and require certain answers from the technical level, in addition to facing questions at their own level. In other words, there is overlap in both directions at this level. Some example issues and questions include the following:

1. An objectively based priority program to provide justification for budget requests
2. A summary assessment of the current status of the network, in graphical and tabular form, based on information from the data base
3. The means for quantitatively determining the effects of lower budget levels, and/or the budget level required to keep the network in some specified state
4. The means for quantitatively demonstrating the effects of deferring maintenance or rehabilitation
5. Estimates of the future status of the network (in terms of average serviceability, condition, safety, etc.) for the expected funding
6. Benefits of a pavement management system, and its major features or "deliverables," etc.

7. Costs of pavement management implementation, including data base development, manpower requirements; system development and installation, training, on-going costs, etc.
8. Implementation of experience of others; documentation of their experience
9. Relationship between pavement management and other management systems, including maintenance
10. Interfacing a pavement management system with transportation management in general

### 3.4.3 Technical Level Users

From a technical perspective, pavement management involves a large number of issues and questions. In addition, the questions and issues faced at the administrative and legislative levels must be appreciated if technical activities are to be meaningful.

The following is an example listing some of these key issues and questions, involving both the network and project levels:

1. Data base design and operation, plus the methods, procedures, and equipment for data acquisition
2. Ensuring the adequacy of the data base
3. Models for predicting traffic, performance, distress, surface friction, etc.—their reliability, consistency, reasonableness, deficiencies, etc.
4. Criteria for minimum serviceability, minimum surface friction, maximum distress, minimum structural adequacy—reasonableness, effects of changes in criteria, etc.
5. Models for priority analysis and/or network optimization
6. Verification of models
7. Relating project (sub) optimization to network optimization
8. Methods for characterizing materials and using results
9. Sensitivity of model analysis results to variations in factors
10. Relationships between vehicle operating costs and pavement characteristics
11. Construction quality control
12. Effects of construction and maintenance on pavement performance
13. Communication between design, construction, and maintenance, within the organizational structure
14. Guidelines for pavement management implementation
15. Relating pavement management to maintenance management
16. Improving the technology of pavement management and making use of implementation projects for this purpose

## REVIEW QUESTIONS

1. Identify the applicability of the various phases of the systems approach, Figure 2.1, on the structure of the project level pavement management system given in Figure 3.1.
2. Describe how the major organizational levels of a federal, state, or local agency are affected by or relate to pavement management.

Chapter 4

# Pavement Management Levels and Functions

## 4.1 THE IDEAL PAVEMENT MANAGEMENT SYSTEM (PMS)

A system in general consists of a set of interacting components that are affected by certain exogenous factors or inputs. In the physical pavement system (structure) the mutually interacting components are usually a surface layer including traffic lanes and shoulders, base layer, subbase layer, and subgrade. The exogenous or outside factors which affect the pavement are environment, traffic, and maintenance. Maintenance is carried out to reduce the rate of deterioration of the pavement from the negative impacts of the traffic and environmental inputs.

A pavement management system on the other hand consists of such mutually interacting components as planning, programming, design, construction, maintenance, and rehabilitation (see Figure 1.3). Exogenous factors affecting a pavement management system include budgets, or information, and nonquantifiable administrative policies.

An ideal pavement management system would yield the best possible value for the available funds while providing and operating smooth, safe, and economical pavements. The minimum requirements of such a system would include adaptability, efficient operation, practicality, quantitatively based decision-making support, and good feedback information. There is no ideal single PMS of course which is best for all agencies. Every agency represents a unique situation with specific needs. Therefore each agency must define carefully what it wants from a pavement management system.

## 4.2 THE NETWORK AND PROJECT LEVELS OF PAVEMENT MANAGEMENT

Pavement management is a process which has two basic working or operational levels: network and project. Figure 4.1 lists the major activities occurring at each level. These activities are discussed in detail in subsequent chapters.

Network level management has as its primary purpose the development of a priority program and schedule of work, within overall budget constraints. Project level work thus comes on stream at the appropriate time in the schedule, and represents the actual physical implementation of network decisions.

Several key points can be made with regard to Figure 4.1:

1. Acquiring and processing input information is the (data base) foundation of pavement management.
2. Criteria to establish when a section is deficient are essential.
3. Deterioration models are required to predict when in time an existing section, or alternative, will become deficient.
4. Alternatives (new construction, rehabilitation, or maintenance) exist at both levels.
5. Alternatives must be analyzed for their technical and economic impact so that priority programs at the network level or the best alternative at the project level can be identified and implemented.

## 4.3 INFLUENCE LEVELS OF PMS COMPONENTS

Four of the major components or subsystems (planning, design, construction, and maintenance plus rehabilitation) have important but changing impacts in terms of a "level-of-influence" concept. This concept which has been used in sectors of industry, such as manufacturing and heavy industrial construction [Barrie 78], shows how the effect on the total life-cycle cost of a project decreases as the project evolves.

Figure 4.2 illustrates the essential features of the "level-of-influence" concept. The lower portion of the figure presents a simplified picture, in bar chart form, of the length of time each major component acts over the life of a pavement. The upper portion shows plots of increasing expenditures and decreasing influence, again over the pavement life. Expenditures during the planning phase are relatively small compared with the total cost. Similarly, the capital costs for construction are a fraction of the operating and maintenance costs associated with a pavement life cycle. However, the decisions and commitments made during the early phases of a project have far greater relative influence on later required expenditures than some of the later activities.

At the beginning of a project, the agency controls all (100 percent influence) factors in determining future expenditures. The question is to build or not to build? A decision not to build requires no future expenditure for the project. A decision to

# Pavement Management Levels and Functions

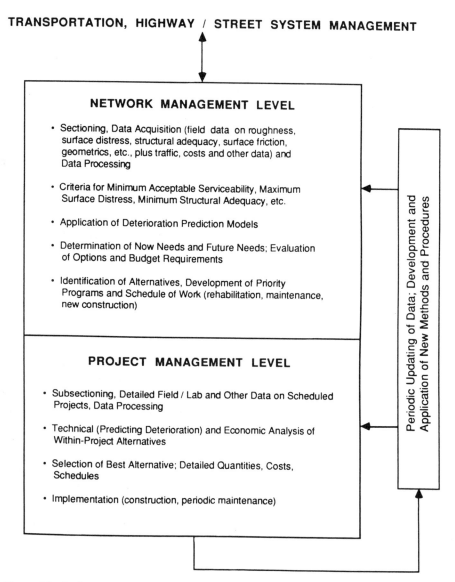

**Figure 4.1** Basic operating levels of pavement management and major component activities.

build requires more decision making, but initially at a very broad level. For example, should it be a flexible pavement or a rigid pavement, and, if rigid, with joints or continuously reinforced? How thick should it be and with what kind of materials? Once decisions are firm and commitments are made, the further level of influence of future actions on the future project costs will decrease.

In the same manner, decisions made during construction, even within the re-

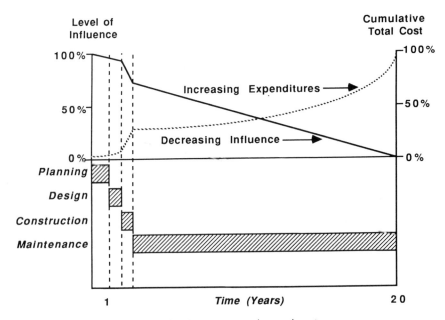

**Figure 4.2** Influence level of PMS subsystems on the total costs.

maining level of influence, can greatly impact the costs of maintaining or rehabilitating the pavement. For example, lack of quality control or substitution of inferior materials may save a few dollars in construction costs, but the extra maintenance costs and user delay costs due to more frequent maintenance activities may consume those "savings" several times over.

With construction completed, attention is now given to maintaining the existing pavement at a satisfactory level. The level-of-influence concept can also be applied to the subsystems of a maintenance management system (MMS). Expenditures during the planning phases of rehabilitation are relatively small compared to the total maintenance cost. However, the decisions and commitments made during early phases of a rehabilitation project have far greater relative influence on what other maintenance expenditures and user costs will be required later.

## 4.4 PAVEMENT MANAGEMENT AT THREE LEVELS

The various activities in a PMS, including decision making, have been categorized into the network level and the project level. The network level can be further divided into the project selection level and the program level. The project selection level involves prioritization to identify which projects should be carried out in each year of the program period. At the program level, budgets are established and general allocations made over an entire network.

This three-level concept is not new and represents the situation that actually occurs in a number of agencies. Also, the terminology sometimes overlaps; for example, in

# Pavement Management Levels and Functions

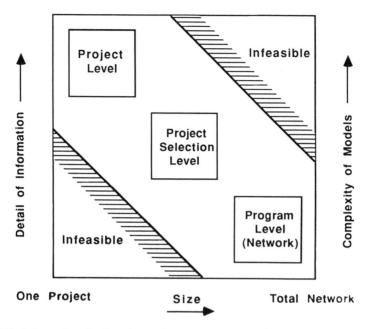

**Figure 4.3** Information detail and complexity of models for a three level PMS.

some papers when "project level" is mentioned, the "project selection level" actually may be meant, and in other cases when the "network" level is mentioned, it is the "program" level which is involved.

If these three levels exist, decisions at each level must interface with one another. For example, the interaction between project selection level and program level is evident in the case of estimating the budget required at the program level because it requires information on the candidate projects.

This three-level concept is illustrated in Figure 4.3. The lower-left triangle represents an area of unreliability because too little information is available for models at the project level, and the upper-right triangle is an area infeasible for modelling due to the size and complexity of the required models.

## 4.4.1 Models at the Three Levels

Several computer programs are available for use at one or more of the three PMS decision levels noted in this section. Some that are closely associated with the Texas State Department of Highways and Public Transportation (SDHPT) will be discussed here.

### 4.4.1.1 Project Level

The PMS models at the project level deal with technical concerns, such as detailed design decisions, for an individual project. As such, they require detailed information on specific sections or subsections of pavement, as shown in the upper left part of Figure 4.3.

The inputs for project level models include load, environmental factors, materials characteristics, subgrade properties, construction and maintenance variables, and costs. What specific data, and in what form, depends on the nature of the models. The typical output from such models would be a set of design strategies that minimize total life-cycle costs, including construction, maintenance, and user costs, while satisfying user, physical, and administrative constraints, such as performance requirements, minimum or maximum thicknesses, and funds availability.

One of the first major working PMS models developed at the project level was the Flexible Pavement Design System (FPS) [Scrivner 68]. This computerized model had several unique features, including the ability to separately calculate deterioration due to traffic and due to environment, and a life-cycle cost analysis subroutine.

Similarly, the Rigid Pavement Design System (RPS) [Hudson 72], was the first available working PMS model relating to rigid pavements at the project level. Subsequently, the System Analysis Method for Pavements (SAMP) [Lytton 75] was developed. It has a broader application than most existing working systems at the project level. Another working project level system with similar features to FPS and SAMP was the OPAC System developed in Ontario [Kher 70]. Any of these system models are still extremely useful guides to any agency setting up a pavement management system at the project level.

### 4.4.1.2 Project Selection Level

The project selection level involves decisions on funding for projects or groups of projects, as opposed to the program level which involves general budget allocation decisions for an entire highway network. The models employed at the project selection level are geared to less detailed data for a set of projects under consideration than ones at the project level, as shown in Figure 4.3. These models, under budget constraint conditions, involve prioritization models based on optimization, near optimization, or other techniques.

### 4.4.1.3 Program Level

The program level involves policy decisions regarding rehabilitation or maintenance for the network as a whole. At this level, allocation of budgets is the major concern, and the models should be designed to optimize the use of funds allocated to rehabilitation and maintenance. The intent should be to consider the status or condition of such as in terms of average serviceability or amount of deficient mileage. In order to do this, data is needed to determine the existing condition of the network as a whole so that the effects of different rehabilitation and maintenance policies and standards can be evaluated.

The network level of pavement management, subsequently described in Part Three of the book, and the examples, incorporates both the project selection and program levels identified in this section.

## 4.5 PMS FUNCTIONS

### 4.5.1 Historical Data Base

The function of a data base is to provide information for decisions at the various levels of a PMS. Less detailed information is needed for models at the network or program level than for models at the project level. In order to develop the wide variety of models required, including those for deterioration or performance prediction, agencies should develop long-term historical data bases.

Long-term and continuous monitoring of deterioration (roughness, surface distress, deflection, etc.) is needed to determine the relative damage attributable to traffic and environmental factors and to predict pavement performance. The National Highway Cost Allocation Study, completed in 1982, has examined the data available for determining causal relationships among traffic use, the environment, and maintenance costs. The FHWA Long-Term Monitoring (LTM) Program was established in an attempt to assess the problems of building a national data base that can develop improved pavement damage relationships [FHWA 82].

The Long-Term Pavement Performance (LTPP) study of the Strategic Highway Research Program (SHRP) should, by the mid to late 1990s, provide the most comprehensive data base in existence.

### 4.5.2 Information Flows

A total PMS functions from the details of the project level to the highest administrative level. At each level, the decisions require varying types and amounts of data, but the flow of information or sequence of actions has a high degree of similarity.

This similarity of information flow was originally illustrated in a NCHRP project [Hudson 79], which identified three basic subsystems: information, analysis, and implementation (see Figure 4.4). The concept is that in carrying out the various activities and making a decision, pertinent information is gathered and the consequences of the available choices are analyzed in the light of this information. Based on this analysis and on other nonquantifiable (perhaps political) considerations and constraints, a decision is made. Once made, the decision is implemented, and the results of the decisions are recorded in the data files or data base and are available to other management levels.

Although budget constraints, decision criteria, and other components of Figure 4.4 are not shown as subsystems, they are, of course, essential to the overall function of the management system. It was considered more useful to identify them separately since they are applied to the outputs of the subsystems. The activities listed within the subsystems of Figure 4.4 are meant to be illustrative and not all-inclusive.

The interface of the network level of Figure 4.4 with the higher level includes "committed" projects that come forward and are submitted as part of the optimized or prioritized program for review and approval. Any such program and its associated costs would likely go forward to the higher level of management as a recommendation, be evaluated with respect to the overall-all transportation policy and programs as well as the sector (i.e., highway, airport) budget allocation, and then be suitably modified if any program revisions are required.

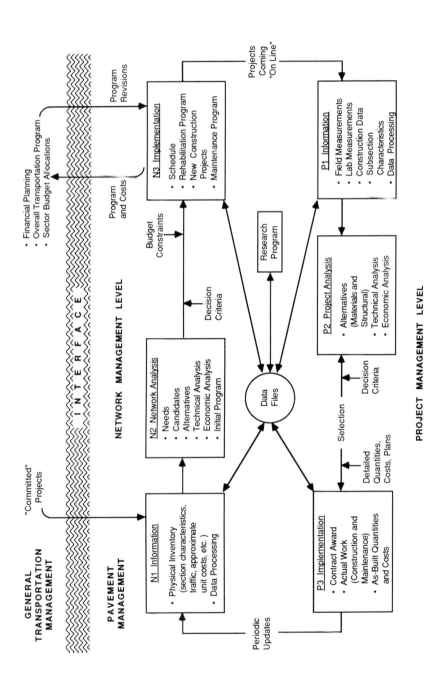

**Figure 4.4** Information flows in a pavement management system. [Hudson 79].

The subsystems shown in Figure 4.4, and key management activities that are applied to the outputs of these subsystems, are discussed in more detail as follows.

### 4.5.2.1 Information Subsystems for the Network Level

This subsystem involves the collection of data which provides the basic foundation for network analysis. The essential activities and types of data collected for this subsystem include the following:

1. Determination of what attributes of the pavement should be measured and/or what types of information should be acquired, and the methods, equipment, etc., needed
2. Identification of homogeneous sections or links in the network
3. Geometric and other inventory characteristics of the sections
4. Traffic measurements or estimates, accidents, etc., for each section
5. Field measurements for structural capacity, ride quality, surface condition, skid resistance, etc., on a sample or mass inventory basis and to a degree of accuracy and/or frequency appropriate to the class of road involved, agency resources, etc.
6. Estimate of approximate unit costs for new construction, rehabilitation construction, and maintenance
7. Identification of available resources (materials, contractor "capacity," physical plant, etc.)
8. Determination of criteria for maximum roughness and surface distress, and minimum structural adequacy and surface friction
9. Identification of "committed" improvements or projects from the overall highway, airport, and/or transportation program
10. Data from as-built projects and maintenance
11. Data processing for input to network analysis subsystem and for transmittal to data files or data bases

### 4.5.2.2 Periodic Updates of Information

Periodic updating of the information subsystem of Figure 4.4 is shown separately even though such updating may be conducted within the subsystem itself. In any case, what is important is the recognition of updating as a key activity and that it applies not only to periodic physical measurements but also to recording changes to the inventory as each project is completed.

### 4.5.2.3 Network Analysis Subsystem

The essential function of the network analysis subsystem is to consider the pavement improvement and/or maintenance needs and to arrive at a program of rehabilitation, new construction, and maintenance. This is accomplished through the following activities:

1. Identification of needs and "candidates" for improvement. Future needs identification requires the application of deterioration or performance prediction models.

2. Generation of alternatives for each candidate project or maintenance section.
3. Selection of program analysis period, discount rate, minimum ride quality levels, etc., for technical and economic analysis; also, identification of what the basis will be for deciding on the final prioritized program (i.e., solely economic or partially economic and partially nonquantitative).
4. Technical analysis of each alternative in terms of estimating performance.
5. Economic analysis of each alternative in terms of calculating life-cycle costs, and benefits or cost-effectiveness.
6. Development of initial program for new construction, rehabilitation, and maintenance.

### 4.5.2.4 Program Decision Criteria and Budget Constraints Applied to Initial Program

The decision criteria and budget constraints applied to the initial program resulting from the network analysis subsystem may simply involve a selection of those projects and that maintenance program which can be done within some available budget. This budget may have been fixed at the higher management level, or several alternative budget levels may be considered.

The projects falling below the budget cutoff would then be put back on the candidate list for consideration the following year, or subsequent years.

Some agencies designate separate budgets for new construction, rehabilitation, and maintenance. As well, some transportation departments allocate budgets by region or district. Whatever the particular practice is, the framework of Figure 4.4 should be applicable.

### 4.5.2.5 Implementation Subsystem

The implementation subsystem of the network management level of Figure 4.4 derives from the previously mentioned application of the decision criteria and budget constraints. It would list the final program and schedule for the new construction and rehabilitation projects, within the analysis or program period, plus the annual maintenance program. In some agencies, this program may be subject to final approval from the higher management level, which has been reflected in Figure 4.4 as a submission of program and costs to this higher level. Program revisions, as also shown in Figure 4.4, may or may not be required.

### 4.5.2.6 Interface Between Network Level and Overall Transportation System Management

Since the major output at the network level is a prioritized or optimized program (subsystem N3 of Figure 4.4), the interface mechanism with the higher level of highway or transportation management should primarily relate to priority programming at this higher level.

The priority programming process as practiced at the broad highway level contains a number of basic elements or steps, as listed in Table 4.1 [TRB 78]. It is primarily based on network level information and analysis, but feedback from project level

**Table 4.1  Steps in the Highway Programming Process [TRB 78]**

1. Project initiation
   (a) technical sources
   (b) nontechnical sources
2. Initial listing
   (a) headquarters
   (b) district
   (c) county
3. Preliminary analysis
   (a) available data and analyses
   (b) planning report
4. Combined listing, first draft
5. Advanced analysis and prioritizing
   (a) technical prioritizing
       (1) sufficiency ratings
       (2) priority ratings
       (3) option-evaluation techniques
       (4) input from other agencies
   (b) nontechnical prioritizing
       (1) political commitments
       (2) legislative mandate
       (3) emergency
       (4) special emphasis
       (5) commitments to other agencies
       (6) system continuity-connectivity
       (7) position in pipeline
   (c) feedback from project planning and development
       (1) development of alternatives/joint development
       (2) environmental analysis
       (3) community and technical interaction
       (4) input from other agencies
6. Combined listing, second draft
7. Financial analysis
   (a) categorical grants
   (b) geographical distribution
   (c) fiscal-year fund projections
   (d) manpower analysis
   (e) financial modifications
8. Preliminary program (projects vs. projected allocations)
9. Executive session
10. Short-range program, first draft
11. Executive and legislative review
12. Short-range program, final draft
13. Scheduling
14. Monitoring
15. Modifying

analysis is also included. This is consistent with the general cyclical flow of information within a PMS, as illustrated in Figure 4.4.

### 4.5.2.7 Projects Coming "On-Line" (From Network Implementation)

This has been separately identified in Figure 4.4 simply to recognize the importance of transforming a project from a network level program to action at the individual project level. Such individual projects would normally come "on line" one or more years before their scheduled construction. This lead time may only need to be one year for certain overlay projects and perhaps several years for a complex project with environmental and other approval requirements.

### 4.5.2.8 Project Level Information Subsystem

This subsystem involves the collection and processing of more detailed data, appropriate to the size and type of project, so that the project analysis and subsequent implementation may proceed. The types of data and component activities may include the following:

1. Identification of homogeneous subsections within the project or section length
2. Field measurements for or estimates of:
   (a) geometrics (lane widths, layer thicknesses, etc.)
   (b) traffic volumes and equivalent axle loads
   (c) structural adequacy, roughness, surface distress, surface friction, etc.
3. Laboratory measurements to determine material properties
4. Acquisition or estimates of unit costs of materials, construction, etc.
5. Identification of criteria or standards for maximum roughness, maximum surface distress, etc.
6. Collection of climatic or environmental data
7. Collection of available data on construction and maintenance variability
8. Data processing for use in the project analysis subsystem and for transmittal to data files

### 4.5.2.9 Project Level Analysis Subsystem

The project analysis subsystem of Figure 4.4 primarily involves design where new construction or rehabilitation projects are concerned. However, the terminology and concepts used also allow for maintenance treatment alternatives to be analyzed. A list of activities for this subsystem would include the following:

1. Generation of current and future materials and layer thickness combination alternatives, plus maintenance treatment alternatives
2. Selection of life-cycle analysis period, discount rate, etc., for technical and economic analysis
3. Technical analysis of alternatives in terms of predicting distress and performance
4. Economic analysis of alternatives to determine life-cycle costs and benefits

# Pavement Management Levels and Functions

### 4.5.2.10 Project Level Decision Criteria and Selection

The decision criteria applied to the various alternatives from the project analysis subsystem, in order to select the best one, may involve both quantitative and nonquantitative factors. These factors should reflect the needs of the network as perceived by the decision maker. A least cost alternative may be selected for example. Or, previous experience and judgment may be combined with the economic-based criterion.

### 4.5.2.11 Detailed Quantities, Costs, and Plans

The documentation of detailed quantities, cost estimates, and plans of the project alternative selected has been separately identified in Figure 4.4. This is not only to show its importance in finalizing the output of the project analysis subsystem but also to stress its importance as input to the implementation subsystem.

### 4.5.2.12 Project Level Implementation Subsystem

This subsystem represents the achievement of a final physical reality from all preceding subsystems of both the network and project levels. Where new construction or rehabilitation is concerned, it includes contract bids and awards, actual work activities, construction control, and finally, documentation of as-built quantities, costs, and geometrics for updating the network information base and for transmittal to the data files.

Where maintenance is concerned, this subsystem would include the actual work performed, quantities, schedules, costs, etc., comprising the application of what is usually termed maintenance management to individual section or project lengths. Maintenance management systems are usually, however, applied to regional or district networks.

### 4.5.2.13 Data Files and Research Programs

The basic subdivision of pavement management into network and project levels is convenient from the point of representing the major levels of decision making. However, the data base or files and the research program can apply to both levels.

Data bases properly designed, operated, and updated are invaluable to efficiently carrying out the activities of both the network and project levels of pavement management.

Research programs, and individual projects within a research program, can involve either or both levels of pavement management. The elements of pavement research management and research implementation guidelines are subsequently discussed.

## 4.6 KEY CONSIDERATIONS IN APPLICATION OF A TOTAL PAVEMENT MANAGEMENT SYSTEM CONCEPT

There are several key considerations in applying a total pavement management system concept, including the following:

1. The need for precise, understandable definitions, but a framework or structure which allows flexibility to use particular practices or methods
2. The need for people in the agencies who have qualifications appropriate to the various activities of pavement management, such as economics, structural analysis, computing, statistics, field measurements, etc.
3. The need for effective incorporation of all the technical, economic, and other factors
4. A well-developed interfacing mechanism between the policy level of transportation management and pavement management at the network level; also, properly coordinated interfacing between the network and project levels of pavement management
5. A well-developed interfacing mechanism between maintenance management and other areas of pavement management

## 4.7 THE FUNCTION OF PAVEMENT EVALUATION

Evaluation is a key part of pavement management because it provides the means for seeing how well the planning, design, and construction objectives have been satisfied. It is used in a broad sense at this point and is directed to measuring and assessing the outputs of a pavement. Part Two subsequently develops the concept of pavement evaluation in much more detail. Figure 4.5 is a schematic representation of the major types of pavement outputs versus time. Some or all of these outputs would be predicted in design, and would then be actually measured as part of the evaluation when the pavement is in service.

In Figure 4.5 the surface distress output has reached a limit of acceptability before any of the other outputs. At this point, the end of the initial service life of the pavement, some rehabilitation measure has been implemented, as shown by the vertical discontinuity. The rehabilitation measure has also been shown to affect the other outputs, such as increased structural adequacy, improved serviceability, improved surface friction, lower maintenance costs, and lower user costs.

The service life of the rehabilitation measure, in the example of Figure 4.5, is then ended by the serviceability reaching a minimum acceptable value. At this point, another rehabilitation measure has been applied and again the other outputs have been affected. Figure 4.5 thus demonstrates that any one or more of the "outputs" of a pavement can reach a limit of acceptability one or more times during the life-cycle or analysis period. The actual measures that can be used for these outputs are subsequently discussed in detail in Part Two.

The function of pavement evaluation in a pavement management system can be summarized as that of measuring and assessing these outputs periodically in order to:

1. Provide data for checking the design predictions and updating them if necessary
2. Reschedule rehabilitation measures as indicated by these updated predictions
3. Improve design models
4. Improve construction and maintenance practices
5. Update network programs

## Pavement Management Levels and Functions

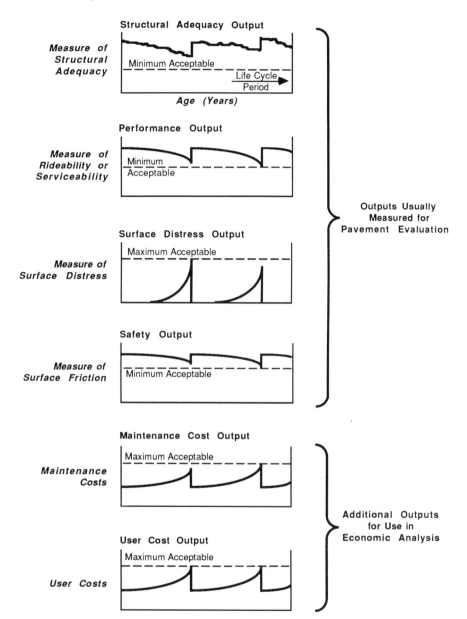

**Figure 4.5** Major types of pavement outputs.

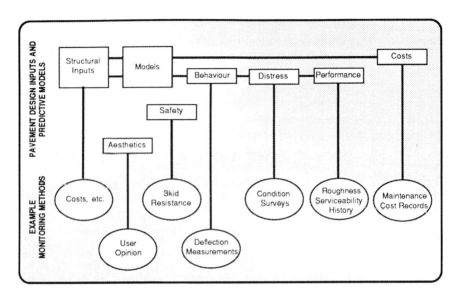

**Figure 4.6** Simplified prediction portion of pavement design and related examples of types of periodic evaluation measurements.

### 4.7.1 The Uses of Evaluation Information: An Illustration

An Illustration to show more explicitly the uses of evaluation information is given in Figure 4.6. It is split into an upper and lower half, the upper half including predictive models along with inputs and outputs, and the lower half showing examples of periodic measurements. These measurements plotted versus time have been portrayed as outputs in Figure 4.5.

The input variables of Figure 4.6 include the physical structure and material strengths, and can be monitored by physical testing and sampling to provide direct information about layer thicknesses and materials properties. Load testing of the total pavement structure is not included because it involves pavement response or behavior of the pavement and not the input information itself.

Safety response could in a broader way be categorized in a general response system; however, it is amenable to treatment and evaluation alone and is normally handled by measuring the surface friction or skid resistance of the pavement.

Costs can be monitored only if records of expenditures, including initial and future (i.e., overlay) construction costs and subsequent pavement related maintenance expenditures are accumulated.

Behavior can be defined as the immediate response of the pavement to load. Thus, deflection tests of all types fall into this category. It should be noted that these load-testing techniques evaluate only the behavioral response of the pavement and not the physical properties directly.

Distress can be defined as limiting response or damage in the pavement. Thus, the accumulated damage that the pavement has suffered is monitored and evaluated.

Because maintenance may have been performed on some of the distress, the evidence of this maintenance in the form of patches and sealed areas should also be monitored. Such monitoring is done routinely by many agencies in the form of condition surveys.

Another category of pavement evaluation of major interest is pavement performance. Because performance is the serviceability history of the pavement, its evaluation implies a time-related accumulation of data. This is best accomplished by periodic measurement or monitoring of the roughness of the pavement. Roughness is directly related to (user) serviceability, as described in detail in Part Two. At the very minimum, evaluation of performance requires two pavement serviceability estimates: one at the time of evaluation and a previous one, usually at the time of construction. A single serviceability measurement at the current time can tell the pavement engineer whether or not the pavement is below the currently desirable level. The change in serviceability from some prior time provides information about the performance history.

A seventh, perhaps relatively less important but complex aspect of pavement evaluation, is aesthetics. How does the pavement "look" to the user and to the designer? This evaluation is totally subjective, and few if any major studies have involved the subject.

The foregoing pavement evaluation measurements interact and there is overlap among them. However, they should not be confused or used interchangeably. For example, the fact that some people evaluate serviceability level using a serviceability equation that includes cracking and patching terms does not mean that the equation provides an adequate evaluation of distress and performance. A serviceability equation is primarily a performance evaluation tool. Behavioral measurements can be used to estimate inputs only in conjunction with some type of theory or model.

### 4.7.2 User-Related Evaluation versus Engineering Evaluation

Among the types of pavement evaluation, most agencies consider the following four as most important: serviceability, structural adequacy, surface distress, and safety. Safety is primarily in terms of surface distress but can include such factors as deep ruts which may affect steering conditions, hydroplaning potential, and icing potential. It is clearly a user-related measure. The other three types of evaluation, however, have often been confused and erroneously interchanged. They should be considered in terms of:

1. Functional behavior, using the serviceability-performance concept
2. Engineering or mechanistic types of evaluation of the pavement structure

In general, serviceability-performance is concerned with the overall function of the pavement; that is, how well it performs its function as a riding surface for vehicular traffic. By and large, this is also the area of main concern to the pavement user.

On the other hand, the mechanistic evaluation of pavement structures is of vital interest to engineers. An understanding of the interrelationship between these two types of evaluation is also vital. Whereas serviceability is a measure of present condition, the mechanistic or structural evaluation is used to estimate the future response of the pavement to load. The relationship can be illustrated by a weak, cracked pavement today with high serviceability, but the prospects are for a rapid loss in serviceability.

### 4.7.3 Pavement Evaluation with Respect to User Costs

Decisions to improve pavements in practice are almost always based on structural, serviceability, distress, safety, or maintenance cost considerations. However, excessive user costs arising from surface characteristics can in some situations be an equally valid criterion for improvement needs and can in fact form the basis for determining the priorities of such needs.

Most past and indeed current practice has been to consider only capital, maintenance, and engineering or administrative costs, with the implicit assumption that user costs do not vary with level of serviceability, surface distress, and extent and time of the improvement. In other words, all strategies are assumed to result in equal user costs. In fact, such user costs may vary significantly and they should not only be evaluated, but also used in the economic analysis of alternative strategies, as subsequently discussed in more detail in Part Four.

## 4.8 SUMMARY

Pavement management is a broadly based process which encompasses the entire set of activities required to provide and maintain pavements. Several important concepts presented in this chapter can be summarized as follows:

1. Pavement management has two basic operating levels of activities and decision making: network and project. In some agencies, the network level incorporates a project selection and a program level.
2. Four of the major subsystems, planning, design, construction, and maintenance, have different degrees of influence over the life of a pavement, with planning having a high initial influence and maintenance having a higher degree of influence as the pavement ages.
3. A data base of historical long-term pavement performance represents a very important function and need. It provides the basis for developing a wide variety of models, including those for pavement deterioration.
4. Proper information flows also represent a very important function in pavement management for both the network and project levels, and between them.
5. Pavement evaluation represents still another key function in that it represents the means for checking how well design, construction, and maintenance objectives have been satisfied, for updating schedules and other information, and for decision support.

## REVIEW QUESTIONS

1. Describe how the level of influence concept applies to pavements.

## Pavement Management Levels and Functions

2. Describe the relationship between a three-level structure of pavement management.
3. Describe the information flows at the project and network level.
4. What are the key considerations or requirements for total pavement management?
5. Clearly, and in summarized form, describe the difference between pavement distress and performance.

Chapter 5

# Using PMS as a Research Planning and Technology Improvement Tool

## 5.1 IDENTIFYING RESEARCH NEEDS

Some of the earliest work in pavement management systems development was done to provide a rational framework for organizing and coordinating existing knowledge on pavements and for projecting future research needs. That purpose is still valid today. Unfortunately, because of the many details involved in pavement design, construction, and maintenance, and the availability of a large amount of technology, agencies often find themselves working at cross purposes, internally, without proper coordination on pavement research.

When NCHRP Project 123 began [Hudson 70], one of the primary objectives was "to delineate additional profitable areas of research in the design, construction, maintenance and economics of pavements." In current pavement management terms it could be said that the objective was to provide a systematic plan of research for continued improvement of pavement design, construction, and maintenance technology. A subsequent NCHRP systems project [Hudson 79] also showed that a pavement management framework provides an ideal way of identifying and structuring research needs. More recently, it has been shown that a pavement management framework is very suitable for identifying research issues and needs, and for the innovation that should result [Hudson 92].

The remainder of this chapter addresses the use of a PMS in research planning and technology improvements, while Chapter 34 (Part Five) subsequently considers the management aspects of implementing research results within the context of a PMS. Then, Chapter 43 (Part Seven) considers research and innovation in terms of future direction.

## 5.2 SYSTEM PARAMETERS AND THE STATE OF THE ART

If the pavement "system" and its subsystems were completely understood, a perfect, invariate set of parameters that defined the system could be developed. Each of these parameters would be a function of space, time, geometry, and other variables. However, there are knowledge gaps, which should decrease in the future but will likely exist for quite some period of time. New or improved knowledge will come from continued study, the development of better models, the application of these models, and the feedback of information to improve the models.

At the present time, the parameters that must be considered in a pavement management system are highly dependent on the state of the art for:

1. The model being used
2. Past experience on which to base knowledge of pavement behavior and the significant factors involved
3. The quality of the instrumentation or measurement techniques available to determine the parameters
4. The quality and extent of the information and data base available
5. The inherent variability governing the amount of data required to define the parameters adequately

These five factors constitute much of the problem in developing improved pavement management systems.

The approach of isolating the various subsystems or factors in pavement management in order to address research needs can give misleading results. For example, trying to model only the effect of environment on pavement deterioration would be incomplete because there would likely be an interaction of environment with load applications.

Another example of subdividing the problem has been the approach of developing design and other methods for immediate use by one part of the agency while another, usually in the research section, is trying to solve the problem in terms of a sophisticated theory to give perfect answers. If this theory ignores or does not build on what is currently being used, the effort is wasteful and will not likely lead to a comprehensive, efficient, and economical method.

In reality, the problem will ultimately be solved in a cyclic way, as shown in Figure 5.1, and improvement will come in gradual steps. Quantum increases would be ideal but not likely. The process of Figure 5.1 makes use of other research, methods, and information. Also, the current state of the art should be used in the initial, perhaps

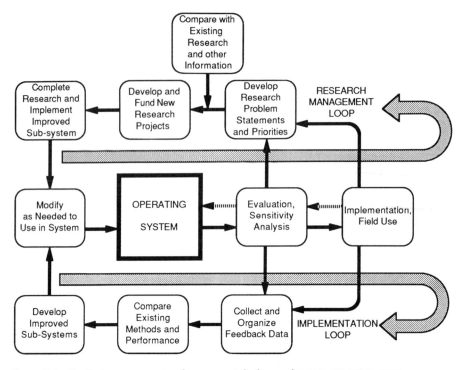

**Figure 5.1** Cyclic improvements of pavement design and management system.

crude, systems model. Sensitivity analyses can be very useful in continuing improvements.

Thus, the way model building, selection of parameters, and the entire system development relate to each other begins to become apparent. In some components, such as traffic, it seems easy to define the significant parameters. However, there are questions as to the form in which the data are to be provided in the model and the way they should be summarized. For example, the AASHO Road Test models involved equivalent 18-kip (80 kN) single-axle loads; however, the original data also includes actual vehicle loads, placement, and other factors such as tire pressure and tire width.

Environmental variables have historically been rainfall, temperature, and depth of frost penetration. However, except for a few theories such as those for slab restraint, the models with environmental variables have involved crude correlations. The use of these correlations causes problems in developing general models because the experiment and the data used to develop the correlations were basically applicable only to a particular situation or locale.

## 5.3 FUTURE ADVANCES IN PAVEMENT MANAGEMENT

Advances in pavement management will likely occur through one or more of the following:

1. Continuing incremental improvements in current technology
2. More widespread use of pavement management by public agencies and the implementation experience acquired
3. New equipment and methods, and their automation
4. Application of new technologies

The body of knowledge in the pavement area has largely been built through incremental improvements in the technology rather than spectacular breakthroughs. This will undoubtedly continue, ranging, for example, from more powerful microcomputers and better software to improvements in materials, construction, and maintenance technology.

As more public agencies implement pavement management systems, the experience base will grow and should in itself contribute to advances in the technology. For example, implementation experience to date has already demonstrated that near-optimization techniques, incorporating a heuristic, marginal cost-effectiveness approach, are quite applicable to the priority programming of rehabilitation improvements for pavement networks, as an alternative to mathematical programming for the optimization [Haas 85].

Many innovations in automated equipment and methods offer major potential for improving almost all aspects of pavement management technology, as subsequently discussed in Part Seven on "Looking Ahead." These include, for example, a very promising new technology in construction involving a different method for accomplishing asphalt compaction [Abdelhalim 87], and the application of knowledge-based expert systems technology. A comprehensive application example of the latter [Hajek 87a and 87b] includes a procedure for selection of pavement preservation treatments in Ontario. A major conclusion of the work was that it should be applied to well-defined areas in the depth and detail required for practical application, commonly available and well-documented software should be used whenever possible, and the expert system should be integrated with the existing pavement management activities of the agency.

However, despite the tremendous attention and expectations for the applicability of expert systems in the pavement field, some caution must also be expressed. They are not a panacea for all problems [Lamb 87] and a lot of hard work will be required to develop useful, practical applications. This observation applies to all new technologies. Further discussion on this subject is also subsequently contained in Part Seven.

### 5.3.1 Effects of Large-Scale Research Programs

The best example of a large-scale research program which should have an impact on pavement management is the Strategic Highway Research Program (SHRP). This effort ($150 million over five years) is generally directed to the areas of pavement performance, asphalts, maintenance cost-effectiveness and control of snow and ice, and concrete and structures.

Pavement management as a process per se will not likely be affected by these research efforts. But they should have a major impact on the technology of the process, particularly with regard to the following:

1. Field data acquisition (improvements in test methods, data processing and data interpretation)
2. Performance prediction (substantial improvements in deterioration modeling)
3. Maintenance and rehabilitation alternatives:
   (a) Materials (better methods of specifying and evaluating)
   (b) Treatments (improved capability of evaluating effectiveness)

Two major caveats should be placed against the impacts of these research programs. First, the results will not provide complete solutions to the massive number of technical problems existing in the pavement field, even though they should make substantial contributions. Second, a lot of technology and knowledge is available today which is not being used as effectively as possible. As well, major efforts will be required before the SHRP results can be put into practice.

### 5.3.2 Evolution of Pavement Management

Pavement management, as a process, will not likely evolve substantially from its present state for at least the next decade. The reasons are that there is a good acceptability among agencies and the major needs are seen to be in technology improvements within the process (as recognized by such initiatives as SHRP) rather than the process itself. Some of the prospects for future evolution are further discussed in Part Seven.

## 5.4 TOWARD REALIZING FUTURE PROSPECTS FOR PAVEMENT MANAGEMENT

It is beyond the scope of this chapter to develop a "blueprint" for realizing the future prospects of pavement management. That would ideally be accomplished in an effort similar to that involved in developing an AASHTO pavement design guide. While numerous technical papers plus workshop and symposium results can provide substantial assistance, a major amount of research is needed for codifying the next major level of pavement management and for significantly improving the state of technology. Again, Part Seven identifies some of the major opportunities for realizing these future prospects.

## 5.5 ESTABLISHING PRIORITIES

One of the important parts of research management is that of establishing priorities for work to be done. Almost no research budget is adequate for attacking all perceived problems. Too often in the past, priorities have been set by the main interests or abilities of existing research staff rather than by needs. This can be overcome to a large degree by identifying the key issues based on a well-developed working pavement management system. Sensitivity analyses can be run with the working system to determine the areas or parameters in the model that seem to affect the output of the system most. These results can be compared with estimates of the accuracy with which the parameters or models are known or can be determined. By combining this information, a priority list of important factors can be determined. This priority list can be compared to research costs and potential payoff or benefit to establish actual research program priorities.

## 5.6 IMPLEMENTING RESEARCH RESULTS

Proper implementation of any research results, as discussed further in Part Five, is best begun at the time that the research is first formulated. This implementation should directly involve the research "customer," e.g., the operating agency, and it requires a proper understanding of the problem. Certainly it may be appropriate to have people specifically charged with implementation of results even within the staff of the pavement management system. However, these people must work closely with the research team and with the operating team of the management system, and they cannot be successful if they are working independently of these groups. Such an implementation group can be responsible for preparing the necessary manuals, documents, and forms for putting the revised system into practice.

The pavement management system thus provides the required organizational structure for both defining research needs and providing a mechanism for implementation of research findings.

Throughout this book, major issues and research needs are inherent or explicitly identified for the various activities, methods, and examples presented. While it is not possible or perhaps even useful to discuss all the specific, detailed research needs and priorities, because they can change very rapidly as innovations and findings are brought forth by the pavement research community, the book concludes with Part Seven on "Looking Ahead."

## REVIEW QUESTIONS

5.1 List several reasons why the state of the art in pavement management should continue to improve.
5.2 Discuss the potential for improving the pavement management process.

# References to Part One

[AASHO 52]        American Association of State Highway Officials, "Public Roads of the Past: 3500 B.C.–1800 A.D.," AASHO, 1952.

[Abdelhalim 87]   Abdelhalim, A. O., W. A. Phang, and R. C. G. Haas, "Realizing Structural Design Objectives Through Minimization of Construction-Induced Cracking," Proc., Sixth International Conference on Structural Design of Asphalt Pavements, University of Michigan, Ann Arbor, July 1987.

[Barrie 78]       Barrie, D., and B. Paulson, *Professional Construction Management*, McGraw-Hill Book Company, New York, 1978.

[CGRA 65]         Canadian Good Roads Association, *A Guide to the Structural Design of Flexible and Rigid Pavements in Canada*, CGRA, Sept. 1965.

[Conway 67]       Conway, R., W. L. Maxwell, and L. W. Miller, *Theory of Scheduling*, Addison-Wesley, 1967.

[Dreyfus 65]      Dreyfus, F. E., *Dynamic Programming and the Calculus of Variations*, Academic Press, 1965.

[FHWA 82]         Federal Highway Administration, "Long-Term Pavement Monitoring Program Data Collection Guide," Program Management Division, Office of Highway Planning, FHWA, Washington, D.C., April, 1982.

[FHWA 87]         "Our Nation's Highways: Selected Facts and Figures," Federal Highway Administration, No. 7558-106, 1987.

# References to Part One

| | |
|---|---|
| [Gass 64] | Gass, S. I., *Linear Programming Methods and Applications*, McGraw-Hill, 1964. |
| [Haas 70] | Haas, R. C. G., and B. G. Hutchinson, "A Management System for Highway Pavements," Proc., Australian Road Research Board, 1970. |
| [Haas 73] | Haas, R. C. G., W. A. McLaughlin, and V. K. Handa, "Systems Methodology Applied to Construction," Proc., National Conference on Urban Engineering Terrain Problems, Montreal, May 1973. |
| [Haas 77] | Haas, R. C. G., et al., "Pavement Management Guide," Roads and Transportation Association of Canada, 1977. |
| [Haas 78] | Haas, R. C. G., and W. R. Hudson, *Pavement Management Systems*, McGraw-Hill, 1978. |
| [Haas 85] | Haas, R. C. G., M. A. Karan, A. Cheetham, and S. Khalil, "Pavement Rehabilitation Programming: A Range of Options," Proc., First North American Pavement Management Conference, Toronto, March, 1985. |
| [Hajek 87a] | Hajek, J. J., "Life-Cycle Pavement Behavior Modeling Using a Knowledge-Based Expert System Technology," Doctoral Dissertation, University of Waterloo, 1987. |
| [Hajek 87b] | Hajek, J. J., G. J. Chong, R. C. G. Haas, and W. A. Phang, "ROSE: A Knowledge-Based Expert System for Routing and Sealing," Proc., Second North American Conference on Managing Pavements, Toronto, November 2-6, 1987. |
| [Hall 62] | Hall, A.D., *A Methodology for Systems Engineering*, Van Nostrand, 1962. |
| [HRB 62] | Highway Research Board, "The AASHO Road Test: Report 5—Pavement Research," HRB Special Report 61-E, 1962. |
| [HRB 71] | Highway Research Board, "Structural Design of Asphalt Concrete Pavement Systems," HRB Special Report 126, 1971. |
| [Hudson 68] | Hudson, W. R., F. N. Finn, B. F. McCullough, K. Nair, and B. A. Vallerga, "Systems Approach to Pavement Systems Formulation, Performance Definition and Materials Characterization," Final Report, NCHRP Project 1-10, Materials Research and Development, Inc., March 1968. |
| [Hudson 70] | Hudson, W. R., B. F. McCullough, F. H. Scrivner, and J. L. Brown, "A Systems Approach Applied to Pavement Design and Research," Published jointly by the Texas Highway Department, Center for Highway Research of The University of Texas at Austin and Texas Transportation Institute of Texas A&M University, Res. Rept. 123-1, March 1970. |
| [Hudson 72] | Hudson, W. R., R. K. Kher, and B. F. McCullough, "A Working Systems Model for Rigid Pavement Design," *Research Record 407*, Transportation Research Board, Washington, D.C., 1972. |
| [Hudson 73] | Hudson, W. R., and B. F. McCullough, "Flexible Pavement Design and Management: Systems Formulation," NCHRP Report 139, 1973. |
| [Hudson 79] | Hudson, W. R., R. Haas, and R. Daryl Pedigo, "Pavement Management System Development," NCHRP Report 215, November 1979. |

| | |
|---|---|
| [Hudson 92] | Hudson, W. R., and R. C. G. Haas, "Research and Innovation Toward Standardized Pavement Management," ASTM, STP 1121, 1992. |
| [Hutchinson 68] | Hutchinson, B. G., and R. C. G. Haas, "A Systems Analysis of the Highway Pavement Design Process," Research Record 239, Highway Research Board, 1968. |
| [Kher 70] | Kher, Ramesh, and W. A. Phang, "OPAC: Economic Analysis Elements," Research Report 201, Ontario Ministry of Transportation and Communications, May 1970. |
| [Künzi 66] | Künzi, H. B., W. Krelle, and W. Oettli, *Nonlinear Programming*, Blaisdell, 1966. |
| [Lamb 87] | Lamb, J., "Expert Systems: The Bubble Bursts," New Scientist, January 1987, p. 52. |
| [Lytton 74] | Lytton, R. L., and W. F. McFarland, "Implementation of a Systems Approach to Pavement Design," Research Record 512, Transportation Research Board, 1974. |
| [Lytton 75] | Lytton, R. L., W. F. McFarland, and D. L. Schafer, "Flexible Pavement Design and Management Systems Approach Implementation," NCHRP Report 160, 1975. |
| [Moder 64] | Moder, J. J., and C. R. Phillips, *Project Management with CPM and PERT*, Rheinhold, 1964. |
| [NA Conf. 85] | First North American Pavement Management Conference, Proc., Published by Ministry of Communications of Ontario, Downsview, Ontario, 1985. |
| [NA Conf. 87] | Second North American Conference on Managing Pavements, Proc., Published by Ministry of Communications of Ontario, Downsview, Ontario, 1987. |
| [Scrivner 68] | Scrivner, F. H., W. M. Moore, W. F. McFarland, and G. R. Carey, "A Systems Approach to the Flexible Pavement Design Problem," Texas Transportation Institute, Res. Rept. 32-11, 1968. |
| [Stark 72] | Stark, R. M., and R. L. Nicholls, *Mathematical Foundations for Design: Civil Engineering Systems*, McGraw-Hill, 1972. |
| [TRB 78] | Transportation Research Board, "Priority Programming and Project Selection," NCHRP Synthesis of Highway Practice 48, 1978. |
| [Wiley 66] | Wiley, C. R., *Advanced Engineering Mathematics*, McGraw-Hill. 1966. |
| [Wilkins 68] | Wilkins, E. B., "Outline of a Proposed Management System for the CGRA Pavement Design and Evaluation Committee," Proc. Canadian Good Roads Association, 1968. |

/ Part Two

# Data Requirements

Chapter 6

# Review of Pavement Management Data Needs

## 6.1 CLASSES OF DATA REQUIRED

One objective of a pavement management system is to coordinate all activities required for providing pavement structures in a cost-effective manner. These activities, for virtually all segments of the highway or airfield agency, have impacts of varying degrees on a comprehensive pavement management system. Support of these activities requires a broad data base involving each segment of the agency. The data base should include pavement condition and performance, among the various items.

While the focus of many existing pavement management systems (PMS) is on condition and performance of the surface and the structure, a comprehensive PMS uses data from a variety of sources. The classes of data needed include the following [Haas 91]:

- Section Description
- Performance Related Data
- Historic Related Data
- Policy Related Data
- Geometry Related Data
- Environment Related Data
- Cost Related Data

```
Section Description                R + M      Geometry Related Data
                                              • Section dimensions        R
Performance Related Data                      • Curvature                 R
• Roughness                        R          • Cross slope               R
• Surface distress                 R + M      • Grade                     R
• Deflection                       R          • Shoulder / curb           R + M
• Friction                         R + M
• Layer material properties        R          Environment Related Data
                                              • Drainage                  R + M
Historic Related Data                         • Climate (temperature,     R
• Maintenance history              R + M        rainfall, freezing)
• Construction history             R + M
• Traffic                          R + M
• Accidents                        R + M      Cost Related Data
                                              • Construction costs        R
Policy Related Data                           • Maintenance costs         R + M
• Budget                           R + M      • Rehabilitation costs      R
• Available alternatives           R + M      • User costs                R
  (maintenance & rehabilitation)
```

R:   data used primarily for rehabilitation          R + M:   data for both uses
M:   data used primarily for maintenance

**Figure 6.1** Major classes and component types of pavement data [Haas 91].

Figure 6.1 lists the foregoing classes and their components, and whether they are primarily applicable to maintenance or rehabilitation, or both.

All but the policy and cost related data classes provide background information required for the analysis and modeling of pavement performance. Comprehensive pavement management requires that data from each source be readily accessible to the pavement management staff. In large agencies, such as state transportation departments, each class of data may be the responsibility of a different section. Hence, there is a need for effective coordination and cooperation. In smaller agencies these functions may be handled by a staff of one or two engineers and technicians. However, organizing, acquiring, and recording the data in a systematic and accessible manner is necessary in all cases.

The data base is a central feature of a PMS as shown in Figure 6.2. The data base serves as the repository of the information required to support virtually all decisions in addition to those concerning maintenance and rehabilitation shown in Figure 6.1. Moreover, the quality of the data base will dictate the value of the PMS.

It has been common practice to maintain these data elements in a disjointed fashion, for example, construction records are maintained by the Construction Division, Planning keeps traffic records, and Operations keeps maintenance data. Under a PMS the data base function can be centralized so that all the concerned divisions have ready access to the needed data and duplication is reduced.

# Review of Pavement Management Data Needs

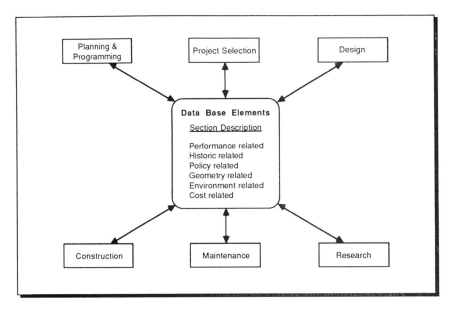

**Figure 6.2** Data base as a central feature of the PMS.

## 6.1.1 Uses of Pavement Management Data

Data for pavement management is used at the network and project levels. The network level data finds its end use in the priority programs that are implemented, plus budgeting and financial planning, Project level data is used for the engineering associated with specific sections or projects.

Some of the typical network and project level uses for the types of pavement data identified in Figure 6.1 are listed in Table 6.1, which has been based on [Haas 91].

## 6.2 THE IMPORTANCE OF CONSTRUCTION AND MAINTENANCE HISTORY DATA

In order to fulfill its purpose, a PMS must follow through from planning and programming design to implementation, including construction and periodic maintenance. Construction, of either a new pavement or rehabilitation of an existing pavement, converts a design recommendation into physical reality. Loss of performance, eventually leading to the need for rehabilitation, is identified within the ongoing process of data collection and evaluation. Such evaluation of pavement performance is also used for determining the current status of the pavement network.

Pavement data, collected over time, provides the basis for developing, updating, and assessing pavement performance models used in planning and programming and in design. Essential to such model development is data on the construction and maintenance of the pavement. Pavement construction data includes information on the as-

**Table 6.1 Typical Uses of Pavement Management Data [Haas 91]**

| Data Item | Network Level | Project Level |
|---|---|---|
| **1. PERFORMANCE RELATED** | | |
| Roughness | a) Describe present status<br>b) Predict future status (deterioration curves of roughness vs. time or loads)<br>c) Basis for priority analysis and programming | a) Quality assurance (as-built quality of new surface)<br>b) Create deterioration curves<br>c) Estimate overlay quantities |
| Surface distress | a) Describe present status<br>b) Predict future status (deterioration curves)<br>c) Identify current and future needs<br>d) Maintenance priority programming<br>e) Determine effectiveness of alternative treatments | a) Selection of maintenance treatment<br>b) Identify needed spot improvements<br>c) Develop maintenance quantity estimates<br>d) Determine effectiveness of alternative treatments |
| Surface Friction | a) Describe present status<br>b) Predict future status<br>c) Priority programming<br>d) Determine effectiveness of alternative treatments | a) Identify spot or section rehabilitation requirements<br>b) Determine effectiveness of alternative treatments |
| Deflection | a) Describe present status<br>b) Predict future status (deterioration curves)<br>c) Identify structural inadequacies<br>d) Priority programming of rehabilitation<br>e) Determine seasonal load restrictions | a) Input to overlay design<br>b) Determine as-built structural adequacy<br>c) Estimate remaining service life<br>d) Estimate remaining load restrictions |
| Layer Material Properties | a) Estimate section-to-section variability<br>b) Develop basis for improved design standards | a) Input to overlay design<br>b) Provide as-built records |
| **2. HISTORIC RELATED** | | |
| Maintenance History | a) Maintenance programming<br>b) Evaluate maintenance effectiveness<br>c) Determine cost-effectiveness of alternative designs and treatments | a) Identify problem sections |

## Table 6.1 *continued*

| Data Item | Network Level | Project Level |
|---|---|---|
| Construction History | a) Evaluate construction effectiveness<br>b) Determine cost-effectiveness of alternative designs and construction practices<br>c) Determine need for improved quality assurance procedures | a) Provide as-built records<br>b) Provide feedback to design |
| Traffic History | a) Priority programming<br>b) Input to estimate general performance/distress trends | a) Input for pavement design<br>b) Identify traffic handling methods<br>c) Estimate remaining service life |
| Accident History | a) Develop countermeasures<br>b) Priority programming | a) Identify high-risk sites<br>b) Develop countermeasures |
| **3. POLICY RELATED** | | |
| Budget | a) Priority programming<br>b) Selection of management strategies | a) Determine cost limitations |
| Available Alternatives | a) Selection of management strategies<br>b) Priority programming | a) Economic evaluation<br>b) Life-cycle cost comparisons |
| **4. GEOMETRY RELATED** | | |
| Section Dimensions | a) Develop general policy or standards | a) Determine section constraints |
| Curvature | a) Develop general policy or standards | a) Determine section constraints<br>b) Assess safety |
| Cross Slope | a) Develop general policy or standards | a) Assess drainage<br>b) Assess safety |
| Grade | a) Develop general policy or standards | a) Assess drainage<br>b) Assess safety |
| Shoulders/Curbs | a) Develop general policy or standards | a) Assess safety<br>b) Assess drainage |
| **5. ENVIRONMENT RELATED** | | |
| Drainage | a) Evaluate general network performance | a) Evaluate section performance |
| Climate | a) Evaluate general network performance | a) Evaluate section performance |

**Table 6.1** *continued*

| Data Item | Network Level | Project Level |
|---|---|---|
| **6. COST RELATED** | | |
| New Construction Costs | a) Priority programming<br>b) Selection of network investment strategies | a) Economic evaluation<br>b) Selection of strategy |
| Maintenance Costs | a) Priority programming<br>b) Selection of network maintenance strategies | a) Evaluation of maintenance effectiveness<br>b) Selection of maintenance sections |
| Rehabilitation Costs | a) Priority programming<br>b) Selection of network rehabilitation strategies | a) Economic evaluation<br>b) Selection of rehabilitation strategies |
| User Costs | a) Priority programming<br>b) Selection of management strategies | a) Economic evaluation<br>b) Selection of mitigation strategies |

built quality of the materials, such as the results of concrete flexural strength tests and asphalt concrete densities. Large variability of construction quality will result in poor performance, compared to pavements with uniform quality. Pavement maintenance data includes records of all maintenance activities that can affect the performance of the pavement such as crack sealing, patching, and surface seals. A high level of maintenance makes it possible to extend the life of the pavement beyond the expected design life.

## 6.3 THE IMPORTANCE OF PERFORMANCE RELATED PAVEMENT EVALUATION

Evaluation of all the pavement data items in Figure 6.1 is important. Of key concern is the data related to pavement condition, particularly that related to the road or airfield users. Users rate the pavement either consciously or subconsciously during every vehicle ride or ground operation of an aircraft.

The major purpose of performance related pavement evaluation is to determine the current condition of the pavement structure. Four key measures can be used to characterize or define the condition of the pavement (see Figure 6.1):

1. Roughness (as related to serviceability or ride comfort)
2. Surface distress
3. Deflection (as related to structural adequacy)
4. Surface friction (as related to safety)

In a complete sense, a "good" pavement provides satisfactory riding comfort to its users, does not require extensive maintenance for the repair of distress, is structurally adequate for the traffic loads, and provides sufficient friction to avoid skidding accidents. There are many ways of carrying out pavement evaluation, and unfortunately there is a multiplicity of terminology, much of it agency specific. However, such organizations as the American Society for Testing and Materials (ASTM) and the American Association of State Highway and Transportation Officials (AASHTO) have developed or are in the process of developing standard definitions. The reader should consult their most recent publication(s) for clarification. Rather than reproduce a thesaurus of terms in this book, every attempt is made to conform with ASTM or AASHTO standard usage, and/or to clearly indicate the meaning of the term.

The distinction between distress, roughness, structural adequacy, and surface friction of the pavement is particularly important. Distress is the physical deterioration of the pavement surface, such as potholes, cracking, and rutting, and it is generally but not necessarily visible. Roughness is derived from the longitudinal profile of the pavement surface and affects ride comfort or quality. Under the premise that pavements are constructed for the users, then roughness is the primary operating characteristic of pavements that affects users, and roughness therefore defines the serviceability or functional response of the pavement. Structural adequacy is the ability of the pavement to carry loads without resulting in undue distress. The safety of the pavement surface is primarily related to the surface friction or skid resistance of the pavement, but it can also be affected by severe rutting or potholes.

These four measures or characterizations, along with maintenance and user costs can be viewed as representing the "outputs" of the pavement, that is, they are the variables that can be measured to determine whether or not the pavement is behaving satisfactorily. These outputs would originally be predicted at the design stage and then periodically evaluated while the pavement is in service. The service life of the pavement would be reached when any one of the measures reach a minimum (or maximum, depending on the measure) acceptable level. At that point, if sufficient funds are available, rehabilitation would be carried out and a new service cycle would begin.

## 6.4 OBJECTIVITY AND CONSISTENCY IN PAVEMENT EVALUATION

The uses of pavement data, and its evaluation, have been listed in Table 6.1. In essence, pavement evaluation serves the planning and design activities of pavement management and can also serve construction and maintenance activities.

When evaluating existing pavements consistent and repeatable quantification of condition is an essential requirement. Many pavement evaluation schemes rely on the judgment and opinion of a human rater. While this may provide useful insight into condition, such evaluations lack uniformity and generally lose meaning over time as the attitude and ability of the rater change and/or new personnel are added. Whenever possible, evaluation should be performed objectively using calibrated instruments.

The primary need of an engineering evaluation of pavement condition is consistency of the data base across time and space. Frequently, agencies with large pavement networks rely on several teams for evaluating the pavements. Without detailed instructions and training in the evaluation process, there will be inconsistencies among ratings that invalidate the data. The problem is even more acute when the data is examined across time to evaluate pavement performance.

Thus engineering evaluation of pavements requires a well-documented set of practices and procedures, plus good training. For example, in the measurement and evaluation of pavement distress, the only practical method of survey for many agencies is with trained observers. In these cases, the observers should be given explicit instructions on methods of performing the survey.

## 6.5 COMBINING PAVEMENT EVALUATION MEASURES

Detailed pavement condition information is not necessary at all levels of pavement management. For example, senior administrators may only be interested in summary descriptions of the present status of individual sections, or the network as a whole.

In these cases, a tool is needed to summarize or aggregate the individual measures into a statistic for identifying the overall quality or condition. Several approaches or models for an overall pavement quality index have been used. Generally, they are based on adding or multiplying together the individual measures of pavement condition. Weighting factors are assigned to each measure to reflect their relative importance. Combined measures are further discussed in Chapter 12.

## REVIEW QUESTIONS

1. Identity the benefits of an engineering evaluation of pavement conditions.
2. What are the requirements for an engineering evaluation of pavement condition?
3. What is the purpose of a combined index of pavement condition?

Chapter 7

# Inventory Data Needs

## 7.1 INVENTORY DATA NEEDS

The management process requires an inventory of the existing facilities. Depending on the requirements of the agency, the level of detail in the inventory will vary. However, all pavement management systems need at least a rudimentary inventory of the pavement network. This inventory is essential for the development of a PMS. The inventory data base provides the pavement manager an accounting of the extent of the network, the types of pavements, their geometry, traffic levels, environmental conditions, etc. Essentially, it incorporates information on what exists now, plus past information on such items as maintenance and construction performed.

Developing the inventory of the pavement system can be a time-consuming task. The basic steps in the inventory include the selection of the variables or data elements that will be used in the inventory, definition of the pavement section description or referencing method, and assembling and recording the data into the database. Figure 6.1 which lists all the major classes and component types of data, including inventory data, can provide initial guidance for the selection.

## 7.2 TYPES OF INVENTORY DATA

The inventory of the network generally refers to the permanent features of the highway or airfield pavements. Because a wide variety of data elements are candidates for inclusion, a compromise between the level of detail desired and practical cost considerations for collecting and managing the data will usually be required. Some large agencies compile extensive and detailed inventories while other agencies operate with minimum data.

There are several major classes of inventory data to be considered in pavement management, and they would, for many agencies, include the following:

- Section reference and description
- Geometry
- Pavement structure
- Costs
- Environment (weather) and drainage
- Traffic

The geometry of the pavement refers to its classification and physical features of the pavements. Pavement structure defines the thickness and material types of the layers. Cost data includes construction, maintenance, rehabilitation, and user costs. Environment data identifies those local conditions that can influence pavement performance. Traffic data changes annually but it relates to the other data items and is therefore frequently incorporated into inventory data as a matter of convenience and relevance.

### 7.2.1 Selection of Pavement Management Sections

Highway agencies frequently have multiple methods of referencing the location of pavement sections. Historically, different sections or divisions in the agency have data collection and use needs that are not totally compatible with the needs of other divisions. For example, the planning section may use traffic control sections for the collection and storage of traffic data; the construction section may use a construction project numbering scheme; and operations may use a route milepost method for scheduling maintenance operations. A PMS needs to coordinate these functions. The PMS must use a permanent referencing system and the location methods used by all other divisions will be changed or cross referenced. Therefore, one of the first tasks required for the development of a PMS is to establish a common referencing method to identify pavement sections within the network. This can be one of the existing methods in the agency.

#### 7.2.1.1 Referencing Method

The development of the referencing system should start with a review of the current practices of the agency. Each of the divisions responsible for collecting pavement data, or associated data, should be identified and reviewed. For pavement management one of the existing methods may be selected or a new method may be defined. In

## Inventory Data Needs

either case, plans should be made for coordinating the different referencing methods to permit a free interchange of data. There are four basic methods of referencing pavement sections: route-milepost, node-link, branch-section, and geographic information system (GIS).

The route milepost system is widely used in state highway agencies. Each route is given a unique name or number and the starting point of the route is defined. Then mileposts are sequentially numbered along the length of the route.

In the node-link method, key points in the network are defined as nodes and the sections between these nodes define the links. Nodes are usually defined at intersections, boundaries, and points of change in the pavement characteristics, such as a change in surface type.

The branch-section method is used in the PAVER pavement management system developed by the Corps of Engineers [Shahin 79] for airfield pavements and extended for use by city and county highway agencies. General features of the pavement network are defined as branches and homogeneous units of the branches are defined as sections. In airfield pavements, a runway or apron may be defined as a branch, and homogeneous areas within the runway or apron would be defined as sections. In highway work, routes or streets may be defined as the branch and homogeneous sections within the route would be defined as sections.

Geographic information systems (GIS) use a coordinate system to define the location of each feature of the network. Connecting relationships between feature coordinates define the routes or branches. The geometric layout of the pavements can be completely defined with the coordinate system. Although definition of the coordinates is a time-consuming task to initially establish the data base, once the coordinates are defined and stored, the computer can be used to generate maps showing selected pavement features and new features can be readily added to the data base. The use of GIS based systems will probably expand as the cost of computers decreases and software is developed to support the use of GIS.

In 1990, the Federal Highway Administration (FHWA) developed an advanced PMS course. It includes a description of their support of the GIS development in terms of the Geographic Road Information Display System (GRIDS). Several state and local agencies have developed, or are actively developing GIS, for the management of their infrastructure, including the highway network and pavement management. Figure 7.1 demonstrates a feasible means for interacting PMS with GIS [NC DOT 88]. Note that the primary function identified for the GIS is the production of maps for a visual characterization of the pavement management reports.

### 7.2.1.2 Section Definition

Once the referencing method is established, the specific pavement sections must be defined for use in the data base. The methods used vary widely between agencies. Sections can be defined to have either uniform characteristics or fixed length.

The advantage of the fixed length approach is the simplicity of the data structure and the ease of locating sections in the field during data collection. The disadvantage of this approach is that pavement structure characteristics can change within the length of the section. For example, the Arizona Department of Transportation uses the route

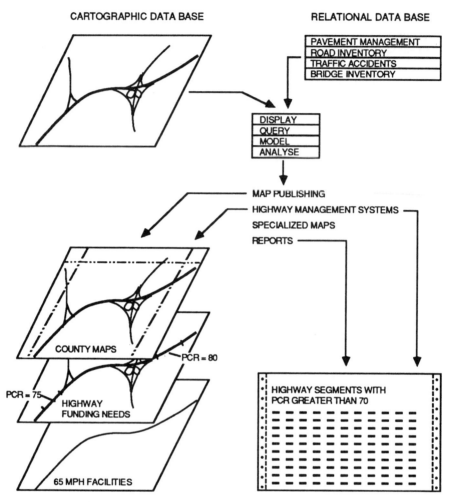

**Figure 7.1** Interaction of geographic information system and pavement management, based on a North Carolina DOT feasibility report [NC DOT 88].

milepost system for referencing sections. Data is collected at each milepost in the system. Management decisions are based on construction projects, yet these projects rarely begin or end at an exact milepost. Thus, the location of changes in the pavement structure, which can affect performance, are not identified in the data base.

The advantage of selecting pavement sections with homogeneous characteristics is ease of analysis of the data. The condition of each pavement section can be evaluated and maintenance or rehabilitation plans can be formulated. Section boundaries are usually defined based on selected control parameters, such as construction contract beginning or end, traffic levels, maintenance districts, intersection with another major facility, change in pavement type or geometry (number of lanes), etc. [RTAC 77].

# Inventory Data Needs

The length of the section can vary from a few hundred feet to several miles. The disadvantage of this approach is that data collection crews can have difficulty in identifying the boundaries between the pavement sections in the field. Also, in this approach, it may be necessary to subdivide the pavement sections in the future if rehabilitation projects change the pavement structure.

The definition of pavement sections for local agencies with many residential streets and cul-de-sacs is frequently difficult as there are many irregular shapes and short segments. A number of cities, such as Mesa, Arizona, handle this problem by treating all of the local and collector streets in a residential area as a homogeneous section. The general location of the residential areas is defined with coordinates. The management unit, or pavement section, is the total square yards of local streets and collectors within each of the residential areas.

The entire structure of the pavement management data base and all subsequent data collection and analyses are affected by the definition of pavement sections. Therefore, it is extremely important to have a method of defining and location referencing these sections that is workable, consistent, and amenable to updating.

## 7.2.2 Geometry Inventory

The geometry inventory defines the physical characteristics or features of the pavement sections. It can include, for each pavement section, the following:

- Location reference and length of section
- Classification (i.e, freeway or interstate, arterial collector, arterial, local out, runway, taxiway, apron, parking area)
- Number of lanes, widths, shoulder type and width, grade, cross slope, curvature, presence and dimensions of curbs

Geometric characteristics of the pavement, such as number of lanes and widths, shoulder type and widths, grade, cross slope, and curvature can be included in the inventory for basic planning information and to indicate whether the existing geometry meets current standards. This data is generally used in planning major rehabilitation projects to determine if reconstruction is required based on geometric consideration.

## 7.2.3 Pavement Structure Inventory

The pavement structure inventory is basically a record of the construction history of the project. For some agencies the only pavement structure data available is the type of surface. This is inadequate for a good PMS. At the other extreme are systems that contain complete construction details of the pavement construction history. A data base of this type would include the thickness, type of material, and year of construction for each layer in the pavement structure. In addition, results of construction quality control tests can provide such data on the quality of the materials such as the densities of the compacted layers, compressive strength of concrete, Marshall stability of the asphalt concrete layers, etc. Summary statistics, such as the mean, standard deviation, and coefficient of variation, can be used to capture the material quality data.

The pavement structure inventory should also have records of major maintenance or reconstruction projects, including type of treatment and layer thickness.

### 7.2.4 Cost Data

The cost inventory should include data on the cost of new construction, maintenance, and rehabilitation. It may also include user costs. Construction and rehabilitation costs can be compiled from records, estimates and surveys of recently completed projects. These costs should be updated on a regular basis, at least annually. If the agency has implemented a maintenance management system, average maintenance costs can be determined by analyzing the data records. Otherwise, maintenance costs must be estimated based on the expected performance of the maintenance crews and the condition of the pavements. User costs are estimated based on the traffic volumes, condition of the pavements, and models of vehicle operating cost.

### 7.2.5 Environmental (Weather) and Drainage Data

Environmental conditions can have a serious effect on the performance of pavements. When these conditions vary significantly across the geography of the agency's jurisdiction, a record of the local environmental conditions can assist the pavement manager in predicting performance and in the selection of pavement rehabilitation strategies. There are several measures that can be used as an index of environmental conditions, such as the Thornthwaite index, freeze-thaw cycles, freezing index, seasonal rainfall, or a regional factor developed by the agency.

Drainage and shoulder characteristics can have a direct impact on pavement performance and will also affect the selection and cost of maintenance and rehabilitation treatments. While there are many techniques of ensuring adequate or good drainage, the data characterizing drainage is usually either in terms of a porosity or permeability value, or subjectively recorded in terms of good, fair, or poor.

### 7.2.6 Traffic Inventory

Traffic data is required in pavement management for the prediction of performance and the assignment of priorities during the selection of rehabilitation projects. For the selection of projects, a measure of traffic volumes is required. For highway agencies the average annual daily traffic (AADT), with a breakdown into percent passenger vehicle and percent trucks, is a common measure of the total traffic on the section.

Performance modeling, on the other hand, requires an estimate of the heavy vehicle traffic that generates the majority of the distress. For highway pavements, the usual measure is the total 18-kip equivalent single-axle loads ($ESAL_{18}$) can be used to estimate the quantity of vehicles that damage the pavements or that the pavement has carried or is expected to carry. Traffic growth rates should be included for both the AADT and the $ESAL_{18}$. For airfield pavements, records should be maintained on the total numbers of movements of each aircraft class.

## 7.3 COLLECTING AND PROCESSING INVENTORY DATA

Once the data elements for the inventory and the pavement sections are defined, the sampling plan put into place, and the resources allocated, the actual data can be collected or assembled. Depending on the sophistication of the agency, this task can be relatively simple or very time consuming. Generally, most of the information required for the inventory should be located in the historical records of the agency. If computer records are available, assembling the data base may be as simple as reformatting the records into that required for the pavement management data base. However, agencies without a PMS may not have computerized records. In this case, hard copy records and manual files must be used for developing the data base.

Generally the construction history begins with the as-built plans. These can provide data on the length and width of the pavement project as well as the type of materials, thicknesses, and year of construction of each layer. There are cases, especially in local agencies, where a pavement structure may have evolved over time without being designed and without proper construction records. In these cases it may be necessary to rely on the recollection of people who have been with the agency for a long time to get an estimate of the construction history.

In some cases, the construction history will not be available in any form. If that occurs, the pavements can be cored or trenched to examine the structure. Generally, it is not necessary to have a separate program of coring for establishing the pavement construction history. The data can be collected as part of a structural evaluation, or for other reasons. For example, one agency worked with the utility company to develop a form that work crews can fill in whenever they cut a trench across a pavement. This form captures the thickness and material type for each layer.

Any data should be collected and processed systematically. Data collection forms should be designed for understandability, ease of use, and precise recording. If the data assembly is being performed in the office, the forms may be on a computer screen and the clerk assembling the data can enter information directly into the computer. If the data is being assembled away from the office, paper forms should be designed that will allow direct computer entry. Laptop or notebook computers that can be carried into the field also provide a viable alternative for recording data and eliminate the need to transcribe the data in the office.

The inventory data forms the basis for the pavement management system. Assembling this data can be a time-consuming and costly task, but it is essential and worthwhile.

## REVIEW QUESTIONS

1. What elements should be included in a pavement inventory?
2. What are the relative advantages and disadvantages of the fixed length versus the uniform characteristics methods for defining pavement management sections?
3. What is the single most important aspect of defining a pavement referencing system?

Chapter 8

# Pavement Performance

## 8.1 THE SERVICEABILITY-PERFORMANCE CONCEPT

The evaluation of pavement performance involves a study of the functional behavior of a section or length of pavement. For a functional or performance analysis, information is needed on the history of the riding quality of the pavement section for the time period chosen and the traffic during that time. This can be determined by periodic observations or measurements of the pavement riding quality coupled with records of traffic history and time. It is this history of deterioration of the ride quality or serviceability provided to the user, that defines pavement performance as shown on Figure 8.1 [Carey 60].

Until a measure of pavement serviceability was developed in conjunction with the AASHO Road Test [HRB 62], inadequate attention was paid to evaluation of pavement performance. A pavement was considered to be either satisfactory or unsatisfactory (i.e., in need of repair or replacement). Pavement design technology did not directly consider performance. Design engineers have varied widely in their concepts of desirable performance. For example, at one extreme an engineer asked to design a pavement for a certain expected traffic level for 20 years might consider the job properly done if little or no cracking occurred during the design period. On the

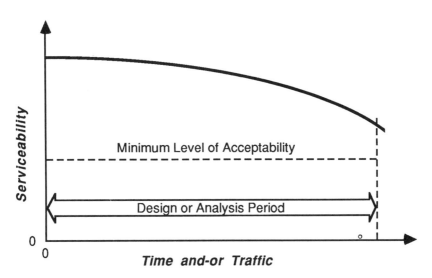

**Figure 8.1** Deterioration of ride quality or serviceability over time.

other hand, a second designer might be satisfied if at the end of the design period, the pavement had reached a totally unacceptable level of serviceability.

Many design methods involve determination of the pavement thickness required so that computed stresses or strains do not exceed some specified level. While it is clear that cracking distress will occur if the pavement is over stressed, information relating distress to functional behavior was not available prior to the AASHO Road Test. A method of performance evaluation was badly needed at the time of the Road Test, and it was fulfilled with the "serviceability-performance concept" developed by Carey and Irick [Carey 60]. They pointed out that serviceability must be defined relative to the purpose for which the pavement is constructed; that is, provide a smooth, comfortable, and safe ride. In other words, the measurement should relate explicitly to the user, who is influenced by several attributes of the pavement, including:

1. Response to motion as characterized by the particular pavement-vehicle-human interaction for a particular speed
2. Response to appearance, as characterized by such factors as cracking, patching, color, shoulder condition, etc.

### 8.1.1 Functional Pavement Rating

Definition of the functional behavior of the pavement as the ability to provide a smooth, comfortable, and safe ride required the development of a rating method to characterize these attributes, which depend on the user's perception of the adequacy of the level of service of the pavement. Thus, user opinions must be measured in order to rate the serviceability of the pavement.

A rating procedure requires construction of a scale for quantifying serviceability. Teachers often rate students on a scale of 0 to 100 percent; amateur golfers are rated by an arbitrary system called a handicap, which is derived as a percentage of their average scores over par for a period of time. Many arbitrary scales are used in a wide variety of applications. Development of a scale for rating pavements is complicated by the interaction between the vehicle occupant, the vehicle parameters, and the roadway characteristics.

Three basic requirements for the development of an arbitrary pavement rating scale have been defined [Hutchinson 64]:

1. Development of a suitable mathematical model to characterize pavement roughness
2. Development of a suitable model of highway vehicle suspension characteristics that may be used with the pavement roughness characteristics model for predicting the response of the vehicle
3. A quantitative knowledge of the response of humans to motion

These requirements imply that the user's perception of pavement quality depends primarily on the vertical acceleration experienced driving. Vertical acceleration in turn depends on the roughness characteristics of the pavement, the vehicle mass and suspension parameters, and the travel speed. But the pavement engineer is also interested in the characteristics of the surface. The vehicle serves as a filter which complicates the relationship between the user perception and the surface characteristics.

## 8.1.2 Development of the Serviceability Index

The WASHO Road Test [HRB 55] in the early 1950s showed the difficulty of establishing a failure condition for pavement sections. Subsequently, the idea of subjective average pavement ratings [Carey 60] to measure serviceability was based on the following five fundamental assumptions:

1. Highways are for the comfort and convenience of the traveling public. Stated another way, a good highway is one that is safe and smooth.
2. Users' opinions as to how they are being served by highways is by and large subjective.
3. There are, however, characteristics of highways that can be measured objectively and that when properly weighed and combined are in fact related to users' subjective evaluation of the ability of the highway to serve them.
4. The serviceability of a given highway may be expressed by the mean evaluation given by all highway users. Honest differences of opinion preclude the use of a single opinion in establishing serviceability ratings. The mean evaluation of all users, however, should be a good measure of highway serviceability.
5. Performance is assumed to be an overall appraisal of the serviceability history of a pavement. Thus, it is assumed that the performance of a pavement can be described if one can observe its serviceability from the time it was built up until the time its performance evaluation is desired.

**Pavement Performance**

Based on these fundamental assumptions, the Present Serviceability Index (PSI) measure was developed and used at the AASHO Road Test [HRB 62]. A similar, widely used technique was developed by the Pavement Design and Evaluation Committee of the Canadian Good Roads Association (currently the Transportation Association of Canada) in the late 1950s and early 1960s [CGRA 59, 65, 67, 71]. It was termed Riding Comfort Index (RCI).

In the AASHO and Canadian Studies, panels of raters drove over a number of pavement sections. Certain rules were established for the rating sessions [Carey 60, CGRA 59, 65]. Each panel member records an independent opinion on the type of form shown on Figure 8.2. The AASHO terminology for each such rating is *Individual Present Serviceability Rating*, with the mean of the individual ratings termed the *Present Serviceability Rating* (PSR). The Canadian equivalent was originally termed the *Present Performance Rating* but was changed in 1968 to *Riding Comfort Index* (RCI) to denote more explicitly the evaluation of pavement riding quality only [Wilkins 68].

The major difference between the two approaches, as shown by comparing Figures 8.2.a and Figure 8.2.b, is the construction of the scales. There are five descriptive cues in each; however, the Canadian scale has 10 numerical categories whereas the AASHO scale has 5 such categories. Both methods emphasize that only the descriptive words are used by the raters when evaluating pavements. The numerical rating is scaled from the forms during data reduction.

It is impractical and expensive to do serviceability ratings on an entire pavement network. Consequently, considerable effort has gone into correlating various (objective) high speed and efficient mechanical measurements with the (subjective) ratings on samples of the network. The purpose is to then use these correlations for estimating serviceability from measures of pavement roughness.

## 8.2 CHARACTERIZATION OF PAVEMENT ROUGHNESS

Studies at the AASHO Road Test [Carey 60] indicated that about 95 percent of the information about the serviceability of a pavement is contributed by the roughness of the surface profile. At about the same time, it was stated [Hveem 60] "there is no doubt that mankind has long thought of road smoothness or roughness as being synonymous with pleasant or unpleasant." But, the effects of a given degree of roughness vary with the speed and characteristics of the vehicle and tolerance of the vehicle passenger or driver.

### 8.2.1 Roughness Defined

Pavement roughness is a phenomenon experienced by the passenger and operator of a vehicle or airplane traveling over the surface. It is common to view roughness in terms of the *distortion of the pavement surface* which contributes to an undesirable or uncomfortable ride. This definition requires a measurement and analysis method for quantifying distortions of the pavement surface. Once the measurement and analysis method is selected, individual agencies can establish interpretation scales to determine the severity of the level of roughness.

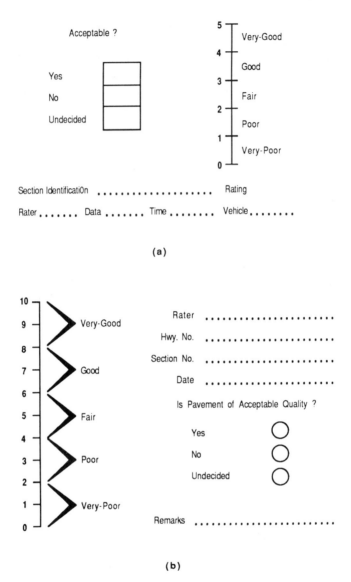

**Figure 8.2** Evaluation forms for individual, subjective pavement ratings: (a) Individual Present Serviceability Rating (PSR) form used at AASHO Road Test; (b) Present Performance Rating (now Riding Comfort Index) form developed by the Canadian Good Roads Association.

## 8.2.2 Components of Roughness

An evaluation of the entire pavement surface is required to define roughness completely. However, for most purposes, roughness can be divided into three profile components of distortion: transverse, longitudinal, and horizontal. Of particular interest are the variations in profile that impart acceleration to the vehicle or occupant and thus influence comfort and safety.

Distortions of the pavement surface can generate both vertical and lateral acceleration in the vehicle. Vertical acceleration is the major contributing factor to occupant comfort and derives from longitudinal distortion of the pavement profile. Lateral accelerations are the result of vehicle roll and yaw. Roll results from rotation about the longitudinal axis of the vehicle while yaw is the rotation about the vertical axis. The curvature of the roadway, which contributes to yaw, is normally handled through good geometric design. Roll results from differential transverse pavement elevations. Under severe conditions, it can impart an undesirable level of vertical acceleration.

Approximately 70 percent of vehicles travel in a well-defined wheel path with the right wheel located 2.5 to 3.5 feet from the pavement edge. The wheel tracks of automobiles and trucks are approximately 6 and 7 feet apart, respectively. Therefore, line measurement of the longitudinal profile in the wheel path provides the best sample of road surface roughness. Furthermore, comparison between the two wheel paths can provide some measure of the transverse variations that affect roll. For airfield pavements the longitudinal profile measurements should be performed in the wheel path of the aircraft using the facility.

Basing the definition of pavement roughness as distortions in the pavement surface generates the conclusion that road roughness evaluation *requires* measurement of the longitudinal profile of the pavement in the vehicle wheel path. For engineering interpretation, the measurements are usually handled with a mathematical model that generates a summary statistic for the length of pavement being evaluated. A variety of such summary statistics have been used, ranging from power spectra to some type of roughness index.

In general, most passenger cars exhibit ride characteristics that are relatively similar. With the limitation of relatively fixed vehicle parameters, ride is thus a function of the car excitation generated by the various combinations of road profile and vehicle speed. Most drivers have experienced the sensation of either slowing down or speeding up to improve the ride on a particular road. This indicates that the road has a wave length content that, at a particular speed, produces an excitation in the car at one of its resonant frequencies of about 1 or 10 Hz. The relationship between wavelength, car speed, and car resonant frequency is shown in Figure 8.3. Any surface wave length can cause an excitation at one of the car's resonant frequencies depending on speed. If the amplitude of that wavelength is large, the ride will be noticeably affected.

There are many devices used for roughness evaluation by highway and airport agencies that do not measure the profile of the pavement. Rather, they measure the response of a vehicle to the roughness of the road, and are therefore sensitive to the vehicle characteristics and measurement speed. As long as these two parameters are constant, such devices produce a useful index of roughness. However, the limitations of these devices need to be fully understood, especially when analyzing their output over time, or when comparing the output of several devices.

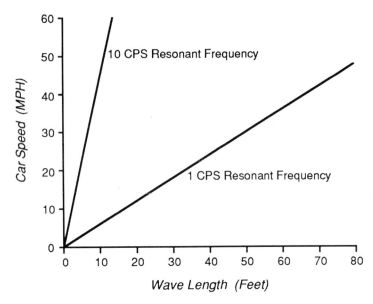

**Figure 8.3** Relationships among resonant frequencies of cars, car speed, and pavement surface wavelength.

## 8.3 EQUIPMENT FOR EVALUATING ROUGHNESS

Even though the concept of the functional performance of pavements was developed at the AASHO Road Test in the late 1950s, the need to evaluate roughness of pavements was recognized in the 1920s, as reported in *Public Roads* (February 1941). Roughness is the primary component of serviceability and a large number of different roughness evaluation methods or devices have been correlated with panel ratings of Present Serviceability Rating (PSR) or Riding Comfort Index (RCI). These devices may be categorized as profile devices, profilographs, and response type devices.

Mechanical equipment or devices for measuring true profiles were not available at the time of the AASHO Road Test. Therefore, an understanding of the pavement roughness measurements at the Road Test is necessary for relating current measurements to the pavement performance models in the AASHO design procedures. Initially pavement roughness was measured with the AASHO slope profilometer; however this device was cumbersome and complex. Hence the engineers at the Road Test developed a simplified clone known as the CHLOE profilometer that was to be used by states after the Road Test.

The Road Test profilometer is diagrammed in Figure 8.4. It recorded the angle, $\theta$ at 1 foot intervals along the pavement section, while being towed at a speed of 5 miles per hour, where $\theta$ is the angle between the line that connects the centers of the support wheels of the profilometer and the tow vehicle, and the line that connects the centers of the two small wheels on the profilometer. The distance $XL_1$, is long enough (25.5 feet) that the line between the wheel centers is approximately parallel to a perfectly smooth pavement surface. The small wheels on the profilometer are

**Pavement Performance**

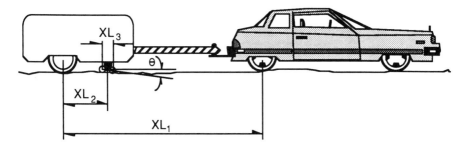

**Figure 8.4** Schematic of the AASHO Road Test profilometer

close enough (0.75 foot apart) that the line between their centers is parallel to the tangent to the road surface at the point midway between them.

Because the range of the angle $\theta$ rarely exceeds $\pm 3$ degrees, the Road Test profilometer recorded the radian measure of the angle. Slope variance is thus computed as the variance in angle $\theta$ using the standard variance equation defined in statistics textbooks. The CHLOE profilometer had good repeatability. However, the slow operating speed and inability to measure wavelengths longer than 12 feet eliminated the use of the CHLOE for regular roughness measurements.

### 8.3.1 Pavement Profile Measurement Devices

Pavement profile may be measured in the field and evaluated or summarized by computer, or it can be processed through a mechanical response type device to be discussed later. Profile measurements must be continuous or closely spaced points to capture the influence of distortions on ride quality. ASTM specifies a maximum spacing of 1 foot between each measurement [ASTM 91]. The most straightforward technique for measuring the profile of a pavement is with precision rod and level surveys. However, this is time consuming, costly, and limited to the evaluation of short lengths of pavements. Therefore, considerable research has been performed for the development of pavement profiling devices.

#### 8.3.1.1 Face Dipstick

One of the simplest devices for measuring the profile of a pavement is the Face Dipstick [Donnelly 88]. Originally developed for evaluating the evenness of building slabs, it consists of an accelerometer mounted on a frame with a set of contact feet 12 inches (300 mm in the metric version) apart (Figure 8.5). A handle is mounted on the frame for "walking" the Dipstick along the profile path by pivoting on the front foot and rotating through 180 degrees. The microcomputer mounted on the Dipstick records the data and can compute simple roughness summary statistics. An accelerometer measures the slope of the frame. Knowing the frame and the distance between the feet allows determination of the change in elevation between the feet. The reported accuracy is 0.0015 inch per reading [Donnelly 88].

Production rate of the Dipstick can be up to 900 ft/hr or more in one wheel path. Advantages of the Dipstick include relatively low initial cost and operational simplicity.

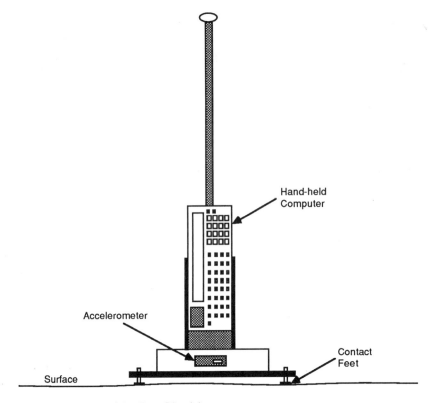

**Figure 8.5** Schematic of the Face Dipstick.

Although it is quicker than precision rod and level surveying, the primary disadvantage is the low operating speed and the need to close the facility while the measurements are performed. The Dipstick is mainly applicable to the evaluation of short pavement sections, or for calibration of response type devices.

### 8.3.1.2 TRRL Profilometer (High Speed Road Monitor)

The Transport and Road Research Laboratory (TRRL) of England developed a profilometer in the mid-1970s based on the back-site fore-site rod and level surveying method [Dickerson 76]. As shown in Figure 8.6, four lasers are used to measure the distances to the pavement surface. At the start of a run, an initial set of measurements is recorded to establish the datum. Then, as the vehicle moves forward, each laser measures the distance between the sensor and the pavement surface at the location of the preceding sensor measurement. Thus, back-sites and fore-sites are continuously recorded and the profile of the road is determined. The TRRL profilometer can operate at variable speeds up to the normal travel speed on freeways.

**Pavement Performance**

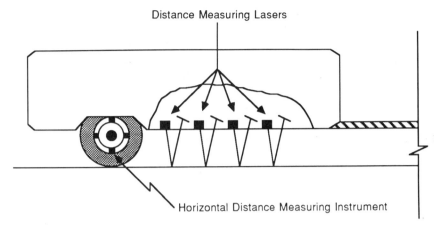

**Figure 8.6** Schematic of TRRL profilometer (high-speed road monitor).

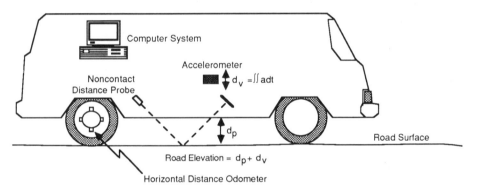

**Figure 8.7** Concept of inertial profile.

### 8.3.1.3 Inertial Profilometers

There are several devices that use the principle of inertial profilometry to measure the profile of the pavement surface. Inertial profilometry requires the following four basic components (as shown on Figure 8.7):

1. Device to measure the distance between the vehicle and the road surface
2. An inertial referencing device to compensate for the vertical movement of the vehicle body
3. A distance odometer to locate the profile points along the pavement
4. An on-board processor for recording and analyzing the data

Several types of transducers are used for measuring the distance between the pavement surface and the vehicle, $d_p$. The inertial reference device is usually either a mechanical or electronic accelerometer mounted to represent the vertical axis of the vehicle. As the vehicle traverses the pavement, vertical acceleration of the vehicle body is measured with the accelerometer. The accelerations are integrated twice to quantity the vertical movement, $d_v$, of the vehicle body. These movements are added to the distance measurements, $d_p$, to obtain elevations of the pavement profile. Some devices record the actual profile of the pavement surface while others process the data on board and only record a roughness summary statistic.

#### 8.3.1.3.1 Surface Dynamics Profilometer

The concept of inertial profilometry was originally applied by the General Motors Research Lab for the evaluation of vehicle test tracks [Spangler 62, 64]. The original device was known as the General Motors Research (GMR) or Surface Dynamics Profilometer (SDP). It used electromechanical potentiometers connected to road-following wheels to measure the distance between the vehicle body and the pavement surface. Data was recorded on an analog computer which filtered the profile signals before recording them. In 1966, K. J. Law Engineers Co. was licensed to manufacture the SDP. Over the years there have been many improvements in the device including the use of on-board digital computers and noncontact distance measuring sensors. In its current form the Law Profilometer, Model 690D, can measure pavement profiles at intervals of 6 inches, in both wheel paths, at highway speeds. In addition to recording pavement profile data, the on-board computer can compute a variety of roughness summary statistics. The 690D meets the criteria of ASTM E 950 [ASTM 91]. While it does have a high initial cost and is complex, there are major advantages, including the following:

1. Determination of actual profiles
2. Ability to handle large amounts of data automatically
3. Flexible operating speed
4. Ability to detect and analyze long wavelengths (especially important for airport runways and high-speed highway traffic)
5. Excellent repeatability

#### 8.3.1.3.2 FHWA Profilometer

The Federal Highway Administration sponsored research at the University of Michigan which produced an inexpensive device for measuring pavement profile and rut depth [Donnelly 88]. The PRORUT device developed on this project uses the principles of inertial profilometry to obtain precise pavement profiles. Low cost computer equipment and distance sensors are used to minimize the cost of the data collection and processing equipment. Only one prototype device was produced under the research contract.

#### 8.3.1.3.3 APL Profilometer

The Longitudinal Profile Analyzer (APL) was developed by the French road research laboratory, LCPC, and is distributed by MAP S.A.R.L [MAP 90]. It is a single wheel

# Pavement Performance

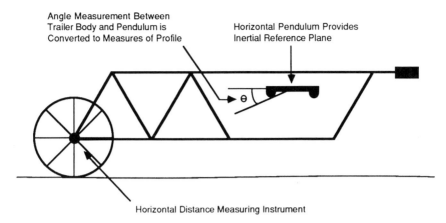

**Figure 8.8** Schematic of the APL profilometer.

and instrumented trailer device with a horizontal pendulum used to establish the inertial reference plane (see Figure 8.8). As the trailer wheel moves up and down in response to the road roughness, the angle between the pendulum and the wheel frame is measured and converted to a vertical distance measurement which is recorded at specified distance intervals. Due to the mechanical nature of the device, measurements must be performed at constant speed and the response is quite sensitive to the speed. Measurement of the profile distortions that are significant for highway pavements requires operating the APL at approximately 13 mph.

#### 8.3.1.3.4 Low Cost Profile Based Devices

There are two devices that use relatively low cost sensors and computers for evaluating the profile of the road and then computing and recording roughness summary statistics: the Law Model 8300 Roughness Surveyor and the South Dakota Profiler [Donnelly 88]. Both of these devices use ultrasonic sensors to measure the distance from the vehicle body to the pavement surface. An accelerometer is used to compensate for the vertical motion of the vehicle body. Microcomputers are used to analyze the data and store a roughness summary statistic. These devices do not provide the accuracy of the Law Model 690D. However, their advantages include the ability to operate at highway speeds and relatively low cost. Disadvantages include limits on accuracy and potential operational problems on open textured pavements. It appears the advantages outweigh the disadvantages in that a significant number of states and other agencies are currently using or evaluating South Dakota profilers.

### 8.3.2 Profilographs

Rolling straight edge devices, or profilographs, have been widely used for evaluating the smoothness of concrete pavements during construction. There are several designs of devices operating on a similar principle, as shown in Figure 8.9 (Rainhart Profi-

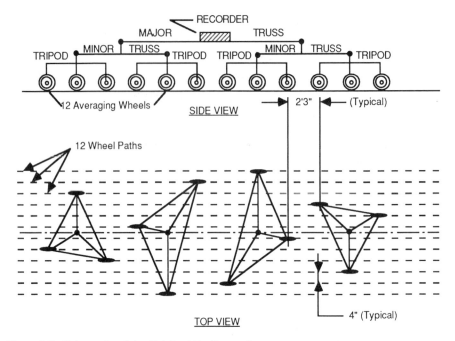

**Figure 8.9** Schematic of the Rainhart Profilograph.

lograph). These devices consist of bogie wheel sets at the front and rear, a recording wheel at the center and a strip chart recorder for capturing the movement of the recording wheel relative to the bogie wheels. The recording wheel is free to move vertically as the device passes over bumps in the pavement. Strip chart records of the road roughness are analyzed and the output is reported in inches of roughness per mile.

Profilographs have been developed by the California and Texas highway departments and the Transport and Road Research Laboratory in England. The number and arrangement of the bogie wheels and the length of the device are the primary design differences between the different types. Advantages of profilographs include low initial cost, simple operations, and good repeatability. The disadvantages include low operating speed and inability to measure roughness at wavelengths equal to integer multiples of the wheel base.

### 8.3.3 Response Type Measurements

There are two basic designs of response type road roughness measuring systems (RTRRMS) or devices: those measuring the displacement between the vehicle body and axle, and those which use accelerometers to measure the response of the vehicle axle or body. In reality, these devices measure the response of the vehicle to the roughness of the road; hence, the term *RTRRMS* to describe this class of measuring equipment [Gillespie 80]. Due to their low cost, simple design, and high operating

## Pavement Performance

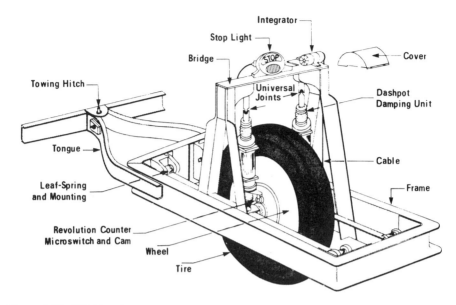

**Figure 8.10** BPR Roughometer.

speed, these devices have been widely used by highway agencies to collect roughness data for pavement management systems.

### 8.3.3.1 Mechanical RTRRMS

One of the earliest RTRRMS was the Bureau of Public Roads (BPR) roughometer. The original device was mounted in an automobile and subsequently converted to a trailer type device, as shown in Figure 8.10. The parameters of the trailer, such as the mass, shock absorbers, springs, etc., were selected to represent one quarter of a passenger car. Roughness was measured as the differential movement between the wheel axle and the trailer body, and was dependent on the trailer characteristics and travel speed.

The most popular RTRRMS device in the United States is the Mays Ride Meter (MRM) or Maysmeter, developed for the Texas Highway Department in the 1960s [Phillips 69]. As shown in Figure 8.11, the Maysmeter consists of a rod attached to the vehicle axle and to a transmitter mounted on the vehicle. The amount of movement relative to the body is measured by the transmitter and the signal is sent to a recording device. Originally, the recording device was a strip chart recorder which ejected a length of paper in proportion to the measured roughness; distance was recorded with an automatic event marker driven by the vehicle's odometer. Subsequently, electronic devices, including laptop computers, have been developed for recording Maysmeter output. Originally, all Maysmeters were mounted in automobiles. However, since the output is sensitive to the vehicle characteristics, such as shock absorbers and mass,

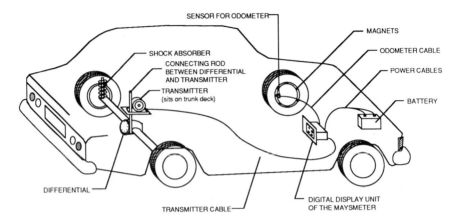

**Figure 8.11** Maysmeter

some agencies mount the transmitter in two wheel trailer versions to improve the time stability of measurements.

The Portland Cement Association (PCA) developed a RTRRMS that is conceptually similar to the Maysmeter. The basic PCA meter design has been adopted and manufactured by several agencies and goes under such names as the Cox and Wisconsin meter. The basic difference between the PCA meter and the Maysmeter is that the former uses a series of counters to measure the number and magnitude of the vertical deviations between the vehicle body and axle.

A wide variety of other displacement type RTRRMS's have been developed both internationally and by local highway agencies. Examples include:

1. The bump integrator, based on the BPR design, used by the TRRL
2. The IJK meter developed by the Iowa Highway Department to simulate the response of the PCA meter while eliminating some of its operational problems
3. The NASRA meter developed in Australia [Potter 78]

Regardless of the specific design of the sensor and transmitter used in a particular RTRRMS device, they all are sensitive to the vehicle characteristics and operating conditions. Therefore, these devices must be calibrated to a roughness standard to produce time stable results.

### 8.3.3.2 Accelerometer-Based RTRRMS

One of the approaches to reduce the dependence of the measurements on the characteristics of the host vehicle is to use an accelerometer as the primary motion sensor. The accelerometer can be mounted on either the vehicle axle or body. The Automatic Road Analyzer (ARAN), Portable Universal Roughness Device (PURD), Bruel & Kjear Road Roughness Rating System, and the Tech West Photologging Equipment use accelerometers mounted on the vehicle axle [Donnelly 88]. The PURD is mounted in a special trailer while the other devices are mounted on the axle of the host van.

**Pavement Performance**    91

The Slometer, developed for the Texas Department of Highways and Public Transportation, has the accelerometer mounted in the trunk of the vehicle [Babb 83, Walker 87]. The Dynatest 5000 has an accelerometer mounted on the vehicle chassis [Donnelly 88].

The axle mounted accelerometers are not as sensitive to the vehicle parameters as the displacement type devices. Movement of the axle in response to road roughness depends on the amount of tire distortion and the upward vertical force generated when the tire hits a bump and the downward vertical force of the vehicle suspension. If the force of the suspension on the axle is greater than the upward force generated by the bump, then the tire maintains contact with the pavement so the axle provides a reasonable tracking of the pavement surface. The output of the accelerometer can be integrated twice to obtain an estimate of the vertical axle movement. However, this integration process can magnify the effect of undesired noise in the signal. Generally the axle mounted RTRRMS's use a measure of the root-mean-square acceleration of the axle to quantify pavement roughness.

The Slometer [Babb 83, Walker 87] is intended to be a low cost roughness measuring device that is not affected by vehicle parameters. Components of the device include an accelerometer mounted in a vehicle, a microcomputer, and an operator console. The unit records the vehicle body vertical acceleration as it travels along the pavement surface. A unique feature is the software which analyzes the vertical acceleration data in a manner that isolates the effects of vehicle parameters and removes them from the resulting roughness summary statistic. Using a correlation between roughness and Serviceability Index (SI), the device can provide an SI output.

The Dynatest 5000 uses an accelerometer mounted on the vehicle chassis, and its output is the root mean square vertical acceleration of the chassis. Because this output is sensitive to the vehicle characteristics, the device is supplied with calibration equipment and procedures. The calibration equipment consists of a dual beam leveling system to obtain the slope variance of the test section.

## 8.4 A UNIVERSAL ROUGHNESS STANDARD

The equipment for evaluating pavement roughness measures either the profile of the pavement or the response of the vehicle to pavement roughness. In the first case, a method is needed to reduce the data into a meaningful statistic that represents the roughness of the pavement surface. In the second case, a method is needed to correlate the response measurement to a common scale, so that the output of different devices can be compared and to provide a time stable measure of pavement roughness; i.e., the calibration process. Since calibration requires a time stable measure for establishing the reference scale, the approach used in the following sections is to first present the various algorithms that can be used to generate a roughness summary statistic. Then the way in which these statistics can be used to calibrate response type systems is described.

It should be noted that the approach also involves development of a roughness evaluation scale that is independent of user quality assessment. Once the scale is determined, then subjective criteria can be established to determine the roughness levels that correspond to various levels of riding quality or comfort.

### 8.4.1 Roughness Summary Statistics

Pavement profile data is elevations at discrete intervals along a pavement surface. Raw profile data cannot be readily used by the pavement engineer. It must be processed or filtered in some manner to produce a meaningful representation of the pavement roughness. Originally, digital filtering techniques were used in an attempt to extract wavelength and amplitude information from the profile data using power spectral density analysis techniques [Hutchinson 65]. While this approach provides useful information on the specific components of pavement roughness in the profile data, the technique has not been widely applied. A 1984 review of the techniques for computing a roughness summary statistic found three types in common use [Hudson 84]:

1. Quarter-car simulation
2. Root-mean-square vertical acceleration (RMSVA)
3. Slope-variance (SV)

Additional summary statistics that have found some degree of use include mean absolute vertical acceleration (MAVA) and Straight Edge Index (SEI), as pointed out in a 1983 study [Joseph 83]. As well a "profile index" was introduced in 1985 as the result of research performed for the National Cooperative Highway Research Program [Janoff 85].

#### 8.4.1.1 Quarter-car Simulation

The concept of quarter-car simulation as a method for analyzing pavement profile data was originally an attempt to simulate the output of the BPR roughometer. Subsequently, vehicle simulation studies at the University of Michigan demonstrated that full-car and half-car simulation models do not provide an advantage over the quarter-car simulation with respect to the calibration of RTRRMS devices and are computationally much more complicated [Gillespie 80].

The parameters of the quarter-car are shown in Figure 8.12. They include the sprung mass of the vehicle body; the suspension spring and damper (shock absorber) constants; the unsprung mass of the suspension, tire, and wheel; and the spring constant of the tire. Theoretical correctness would require a damper constant for the tire; however, practical application generally ignores this term. Mathematically the behavior of a quarter-car can be described with two second order equations:

$$M_s \ddot{Z}_s + C_s(\dot{Z}_s - \dot{Z}_u) + K_s(Z_s - Z_u) = 0 \qquad (8.1)$$

and

$$M_s \ddot{Z}_s + M_u \ddot{Z}_u + K_t(Z_u - Z) = 0 \qquad (8.2)$$

# Pavement Performance

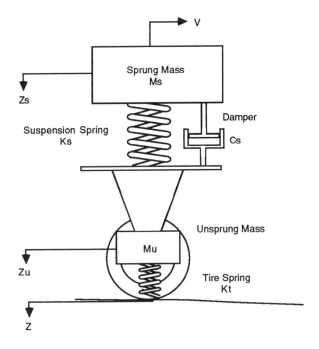

Quarter-Car Model

**Figure 8.12** Quarter-car model [Gillespie 80].

where

$Z$ = road profile elevation points
$Z_u$ = elevation of unsprung mass (axle)
$Z_s$ = elevation of sprung mass (body)
$K_t$ = tire spring constant
$K_s$ = suspension spring constant
$C_s$ = shock absorber constant
$M_u$ = unsprung mass (axle)
$M_s$ = sprung mass

The double dot notation above the elevation terms represents acceleration while the single dot represents velocity.

Since RTRRMS devices generally measure the movement between the vehicle axle and body, simulation requires calculation of the difference in elevation between the body and axle in response to the road profile and forward motion of the vehicle. This is accomplished by integrating the difference in the velocities between the sprung and unsprung mass; producing the quarter-car statistic, QCS:

$$\text{QCS} = \frac{1}{C} \int_0^T |\dot{Z}_s - \dot{Z}_u| \, dt \qquad (8.3)$$

**Table 8.1 Quarter-Car Model Parameters [Gillespie 80]**

| Vehicle Type | $K_1$ (Sec$^{-2}$) | $K_2$ (Sec$^{-2}$) | $M$ | $C$ (Sec$^{-1}$) |
|---|---|---|---|---|
| HRSI Reference | 653 | 62.3 | 0.150 | 6.0 |
| BPR Roughometer | 667 | 113.3 | 0.167 | 5.0 |

The term $C$ represents either the total time required to traverse the section of road being simulated, $T$, or the length of the section, $L$. If the time factor is used to normalize the quarter-car statistic, the calculation results in an average rectified velocity, while a distance base yields the average rectified slope.

There are several acceptable numerical techniques for the solution of equation 8.3. However, the linear nature of the equations permits an exact solution with the state transition matrix method [described in Gillespie 80].

Historically, two sets of vehicle parameters have been used for computing quarter-car statistics for the calibration of RTRRMS devices. A set representing the original BPR Roughometer trailer was used for several years, until research at the Highway Safety Research Institute (HRSI) produced an updated set of vehicle parameters, as given in Table 8.1 [Gillespie 80]. The World Bank recommends the HSRI vehicle parameters and have termed the quarter-car statistic computed as the International Roughness Index, IRI [Sayers 86].

Although the mathematical base for quarter-car simulation is somewhat complex, computer programs are readily available for performing the calculation. The MAPCON suite of programs available from the MACTRANS computer center at the University of Florida contains the needed algorithms [Zaniewski 86].

### 8.4.1.2 Root Mean Square Vertical Acceleration (RMSVA)

The search for an improved statistic to calibrate RTRRMS led to the development of a simple power spectral density analysis method for analyzing pavement profile data [McKenzie 78, 82]. It computes the vertical acceleration of a point moving between discretely measured profile measurements, as well as the root-mean-square value of the individual accelerations.

The process assumes a set of equally spaced profile points, $Y_1, Y_2, \ldots Y_n$, with a horizontal distance between each point of $\Delta S$. RMSVA is the root-mean-square ratio of the change of adjacent profile slopes to the horizontal distance between points. If a set of adjacent points, $Y_A, Y_B, Y_C$ is considered, as shown in Figure 8.13, then the following calculations can be carried out.

$$\text{Slope at } A = \frac{Y_A - Y_B}{\Delta s} = |\theta_1| \quad \text{since } \theta_1 \text{ is small.} \quad (8.4)$$

$$\text{Slope at } B = \frac{Y_C - Y_B}{\Delta s} = \theta_2. \quad (8.5)$$

$$\text{Change of slope} = \theta_2 - (-\theta_1) = \theta_2 + \theta_1 = \Delta\theta \quad (8.6)$$

$$= \frac{Y_C - Y_B + Y_A - Y_B}{\Delta s}.$$

**Pavement Performance**

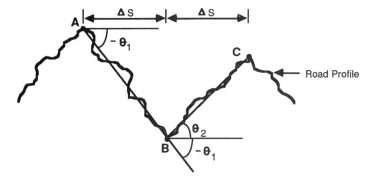

**Figure 8.13** Representation of pavement profile for root-mean-square vertical acceleration (RMSVA) analysis.

Ratio of change of slope to the distance is given by

$$\left(\frac{\Delta\theta}{\Delta s}\right)_B = \frac{(Y_C - Y_B) - (Y_B - Y_A)}{\Delta s^2}. \tag{8.7}$$

$\dfrac{\Delta\theta}{\Delta s}$ is the estimate of the second derivative.

$$\text{Thus, RMSVA} = \left[\sum_{i=2}^{n-1} \frac{\left(\frac{\Delta\theta}{\Delta s}\right)_i^2}{n-2}\right]^{0.5}$$

$$= \sum_{i=2}^{n-1} \left[\frac{\{(Y_{i+1} - Y_i - (Y_i - Y_{i-1})\}^2/\Delta s^4}{n-2}\right]^{0.5} \tag{8.8}$$

where

$n$ = number of profile elevations
$\Delta s$ = sample length (horizontal distance between points)
$Y_i$ = profile elevation at point $i$

It may be noted that RMSVA contains second derivatives of the profile which are invariant to the influence of underlying pavement geometry; hence, they exhibit properties of stationarity. Computer programs have been developed for calculation of RMSVA [Queiroz 81, Zaniewski 86].

Because RMSVA can be calculated for a wide spectrum of base lengths, $\Delta s$, it is a type of power spectral density analysis. The use of RMSVA in correlating with Maysmeter results is illustrated in a 1982 study where base lengths ranging from 0.5 foot to 64 feet were used [Sriniwarat 82]. Multiple regression analysis showed that a combination of RMSVA at 4 and 16 feet base lengths, $RMSVA_4$, and $RMSVA_{16}$, correlated quite well with the various Maysmeters, with an average $R^2$ of .96. The study yielded the following formula for a reference Maysmeter, MO:

$$MO = -20 + 23\text{RMSVA}_4 + 58\text{RMSVA}_{16}. \tag{8.9}$$

The MO statistic is in units of in/mile for RMSVA values in ft/sec$^2$. Since the MO statistic is used for calibration of the Texas Maysmeters, it has been related to the Present Serviceability Index, PSI, as follows:

$$\text{PSI} = 5e^{-c(\ln(aMO)/d)} \tag{8.10}$$

where $a$, $c$, and $d$ are constants derived from regression analysis, and independent of any particular Maysmeter. In a complementary 1982 study [McKenzie 82] these constants were found to be 32, 9.387, and 8.493, respectively.

### 8.4.1.3 Slope Variance

Because the slope variance measure (SV) of pavement roughness was generated by the profilometer at the AASHO Road Test [HRB 62], any attempts to relate current roughness measures to SV should refer back to the Road Test data base. However, CHLOE profilometers are no longer available or suitable for routine use. As a surrogate, a mathematical simulation of the CHLOE, operating on measured profile points, was developed [Hudson 84]. The CHLOE slope variance is computed as

$$SV = \frac{\sum_{i=1}^{n} X_i^2 - \frac{1}{n}\left(\sum_{i=1}^{n} X_i\right)^2}{n - 1} \tag{8.11}$$

where

$SV$ = slope variance
$X_i$ = the $i^{th}$ slope measurement
$n$ = total number of measurements

The major problem with simulation of slope variance is determination of the slopes. Modern profile measurement techniques generally make measurements at fixed spacings that are different from those of the CHLOE profilometer. In order to develop the simulation model, it was assumed the elevation between the measured profile points could be represented with a straight line approximation. The elevations under the tow vehicle wheel, support wheel, and measurement wheels were evaluated and used to compute the slope angle. This angle was then used in equation 8.11 to compute the slope variance.

### 8.4.1.4 Profile Index

An NCHRP project in the mid-1980s was directed to:

- Developing a scale that accurately reflects the public's perception of pavement roughness
- Developing transformations that relate pavement profiles to this scale
- Showing how RTRRMS output relates to this scale [Janoff 85]

**Pavement Performance**

The researchers selected a power spectral density approach for the calculation of roughness summary statistic, termed the profile index, PI, from pavement profile data.

PI is defined as the square root of the mean square of the profile height. The rules for calculating the Pi from an unfiltered record of the true pavement profile are described as follows [Janoff 85]:

1. Find the average value of profile height and subtract it from the original record of profile height.
2. Square the new profile record and find the average value over the complete record.
3. Find the square root of the "mean square" calculated in the vertical plane.

By these rules, PI is sensitive to very long wave length deviations in the pavement profile. To overcome this problem the profile is filtered to remove undesirable components of the profile before calculating PI.

During the research, 52 pavement profiles measured with the Ohio Department of Transportation digital profilometer (Law Model 690D) were processed. The actual method used to compute PI involved processing each profile with a series of one-third octave bands as follows:

1. Find the Fourier coefficients of the pavement profiles with a Fast Fourier transform program using the Singleton procedure [Singleton 67].
2. Find the mean square profile height of: (a) the sum of the right and left wheel path profile (vertical component) and (b) the difference between the wheel path profiles (roll component).
3. Find the sum of the mean square profile of each wheel path.
4. Sum the values of the mean square profile height over each one-third octave band and take the square root.

The details of this processing profile measurements are available [Appendix D of Janoff 85].

### 8.4.2 Selection of a Roughness Summary Statistic

Selection of a roughness summary statistic for the description of pavement profile data should be based on the needs of the user agency. However, these needs should also be examined from a broad base. For example, the Federal Highway Administration (FHWA) in the United States has not mandated the use of a specific technique for roughness evaluation, but it does require the states to provide roughness information for the Highway Performance Monitoring System. In order to maximize the utility of this information, the states should be providing information in a consistent manner. Similarly, local agencies should be aware of these practices since much of their funding comes from the states. If they use measures that are compatible with state practices, a competitive advantage in the solicitation of funds can exist.

To a large extent, many third world or developing nations are approaching uniform roughness evaluation due to encouragement from the World Bank. In particular, the International Roughness Index (IRI) has been adopted by a number of countries.

Regardless of the statistic selected, it is important to make roughness measurements in a consistent and uniform manner. Due to the mechanical nature of response type devices, they cannot be depended upon to provide consistent measures over time and between vehicles. However, the RTRRMS devices are in widespread use due to their low cost. Thus, the need for calibration with a consistent and stable measure exists.

The calibration process should yield pavement roughness measurements which can be meaningfully interpreted across time and between agencies. The measurement of roughness on a uniform or universal scale should not be confused with the interpretation of the data. Each agency can and should develop standards for what is an acceptable or unacceptable level of roughness on the standard scale.

### 8.4.3 RTRRMS Calibration Procedures

Calibration is the process of determining and quantifying the operation of any specific RTRRMS in relation to a stable measure. The output of a RTRRMS is dependent on many different factors related to the host vehicle (e.g., suspension system, tire pressure, weight, wheelbase, etc.) and the instrument itself. Therefore, calibration is necessary to convert the output of a particular RTRRMS to a standard or universal measure.

The calibration procedure requires several steps that need to be carefully followed:

1. Establishing a calibration course:
   a. selection of calibration sections
   b. obtaining detailed profile data for each calibration section
   c. analysis of the profile data to determine the calibration statistic, such as the IRI or MO
2. Calibration of RTRRMS systems:
   a. data collection
   b. analysis
3. Calibration control:
   a. periodic measures with RTRRMS systems on a subset of the calibration sections or on separate control sections
   b. data analysis
   c. recalibration of RTRRMS's when needed

#### 8.4.3.1 Selection of Calibration Sections

The number of sections selected, ranging from very rough to very smooth, should generally be between about 15 to 30. True profile of each section is then measured, using for example a Law Profilometer, the Face Dipstick, or rod and level. The elevations from the profile measurements are converted to the selected standard statistic, such as IRI or MO, representing the "true" section roughness. The RTRRMS is then operated and a Roughness Output (RO) is obtained for each section. The standard statistics and the RO's are related through a regression equation, which can

# Pavement Performance

then subsequently be used for estimating the standard roughness statistics from the output of the RTRRMS.

### 8.4.3.1.1 Number of Calibration Sections Required

In order to give statistical significance to the calibration equation, a minimum of 15 sections should be used. If a Law Profilometer is available, it is worthwhile to measure some extra, reserve calibration sections which can be substituted for the primary sections if needed.

### 8.4.3.1.2 Range of Roughness

It is important that selected calibration sections have a wide range of roughness from smooth to moderate roughness to rough. For selecting sections the serviceability concept can be used; i.e., smooth sections would have a PSR of more than 3.5, moderately rough would be in the range of 2.5 to 3.5, and rough sections would have a PSR of less than 2.5. There should be approximately one third of the sections in each of the roughness categories.

### 8.4.3.1.3 Length and Grade of Section

A calibration section should be straight with no curves or turns, and should have as flat a grade as possible. Lengths of 1,000 feet (320 m) have been commonly used in the United States, Canada, and many other parts of the world.

### 8.4.3.1.4 Safety Considerations

If at all possible, each calibration section should be located on a lightly trafficked and sufficiently wide roadway. For rod and level or Dipstick surveys, a crew will be occupying one entire lane width. Since the profile survey needs to be repeated every 6 to 12 months, it is also important to locate sections on lightly trafficked roads. If a section must be located on a higher traffic facility, it should have at least two lanes in each direction. Extra care should be taken to protect the survey crew in terms of traffic rerouting signs, flagging, and traffic cones.

### 8.4.3.2 Profile Measurements and Analysis

Research has demonstrated that profile measurements at 2-foot intervals are adequate for calibration of RTRRMS. The Law Profilometer can measure profile points at 6-inch intervals, the Dipstick at 12-inch intervals. Due to the labor required to perform rod and level surveys, a 2-foot interval is recommended.

The Law Profilometer and Face Dipstick are supplied with microcomputers programmed to compute various roughness summary statistics that can be used for calibration. Rod and level data can be analyzed with the MAPCON program available from the MACTRANS center at the University of Florida [Zaniewski 86].

### 8.4.3.3 Calibration of RTRRMS Systems

The calibration process requires at least five replicate measures of the roughness on each calibration section with each RTRRMS. The mean values are then used in a linear regression analysis with the calibration statistic, such as IRI or MO. A variety of microcomputer packages and spread sheet programs are available for this purpose. In performing the regression analysis, the output, RO, from the RTRRMS (i.e., mean values of roughness) is the independent variable and the calibration statistic is the dependent variable, for example:

$$RO = a + bMO \pm e \qquad (8.12)$$

In actual use, equation 8.12 is turned around such that the output from the RTRRMS, RO, will be used to estimate the calibration statistic, for example:

$$\widehat{MO} = (RO - a)/b \qquad (8.13)$$

The cap on MO indicates that it is a value estimated from the independent variable.

### 8.4.3.4 Calibration Control

RTRRMS output changes over time. Thus, it is necessary to periodically test each RTRRMS. The purpose of calibration control is to determine when any RTRRMS's response has changed sufficiently to warrant recalibration.

Calibration control requires two rough and two smooth sections. These sections should have zero grade, be on tangents, and have homogeneous roughness throughout their length. The sections should be on lightly trafficked roads to permit constant speed operation of the RTRRMS. Initially, the RTRRMS is operated 25 times on each section and the mean, $\bar{x}_{ci}$, and standard deviation, $S_{ci}$, are computed for each section. The subscript $c$ indicates these are control values, and the $i$ indicates section number. The RTRRMS may now be used to obtain road roughness measurements.

Biweekly or every 2,000 miles of operation, the calibration of the instrument should be tested by obtaining five repeat runs on one smooth and one rough test section. The mean for the control test runs, $\bar{x}_{ti}$, is computed for each section and compared to the control mean and standard deviation using the Student's test. If the test shows the RTRRMS is producing acceptable results at the specified level of significance, it can continue to be used for road roughness measurements.

When the control test indicates a problem, then careful testing is required to determine if the problem lies with the data reduction, the RTRRMS, or if the properties of the control section have changed. The first step is to review the data to make sure it was transcribed correctly and the mean is correct.

If the data is not the source of the problem, then five additional measures are made on the section where the problem was indicated and the mean is compared to the initial control mean. If the new test shows the RTRRMS is acceptable, then it can be returned to service, with the assumption that the prior test was the result of the random variation.

If the second test indicates there is still a problem, then five repeat runs are obtained on the three other control sections, and the means are computed and compared to the control data. If these indicate the system is in calibration, then the RTRRMS is assumed to be operating correctly and the characteristics of the control section, which should no longer be used, have changed.

If the data from any one of the last three control sections indicates a problem, then the RTRRMS should be recalibrated.

**Pavement Performance**

The calibration and control of the output of RTRRMS devices is a time-consuming and laborious process. However, the alternative to proper calibration is the collection of unreliable roughness information.

## 8.5 RELATING ROUGHNESS TO SERVICEABILITY

The primary use of objective roughness measurements is to estimate the pavement serviceability, which is a subjective (i.e., user) rating of pavement ride quality. The first and most widely used method for this purpose was the development of the Present Serviceability Index, PSI, at the AASHO Road Test [HRB 62]. The original functional form of the PSI equation is

$$\text{PSI} = C + (A_1 R_1 + \ldots) + (B_1 D_1 + B_2 D_2 + \ldots) \pm e \quad (8.14)$$

where

$C$ = coefficient (5.03 for flexible pavements and 5.41 for rigid pavements)
$A_1$ = coefficient ($-1.91$ and $-1.80$ for flexible and rigid respectively)
$R_1$ = function of profile roughness [$\log(1 + SV)$], where $SV$ = mean slope variance obtained from the CHLOE profilometer
$B_1$ = coefficient ($-1.38$ for flexible and 0 for rigid)
$D_1$ = function of surface rutting ($RD$), where $RD$ = mean rut depth as measured by simple rut depth indicator
$B_2$ = coefficient ($-0.01$ for flexible and $-0.09$ for rigid)
$D_2$ = function of surface deterioration ($C + P$), where $C + P$ = amount of cracking and patching, determined by procedures developed at the AASHO Road Test
$e$ = error term

Given this general form of equation, it is necessary to determine the coefficients for a particular set of input variables. This was done at the Road Test for several sets of variables [Carey 60]. It is important to understand that the result is a best-fit equation based on all observed data used in the equation. Other variables were candidates for inclusion in the equation but they added no significance in predicting PSI.

As well, a regression equation is not a causative relationship and covariance between terms can account for very small coefficients on a variable that alone is only slightly less well correlated with the dependent variable. For example, if the observed roughness in a pavement is due to cracking, the two factors are correlated. Consequently, once the roughness term is included in the equation, little variation remains to be explained by adding the cracking terms and thus the coefficient is small. This does not indicate a lack of concern for cracking, but if users of the equation alter it arbitrarily because they intuitively "feel" that cracking is more important, then erroneous and unpredictable results will occur.

### 8.5.1 Evolution of the Serviceability-Roughness Concept

Any change in measurement methods or units will result in a modified equation. This can be done either by performing an entirely new regression if all data is available, or by comparing the old measurement to the new and making an appropriate substitution in the equation. For example, at the AASHO Road Test [HRB 62] the BPR roughometer output, $R$, in inches per mile, was correlated to slope variance, $SV$, and substituted into equation 8.14. The resulting equation for rigid pavements was

$$PSI = 5.41 - 1.80 \log (0.40R - 30) - 0.09 \sqrt{C + P} \qquad (8.15)$$

It must be emphasized that any PSI model, such as that represented by equation 8.14, is not an end in itself. Carey and Irick made this quite clear in pointing out that it is intended only to predict PSR to a satisfactory approximation [Carey 60]. Unfortunately, this intent and use of the concept is sometimes forgotten. As well, some engineers are inherently "hostile" to the concept of a completely subjective evaluation as represented by PSR. They prefer to evaluate their structures by measurable physical criteria, which is fine per se, but forgets the user for whom the structure was built.

The PSI equation was developed by multiple regression techniques, as previously noted. That is, a set of physical measurements were related to the subjective, user evaluations in terms of the mean panel rating values, PSR. Although these physical measurements include condition or distress data (i.e., mean rut depth plus cracking and patching), it is roughness that provides the major correlation variable (i.e., correlation coefficients between PSR and PSI are increased by only about 5 percent after adding in the condition data). Thus, whenever PSI is calculated from physical measurement data, this is really only an estimate of PSR; that is,

$$PSI = PSR \pm e \qquad (8.16)$$

where $e$ is an error term. In other words, PSI and PSR are not two different ways of obtaining pavement serviceability. PSI represents a means of using objectively obtained data to estimate a subjectively based parameter [Carey 62, Haas 71].

The original Canadian studies previously noted [CGRA 67, 71] also tried to relate panel ratings to physical measurement data by multiple regression techniques (roughness data was not included). Although these efforts were relatively successful in explaining performance variations, the regressions were not significant enough as a predictive design tool for many pavement groups. Consequently, a major program was initiated to relate the subjective ratings of Riding Comfort Index (RCI) to roughness measurements, primarily using the RTRRMS devices [RTAC 72]. Figure 8.14 contains example correlations from the Canadian studies, involving a single RTRRMS (Transport Canada's device) and three provinces. Results of these studies included recommendations for calibration and operating procedures for the RTRRMS [RTAC 72, Argue 71, 73].

It should be noted that correlations such as those shown in Figure 8.14 can change significantly among regions and with time. Thus, the foregoing recommendations include periodic recalibration experiments.

The original PSI equation was developed in the late 1950s. Consequently, its applicability may be questioned in view of the regional and time changes that can occur. In fact, a corollary to the concepts of a universal roughness index, presented earlier, would require a slight reinterpretation of the practical application of the serviceability concept. An inherent assumption of the universal roughness scale concept

**Pavement Performance**

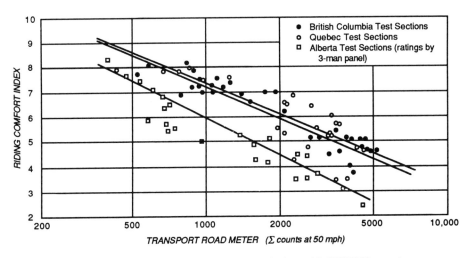

**Figure 8.14** Example correlation of Riding Comfort Index with RTRRMS roughness measurements, from Canadian studies [Argue 73].

is the ability to uniquely define the "true" roughness of a pavement section. Roughness as described by such a scale is not affected by vehicle parameters, attitudes of society, travel speeds, classification of the highway, etc. Conversely, the serviceability concept [Carey 60] is the mean rating of all users of a section of pavement, which can be sensitive to a wide variety of parameters. For example, a pavement with a high roughness may be rated as unacceptable on a high speed road while a pavement with an equal roughness will be rated as acceptable for low traffic speeds.

Under this philosophy, the serviceability concept is a means of scaling the manner in which society rates the quality of a pavement with respect to the true roughness of the surface. There are two major ramifications of this concept.

1. Serviceability should be correlated with "true" measures of pavement roughness. Distress terms should be excluded from the serviceability equations.
2. Serviceability-roughness correlation equations are empirical in nature. Therefore, any serviceability-roughness experiment should be based on proper statistical design where the inference space of the resulting equation corresponds to the range of perceptions of the highway users.

The correlation between the users' evaluation of pavement riding quality or comfort and objective measures of surface parameters should be limited to the roughness of the surface. Distress terms were included at the time of the AASHO Road Test to improve the statistical fit of the serviceability regression equation. These parameters actually contributed very little, however, to the equation and have generated confusion in the engineering community with respect to the nature of the serviceability concept. Limiting the independent variables in the serviceability equation to roughness only focuses the application of the concept on the functional performance of the pavement. Distress evaluation is important in terms of an engineering evaluation of the pavement and should be treated independently of the functional evaluation.

A further extension of this concept is the limiting or minimum acceptable serviceability criterion used to determine when a treatment strategy is required. This criterion, as discussed in more detail in Part Three, should be a function of the class of the highway and the demands of the user society.

The serviceability-roughness relationship should be reestablished whenever there is a change in the conditions comprising the inference space. This principle has been followed in several studies, including Canada [RTAC 72], Texas [Roberts 70], and Arizona [Way 78]. Similar concepts were used in two NCHRP studies [Janoff 85, 88].

### 8.5.2 Development of Serviceability-Roughness Relationships

Since pavements are provided for users, their response, such as the previously noted panel ratings, has to be measured. But the modeling of such subjective opinions or ratings has been developed primarily in the field of psychology and engineers are often unaware of its features and its limitations.

The literature on this subject, termed *psychophysical scaling*, is extensive. Of particular interest to the pavement engineer is the work of Stevens, who classified measurements on the basis of the transformations that leave the scale form invariant [Stevens 59]. Subsequently, it was shown that the considerations presented by Stevens are particularly relevant to the pavement field in terms of the validity of certain statistical manipulations that are performed on evaluation data [Hutchinson 69, Haas 71]. These considerations should be carefully reviewed when devising experiments to relate subjective user opinions to objective mechanical measurements.

Some assumptions are involved in acquiring or modeling user opinions as in the PSR model of the AASHO Road Test or the RCI model of the Canadian studies. These assumptions neglect certain systematic errors as follows:

1. Leniency error (i.e., a rater's tendency, for various reasons, to rate too high or too low)
2. Halo effect (i.e., a rater's tendency to force a particular attribute rating toward his or her overall impressions of the object)
3. Central tendency error (i.e., a rater's hesitation to give extreme judgments, thereby tending ratings toward the mean of the rating panel)

A number of guidelines for constructing rating scales and a discussion of the precautions to be used in interpretation have been presented [Hutchinson 64, Haas 71].

### 8.5.3 Toward Achieving Better Compatibility in Serviceability-Performance Evaluation

Many agencies have put considerable effort into developing, applying, and analyzing serviceability measuring schemes. This is encouraging; however, it has led to a proliferation of methods and data formats, many of which are incompatible with other data. Compatibility has two major aspects:

1. "External" compatibility, relating to whether the results of one agency's work have any quantitative relation or meaning to those of another agency
2. "Internal" compatibility, relating to correlating results, achieving repeatability, etc., within an agency

Achieving better compatibility in pavement performance evaluation is assisted by the following:

1. Performance evaluation of pavements should be on a well-planned basis and be an integral part of the overall pavement management system.
2. The concept of serviceability, and its components, should be clearly understood, as well as the underlying assumptions. Moreover, it should be explicitly recognized that serviceability measures, such as those developed at the AASHO Road Test and in Canada, are not ends within themselves; they exist to estimate the road users' opinions.
3. There are a variety of possible errors in subjective evaluations of serviceability. These can be significant, and it is important that the principles underlying subjective rating scale design and analysis are well understood.
4. Serviceability can be conveniently estimated from a correlation with roughness measurements. But because serviceability is subjective, the correlation and its use are unique to each region and may vary with time. Thus each region should have its own correlation.
5. The problems of internal compatibility can be largely controlled by carefully designed experiments so that proper statistical analyses may be applied to the calibration of the RTRRMS.

## 8.6 APPLICATIONS OF ROUGHNESS DATA

Pavement roughness data fulfills a wide variety of needs in the management function. These needs occur at the network and project levels, for rural and urban road pavements and for airport pavements. The basic technology and considerations for obtaining roughness data are applicable to all needs. However, the level of detail, frequency, operating speeds for roughness measurements, and other factors can vary considerably.

### 8.6.1 Roughness Evaluation at the Network Level

At the network level, roughness data provides a functional evaluation of the pavement surface which can be used in project selection and programming. For example, in California's (CALTRANS) pavement management system, roughness is the primary variable used in assigning priorities for the selection of projects. This is based on the philosophy that the rough pavements are not serving the public in an acceptable manner whereas distressed pavements with a reasonable ride quality are still doing an acceptable job of serving the public. Based on the availability of low cost RTRRMS and their high operating speeds, roughness surveys are carried out by most of the state highway agencies. However, it must be reiterated that the need for calibration of these devices to a common roughness scale remains an important concern at this time.

Due to the way roughness data is used at the network level, the information does not need to be extremely precise. Network roughness surveys are used for overall needs assessments and rehabilitation investment programming purposes. Consequently, the high speed, low cost RTRRMS are well suited for this task.

### 8.6.2 Roughness Evaluation at the Project Level

Pavement performance evaluation data for particular project purposes usually requires precise measurements. Examples of roughness measurement needs at the project level include construction quality control, periodic serviceability estimates, evaluation of rehabilitation options, etc. In order to accomplish this, accurate roughness measurements are desired, which involves the use of slow devices such as the rolling profilographs, or higher speed but more expensive devices such as the Law Model 690D Profilometer.

### 8.6.3 Roughness Evaluation in Urban Areas

Historically, highway agencies have concentrated roughness measurements in rural locations. Some agencies measure roughness on high speed urban routes and limited access highways but few perform roughness measurements on urban streets. This is the result of two factors:

1. The RTRRMS devices used by most highway agencies must be operated at constant speeds, which may not be possible in urban areas.
2. Due to lower traffic speeds in urban areas, pavement engineers have been less concerned with urban roughness.

The first point is valid and is capable of being overcome by speed correction factor for RTRRMS's and with the development of low cost profile based devices such as the Law Model 8300 and the South Dakota Profiler. These latter devices certainly have the potential to measure roughness in the urban environment.

The second point also has some validity in that passengers in a high speed vehicle will experience more discomfort than passengers in slower speed vehicles on pavements of equal roughness. However, this does not limit the utility of roughness measurements in an urban environment [Karan 77]. The primary difference in the application of the measurements would be in the limiting criteria used to define the need for rehabilitation.

### 8.6.4 Roughness Evaluation at Airports

Airport pavements present several particular problems and considerations in the area of performance evaluation, including the following:

1. There are two types of users, passengers and flight crew. It is probably sufficient to consider only the flight crew (in particular, the pilot) because their

response to roughness is more critical (primarily with regard to safety) and because they are responsible for the other class of users.
2. The range of aircraft types presents a much wider variation of pavement-vehicle interaction than that which occurs on roads.
3. The effects of airport pavement roughness are related primarily to safety and undercarriage damage, as contrasted to roads where they are related more to variation in "quality" of ride provided to the user.
4. There has been no widely accepted measure of performance, in terms of the user, yet developed for airport pavements.
5. Long wavelengths are more important on airport runways than on roads because of higher speeds, different wheel and gear configurations, etc.

The effects of excessive airport pavement roughness, with respect to porpoising or skipping, slowed acceleration on takeoff, cockpit vertical acceleration, difficulties encountered by the flight crew in reading instruments, structural damage to aircraft, and so on, have been well summarized [Alford 71]. Aircraft responses, and the possibilities of structural damage, have been considered in a number of publications; e.g. [Lee 68, Yang 72].

During the past few years, RTRRMS devices have come into fairly widespread use for runway roughness evaluation. The Surface Dynamics Profilometer (SDP) was suggested as early as 1971 for measuring airport pavement roughness [Hudson 71], and has subsequently found a significant degree of use. It has the capacity to measure roughness in terms of the various wavelengths and amplitudes that are important to pavement-aircraft-pilot interaction, and it is fast enough to minimize runway closure for measurements. The development of a serviceability type measure for airport pavements, similar in concept to that developed for highways, was also suggested in 1971. Shortly thereafter, it was incorporated into an experiment [Steitle 72] employing a user rating form (with pilots as the users) in evaluating runways at Love Field in Dallas, Texas. Profile measurements were also made with the SDP and it was demonstrated that these could be related to user evaluations.

## REVIEW QUESTIONS

1. Define precisely the difference between pavement serviceability and performance.
2. Define precisely pavement roughness versus ride quality.
3. How are pavement profile measurements used for the evaluation of pavement roughness?
4. Describe the measurement principles of an inertial profilometer.
5. What is the main disadvantage of a response type roughness measuring system?
6. Why is a universal roughness standard necessary?
7. Prepare a tabular comparison of the equipment described in this chapter including key features, operating characteristics, limitations, etc.
8. Working in groups of at least four people, rate the serviceability of 15 pavement sections ranging from smooth to rough. Use statistical analysis to compare the results of the different raters. If possible, obtain roughness measurements from the local highway agency and correlate with the results of the serviceability ratings.

Chapter 9

# Evaluation of Pavement Structural Capacity

## 9.1 BASIC CONSIDERATIONS

There are several reasons for evaluating pavement structural capacity or adequacy and many ways of doing it. Determining structural capacity requires first monitoring or measuring some characteristic of the pavement. It then involves analysis of the resulting data, either on a theoretical or empirical basis, to estimate the load-carrying capabilities and the service life of the pavement under the expected traffic conditions.

In dealing with structural capacity evaluation, the total network of an agency and particular projects or sections within the network need to be considered. For example, allocations of funds will be determined on a network basis, and these decisions determine the availability of funds for any particular project. Therefore, procedures for evaluating the total system at a level adequate to identify structurally deficient sections should exist. Ultimately, of course, detailed structural evaluation should be carried out on each project as part of the rehabilitation design.

This chapter outlines some of the more common methods for nondestructive and destructive structural evaluation. Use of the information for design and maintenance purposes is discussed in Parts Three and Four.

# Evaluation of Pavement Structural Capacity

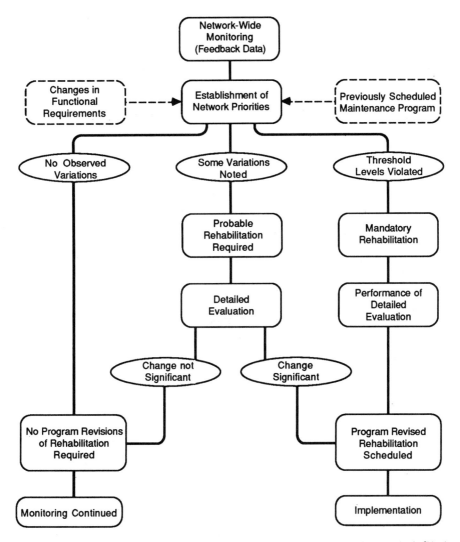

**Figure 9.1** Logistical plan for proceeding from network monitoring to project analysis [Hudson 74].

## 9.1.1 Monitoring the Pavement Network

Structural monitoring on a regular or periodic basis can be accomplished within the scope of a network monitoring plan, applicable basically to all classes of evaluation data as shown in Figure 9.1 [Hudson 74]. Deflection or curvature measurements of the pavement surface under specified loads represent the usual means for such monitoring [McComb 74].

The need for structural evaluation and/or structural deficiencies can be indicated by other types of evaluation, such as surface distress monitoring. As another example, if maintenance or rehabilitation is required because of inadequate surface friction, the wise pavement manager will also carry out a structural evaluation. This may indicate that rather than just providing a thin, skid-resistant surface to restore safety, a thicker overlay for future expected traffic loads could be more cost-effective.

### 9.1.2 Criteria for Detailed Evaluation

In the network evaluation program some threshold criteria are required. They usually take the form of limiting deflection levels for a specified load and expected number of equivalent axle load repetitions. The other types of periodic (i.e., annual, biannual, etc.) evaluation as discussed elsewhere in Part Two, can include serviceability, skid resistance, and distress survey information. Also important are the continuing inspections by maintenance engineers and field people in reporting any unusual or rapid changes in the behavior or condition of the pavements.

### 9.1.3 Functional versus Structural Evaluation

Both functional and structural evaluations of pavements are important. They supplement rather than replace each other. Serviceability observations below the acceptable level are one way to trigger a structural evaluation. The rough pavement may have adequate strength and require only a level-up surface layer, or it may be structurally weak and require replacement or a thick overlay. To complete the interrelationships, the structural evaluation of a pavement should be capable of predicting its service life for the expected load repetitions.

### 9.1.4 Destructive versus Nondestructive Evaluations

Test methods are commonly classified as destructive or nondestructive. Normally, the difference relates to whether or not physical disturbance of the materials occurs. For pavements, destructive evaluation is usually a test pit for sampling and testing the component materials, sometimes in place.

Many effective evaluation techniques involve surface measurement of deflection or curvature combined with small-core drilling to obtain thicknesses and samples of the underlying material for laboratory testing. These are considered nondestructive tests in subsequent discussions, because no major disruption of the pavement structure is required.

## 9.2 NONDESTRUCTIVE MEASUREMENT AND ANALYSIS

Deflection measurement techniques are widely used for the structural evaluation of pavements. In general, these techniques are selected over destructive methods due to lower cost, less interruption to the traffic, less damage to the pavement, and the ability to make a sufficient number of measurements to quantify variability. Alternative methods of structural evaluation such as the spectral analysis of surface waves [Nazarian 86], have been studied but not developed to the point of common implementation.

Deflection measurements require at least a two person crew, traffic interruption of the facility being tested, and the use of a deflection measuring device which can

# Evaluation of Pavement Structural Capacity

be relatively expensive. In addition, there will be mobilization costs to get to the test site. Hence, structural evaluation is a relatively expensive component of the pavement monitoring process. This means that structural evaluation should be carefully planned, including the selection of equipment, the data collection plan, and the data analysis method.

## 9.2.1 Data Collection Plan

The data collection plan should be based on the intended use of the data and variability of the pavement's structural capacity. Moreover, it should be specific with respect to acceptable seasonal testing periods and spatial distribution of the test points along and across the pavement section.

The magnitude of deflection at a specific point on a pavement will vary with environmental conditions. Excess moisture in the base and subgrade during spring thaw conditions will weaken the structure and increase deflection. As well, increasing temperatures will decrease the resilient modulus of asphalt concrete and cause curling or warping of a concrete slab, each resulting in greater deflection.

Pavement design methods treat these seasonal or environmental variations in different ways. Some procedures are based on critical spring deflections while others use summer or "normal" deflections. Thus, the structural evaluation plan should be developed in concert with the analysis method.

The spatial location of the deflection points should be adequate to capture the variability in structural capacity of the pavement. This variability can be due to such factors as nonuniform subgrade materials, varying depth to bedrock, local areas with high moisture, construction variations, etc. Pavements with a high variability should be subjected to a greater number of deflection measurements.

A good approach to sampling involves a two-stage plan. The first stage would have relatively large spacings, and the variability would then be analyzed to determine the number of deflection locations required to quantify the structural capacity to a desired level of reliability. Procedures for this type of analysis have been described [Anderson 74].

This approach may be useful in airport pavement evaluation where the paved area is relatively confined and has to be closed during the pavement evaluation. However, most highway agencies prefer to use a fixed-sampling plan to minimize traffic control problems. Unfortunately, a fixed-sampling plan can result in lack of sufficient data for estimating the variability of the pavement, or an excess of expenses for the collection of unnecessary data. Failure to quantify the variability of the pavement's structural capacity can lead to an inappropriate rehabilitation strategy.

## 9.2.2 Analytical Approaches

There are many analytical approaches to structural evaluation, but they can be grouped into five basic classes:

1. Comparison of measured behavior, usually deflection, to "allowable" deflections based on past performance

2. Comparison of measured behavior against calculated allowable criteria, usually determined from elastic layer analysis and usually in terms of deflection
3. Using an existing design method to estimate remaining life or load carrying capacity with measured behavior as an input
4. Use of measured deflections and layer thickness data to quantify the material properties of each layer using a "back calculation" procedure
5. Combination methods using laboratory material test results in conjunction with the back calculation procedure to provide material properties required for a theoretical analysis of fatigue and measured behavior to provide limiting criteria

The first three approaches have been used successfully under limited conditions. They are hard to adapt to changes in materials, environments, or load limits. The last two approaches offer a more general solution to the structural evaluation problem. However, they are not easy to implement due to the inherent limitations of the currently available mechanistic pavement analysis models.

## 9.3 DEFLECTION DEVICES

Currently, all practical nondestructive evaluations of pavement structural capacity are performed with deflection measuring devices. They can be placed in four broad categories:

1. Static devices which measure the pavement's response to a static load or a single application of a slow moving load
2. Vibratory devices which measure the pavement's response to a vibratory or cyclical load
3. Impulse devices which load the pavement by dropping a known mass through a known distance and measuring the response of the pavement
4. Multimode devices

There have been several comparative studies of the relative capabilities of the various deflection devices. A comprehensive study for the Federal Highway Administration provides much of the basis for the following sections. Table 9.1 summarizes the features of each of the devices.

### 9.3.1 Static Devices

This category includes the following:

- Plate Bearing Tests
- Curvature Meter
- Benkelman Beam
- Automated Deflection Beams
- Curviameter

With the exception of the plate bearing test, these tests measure deflection under truck wheel loads. Curvature Meter data is collected using static loading conditions. Both techniques are labor-intensive, time consuming, and not in widespread use. Benkleman Beam, Curviameter, and automated deflection beams measure deflections under slow-moving wheel loads.

### 9.3.1.1 Plate Bearing Tests

The plate bearing test [Moore 76] is a static deflection procedure. Load is applied through a hydraulic jack on 30-inch diameter rigid plates, stacked on the pavement surface, and reacting against the frame of a stationary reaction frame. It is a slow and laborious test, requiring from 30 minutes to over an hour to complete. For this reason, it is not currently a popular procedure for pavement evaluation. Historically, the plate load test was widely used for airfield pavement design.

### 9.3.1.2 Curvature Meter

The Curvature Meter was developed to estimate the radius of curvature and calculate the corresponding maximum deflection on a pavement surface due to a stationary wheel load [Dehlen 62]. It consists of a long thin aluminum bar with supporting feet at the ends and an Ames dial gauge fixed at the center. The instrument is placed between the dual wheels of a stationary loaded truck and the deflection basin is measured at middle ordinate of a curve with a cord length of 12 inches (0.3 m). The Curvature Meter has not been used on a regular basis by highway agencies in North America.

### 9.3.1.3 Benkelman Beam

The Benkelman Beam is a simple and inexpensive device for deflection measurements. It was developed at the WASHO Road Test [HRB 55] and has been used extensively by highway agencies for pavement research, evaluation, and overlay design around the world.

The Benkelman Beam consists of a simple lever arm attached to a lightweight aluminum or wood frame, as shown in Figure 9.2. Measurements are made by placing the tip of the beam probe between the dual tires of a loaded (usually an 18-kip (80 kN) axle load) truck at the point where deflection is to be determined. As the loaded vehicle moves away from the test point, rebound or upward movement of the pavement is measured by the dial gauge. The equipment is versatile and simple to operate. However, it is slow and labor intensive. In some cases, particularly on stiff pavements, the support legs may be within the deflected area resulting in inaccurate measurements.

### 9.3.1.4 Automated Benkelman Beam

Automated deflection beam devices, which operate on the same principle as the Benkelman Beam, were created to increase the speed of deflection measurements. Deflection beams are mounted on the load vehicle. The beams are positioned and maximum deflection recorded automatically from test point to test point while the operator drives the truck along the pavement.

Table 9.1 Characteristics of Nondestructive Deflection Testing Devices [Hudson 87a]

| Device Name | Type of Unit | Loading Principle | Loading System | Static Load, lb. | Dynamic Force, lbf. | Load Transmitted by | Deflection Measuring Device | Method of Recording Data | Available in USA |
|---|---|---|---|---|---|---|---|---|---|
| Benkelman Beam | Manual Operation | Rolling Wheel | Loaded Truck | (a) | N/A | Truck Wheels | Dial Gauge | Manual | Yes |
| California Traveling Deflectometer | Self-Contained Automated | Rolling Wheel | Moving loaded truck | (a) | N/A | Truck Wheels | Deflection Transducer | Manual, Printed, or Automated | Yes |
| La Croix Deflectograph | Self-Contained Automated | Rolling Wheel | Moving truck loaded with blocks or water | (a) | N/A | Truck Wheels | Deflection Transducer | Manual, Printed, or Automated | No |
| Dynaflect | Trailer Mounted | Steady State Vibratory, Frequency 8 Hz | Counter rotated masses | 2,100 | 1,000 | Two 16" dia. Urethane Coated Steel Wheels | 5 Geophones | Manual, Printed, or Automated | Yes |
| Model 400 B Road Rater | Trailer Mounted (b) | Vibratory Frequency 6–60 Hz | Hydraulic Actuated Masses | 2,400 | 200–3,000 | Two 4" by 7" Pads with 5.5" Center Gap (c) | 4 Geophones | Manual, Printed, or Automated | Yes |
| Model 2000 Road Rater | Trailer Mounted | Vibratory Frequency 6–60 Hz | Hydraulic Actuated Masses | 3,500 | 200–5,500 | Circular Plate 18" dia. (d) | 4 Geophones | Manual, Printed, or Automated | Yes |

| | | Vibratory Frequency 5–80 Hz | Hydraulic Actuated Masses | | | | | |
|---|---|---|---|---|---|---|---|---|
| Model 2008 Road Rater | Trailer Mounted | | | 7,500 | 500–9,000 | Circular Plate 18" dia. (d) | 4 Geophones | Manual, Printed, or Automated | Yes |
| KUAB 50 FWD | Trailer Mounted | Impact | Two Dropping Mass System | 2,000 | 2,700–11,300 | Sectionalized Circular Plate 11.8" dia. (d) | Up to 5 Seismometers | Manual, Printed, or Automated | Yes |
| KUAB 150 FWD | Trailer Mounted | Impact | Two Dropping Mass System | 2,000 | 2,700–33,700 | Sectionalized Circular Plate 11.8" dia. (d) | Up to 12 Seismometers | Manual, Printed, or Automated | Yes |
| Dynatest 8000 FWD | Trailer Mounted | Impact | Dropping Mass | 2,000 | 1,500–27,000 | Circular Plate 11.8" dia. (d) | 7 Geophones | Manual, Printed, or Automated | Yes |
| Dynatest 800 FWD | Trailer Mounted | Impact | Dropping Mass | 2,000 | 6,500–19,000 | Circular Plate 11.8" dia. (d) | 7 Geophones | Manual, Printed, or Automated | Yes |
| Phonix ML 10000 FWD | Trailer Mounted | Impact | Dropping Mass | 1,900 | 2,300–23,000 | Circular Plate 11.8" dia. (d) | 3 or 6 Geophones | | Yes |
| CRSTP Curvimeter | Self-Contained | Rolling Wheel | Moving truck loaded with steel pellets | (a) or variable | N/A | Truck Wheels | | Automated | Yes |

a. 18,000 lbs single axle truck
b. Earlier versions were mounted on vehicles
c. Circular plates are available
d. Plates of other dimensions are available

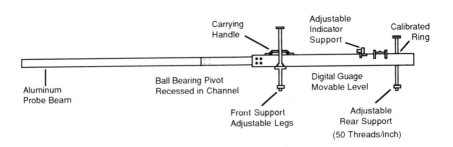

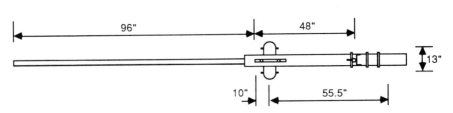

**Figure 9.2** Schematic diagram of the Benkelman Beam, with critical dimensions (Note: 1 in = 25.4 mm).

The La Croix Deflectograph, manufactured in Europe, is a commercially available automated deflection beam. It has been used widely throughout Europe, but not in North America. The vehicle moves at a constant speed of 3 km per hour during testing. Vertical deflections are measured between the twin rear wheels of the loaded axle by means of a sensor rod connected to a reference beam. The operation is fully automated. The unit produces a graphic display and data is recorded on a magnetic tape for computer analysis.

The British Pavement Deflection Data Logging Machine is similar to the La Croix, but is manufactured in the United Kingdom to British standards.

The California Traveling Deflectometer was custom-built for the California Department of Transportation [Caltrans 78]. It is essentially an automated Benkelman Beam. Deflections are measured at 20 feet (6.22 meters) intervals as the vehicle moves at a constant speed of one-half mile per hour (0.8 km per hour).

The CEBTP Curviameter [Paquet 77, CEBTP] was developed in France for continuous deflection measurements on flexible pavements at a relatively higher speed, 11.2 mph (18 km/h). It uses a geophone to measure the vertical acceleration of a

# Evaluation of Pavement Structural Capacity

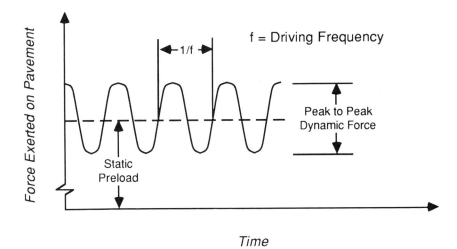

**Figure 9.3** Typical dynamic force output of steady state vibratory devices.

point on the pavement surface between dual test wheels. This measure provides the value of the curvature at that point and, in particular, its maximum value at the peak of the deflection curve. The deflection is obtained by integration of the geophone signal. The forward speed and vertical acceleration measurements are coordinated to obtain a measure of the curvature, or slope, of the deflection basin.

The Curviameter vehicle also carries an electronic deflection beam to measure static deflections similar to the Benkelman Beam. Curviameter deflections have correlated well with the Benkelman Beam deflections [CEBTP].

## 9.3.2 Vibratory Devices

Steady-state vibratory devices, including the Dynaflect, Road Rater, and the U.S. Army Engineer Waterways Experiment Station 16-kip (71 kN) vibrator, produce a sinusoidal force superimposed on a static load, as illustrated in Figure 9.3. The Dynaflect [Scrivner 66, Geolog] and the Road Rater [FDI] are commercially available and widely used nondestructive testing devices in this category.

### 9.3.2.1 Dynaflect

The Dynaflect (Figure 9.4) is a light load, fixed-frequency device. It consists of a force generator and five geophones housed in a small two-wheel trailer [Geolog]. A remote control and readout panel is carried in the towing vehicle, which allows operation from the driver's seat. The load is generated by two counter-eccentric masses rotating at 8 Hz. The peak to peak dynamic load is 1,000 lb distributed between the two load wheels. Five geophones mounted on the trailer tow bar at 12-inch intervals (see Figure 9.4(b)) measure the deflection basin.

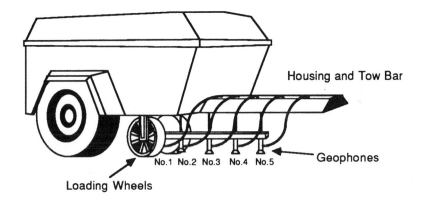

(a) The Dynaflect with Load Wheels in the Test Position

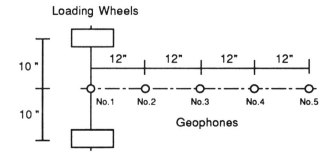

(b) Configuration of Dynaflect Load Wheels and Geophones

**Figure 9.4** Illustration of the Dynaflect (Note: 1 in = 25.4 mm).

### 9.3.2.2 Road Rater

The Road Rater is a vibratory device capable of varying both the load magnitude and frequency. Although three models are produced, they are mechanically similar [FDI]. The major elements of the loading system are shown in Figure 9.5. Magnitude of the static load is varied by transferring the weight of the trailer from the travel wheels to the deflection load plate. To generate the dynamic load, a mass is hydraulically raised and lowered. The deflection load plate provides the reaction point for displacing

**Evaluation of Pavement Structural Capacity**

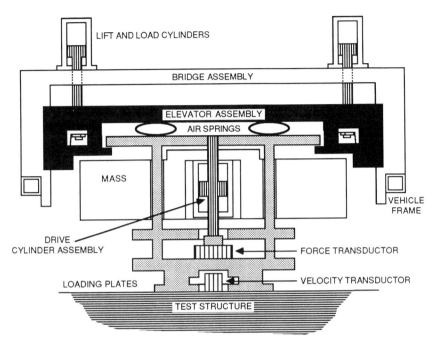

**Figure 9.5** Schematic drawing of the Road Rater static and dynamic loading system [FDI].

the mass. Hence, as the mass is lifted, the deflection force is increased and as the mass is lowered, the deflection force is released. Four velocity transducers are used to measure the deflection of the pavement; one in the center of the load plate and three placed at 1-foot intervals. The three models of the Road Rater differ in the range of dynamic load magnitude and frequency that can be applied to the pavement as follows:

| Road Rater Model | Dynamic Load Range (lb) | Frequency Range (Hz) |
|---|---|---|
| 400 | 200–3,000 | |
| 2000 | 200–5,500 | 6–60 |
| 2008 | 500–9,000 | 5–80 |

### 9.3.3 Impulse Devices

Impulse load deflection measurement devices typically employ a mass falling onto a buffered load plate. These devices are typically referred to as falling weight deflectometers (FWD). Figure 9.6 illustrates the basic principle of an FWD. Variations in the applied peak force levels are achieved by varying the magnitude of the dropping

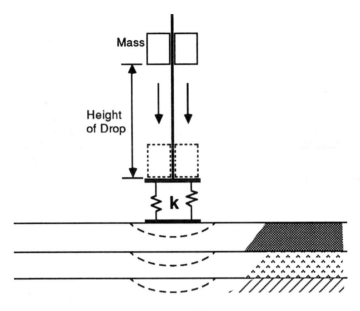

**FWD Loading**

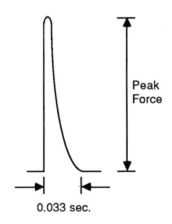

**Typical Force Signal of FWD**

**Figure 9.6** Basic principle of a Falling Weight Deflectometer (FWD).

# Evaluation of Pavement Structural Capacity

mass and the height of drop. Vertical peak deflections are measured by the FWDs in the center of the loading plate and at varying distances away from the plate. They can be plotted as deflection basins. In some opinions, the transient force impulse created by the FWD in the pavement more closely approximates the pulse created by a moving wheel load than either the static or steady-state vibratory load devices. Developed in Europe, these devices have since become popular in the United States and Canada due in large part to their adoption in the mid to late 1980s for the Long-Term Pavement Performance (LTPP) study in the Strategic Highway Research Program (SHRP).

### 9.3.3.1 Dynatest FWD

A schematic of the Dynatest Model 8000 FWD [Dynatest] is illustrated in Figure 9.7. Factory calibrated geophones are used to register peak deflections. The operating sequence is completely automated. Pulse loads between 1,500 to 27,000 lb (7 to 120 kN) are produced with the Model 800 FWD. The Model 800 FWD produces pulse loads between 6,500 to 19,000 lb (30 to 85 kN).

Optional features available with various Dynatest FWD models include: automated pavement temperature sensing, automated air temperature sensing, a geophone extension bar for PCC joint testing, and a video system for accurate, fast PCC joint testing.

### 9.3.3.2 KUAB FWD

The KUAB FWD [KUAB, PTI, 85] unit is trailer mounted and completely enclosed by a metal housing. Doors in the bottom of the unit automatically open to allow the test equipment to be lowered to the pavement surface. This feature protects the mechanical and electrical components from water, oil, dust, etc. The enclosure can be locked to discourage vandalism. The most popular models are the Model 50, with an impulse force range of 2,700 to 11,250 lb (12 to 50 kN), and the Model 150, which has an impulse range of 2,700 to 33,700 lb (12 to 150 kN).

Operation is completely automated. Housed seismometers (deflection transducers) are used to measure peak deflections. They can be calibrated in the field with a micrometer incorporated in the housing. An impact load is applied to the plate by means of a two-mass system. This creates a wider pulse duration than other FWD designs to better represent the stress duration created by trucks. The loading plate is segmented to improve contact with the pavement. The principle of the load system and deflection measurement is illustrated in Figure 9.8.

### 9.3.3.3 Phonix FWD

The Phonix Model ML 10000 FWD [PTA 84, PTA 86] produces a dynamic impulse load range of 2,300 to 23,000 lb (10.2 to 102.3 kN). It includes three or six deflection transducers in adjustable mounts along an 8-foot (2.4 m) raise/lower bar. The electronic control system consists of a microcomputer, control software, and sensors. Options include IBM-PC microcomputer, air and surface temperature measuring devices, and a paint-marking system.

The Phonix FWD trailer features a protective metal housing and covering for the falling weight assembly, which includes a cylindrical mass (in nine sections) of maxi-

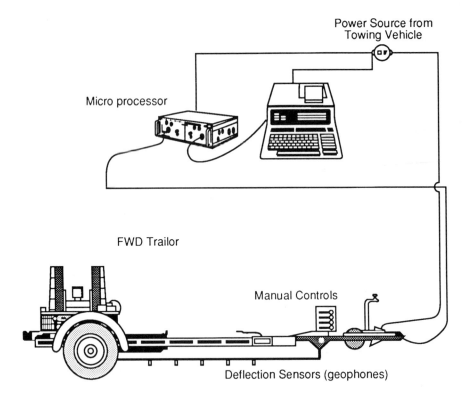

(a) FWD in Operating Position

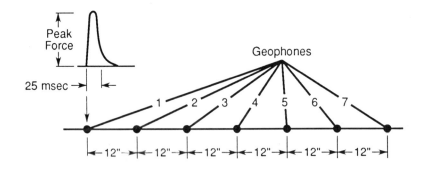

(b) Geophones Configuration

**Figure 9.7** Schematic diagram of Dynatest Model 8000 FWD (Note 1 in. = 25.4 mm) [Dynatest].

# Evaluation of Pavement Structural Capacity

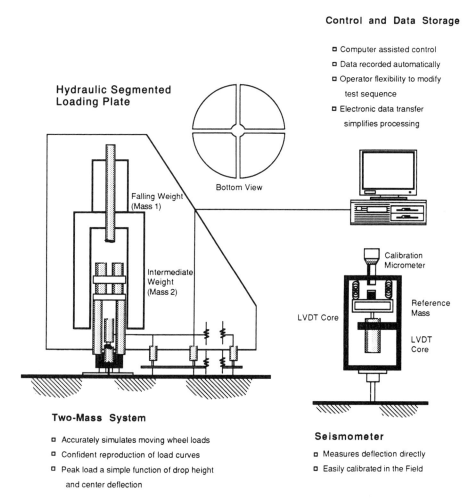

**Figure 9.8** Schematic diagram highlighting key technological features of the KUAB 150 FWD [KUAB].

mum 660 lb (300 kg). The impact is transmitted via rubber pads over the load plate. A schematic of the system is shown in Figure 9.9.

## 9.3.4 FHWA "THUMPER"

The Federal Highway Administration (FHWA) has a multimode load deflection device custom built by Cox and Sons of California. Known as the FHWA "Thumper," it is a research oriented deflection measuring device designed to incorporate many capabilities of other devices. While the "Thumper" is technologically sophisticated and an excellent research tool it has limited application for practicing engineers since it is not commercially available.

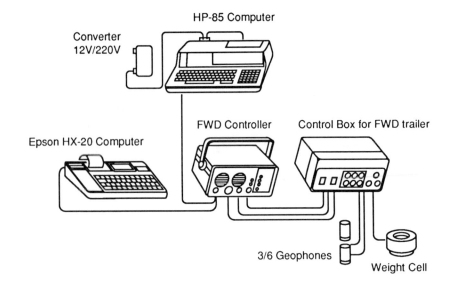

(a) Control and Data Collection Instruments

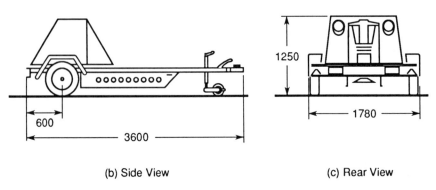

(b) Side View          (c) Rear View

**Figure 9.9** Schematic of the Phonix ML10000 FWD with all dimensions in millimeters [PTA 86].

## 9.4 DESTRUCTIVE STRUCTURAL EVALUATION

Although pavement condition can be evaluated by measurements of surface irregularities, or defects, it occasionally becomes necessary to remove portions of the pavement structure to ascertain just where problems are occurring and why. The term *destructive testing* is applied to these evaluation methods because the original structure of the complete pavement is destroyed with respect to future testing at that particular location. In general, such evaluation procedures are restricted to pavements that show evidence of distress; however, they have been used on test roads [HRB 52, 62] and form an important part of the SHRP LTPP study.

The techniques used depend on the type of information desired, but generally involve cutting into each pavement layer and removing samples for inspection and testing. At times the objective is to obtain undisturbed samples of the various layers.

The actual cross section of the various layers of rutted flexible pavements can be analyzed to determine the behavior of each layer and the functioning of the system. One such study in Kentucky [Drake 59] revealed the subgrade soil had intruded into the water-bound base course material, thus suggesting the need for changes in the gradation of the base course material and modifications of certain construction procedures.

Trenches were cut transversely across flexible pavements at the AASHO Road Test to obtain information concerning the amount of wheel path rutting at the top of each of the component structure layers as well as to obtain information on the existing condition and strength of the materials. It was found that rutting was due principally to decreases in thickness of the component layers, which in turn resulted from lateral movement of the materials. These results, along with density and strength tests on samples of the removed material, provided considerable information on the structural capabilities of the pavement.

Several states have conducted research on degradation of base courses under traffic. Samples removed from the base course layer at various time intervals were tested to determine what increase, if any, in fines has occurred.

Many states excavate and examine isolated trouble spots in pavements to determine the cause of the particular problem and take steps to correct the situation. These individual investigations are rarely reported in publications; in fact, the information rarely goes beyond the group involved in the actual problem. Consequently, the available information concerning destructive testing methods and the attendant results is limited to those occasions where these methods were incorporated into an overall program of evaluation such as at road tests.

The advantages of opening up pavements for detailed investigations below the surface must be weighed against the disadvantages of removing portions of the pavement and replacing it with patches. It is important that all variables that affect pavement performance be evaluated before definite conclusions are reached. Surface defects can be used as general guides to the underlying conditions; however, it is often necessary to determine the true position and cause of failure for a completely reliable analysis.

## 9.5 STRUCTURAL CAPACITY INDEX CONCEPTS

Unlike the serviceability-performance concept for the functional behavior of a pavement, there is no existing unified approach for establishing a structural capacity index. Basically the pavement engineer has two concerns with respect to structural capacity:

1. What is the maximum vehicle load the pavement can withstand without causing excess immediate distress?
2. How many axle load repetitions can the pavement withstand?

The first concern is applicable to airfield pavements where the maximum rated capacity of the pavement must be known to define the class of aircraft which can use the facility. It is also of concern to highway engineers faced with posting load limits during spring thaw periods and with issuing permits for trucks which exceed legal weight limits. Both airport and highway engineers are concerned that a single or relatively limited number of high stress applications will cause excessive damage to the pavement structure.

The development of a structural adequacy index for the repeated load case is complex due to the need to predict future traffic on the pavement structure, especially with respect to the distribution of heavy vehicle axle types and weights. Current theories for pavement analysis make it possible to estimate the remaining life of a pavement, provided several parameters can be quantified:

1. The fatigue behavior of the surface materials
2. The modulus properties of each pavement layer
3. The thickness of each pavement layer
4. The complete traffic history for the pavement section
5. Reliable estimates of the future axle load distributions

There have been attempts to define a structural capacity or structural adequacy index in terms of the same scale as for Present Serviceability Index, PSI (0–5) or Riding Comfort Index, RCI (0–10). One such working method is actually an approximation of fatigue analysis in that it utilizes a maximum tolerable deflection (MTD) versus load repetitions relationship [Haas 83], analogous to a tensile strain versus load repetitions fatigue relationship. Weighting factors transpose the observed deflection into a 0 to 10 scale. A brief description of the method follows.

First, a Truck Factor (TF) and Design Traffic Number (DTN) are calculated in order to estimate the total Equivalent Single Axle Loads (ESAL), using the following equation:

$$\text{TF} = 0.0353 + 0.0003 \left(\frac{\text{AADT}}{2}\right) \times (\text{LDF}) \times \left[\frac{(1 + g)^n + 1}{2}\right] \quad (9.1)$$

where

$\quad$ AADT = annual average daily traffic (2 directions)
$\quad$ LDF = lane distribution factor (1.0 for 2 lane; 0.8 for 4 lane, etc.)
$\quad$ $g$ = the compound annual traffic growth rate
$\quad$ $n$ = the number of years in the analysis period

The minimum value of TF is 0.75 and the maximum is 2.0. The design traffic number or DTN is calculated as follows:

$$\text{DTN} = \left(\frac{\text{AADT}}{2}\right) \times (\text{LDF}) \times (T) \times (\text{TF}) \times \left[\frac{(1 + g)^n + 1}{2}\right] \times (\text{LEF}) \quad (9.2)$$

### Evaluation of Pavement Structural Capacity

where

TF, AADT, LDF, $g$, and $n$ are defined as in equation 9.1.
$T$ = the fraction of commercial traffic
LEF = the load equivalency factor

The calculated DTN is then used to determine the total passes of an 18-kip (80 kN) equivalent single-axle load (ESAL) over the analysis period as follows:

$$\text{ESAL} = \text{DTN} \times 300 \times n. \tag{9.3}$$

The constant 300 represents the effective number of design days in a year. The MTD is then calculated in terms of the Benkelman Beam deflection ($\text{MTD}_B$) using the following relationships [RTAC 77]:

$$\text{MTD}_B = 10.0^{[0.40824 - 0.30103 \times \log_{10}(\text{ESAL})]}$$
$$\text{MTD}_B = 0.1 \quad \text{for ESAL} \leq 47651 \tag{9.4}$$
$$\text{MTD}_B = 0.02 \quad \text{for ESAL} > 10^7$$

This Benkelman Beam criterion is then converted to a Dynaflect value by using the appropriate Benkelman Beam-Dynaflect correlation.

A pavement section is considered to be structurally inadequate if the measured or designed deflection (in terms of a mean value plus two standard deviations, $\bar{x} + 2\sigma$) is greater than the MTD. To evaluate the degree of structural deficiency a Structural Adequacy Index (SAI) is used. The values for SAI range from a "perfect" score of 10 to a totally inadequate score of 0. A value of 5.0 indicates a barely adequate pavement structure (i.e., in most cases, design deflection equal to MTD).

SAI is derived using deduct values (Table 9.2) according to the following procedure:

1. Calculate the difference between the design deflection and MTD.
2. Determine the percentage of deflection measurements (adjusted to a spring value) which exceed the calculated MTD, for a positive difference calculated in 1.
3. Determine the traffic range (low, medium, high).
4. Read the value of density corresponding to the parameters evaluated in steps 1, 2, and 3 from Table 9.2.
5. Subtract the density determined in step 4 from the adequate score of 5 to give the SAI.

For cases where the difference calculated in step 1 is negative, a similar procedure is used. Instead of determining the percentage of deflection readings exceeding the MTD, the percentage of deflection readings less the MTD is determined.

The density value is then determined by reversing the order of the traffic columns so that the "high traffic" column applies to the ranges of low traffic volumes, the "low traffic" column to the high range of traffic volumes, and the "medium traffic" column again to the medium range of traffic volumes. Then, instead of subtracting the corresponding density from 5.0, the density is added to 5.0 to give the SAI.

Table 9.2  Structural Adequacy Point Deduction Table

| Differences (mils)** | Density | | | | | | | | |
|---|---|---|---|---|---|---|---|---|---|
| | Less than 30%* | | | Between 30 and 60%* | | | Greater than 60%* | | |
| | Low Traffic | Medium Traffic | High Traffic | Low Traffic | Medium Traffic | High Traffic | Low Traffic | Medium Traffic | High Traffic |
| 0.00 | 0.0 | 0.0 | 0.0 | 0.0 | 0.0 | 0.0 | 0.0 | 0.0 | 0.0 |
| 0.10 | 0.0 | 0.3 | 0.3 | 0.3 | 0.3 | 0.5 | 0.3 | 0.5 | 0.8 |
| 0.20 | 0.3 | 0.5 | 0.8 | 0.5 | 0.8 | 1.0 | 0.8 | 1.0 | 1.5 |
| 0.30 | 0.5 | 0.8 | 1.5 | 0.8 | 1.0 | 2.0 | 1.3 | 1.5 | 2.5 |
| 0.40 | 0.8 | 1.0 | 2.0 | 1.3 | 1.5 | 2.5 | 1.5 | 2.0 | 3.0 |
| 0.50 | 1.0 | 1.5 | 2.5 | 1.5 | 2.0 | 2.8 | 2.0 | 2.5 | 3.3 |
| 0.60 | 1.3 | 2.0 | 3.0 | 2.0 | 2.8 | 3.3 | 2.5 | 3.0 | 3.8 |
| 0.70 | 1.5 | 2.5 | 3.3 | 2.0 | 3.0 | 3.5 | 2.8 | 3.3 | 4.0 |
| 0.80 | 1.8 | 2.8 | 3.5 | 2.3 | 3.3 | 4.0 | 2.8 | 3.5 | 4.3 |
| 0.90 | 2.0 | 3.0 | 3.8 | 2.5 | 3.5 | 4.3 | 3.0 | 3.8 | 4.5 |
| 1.00 | 2.0 | 3.0 | 4.3 | 2.5 | 3.8 | 4.5 | 3.0 | 4.0 | 4.8 |
| 1.10 | 2.3 | 3.3 | 4.5 | 2.8 | 3.8 | 4.5 | 3.3 | 4.3 | 4.8 |
| 1.20 | 2.3 | 3.3 | 4.5 | 2.8 | 4.0 | 4.8 | 3.3 | 4.3 | 5.0 |
| 1.30 | 2.5 | 3.3 | 4.5 | 3.0 | 4.0 | 4.8 | 3.3 | 4.5 | 5.0 |
| 1.40 | 2.5 | 3.5 | 4.8 | 3.0 | 4.3 | 5.0 | 3.5 | 4.5 | 5.0 |
| 1.50 | 2.5 | 3.5 | 4.8 | 3.0 | 4.3 | 5.0 | 3.5 | 4.8 | 5.0 |
| 1.60 | 2.8 | 3.8 | 4.8 | 2.3 | 4.5 | 5.0 | 3.8 | 4.8 | 5.0 |
| 1.70 | 2.8 | 3.8 | 5.0 | 3.3 | 4.5 | 5.0 | 3.8 | 4.8 | 5.0 |
| 1.80 | 2.8 | 4.0 | 5.0 | 3.5 | 4.8 | 5.0 | 3.8 | 5.0 | 5.0 |
| 1.90 | 3.0 | 4.0 | 5.0 | 3.5 | 4.8 | 5.0 | 4.0 | 5.0 | 5.0 |
| 2.00 | 3.0 | 4.5 | 5.0 | 3.8 | 5.0 | 5.0 | 4.0 | 5.0 | 5.0 |
| 2.10 | 3.0 | 4.5 | 5.0 | 3.8 | 5.0 | 5.0 | 4.0 | 5.0 | 5.0 |
| 2.20 | 3.3 | 4.8 | 5.0 | 3.8 | 5.0 | 5.0 | 4.3 | 5.0 | 5.0 |
| 2.30 | 3.3 | 4.8 | 5.0 | 4.0 | 5.0 | 5.0 | 4.3 | 5.0 | 5.0 |
| 2.40 | 3.3 | 5.0 | 5.0 | 4.0 | 5.0 | 5.0 | 4.3 | 5.0 | 5.0 |
| 2.50 | 3.5 | 5.0 | 5.0 | 4.3 | 5.0 | 5.0 | 4.5 | 5.0 | 5.0 |
| 2.60 | 3.5 | 5.0 | 5.0 | 4.3 | 5.0 | 5.0 | 4.5 | 5.0 | 5.0 |
| 2.70 | 3.8 | 5.0 | 5.0 | 4.3 | 5.0 | 5.0 | 4.5 | 5.0 | 5.0 |

(*) Percentage of deflection readings exceeding the Maximum Tolerable Deflection.
(**) Difference between the design deflection, $(X_{avg} + 2S)$* Spring/Fall Ratio, and MTD

# Evaluation of Pavement Structural Capacity

This method of evaluating structural adequacy gives a "bonus" to pavements whose design deflections are less than the MTD requirement for structural adequacy. Thus, the SAI determined in this manner reflects the degree of both structural inadequacy and structural adequacy. To illustrate this method of determining SAI, two examples (one for negative difference and one for positive difference) are presented as follows.

## Example 1

If a two lane pavement section has a design deflection ($\bar{x} + 2\sigma$) of 2.57, the MTD has been calculated at 1.87, the AADT is 1000, and 45 percent of the adjusted measured deflections exceed the MTD, then:

1. Difference = design deflection − MTD = 0.70
2. 30% ≤ 45% ≤ 60%; medium frequency of exceeded MTD
3. AADT = 1000 (low traffic)
4. from Table 9.2, density = 2.0
5. SAI = 5.0 − 2.0 = 3.0

## Example 2

If a two lane pavement section has a design deflection ($\bar{x} + 2\sigma$) of 1.23, an MTD of 1.65, the AADT is 1000, and 65 percent of the adjusted measured deflections exceed the MTD, then:

1. Difference = MTD − design deflection = 0.42 or approximately 0.40
2. 60% ≤ 65%; high frequency of adjusted measured deflection below MTD
3. AADT = 1000 (low traffic); therefore use "High Traffic" column in Table 9.2
4. From Table 9.2, density = 3.0
5. SAI = 5.0 + 3.0 = 8.0

## 9.6 NETWORK VERSUS PROJECT LEVEL APPLICATIONS OF STRUCTURAL CAPACITY EVALUATION

Structural capacity evaluation can provide a wealth of information concerning the expected behavior of pavements. Due to the expense of data collection and analysis, structural capacity is not currently evaluated at the network level of pavement management by many agencies. Generally, structural evaluation is applied to specific sections at the project level of pavement management.

Nevertheless, structural capacity information, perhaps derived from less intensive sampling than for project level purposes, can be very useful at the network level, particularly if presented in terms of a Structural Adequacy Index, SAI. For example, such states as Utah, Idaho, and Minnesota and such provinces as Alberta and Prince Edward Island have extensive network level structural evaluation data.

## REVIEW QUESTIONS

1. Describe the application of nondestructive data for pavement evaluation.
2. Describe the operating principles and advantages of a falling weight deflectometer.
3. Use an elastic layer program to estimate the modulus of each pavement layer based on the following data:

| LAYER | MATERIAL | THICKNESS | POISSON'S RATIO |
|---|---|---|---|
| Surface | Asphalt Concrete | 4.0 | 0.35 |
| Base | Aggregate Base | 4.0 | 0.4 |
| Subbase | Aggregate Base | 9.0 | 0.4 |
| Subgrade | Fat Clay | – | 0.45 |

Deflection load 9,000 lb on a 6 in radius plate

| Distance from center of the load (in) | Deflections<br>Deflections (mils) |
|---|---|
| 0 | 9.56 |
| 12 | 6.81 |
| 24 | 3.61 |
| 36 | 1.99 |
| 48 | 1.19 |
| 60 | 0.82 |
| 72 | 0.58 |

4. Make a tabular comparison of the characteristics of the deflection measuring devices described in this chapter. Include key features, operational characteristics, limitations, etc.

Chapter 10

# Evaluation of Pavement Distress: Condition Surveys

## 10.1 PRINCIPLES OF SURFACE DISTRESS SURVEYS

Most highway and airport agencies conduct periodic surface distress surveys of their pavements. They measure and evaluate various types of cracking, raveling, disintegration, deformation, and so on. Such surveys are directed in large part toward assessing the maintenance measures needed to prevent accelerated, future distress, or the rehabilitation measures needed to improve the pavement. Thus, surface distress surveys, often simply called distress surveys, are useful for the manager of the pavement, but they are related to the user insofar as distress is frequently the cause of both present and future loss of serviceability. However, distress measurements should not in themselves be taken to represent user response.

Distress surveys should include a reasonable amount of detail. They should identify the distress type, severity, and extent, and in some cases the location. Although this is basic to most, because of the many uses made of the information, an extremely wide variation exists in the manner in which the surveys are obtained, recorded, analyzed, summarized, and stored [Hudson 66]. There is considerable literature available on the various survey methods, including criteria such as that originally contained in HRB Special Report 30 [HRB 57] and the condition survey manuals of various agencies, which have been summarized [Zaniewski 86].

### 10.1.1 Components of Distress Surveys

Distress surveys measure various types, severity, and density, or extent of distress. There is some degree of commonality between the different methods with respect to the components or factors that are usually measured. These often include the following general classes of factors:

1. Surface defects
2. Permanent deformation or distortion
3. Cracking
4. Patching

Several specific distress types exist within each of these classes.

Some agencies include skid resistance and roughness measurements in what they term condition surveys or condition evaluation. Whether or not to include such factors is the choice of each agency. However, the combination of engineering and subjective evaluations in some rating measures is a "mixture of apples and oranges" [Haas 71]. Such a mixture may work reasonably well for a particular agency for rehabilitation decisions, but it is usually devised by only a few individuals in that agency, based on their experience, and therefore has limited applicability or meaning outside the agency.

### 10.1.2 Considerations for the Development of a Distress Survey Procedure

A distress survey should quantify the type, severity, and extent of each of the distress types commonly found on the pavement surfaces within the agency. Keep it simple. Complex survey procedures are typically expensive and provide an excess of information.

The first step is to define how the data is going to be used. The level of detail required in the survey should be designed to meet the needs of the identified uses. Network programming decisions can be made with far less detail than is needed for determining the amount of patching required on a specific section of pavement.

Once the level of detail has been established, the next step is to select or define the types of distresses that will be included in the survey. A Federal Highway Administration project in 1985 found that the practices of state highway agencies with respect to the number of distresses evaluated in their surveys varies from 1 to 20 [Zaniewski 85]. Airfield condition surveys frequently use 19 distress types. The selection of distresses for use in the survey should be made by a committee of engineers familiar with local pavement conditions.

Once the types of distresses are determined, carefully written definitions of the distress appearance should be prepared and supported with photographs. The definition should include instructions for rating the severity and measuring the extent of each distress type. When the types of distress and associated measurement criteria are defined, the survey methodology can be established. Most pavement distress surveys are performed by human observation of the pavement in the field. Automation equipment and procedures that can supplement human observations are being developed, but completely automated systems remain to be developed.

## 10.2 SURVEY METHODOLOGY

Pavement distress surveys may be performed by walking along the pavement section, or from a moving vehicle. Walking surveys provide the most precise data about the condition of the rated pavement sample, and are generally used for airfield evaluation. However, they are time consuming and it is usually not possible to survey the entire pavement surface in a highway network. Consequently, agencies using walking surveys usually only observe a portion or sample of the network.

One of the most important considerations when performing a walking survey is to define the sampling methodology with respect to site and selection of sample units. There are several methods used, including sampling at fixed distance intervals, making a predetermined random selection, or having the survey crew pick a "representative" sample. Statistical theory requires random selection of samples to represent a population. This concept is difficult for some engineers to accept, because they may see a section which has a considerable amount of distress, but the random sample has, for example, recently been patched. However, selecting a sample that is more "representative" of the distress on the pavement section will distort or bias the information about the condition of the pavement network.

Under the rules of random selection some sections will be rated as having less distress than actually exists, but other sections will be rated as having more distress than exists. As a result, random selection of pavement samples for rating will produce a good assessment of the overall condition of the highway network, provided the sample size is adequate.

The size of the sample unit should be determined using standard statistical methods, based on the distress variability and the desired accuracy of the survey. From a practical point of view, however, many agencies select a sample size based on the crew time available and the cost of performing the survey. In this situation, the information on sample size and variability should be used to compute the accuracy of the survey, and in turn determine whether more funds should be allocated to the survey.

Several agencies collect distress data while driving along the section, usually on the shoulder, at slow speed, 4 to 10 mph. The primary advantage of this approach is the ability to cover more or all of the network. The disadvantage is that the quality of the data is not as good as from a walking survey.

The need to perform the distress survey in a consistent manner from one section to the next is extremely important. It is acceptable to combine a slow speed windshield survey with a walking survey provided the same procedure is used on every section in the network, and a random method is used for selecting the sample where the walking survey will be performed.

Some agencies have attempted to observe pavement distress while driving over the section at high speed, 40 mph or more. At high speeds, however, it is not likely that the observer will see, recognize, and record distresses in a consistent manner. If the funding available requires high speed data collection, then roughness measurements should be the primary pavement evaluation tool.

## 10.3 TYPES OF DISTRESS

Development and implementation of a pavement distress survey procedure require a clear definition of the distress types. Fortunately, a wide variety of manuals are available [Shahin 79, Smith 80, Zaniewski 85, SHRP 89]. Such references can be very useful in developing or selecting pavement distress survey procedures for the needs of individual agencies. Frequently distresses are classified as being generated by either traffic, environment, or a combination. This classification assists in the selection of treatments to repair the distress.

Tables 10.1, 10.2, and 10.3 provide example definitions of distress types for flexible, rigid, and aggregate-earth surfaced roads [Shahin 79, ERES 86, Eaton 87]. In general, the names of the distresses are descriptive of their physical appearance. The information in these tables is descriptive. Practical application requires specific definitions of the boundary limits between the severity levels for each distress type and specific techniques for measuring the extent of the distresses.

## 10.4 EXAMPLES OF DISTRESS SURVEY PROCEDURES

Most if not all agencies that have implemented a pavement management system (PMS) incorporate a distress survey procedure within the PMS. These procedures range from very detailed measurement and mapping of specific distress types to procedures which rate the overall condition of the road based on engineering judgment.

### 10.4.1 PAVER Distress Survey

The PAVER distress survey procedure [Shahin 79] is one of the most detailed methods implemented to date. Figure 10.1 shows the PAVER data collection form for flexible pavements. A key feature of the PAVER method is the selection of sample units for inspection. Sample units on flexible pavements are 2,500 ± 1,000 ft$^2$ and on rigid pavements they are 20 ± 8 slabs if the joint spacing is 30 ft or less. When the joint spacing is greater than 30 ft an "imaginary" joint is used to divide the slab and the sample unit consists of 20 ± 8 of the divided slabs. If the purpose of the PAVER survey is to "determine the overall condition of the pavement network," the recommendation is to inspect one or two "representative" sample units. However, if the purpose of the survey is to analyze maintenance and rehabilitation alternatives, then the number of sample units is computed based on the allowable error and the variability of observed distress.

Strict adherence to the PAVER distress survey procedures requires the field crew use a hand odometer for measuring lengths and widths of distressed areas, and a 10-foot straight edge and ruler for measuring vertical deviations in the pavement surface, for example, rutting. If these recommendations are implemented, the resulting information on the condition of the sample unit will be very precise. However, whether or not the information on the condition of the network is accurate depends on how well the survey crew picked the representative samples.

### 10.4.2 Ministry of Transportation of Ontario

Pavement distress survey guides have been published by the Ministry of Transportation of Ontario (formally the Ministry of Transportation and Communications of Ontario) for the evaluation of rigid, asphalt concrete, surface-treated, and gravel-surfaced roads [MTCO 82, 89a, 89b, 89c, 89d]. The survey procedures recommended in these guides are similar with specific distresses described for the different surfaces.

Figure 10.2 is the distress rating form for flexible pavements. The manuals have detailed procedures on the identification of the distress types, severity and extent, including photographs of each distress at various levels of severity. The severity of distress is rated in five categories ranging from very slight to very severe. Extent is also classified in five categories ranging from few (less than 10 percent) to throughout (more than 80 percent).

The evaluation of a section starts with a ride condition rating (the Ontario equivalent to a present serviceability rating) of ride quality performed at the posted speed limit. The rater then drives back over the section at slow speed, 30 mph in rural areas and 20 mph in urban areas, to assess the pavement distress. The rater makes two or three stops per section to examine distress type and severity.

## 10.5 EQUIPMENT FOR DISTRESS EVALUATION

Manual surface distress surveys have several drawbacks including the hazard of being on the road, variability between raters, inadequate sample sizes, and the high cost of the crews performing the slow speed inspection. These factors have encouraged the development of automated techniques for evaluating pavement distress. Available equipment ranges from programming laptop/notebook computers (automated data collection forms) to video and film photography of the pavement surface. However all working methods to date still require some degree of human intervention in the process of identifying and quantifying the distress types. Research and development for equipment to replace the human observer continues. Due to the rapid development of products in this area, those interested in equipment and methods for automated distress evaluation will need to access the most current information on the subject.

A 1987 Federal Highway Administration project [Hudson 87b] evaluated several types of automated distress survey equipment. This research compared the capabilities of two film, one video, and one laser based automated distress survey devices to manual condition surveys. It should be noted that many additional pieces of such equipment are on the market, or are being developed, in various countries around the world.

The film and video based devices capture continuous images of the pavement surface which can be evaluated in the office. This removes the survey crew from the hazardous road conditions, and provides a permanent record of the surface condition. These devices can be very advantageous for data collection on congested urban freeways.

**Table 10.1 Examples of Flexible Pavement Distress Definitions Based on the Metropolitan Transportation Commission Pavement Management System [ERES 86]**

| Distress Types | Description | Severity Levels | Extent Measures |
|---|---|---|---|
| Alligator Cracks | Interconnected cracks caused by fatigue failure due to repeated loads | L  Fine parallel longitudinal cracks<br>M  Network of cracks with some spalls<br>H  Well defined pieces that may rock under traffic and spalled at the edges | Square feet of surface area |
| Block Cracking | Interconnected cracks caused by shrinkage of the asphalt and daily temperature cycling. Size of the blocks ranges from 1 × 1 to 10 × 10 ft. Generally occurs over a wide area of the pavement surface. | L  Any filled crack or nonfilled cracks less than 3/8 in.<br>M  Nonfilled cracks 3/8 to 3 in. wide or non-filled cracks up to 3 in. wide surrounded by light random cracking<br>H  Any crack surrounded by random cracking or nonfilled crack more than 3 in. | Square feet of surface area |
| Distortions | Corrugations, bumps, sags and shoving. Abrupt upward or downward displacements of the pavement surface. Distortions are evaluated relative to the effect on ride quality. | L  Cause vibration in the vehicle but do not require the vehicle to be slowed<br>M  Significant vibration in the vehicle, some reduction in speed is necessary for safety and comfort<br>H  Excessive vehicle vibration requires considerable speed reduction for safety and comfort | Square feet of surface area |

| | | |
|---|---|---|
| Longitudinal and Transverse Cracking | Cracks that are either parallel or transverse to the pavement centerline. Longitudinal cracks are generally related to construction defects and transverse cracks are related to temperature variations and hardening of the asphalt. | See Block Cracking

Linear feet of cracking |
| Patching and Utility Cuts | Repair of the pavement with new material. | L Patch in good condition and does not affect ride quality
M Moderate deterioration, some effect on ride quality
H Severe deterioration, with high severity effect on ride quality

Area of patching |
| Rutting | Depression in the transverse profile of the pavement surface | L Less than 1 in. depth
M 1 to 2 in. depth
H More than 2 in. depth

Square feet of surface area |

**Table 10.2  Examples of Rigid Pavement Condition Distress Definitions Based on the Paver Pavement Management System (Partial listing of the 20 distress types) [Shahin 79]**

| Distress Types | Description | Severity Levels | Extent Measures |
|---|---|---|---|
| Blow-up and Buckling | Slab expansion with insufficient joint width causes upward movement of the pavement at the joint. Rated relative to the effect on ride quality | L  Little effect on ride quality<br>M  Medium severity ride quality<br>H  High severity ride quality | Rate two slabs as distressed for a joint blow-up and as one slab for a crack blow-up |
| Corner Break | Crack intersects the joints at a distance less than or equal to one-half the slab length on both sides. Crack extends vertically through the depth of the slab. | L  Low severity crack with area defined by the break having little or no other cracking<br>M  Medium severity cracking with some medium severity cracks in the area defined by the break and the joint<br>H  High severity cracking and the area between the break and the joint is severely cracked | Extent measured by the number of slabs that are affected |
| Durability or D Cracking | Caused by freeze thaw expansion of large aggregate resulting in a breakdown of the concrete | L  Cracks are tight and cover less than 15% of the pavement area<br>M  D cracks cover more than 15% of the area and are tight or cover less than 15% of the area but display pop outs or can be easily removed.<br>H  More than 15% of the area and pieces are easy to move | Count the number of slabs at the different severity level. |

| | | |
|---|---|---|
| Linear Cracking | Cracks divided the slab into two or three pieces. Can be caused by warp or friction, traffic and repeated moisture cycling. Slabs divided into four or more pieces are counted as divided slabs. | L Nonfilled less than 1/2 in. or any filled crack, no faulting<br>M Nonfilled cracks 1/2 to 2 in. or any nonfilled crack up to 2 in. with less than 3/8 in. faulting or filled cracks with up to 3/8 in. faulting<br>H Nonfilled cracks wider than 2 in. or filled cracks with more than 3/8 in. faulting | Count the number of slabs with each severity level. |
| Polished Aggregate | Caused by repeated traffic applications. The aggregate in the surface becomes smooth to the touch. The measured skid resistance is low. | No severity levels are defined | Count each slab with a polished condition. |
| Pumping | Ejection of material from the slab foundation through joints or cracks. Caused by deflections of the slab by passing traffic loads | No degrees of severity are defined | One pumping joint between two joints is counted as two slabs. |

Table 10.3 Examples of Aggregate Road Surface Distress Types Based on the Cold Regions Research and Engineering Laboratory [Eaton 87]

| Distress Types | Description | Severity Levels | Extent Measures |
| --- | --- | --- | --- |
| Improper Cross Section | Not enough crown for proper drainage | L Small amounts of ponding water<br>M Moderate amounts of ponding water<br>H Large amounts of ponding water, road surface contains severe depressions | Measure linear feet along the center line of the affected area |
| Inadequate Drainage | Evaluate ability of ditches and culverts to carry away water | L Small amounts of water ponding and/or overgrowth and debris in the ditches<br>M Moderate amounts of ponding in the ditches or overgrowth and debris and erosion of the ditches into the shoulders and the roadway<br>H Large amounts of ponding, overgrowth and debris and erosion | Linear feet of affected area measured along the center line |
| Corrugations | Closely spaced ridges and valleys at fairly regular intervals | L Corrugations are less than 1 in. deep<br>M Corrugations 1 to 3 in. deep<br>H Corrugations more than 3 in. deep | Square feet of affected area |

| | | | |
|---|---|---|---|
| Potholes | Bowl-shaped depressions in the road surface. Usually less than 3 ft. in diameter | L Less than 2 ft. in diameter and less than 2 in. deep or less than 1 ft. in diameter and 2 to 4 in. deep<br>M More than 4 in. deep and less than 1 ft. in diameter, or 2 to 4 in. deep and 1 to 2 ft. in diameter, or more than 2 ft. in diameter and less than 2 in. deep<br>H More than 2 in. deep and more than 4 ft. in diameter | Count the number of each severity level in the sample area |
| Ruts | Depressions in the wheel path | L Less than 1 in. deep<br>M 1 to 3 in. deep<br>H More than 3 in. deep | Square feet of affected area |
| Loose Aggregate | Aggregate particles are separated from the soil binder | L Loose aggregate on the road surface<br>M Moderate aggregate berm on the shoulder or less traveled area. A large amount of fine area on the road surface<br>H Large aggregate berm on the shoulder and less traveled area | Linear feet parallel to the road surface |

## ASPHALT OR TAR SURFACED PAVEMENT
## CONDITION SURVEY DATA SHEET FOR SAMPLE UNIT

BRANCH _____  SECTION _____

DATE _____  SAMPLE UNIT _____

SURVEYED BY _____  AREA OF SAMPLE _____

### DISTRESS TYPES

1. Alligator Cracking
2. Bleeding
3. Block Cracking
4. Bumps and Sags
5. Corrugation
6. Depression
7. Edge Cracking
8. Jt. Reflection (PCC)
9. Lane/Shldr Drop Off
10. Long & Trans Cracking
11. Patch & Util Cut Patching
12. Polished Aggregate
13. Potholes
14. Rail/road Crossings
15. Rutting
16. Shoving
17. Slippage Cracking
18. Swell

### SKETCH

### EXISTING DISTRESS TYPES

TOTAL SEVER: L / M / H

### PCI CALCULATION

| DISTRESS TYPES | DENSITY | SEVERITY | DEDUCT VALUE | COMMENTS |
|---|---|---|---|---|
| | | | | |
| | | | | |
| | | | | |
| | | | | |
| | | | | |
| | | | | |
| | | | | |
| | | | | PCI = 100 - CDV = _____ |
| | | | | RATING = _____ |
| DEDUCT TOTAL | | | | |
| CORRECTED DEDUCT VALUE (CDV) | | | | |

**Figure 10.1** PAVER survey form for flexible pavements [Shahin 79].

## FLEXIBLE PAVEMENT CONDITION EVALUATION FORM (MUNICIPALITIES)

Road No. (Street) ______________ Location from ______________ To ______________

Section Length ________ (Km) (m)  Survey Date ________ Traffic Direction ☐ B: Both Directions, N: North Bound S: South Bound, E: East Bound, W: West Bound

Contract No. ______________ Work Project No. ________ Class ☐ F: Freeway, C: Connecting Link, A: Major Artierial M: Minor Artierial, R: Residential

Pavement Condition Rating (PCR) ________ Riding Condition Rating (RCR) ________ Evaluated by ______________

**Riding Condition Rating (At Posted Speed)**

| 10–8 | 8–6 | 6–4 | 4–2 | 2–0 |
|---|---|---|---|---|
| Excellent | Good | Fair | Poor | Very Poor |
| Smooth and Pleasant | Comfort-able | Uncomfort-able | Very Rough and Bumpy | Dangerous at Posted Speed |

| Pavement Distress Manifestation | | | Severity of Distress: Slight | Moderate | Severe | Density of Distress % Extent of Occurrence: Intermittent | Frequent | Extensive |
|---|---|---|---|---|---|---|---|---|
| | | | | | | <20 | 20-50 | >50 |
| | | | 1 | 2 | 3 | 1 | 2 | 3 |
| Surface Defects | Ravelling | 1 | | | | | | |
| | Flushing | 2 | | | | | | |
| | Potholes | 3 | | | | | | |
| | Pavement Edge Breaks | 4 | | | | | | |
| | Manholes & Catchbasins | 5 | | | | | | |
| Surface Deformation | Rippling and Shoving | 6 | | | | | | |
| | Wheel Track Rutting | 7 | | | | | | |
| | Distortion | 8 | | | | | | |
| | Utility Trenches | 9 | | | | | | |
| Cracking | Longitudinal | 10 | | | | | | |
| | Transverse | 11 | | | | | | |
| | Pavement Edge | 12 | | | | | | |
| | Map | 13 | | | | | | |
| | Alligator | 14 | | | | | | |

**Distress Comments** (Items not covered above) ______________

______________

**Recommendation by Evaluator** ______________

______________

| Shoulder Distress Manifestation | | | Severity of Distress | | | | | | Density of Distress % Extent of Occurrence | | | | | |
|---|---|---|---|---|---|---|---|---|---|---|---|---|---|---|
| | | | Right | | | Left | | | Right | | | Left | | |
| Dominant TYPE | √ one | Distress | Sli | Mod | Sev | Sli | Mod | Sev | <2o | 2o-5o | >5o | <2o | 2o-5o | >5o |
| | | | 1 | 2 | 3 | 1 | 2 | 3 | 1 | 2 | 3 | 1 | 2 | 3 |
| Paved Full | | Pavement Edge | | | | | | | | | | | | |
| | | Paved Shoulder Separation | | | | | | | | | | | | |
| Paved Partial | | Cracking | | | | | | | | | | | | |
| | | Breakup and Potholes | | | | | | | | | | | | |
| Surface Treated | | Distortion | | | | | | | | | | | | |
| Primed | | Pavement Edge Curb Separation | | | | | | | | | | | | |

**Maintenance Treatment**

| Pavement | Extent of Occurrence %: <20 | 20-50 | >50 | Shoulder | Extent of Occurrence %: <20 | 20-50 | >50 |
|---|---|---|---|---|---|---|---|
| | 1 | 2 | 3 | | 1 | 2 | 3 |
| Manual Patching | | | | Manual Patching | | | |
| Machine Patching | | | | Manual Spray Patching | | | |
| Manual Spray Patching | | | | Manual Chip Seal | | | |
| Manual Chip Seal | | | | Crack Rout and Seal | | | |
| Machine Chip Seal | | | | | | | |
| Fog Seal | | | | | | | |
| Surface Treatment | | | | | | | |
| Manual Burn & Seal | | | | | | | |
| Crack Rout and Seal | | | | | | | |

**Figure 10.2** Ontario distress survey form for flexible pavement in municipalities [MTCO 89a].

### 10.5.1 PASCO ROADRECON System

PASCO Corporation has worked on the development of a pavement distress survey vehicle since the late 1960s [PASCO 85, 87]. The current device, as shown in Figure 10.3, produces a continuous film strip recording of the pavement surface and a measure of roughness. Photographs are taken at night with controlled amount and angle of lighting. A hairline optical bar can be projected onto the pavement surface to provide a reference line for evaluating rut depth. The vehicle can be operated at speeds up to 50 mph and photograph an area 16 feet wide. Evaluation of pavement distress requires manual interpretation of the photographs. A special viewing table facilitates the data reduction. This equipment has been used in the SHRP Long-Term Pavement Performance Study.

### 10.5.2 GERPHO System

Like the ROADRECON, the GERPHO (Figure 10.4) records a continuous image of the pavement surface on 35 mm film. The GERPHO also uses artificial light to operate at night. Distress information is extracted from the film using a specially designed display table. The equipment has been used extensively in France [MAP 86], with other applications in Spain, Portugal, and Tunisia [DeWilder 85].

### 10.5.3 Automatic Road Analyzer (ARAN)

The Automatic Road Analyzer, ARAN, (Figure 10.5) measures rut depth and transverse profile with ultrasonic sensors, ride/roughness quality with an accelerometer on the rear axle, takes a video picture of the road right-of-way through the windshield, takes a video picture of the pavement surface with a shuttered video camera, and uses an observer-operator set of two keyboards to record and store distress data for up to 20 types, 3 levels of severity and 5 levels of extent [HPI 85]. Rut depth is measured with seven ultrasonic sensors on 12-inch (30.5 mm) centers, mounted across the front of the vehicle. Additional sensors and bar extensions can extend the rut bar to 12 feet (3.1, 3.4, or 3.7 m).

### 10.5.4 Laser Road Surface Tester (RST)

The laser Road Surface Tester (RST), as shown in Figure 10.6, was developed by the Swedish Road and Traffic Research Institute. It uses laser technology to measure crack depth and width, rut depth, longitudinal profile, macrotexture, cross profile, and distance. A "windshield" distress survey can be performed by one of the operators to identify types of cracking and other distresses [Novak 85]. An on-board microcomputer integrates the sensor signals with the accelerometer and distance transducer, averages the data into management sections, and provides the processed data in real time.

## 10.6 PAVEMENT DISTRESS INDEXES

In order to perform reliable surveys of pavement distress, the extent and severity of each distress type must be identified. However, in managing pavements and allocating

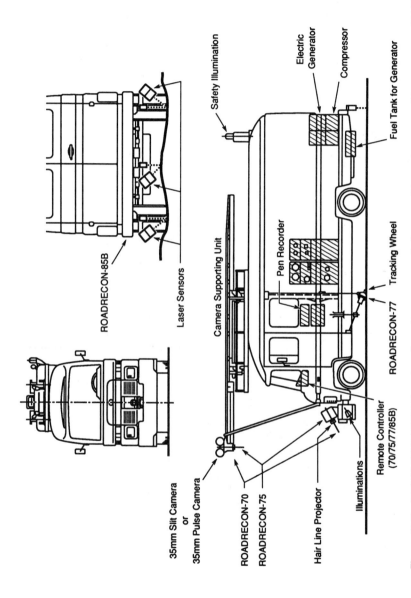

**Figure 10.3** PASCO ROADRECON system featuring automated photographic equipment and laser sensors [PASCO 87].

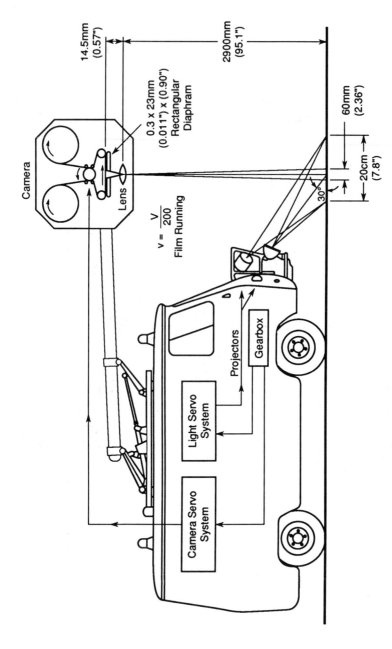

**Figure 10.4** Schematic illustration of the principles of GERPHO's automated photographic system [MAP 86].

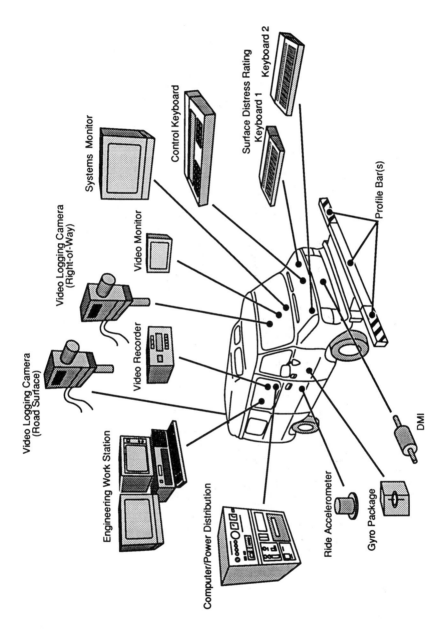

**Figure 10.5** Schematic illustrating various components of HPI's Automatic Road Analyzer (ARAN) [HPI 85].

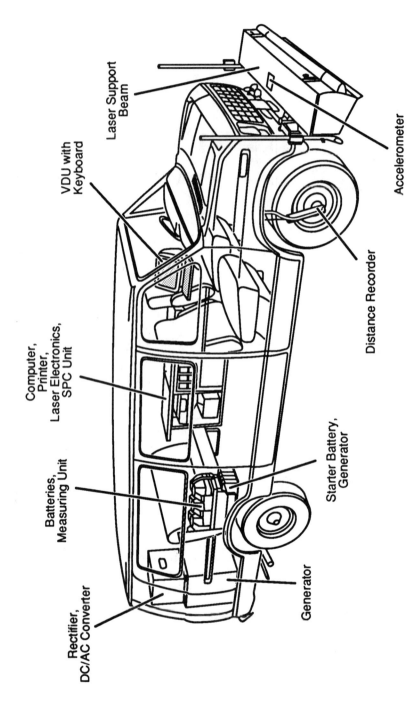

**Figure 10.6** Schematic illustrating various components of the laser Road Surface Tester (RST) [Novak 85].

# Evaluation of Pavement Distress: Condition Surveys

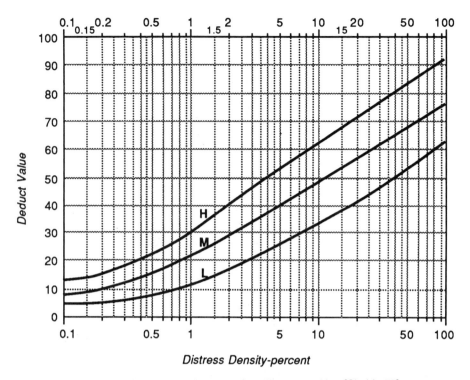

**Figure 10.7** Example of a deduct value curve for alligator cracking [Shahin 79].

resources, difficult decisions must be faced for the selection of projects. Is it more important to repair a section with alligator cracking or rutting? What combinations of extent and severity of the different distresses indicate one pavement section is in a worse state than another pavement with a different set of distresses? Combining the various distresses into a composite index is useful for comparing sections but it is a subjective process.

An early procedure for determining a composite index of pavement distress involved the use of deduct values in Washington State [Le Clerc 71, 73]. This approach was further developed for PAVER and has been implemented in other systems derived from the PAVER method, such as the Municipal Transportation Commission, MTC, pavement management system in the San Francisco Bay Area [ERES 86].

In the deduct value approach, a perfect pavement is assigned an index of 100. The index is decreased by a cumulative deduct value score based on the level and severity of the observed distress. In the PAVER system deduct value curves, such as shown on Figure 10.7 for alligator cracking, were developed for each of the distress types. The abscissa of the graph is the density or extent of distress. Each graph contains three curves corresponding to the severity of the distress. The ordinate is the deduct value. Based on the distress survey, the deduct value for each combination of distress type and severity level is estimated from the charts. The total deduct value is computed by adding the individual distress type deduct values.

On severely distressed pavements with multiple distress types, the total deduct value can actually exceed 100. Thus, under the philosophy that a pavement with two distress-type severity-level combinations which each have a deduct value of 35 is not in as bad a state as a pavement with a deduct value of 70 for one distress-type severity-level combination, a series of curves were established for correcting the total deduct value, as shown in Figure 10.8. Using the total deduct value and the number of distress-type severity-level combinations with a deduct value greater than 5, the corrected deduct value is determined and subtracted from 100 to define the composite distress index (called the pavement condition index, PCI, in PAVER). The PCI is computed for each sample unit observed for a pavement section and the values for the sample units are averaged to define the PCI for the section. Figure 10.9 associates a qualitative statement of the pavement distress state for PCI ranges.

Although the PAVER concept for a composite distress index is widely applied, it should be remembered that the deduct value curves were developed for a specific set of definitions for distress type and severity level. If the user agency elects to modify these definitions, especially with respect to severity level, then the deduct value curves should be carefully examined and modified in an appropriate manner. For example, in PAVER a rut depth must be greater than 1 inch to be classified as high severity. If an agency lowers this boundary to 3/4 inch, then the deduct curve for high severity rutting should also be modified.

## 10.7 APPLICATIONS OF DISTRESS DATA

Pavement distress data has long been recognized by pavement engineers as an important parameter for quantifying the quality of a pavement surface. It is important at both the network and project levels of pavement management systems, although the level of detail required for each application is considerably different. In both cases, the pavement distress information is useful in selecting appropriate treatments, as subsequently described in more detail in Part Three.

At the network level, the concern is with determining what class of treatment is required, for example, continued routine maintenance or resurfacing. A properly formulated composite pavement distress index is usually adequate for this purpose. This index can be determined from either a relatively limited sample of each pavement section or with a windshield survey from a slow-moving vehicle.

The project level requires an estimate of the specific extent and methods of pavement repair, such as to patch a certain area of alligator cracking. This requires a greater level of detail in the distress survey. However, manual survey methods are generally not economical for collecting the quality and quantity of information required at the project level for an entire network.

An indication of the importance of distress data is the fact that distress indexes are used as the only measure of pavement quality in many pavement management systems. This is particularly true for systems used by local governments and in urban

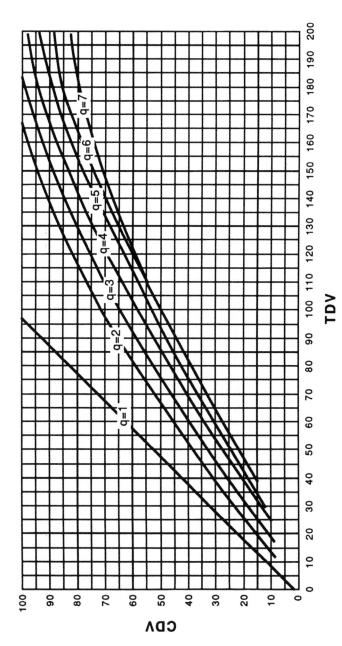

**Figure 10.8** Corrected deduct value curves.

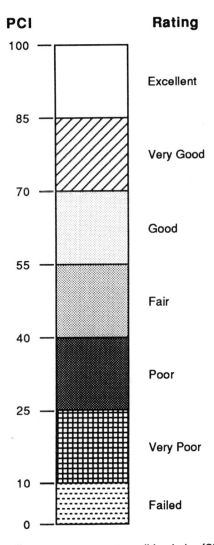

**Figure 10.9** Pavement quality versus pavement condition index [Shahin 79].

areas where roughness measurements are not performed due to equipment availability, cost, relative applicability for lower speed facilities, etc.

Pavement distress is also broadly used as the only quality measure of airfield pavements. Factors contributing to the use of distress, in preference to roughness, for the evaluation of airfield pavements include:

1. Relatively limited surface area permitting the collection of more detailed distress data and therefore an improved ability to rate pavement quality
2. Support of the military and FAA for the PAVER method

# Evaluation of Pavement Distress: Condition Surveys

3. The lack of development of a serviceability concept comparable to the highway application
4. The inability of aircraft to alter speed and direction, particularly on runways
5. The damage potential should an airplane wheel hit severe distress
6. The damage potential if loose pavement pieces are ingested into an aircraft engine

These factors do not mean roughness is unimportant on airfields. In fact, rough pavements may contribute to the fatigue of structural components in the aircraft. Further development and implementation of roughness evaluation procedures for airfields could provide an important improvement to the state of the art in airfield pavement management systems.

## REVIEW QUESTIONS

1. The best way to gain experience with distress evaluation is to participate in performing a condition survey of several distress pavements. Using one of the forms in this chapter, evaluate the condition of several pavements with a range of distress conditions. This exercise is best performed in groups so the results of the different individuals can be compared. The sections selected for this evaluation should be low volume roads for safely reasons. Perform a serviceability rating of these pavements and examine for a correlation between roughness and distress.
2. Examine the literature for definitions of distress types. Compare the definitions used by different agencies for the same distress type.
3. Review the definitions of severity levels for distresses in the PAVER distress identification manual. Discuss the applicability of these definitions for local conditions.
4. Develop a simple condition survey form of your own. Be sure to include all three major elements of a distress survey.

Chapter 11

# Evaluation of Pavement Safety

## 11.1 MAJOR SAFETY COMPONENTS

The evaluation of pavements for safety usually considers only slipperiness, in terms of skid resistance which is called surface friction by some agencies. There are, however, several components of safety evaluation, including:

1. Skid resistance
2. Ruts (as they relate to accumulation of water and the dangers of hydroplaning or ice accumulation)
3. Light reflectivity of the pavement surface
4. Lane demarcation
5. Debris or foreign objects (especially relating to airport pavements)

Deviations in the transverse and longitudinal profile of the pavement surface can generate a hazard to traffic, especially if they result in water standing on the pavement surface. Roughness can also create driver fatigue and therefore contribute to a safety problem. Since the evaluation of roughness and pavement distortions has been treated in previous chapters, it will not be further addressed in this chapter.

There are situations where any one of the foregoing factors can provide a sig-

# Evaluation of Pavement Safety

nificant safety hazard, but pavement slipperiness is by far the most common and important factor. Consequently, it is specifically addressed in the following sections.

## 11.2 SKID-RESISTANCE EVALUATION

The phenomenon of skidding involves a very complex interrelationship among pavement factors, contamination of the pavement surface by oil and water, vehicle parameters (mainly tires), and vehicle operation factors. Nevertheless, due to the importance of skid resistance on traffic safety, research has been successful in developing an understanding of the phenomenon and in developing measurement techniques and evaluation procedures. As a result, the area of skid-resistance evaluation is perhaps better understood than performance evaluation.

## 11.3 BASIC CONCEPTS OF SKID RESISTANCE AND USES OF MEASURED VALUES

The coefficient of friction, $\mu$, in physics, is calculated by dividing the frictional resistance to motion in the plane of the interface, $F$, by the load acting perpendicular to the interface, $L$. It is dependent on the contact area and is therefore not a suitable representation of the actual interaction between the tire and pavement surface. Since skid resistance involves the interrelationship between pavement characteristics, vehicle parameters, and vehicle operation, skid resistance per se is not a fundamental pavement characteristic. In reality the pavement characteristic which dominates the measurement of skid resistance is the micro and macro texture of the pavement surface. In physics such frictional characteristics are referred to as roughness. This term should be avoided in the discussion of pavement friction characteristics because in the pavement field roughness is more generally associated with the influence of the pavement surface on ride quality. Although research has been performed on the measurement and application of pavement texture for predicting skid resistance, this technique apparently has not been implemented.

In recognition of the fact that the friction parameter measured to evaluate pavements is not the traditional $\mu$, pavement engineers usually prefer the term friction factor, calculated as

$$f = \frac{F}{L}. \tag{11.1}$$

Because it is not correct to ascribe a particular $f$ to a pavement without specifying all the tire, speed, temperature, water film thickness, and other conditions that influence it, standards have been developed for skid resistance measurement. The best-known standard is the locked wheel skid trailer described in ASTM E274. Measurements made in accordance with this standard are termed *skid numbers*, SN, calculated as

$$SN = 100f. \tag{11.2}$$

Friction evaluation of airfield pavements is frequently carried out with a device, such as a Mu meter, that measures the force required to hold a wheel at a fixed angle to the direction of travel.

Skidding accidents occur not only by direct forward sliding (as in an emergency stop) where all wheels are locked but also by jackknifing (where only one or one set of wheels lock) and by breaking away or sliding off curves. Most such accidents occur under wet or icy conditions. Thus, skid-resistance measurements are generally conducted on wetted pavements.

Skid-resistance measurements taken by a standard procedure, such as ASTM E274, do not indicate the precise value available to a particular vehicle and driver. This varies with the type of tire, the amount of tire wear, and the maneuver (acceleration, deceleration, lane changing, etc.), in addition to the drainage and texture characteristics of the pavement surface. SN values obtained by a standard method represent the relative ability of a pavement to serve the frictional needs of traffic.

The large amount of literature available in the area of skid resistance to 1972 was well summarized in NCHRP Synthesis Report 14 [HRB 72]. This report considers not only the various techniques of measurement but also skid-resistance requirements and various design or remedial actions that may be taken. Also, it has been pointed out that skid-resistance data can be used for the following pavement management purposes:

1. Identifying areas of excessive slipperiness
2. Planning maintenance
3. Evaluating various types of materials and new construction practices

Committee E 17 of ASTM has continued to support the development of skid resistance measurements. In particular, ASTM Special Technical Publications 929 and 1031 document recent developments in the evaluation of pavement safety [Potlinger 86, Meyer 90].

## 11.4 METHODS OF MEASURING SKID RESISTANCE

All pavement skid-resistance measurements are empirically based. Results produced by the various devices and test methods are meaningful only when evaluated relative to an empirical base of evidence developed over time by using a standard test method. Thus, it is extremely important to follow the standard methods for the calibration and use of the device. The following discussion only provides an overview of the test methods; it is not a substitute for reading and following the specifications for the standard test methods.

### 11.4.1 Skid Measuring Equipment

ASTM E274 is the most widely used method for measuring skid resistance in North America. The method requires the use of a calibrated locked-wheel skid trailer. The major components of the device are:

## Evaluation of Pavement Safety

1. The tire, which must meet ASTM standards
2. The brake for locking the wheel
3. A water tank and distribution nozzle for wetting the pavement with a uniform-specified film of water
4. A force transducer for measuring the force required to pull the trailer when the wheel is locked

After the test wheel is locked and has been sliding for a suitable distance, the force is measured and the skid number, SN, is calculated for that section of the pavement. Many agencies in North America have skid trailers.

Automobile-mounted methods for measuring skid resistance are seemingly logical but they are also dangerous if all four wheels are locked. By using front wheel or diagonal wheel braking only, the instability is reduced [Horne 70]. A major disadvantage of automobile methods is the variability in results caused by vehicle effects (differences in suspension, weight, tires, condition, load distribution, air drag, etc.). Also, a separate water supply for wetting the pavement can significantly increase the costs of such tests.

Portable field testing methods are perhaps best represented by the well-known pendulum-type device developed by the British Transport and Road Research Laboratory [Biles 62] and covered by ASTM E303. It involves dropping a spring-loaded rubber shoe attached to a pendulum. The results are reported as *British Pendulum Numbers* (BPN). Major advantages of this device include its applicability to laboratory testing and areas such as intersections, its simplicity, low cost, and transportability; however, its use in the field requires the diversion or stopping of traffic. Also, because the shoe contacts the pavement at a relatively low speed, the results do not correlate well with locked-wheel trailer test results conducted at 40 mph.

There are several devices that measure skid resistance in the "slip mode." This refers to the phenomenon that occurs when a wheel is gradually braked, with increasing friction factor, to a point of "critical slip" beyond which the wheel locks and the friction factor drops. At the critical slip point, the friction factor is higher than when the wheel is locked. This has practical significance to the development of automatic brake control systems. NCHRP Synthesis Report 14 illustrates that the ratio of $f_{max}$ (i.e., the friction factor at critical slip) to $f_{lock}$ (i.e., the friction factor under locked-wheel conditions) varies with surface texture, friction force, and temperature [HRB 72].

There are also several devices that measure skid resistance in the "yaw mode," where the wheels are turned at some angle to the direction of travel. The side or cornering force is measured and it (as well as the side friction factor) peaks at some yaw angle. Measurement of pavement friction in this manner presents problems similar to those in slip mode measurement. The critical yaw angle is subject to the same variations as critical slip. It is desirable to use an angle that is relatively insensitive to variations in surface characteristics and operating conditions.

A fairly simple version of a yaw-mode device is the Mu meter, currently being used primarily for the evaluation of runways. It was originally developed and tested in Britain and uses two yawed wheels, at 7.5°, which provides balance. The Mu meter measures the force required to maintain the yaw angle of the tire.

Another, more sophisticated device called the Sideways-Force Coefficient Routine Investigation Machine (SCRIM) was developed by the British Transport and Road Research Laboratory in Britain [WDM 72]. It uses a yaw angle of 20° on the test wheel, which can be lifted clear of the road when not in use. The friction-factor measurements are reported as sideways-force coefficient numbers (SFC), where

$$SFC = 100f \tag{11.3}$$

which is numerically the same as the skid number, SN, reported with locked-wheel skid trailer tests according to ASTM E274.

The major advantages of a device such as SCRIM include the continuous record of skid resistance and high allowable operating speeds. The major disadvantage of the SCRIM device is the relatively high initial cost.

### 11.4.2 Pavement Texture

A unique approach to estimating skid resistance indirectly was originally developed by Schonfeld. It involves the use of color stereophotographs to analyze the "texture elements" of the surface in terms of a "texture code number" [Schonfeld 70]. This has been correlated with skid tests using the ASTM skid trailer. The correlation was reasonably good in that the standard deviation of SN estimate varied from about $\pm 3$ (for 30 mph) to $\pm 2.2$ (60 mph), over the SN range tested.

Although the Schonfeld technique is too slow in manual form for routine application, the correlation between texture and skid resistance is an important finding. Since the ability exists to measure pavement texture from a vehicle moving at normal highway speeds, further research into developing an automated method, perhaps using image analysis techniques, to analyze Schonfeld's texture elements could be well worthwhile.

### 11.4.3 Correlations Among Skid Measurements

Unfortunately, correlations among various skid measurement devices are not usually very good, largely because each device measures a different aspect of the frictional interaction between vehicle and pavement. Although some correlations may be reasonably good, over a limited range of conditions, they "should be considered fortuitous, rather than as a fulfillment of a justified expectation" [HRB 72]. The inability to form stable correlations between skid measures emphasizes their empirical nature. It has also hindered the development of criteria for the minimum allowable skid resistance of a pavement.

## 11.5 CHANGE OF SKID RESISTANCE WITH TIME, TRAFFIC, AND CLIMATE

Skid-resistance evaluation, especially for the purpose of assessing future rehabilitation needs, should consider changes on a time and/or traffic basis, as well as on a climatic

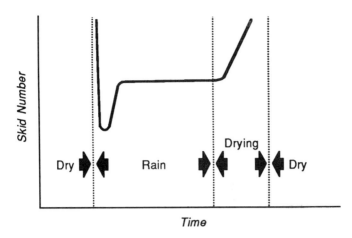

**Figure 11.1** Example of change in skid resistance due to rain [HRB 72].

effect basis. The latter can involve both short and longer periods of time (i.e., rainfall or icing of a short duration versus seasonal climatic changes).

These considerations of time/traffic/climate-based changes in skid resistance require periodic measurements. Various changes in the nature of the pavement surface should be recognized as possible contributors to such skid-resistance changes including:

1. Porosity
2. Wear (i.e., due to studded tires)
3. Polishing of surface aggregate
4. Rutting (due to compaction, lateral distortion, or wear)
5. Bleeding or flushing at the surface
6. Contamination (rubber, oil, water, etc.)

On a short-term basis, skid-resistance changes can occur rapidly, usually because of rainfall, as illustrated schematically in Figure 11.1. On a somewhat longer, seasonal basis, skid resistance may fluctuate as shown in Figure 11.2. On a still longer basis of, say, several years or several million vehicle passes, most pavements show a continual decrease of skid resistance, as illustrated in Figure 11.3.

Predictions of skid-resistance changes with time and/or traffic can be made in either of two basic ways:

1. Extrapolate an existing set of data, acquired over some period of time, into the future.
2. Conduct laboratory experiments either at the initial design stage or prior to designing rehabilitation measures (polishing characteristics of the aggregates, texture, shape, size, laboratory track tests on simulated mixtures, etc.). This approach essentially provides only a qualitative link to estimates of actual skid resistance.

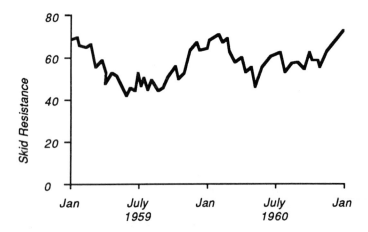

**Figure 11.2** Example of seasonal change in skid resistance [HRB 72].

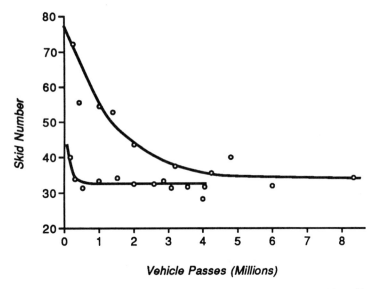

**Figure 11.3** Example of skid resistance loss of two pavement sections with traffic applications [HRB 72].

## REVIEW QUESTIONS

1. Discuss the role of safety evaluation in a pavement management system.
2. Discuss the variability in skid measurements over time. What causes the variation?

Chapter 12

# Combined Measures of Pavement Quality

## 12.1 CONCEPT OF COMBINED MEASURES OF PAVEMENT QUALITY

Pavement evaluation is generally directed toward the following objectives:

1. Selection of projects and treatment strategies at the network level
2. Identification of specific maintenance requirements at the project level

Each of these objectives requires pavement evaluation information to greater or lesser degrees of detail. In the case of lesser detail, aggregation of the individual measures comprising the information may be sufficient. Such aggregation represents a composite or combined measures of quality of the pavement. But the obvious question is how to combine these measures into a reasonable composite index that indicates the overall quality of the pavement.

## 12.2 REASONS FOR USES OF A COMBINED INDEX

The basic reason for a combined index is its use as a communications tool to convey summary information to senior administrators, elected officials, and the public. Figure 12.1 indicates the level of detailed information that is appropriate to different levels

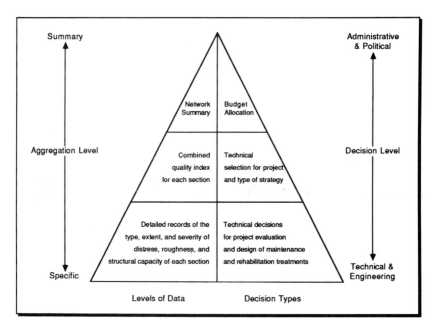

**Figure 12.1** Pavement data aggregation and level of decision.

of activities and decisions in the pavement management process. At the first level involving specific activities (ie., the disaggregated level) and specific technical or engineering decisions, detailed information on distresses, roughness, and structural capacity of each section is required. The second in the diagram involves aggregation and decisions for network management purposes. At this level, composite measures are useful for determining priorities for the selection of project strategies. Composite measures are particularly useful at the third level, which involves administrative and political decisions. People at this level are faced with large amounts of information and therefore aggregation of data to portray overall quality of the entire pavement network, plus future projection of the quality as related to budgets, is of primary importance.

## 12.3 METHODS OF DEVELOPING A COMBINED INDEX

There is no single engineering or mechanistic formula for establishing a combined pavement quality index. Techniques for the development of equations have been derived from the social sciences field, or in a subjective empirical way.

The Pavement Condition Index (PCI) defined in PAVER [Shahin 79] and the MTC systems [ERES 86] are examples of a combined index for a particular deterioration measure, in this case, surface distress. However, combined indexes are generally considered to involve two or more such measures, like the "Overall Pavement Index"

(OPI), suggested by [Baladi 90]. This type of index represents an overall aggregation of the different measures of pavement condition, for example:

$$\text{OPI} = W_1 C_1 + W_2 C_2 + W_3 C_3 + W_4 C_4 \tag{12.1}$$

where

$\text{OPI}$ = overall pavement condition index
$W_i$ = weighting factor for pavement condition measure $i$
$C_i$ = value for pavement condition measure $i$

The Delphi technique is one tool that can be used to establish a composite pavement quality index. This is essentially a structured technique for capturing expert opinion. It involves selection of a panel of experts who are asked to rank a series of hypothetical situations, such as those used to maintain pavements with different combinations of distress, structural capacity, and roughness. The initial rankings are analyzed and the results are redistributed to the panel for reassessment. An eventual convergence toward the mean is sought and the information can then be statistically analyzed to develop an equation for a composite quality index.

Regardless of the technique used to perform the survey of expert opinions, they all require the development of a specific questionnaire. One of the problems in developing the questionnaire is structuring the questions in a manner that is amenable to statistical analysis. The rational factorial approach [Fernando 83], for example, uses the principles of experimental design for the development of the questionnaire and analysis of the results. Further details on the equation developed from the approach, and its application, are provided in Chapter 19.

The Pavement Quality Index, PQI, [Karan 83] is another approach for statistically capturing information from an expert panel. It was developed from an analysis of 40 sections rated for Riding Comfort Index (RCI), Structural Adequacy Index (SAI), and Surface Distress Index (SDI), each on a scale of 0 to 10. Sections were selected to cover a wide range of the basic performance parameters (RCI, SDI, SAI) for three flexible pavement types, granular base, full depth asphalt, and soil cement base.

Two station wagons of similar ride and size were used to carry two panels of four raters each on a visual inspection and ride of the 40 sections, following a training session. On the last day, replicate ratings were made on five sections by five raters.

Pavement Quality Rating Forms, Figure 12.2, were provided for each section. These forms also contained information regarding traffic and deflection magnitudes as well as the RCI, SDI, and SAI.

The POI rating data was first analyzed to check for systematic errors (leniency error, halo effects, and central tendency effects), which were found to be insignificant.

Analysis of variance (ANOVA) techniques were then used to test for source of variation in the data, between panels, within panels, location of drivers versus others and within others, and between sections. The only significant source of variation was due to sections, which should be expected.

Since no systematic errors were found, the raw data was used in regression

SECTION NUMBER:

LOCATION:        21:22              MP 5.36 (8.63 km) to MP 6.36 (10.23 km)

MILEAGE TIE:     MP 0.0 = Jct. 53

PAVEMENT TYPE:   GB

LAYER THICKNESS: 4 in. AC, 2 in. ABB, 6 in. BASE

AGE: 18 YEARS        DTF: 4.8        ESAL (x $10^5$): 2.4

$\bar{d}$: 0.039 in.        MTD: 0.056 in.        RCI: <u>5.5</u>

SDI: <u>6.5</u>        SAI: <u>5.2</u>

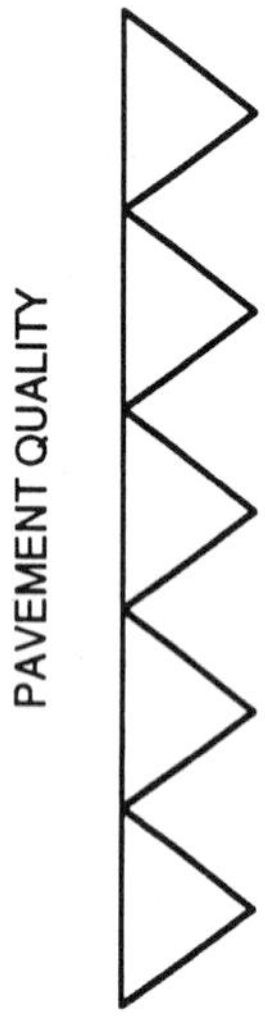

PAVEMENT QUALITY

**Excellent** (Pavement Like New)

**Good** (Many Years of Service Life)

**Fair** (Close to or Needing Rehabilitation)

**Poor** (Should Have Been Rehabilitated in the Last Couple of Years)

**Extremely Poor** (Should Have Been Rehabilitated Many Years Ago)

Is Pavement of Acceptable Quality?  _____  Yes

_____  No

_____  Undecided

COMMENTS: _____

**Figure 12.2**  Sample PQI rating form [Karan 83].

## Combined Measures of Pavement Quality

analyses to develop a POI model. Several transformations of the data were evaluated; however, resulting in a final model:

$$PQI = 1.1607 + (0.596 \cdot RCI \cdot SDI) + (0.5264 \cdot RCI \cdot \text{Log } 10 \text{ SAI}) . \quad (12.2)$$

This model has a standard error of estimate of 0.79 and an $R^2 = 0.76$.

An analysis of acceptance of the pavement quality was also conducted, based on the acceptable/unacceptable response of the rater for each section. It showed a minimum acceptable PQI level of 4.7.

According to a survey carried out by Baladi, a number of states and provinces have developed composite and combined indexes [Baladi 90]. For example, Minnesota uses a PQI formulation, similarly to that for Alberta, which combines a surface rating, ride rating, and structural adequacy rating. New Hampshire's combined index uses roughness plus distress, as does Florida.

### 12.4 PRECAUTIONS TO THE USE OF COMBINED INDEXES

The key to the development of a combined or composite index of pavement quality is to recognize the subjective nature of the problem and the associated techniques for quantifying subjective information. The methodology may be transferable, but not the specific models. Thus, the actual development of an index should be calibrated for each agency.

Due to the aggregate nature of a combined pavement quality index, confusion with interpretation and application can occur. For example, a combined index may be used at the network level for the selection of a general treatment strategy. But at the project level, certain specific information which is masked in aggregation, such as particular distresses, may require the selection of a different treatment. This does not imply composite indexes are inapplicable at the project level, but it does imply that their primary use is to convey summary information at the network level.

### REVIEW QUESTION

1. Write an equation for combining the distress and serviceability measures for a composite pavement quality index. Put the equation in a spread sheet program and test the sensitivity of the equation to different weighting parameters. Using the pavement condition and serviceability rating data collected as directed in previous chapters, determine the composite quality index from your equation.

Chapter 13

# Data Base Management

## 13.1 DATA BASE CONSIDERATIONS

Data in and of itself is a useless commodity. Only through its analysis and application can the benefits of pavement management systems be realized. Thus, while the collection of data is a necessary component of a pavement management system, merely having the data is inadequate. Data access is the key to effective pavement management. Some people distinguish between "data" (values that are recorded in a data base) and "information" (data that is understood by some user). This distinction is unnecessary and the meaning of the terms can be determined from the content.

When pavement management systems were introduced, there was some reluctance on the part of practicing engineers to use the systems because they "didn't want a computer telling them what to do." In response, proponents of PMS stressed the conceptual foundations of pavement management and assured engineers that manual systems could be implemented. However, this was before the revolution in microcomputers that are affordable by even the smallest highway agencies. The concept of implementing a manual PMS at this time would be like walking from New York to Los Angeles. It can be done but it is not very time efficient. The computer is a valuable tool for data storage, processing, analysis, and reporting. The value of this tool cannot be overstated, especially when computers are paired with modern software which

## Data Base Management

removes much of the pain and tedium association with using computers only a few years ago. A computerized data base system has several advantages over paper-based records keeping, including [FHWA 90]:

1. Data is stored in a compact space.
2. Storage and retrieval of the data are much faster than a manual method, permitting the data to be updated on a regular basis and facilitating the use of the information.
3. There is centralized control of information.

Centralized control of the data base has several advantages to the agency. However, the advantages of centralization can only be realized if the system is properly designed and maintained. Advantages of a centralized system are [FHWA 90]:

1. Redundancy is reduced by having each piece of data stored in only one place. This also avoids inconsistencies when updating the data files.
2. Data can be shared by various applications. Frequently, the data needed for pavement management is collected by different divisions within the agency. Having a centralized data base ensures that all divisions will have access to the needed information.
3. Standards can be enforced, in terms of data formats, naming, and documentation.
4. Security restrictions can be applied to control the flow of information and the updating of the information.
5. Data integrity can be maintained by controlling the updating of the data base and use of integrity checks whenever an update is made.
6. Conflicting requirements can be balanced to optimize the data base for the agency rather than the individual user. This is particularly important for sharing data between divisions.

One of the most important considerations in the development of a data base management approach is the spatial and temporal identification of the data. Spatial identification requires being able to physically relate the data to the location on the pavement network. Spatial referencing is accomplished through the section definition process.

In the Federal Highway Administration Advanced Course in Pavement Management Systems [FHWA 90], the location referencing system is identified as the most important component of a pavement management data base. A location referencing system is distinguished from a referencing method as the system includes all office and field procedures required to identify the location of all pavement data elements. The location reference method is a subset of the location reference system. Procedures must be established within the location referencing system to ensure the integrity of the system once it is in the application.

Temporal identification defines when the data was collected and placed in the data base. This point is frequently overlooked as the project engineer will assume the data is the most recent available and will simply work with it. However, unless the data is properly managed, the information can become stale and out of date. The need for date stamping is particularly crucial in large organizations where different divisions in the highway agency will have responsibilities for different elements in the data base.

## 13.2 TYPES OF DATA BASES

Due to the wide variety of data required for effective pavement management, many agencies maintain the data in separate, but relationally fixed data files. Typically, separate data files are maintained for:

1. Construction history, including major maintenance and rehabilitation
2. Traffic
3. Maintenance
4. Pavement quality measures (roughness, distress, skid, structural evaluation)
5. Special studies, generally data collected for rehabilitation design and research

Each of these files needs to be coordinated with respect to the spatial location of the data. One problem which frequently occurs in developing a coordinated data base is that different divisions of an agency will have responsibility for collecting data elements in the various files. Each division may have a distinctively different method for spatial reference. For example, traffic data may be stored by traffic count stations, construction data may be stored by a project reference number, and quality measures may be stored by PMS section numbers, etc. This can be a very difficult and time-consuming problem to overcome. Either all data collection units will need to change to a unified pavement referencing method or algorithms need to be developed for translating between the various referencing methods. The first approach is more difficult to implement due to the need to make people change the way they do things. The second approach can be difficult to implement and has the danger that one of the data collection units will change their referencing method leading to the need to alter the conversion algorithms. However, the advantage of making the data management "transparent" to the users is very appealing and easier to implement across the agency.

## 13.3 DATA BASE ISSUES

Regardless of the types of data that are stored in the data base, it must have integrity, accuracy, validity, security, and documentation. The data base management system must include mechanisms for ensuring these characteristics are preserved, otherwise the data will be worse than having no data at all. These terms are defined as follows [FHWA 90]:

1. Integrity—two pieces of data that represent the same fact must be equal, e.g., a road name must be spelled the same whenever it occurs in the data base.
2. Accuracy—values in the data base must represent, as closely as possible, the

# Data Base Management

actual situation at the indicated location and time. Generally, accuracy is a data collection consideration.
3. Validity—values in the data base must be correct and should pass logic and limits tests.
4. Security—three issues are involved: proper access to the data base, controls on changes to the data base, and prevention of the permanent loss of the data.
5. Documentation—each of the fields in a data base record should be documented with respect to the meaning, unit of measure, format, source, use, relationships, and security considerations.

A final issue which must be addressed in the development of a data management plan is the treatment of archival or historical data. Maintaining historical performance with the current data set clutters up the file and can increase the time required for a data analysis. However, the advantage is that the entire data set is used on a periodic basis and will not become "lost." If the selected data base management requires separate archive files, the data base manager must be cognizant of the value of the archive information and take appropriate steps to maintain and preserve the files. The FHWA Advanced Pavement Management training course [FHWA 90] identifies the following types of data with associated needs for maintaining "backup" copies of the data:

1. Scratchpad Data—This is working data that is not important to the central data management function and there is little concern for losing the data. A single backup copy updated every 5 to 10 days should be sufficient. This class of data is generally a subset of the centralized data base that an engineer has assembled for the analysis of a specific project.
2. Important Data—This data is important but not critical for the agency. The backup procedure is stricter than for the scratchpad data but not as stringent as is required for critical data. An archive copy and two backup copies are sufficient. The backups can be made on alternating work days and the archive copy can be made once per week. The archive copy should be stored off site, allowing for recovery from a catastrophic event such as fire or flood.
3. Critical Data—This data is critical for the operation of the agency and cannot be readily recreated. For example, historical pavement performance records are critical for evaluating the pavement design models and cannot be recreated. Backups of these files should be created whenever the data base is updated and an archive copy, stored off site, should be created before the original records are destroyed.

## 13.4 COMPUTER ISSUES

The rapid development of microcomputers and minicomputers has given engineers a wealth of options for data handling and analysis. The conventional process of computer system selection required a definition of the data processing and storage needs, identification of the software that could be applied to the problem, and then selection of the appropriate hardware. While this approach is conceptually correct, it ignores the fact that most agencies are currently using computers and frequently institutional barriers exist which affect the flexibility of computer selection.

The pavement management engineer needs computer support, both hardware and software, that promotes efficient operation, has the flexibility of ready access to all the data, and can generate a wide variety of reports. The flexibility aspect is particularly important for it is impossible to anticipate all of the data applications when the PMS is initially developed. There is nothing more frustrating to an engineer or administrator than to know a data base contains some required information, but the structure of the software prevents access in a convenient manner.

Modern microcomputers and associated software have eliminated many of these frustrations. High-speed 32-bit processors connected to large fixed disks are generally powerful enough for all but the most complicated analysis of large highway agency data bases. Very successful pavement management systems have been implemented on microcomputers for highway agencies with over 10,000 miles of roads. If the trends in computer hardware continue at the current pace, it is likely that microcomputers of the future will be capable of completely handling the pavement management needs of all highway agencies.

The primary advantage of the microcomputer, in addition to the relatively low initial cost, is that computer hardware and software developers have remained focused on the personal nature of these computers. This has resulted in software that can be widely used by engineers without specific training in computer programming. This is especially true in data base management programs that can be used for processing, storing, and analyzing pavement data with a minimum of computer expertise. In essence, microcomputers provide the flexibility of analysis and reporting required for a successful pavement management system.

As microcomputers have become more powerful, the division between microcomputers and minicomputers has become fuzzy. One criterion would consider the microcomputer to be a personal computer and relegate minicomputers to the role of serving multiple users operating from dumb terminals. However, even this definition is losing clarity as microcomputers become file servers for local area networks.

Mainframe computers are even more powerful than minicomputers. They are also more difficult to use. Historically, the mainframe was so expensive that it was assumed the computer user would be a specialist capable of producing a detailed set of instructions required to perform an analysis. This approach limits the flexibility in using the data bases on a mainframe computer.

In large agencies, where different divisions are responsible for the various central data base elements, a mainframe computer can be a very efficient repository of the data. Frequently, the mainframe is interfaced with microcomputers to provide ready access to the data and the flexibility of analysis. Use of the mainframe computer as the main data base repository promotes centralization of the data base for the agency.

The flexibility limitations of the mainframe computers may be overcome in the future as computer scientists integrate software between the various hardware formats. For example, there is one computer package that uses the concepts of a sequential query language to permit unified operational integration of computer code and data across microcomputers, minicomputers, and mainframe computers. While compatibility across hardware platforms is conceptually feasible, it is currently in the development stage and it should be tested by the agency before a commitment is made to a particular platform.

In essence, the specific hardware and software selection is not a critical issue, provided the data management, analysis, and reporting can be performed in a convenient and flexible manner.

The rapid developments in both computer hardware and software inhibit any discussions of particular equipment or programs. Based on current technology, the majority of pavement management functions can be performed on microcomputers for the vast majority of highway and airport agencies. Furthermore, the future capability of both microcomputer hardware and software will improve, thereby giving pavement management engineers greater capability.

## 13.5 EXAMPLES OF DATA BASE MANAGEMENT

The Arizona Department of Transportation, ADOT, pavement management system began in 1969 with the purchase of a Dynaflect. Since that time there has been a continuous evolution of the development of the system. In this development, extensive detail was paid to the establishment of the data bases used by the system and the engineers in the pavement management division. Data is maintained on both the department's mainframe computer and on microcomputers. The mainframe serves as a central storage facility and the network optimization computer program directly accesses these data bases for the development of the initial five-year program for the state. The microcomputer data bases are used for the daily operations of the pavement management staff. In addition, a procedure has been developed to provide a direct interaction between the pavement management data bases at the network and project level analysis for overlay design.

ADOT maintains three data bases on microcomputers: the pavement management data, the construction history, and the deflection data. These data bases are maintained with the dBase III$^+$ data base manager. The files are indexed to permit rapid retrieval of the information.

The pavement management data base contains information on the route identification, traffic level and growth rate, maintenance costs, and pavement condition data. The pavement condition data include cracking, roughness, and skid measurements. These condition measures are maintained for each year of observation. Thus, the engineers have the ability to review the data across time when considering a pavement for a rehabilitation action. All of the data is maintained on a milepost basis, with one record per milepost for two lane highways. For four-lane divided highways, observations are collected at each milepost for each direction.

The pavement construction data base contains a record for each project that has been performed by the department. The location of each project is identified by the route number and the beginning and ending milepost, recorded to a hundredth of a mile. The type of material and thickness of each pavement layer are recorded for each construction project.

The deflection data base contains records of all the data collected with the Dynaflect and Dynatest Model 8000 falling weight deflectometer. The location of the measurements are identified by route number and fractions of a milepost, lane, direction, and date.

The ADOT data base is a good example of a data base that is centrally managed on a mainframe computer, but is commonly accessed and analyzed with a microcomputer. Historical pavement condition measures maintained on the system permit periodic evaluation of the overall condition of the network and the development of performance models.

## REVIEW QUESTIONS

1. Estimate the size of a data file required to maintain a pavement inventory and condition data base. This will require defining the data elements for the system and the number of pavement sections. Visit a computer store and determine the cost of a computer system that could manage this quantity of data.
2. Using a data base management program on a microcomputer, generate a data base with at least 200 records. Use the program to generate summary statistics about the system.

Chapter 14

# Describing the Present Status of Pavement Networks

## 14.1 USES OF PRESENT STATUS INFORMATION

Invariably, one of the first questions people ask is, "What is the present condition or status of a problem area or facility?" When this question is applied to a network of pavements, the pavement management engineer must be able to respond with specific factual information. In general, the pavement management engineer must be responsive to queries from upper management concerning the status of the pavement network, and to queries from the design and operations divisions concerning specific locations of pavement sections which require attention. In addition, there must be adequate response to queries from the public, frequently concerning the condition of a specific section of pavement. The ability to respond to these queries depends on how well they were anticipated in the development of the data base and the flexibility of the computer software to be responsive to specific queries.

There are several useful formats for expressing the status of the pavement network. These include:

1. Color coded maps indicating in a categorical manner, the condition of all pavements in the network
2. Graphical representations of pavement quality

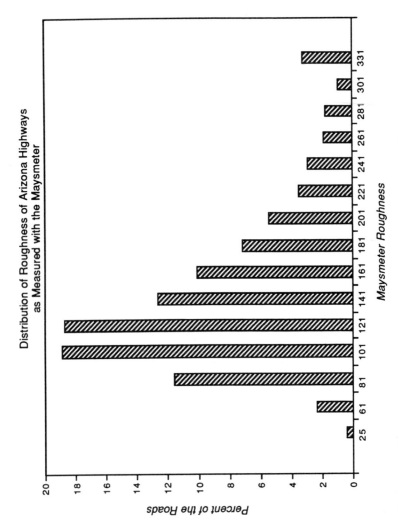

**Figure 14.1** Example distribution of roughness for Arizona highways in 1989.

**Describing the Present Status of Pavement Networks** 175

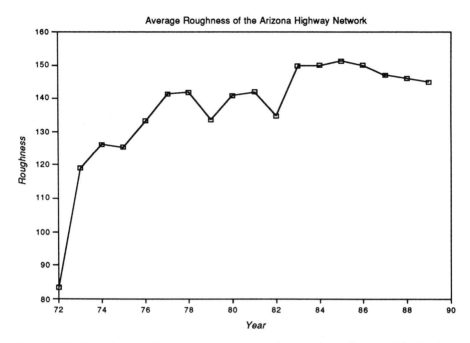

**Figure 14.2** Example use of pavement management data used to define trends in the data.

3. Tabular summaries, e.g., the percentage of each pavement class in specific condition ranges

Geographic information systems (GIS) are particularly adept at producing network summaries on maps. They are a strong visual aid for the representation of the condition of a pavement network. Such maps can also be produced manually using tabular data, outline maps, and colored markers. This can be a time-consuming process, but is worthwhile, especially for presentation to the public, elected officials, and top management.

Graphical representation of pavement quality is another important visual tool. Histograms can show the percentages of pavement in good, fair, and poor condition. These can be further broken down by highway class, airport feature, political jurisdiction, etc. Comparison of the histograms from one year to the next can readily demonstrate changes in quality of the pavement network.

Frequently, the pavement management engineer will ask for information on a specific pavement section. In this case, tabular data can be the most effective tool. Three types of tables are particularly useful in reports:

1. Sequential listing of all pavement sections based on the pavement quality index
2. Sequential listing by pavement quality but first sorted by selected treatment strategies
3. Listings sorted by the common street highway or airport feature name

## 14.2 EXAMPLES OF PAVEMENT CURRENT STATUS AND TREND REPORTS

The value of having data rests in the ability to use the information effectively. Figures 14.1 and 14.2 are examples of the types of analysis that can be prepared for management from the pavement management data base. Figure 14.1 shows the distribution of pavement roughness in Arizona in 1989. Figure 14.2 shows how the data can be used to track trends in the overall condition of the pavement over time. Each of these figures was generated in a few minutes by issuing simple dBase III$^+$ commands and transferring the selected information to Quatro Pro to produce the graphs. Other graphs, such as a breakdown of the data for type of facility, regional factor, or traffic level, could have just as easily been generated. The key point is that having the data on a readily accessible data base gives the analyst a great deal of flexibility in the development of reports and various summary statistics.

### REVIEW QUESTION

1. Discuss separately the types of summary statistics that would be useful for the legislative, administrative, and technical users of pavement management.

# References to Part Two

[Alford 71]   Alford, W. T., "Airport Pavements: An Airline Pilot's Viewpoint," Technical Bulletin No. 10, American Concrete Paving Association, 1971.

[Anderson 74]   Anderson, V. A., and R. A. McLean, *Design of Experiments, A Realistic Approach*, Marcel Dekker, Inc., 1974.

[Argue 71]   Argue, G. A., et at., "Evaluation of the Car Road Meter: A Device for the Rapid Measurement of Pavement Roughness," Proceedings Roads and Transportation Association of Canada, 1971.

[Argue 73]   Argue, G. A., "A Canadian Evaluation Study of Road Meters," Highway Research Board Special Report 133, 1973.

[ASTM 91]   American Society for Testing and Materials, "Standard Test Method for Measuring the Longitudinal Profile and Vehicular Traveled Surface with an Inertial Profilometer," ASTM Standard E 950-83, Annual Book for ASTM Standards, Section 4, Volume 04.03, 1991.

[Babb 83]   Babb, L. V., "Design Enhancement of a Low Cost Road Roughness Measuring Device," Report No. FHWA/TX-83/05-279-2F, Federal Highway Administration, Washington, DC, Jan. 1983.

: Data Requirements

[Baladi 90] Baladi, Gilbert Y., "Analysis of Pavement Distress Data, Pavement Distress Indices and Remaining Service Life," Module A-1 of FHWA Course Notes on Advanced Pavement Management Systems, FHWA, 1990.

[Biles 62] Biles, C. G., B. E. Sabey, and K. H. F. Cardew, "Development and Performance of the Portable Skid Resistance Tester," ASTM Special Technical Publication No. 326, 1962.

[CALTRANS 78] "Methods of Test to Determine Overlay Requirements by Pavement Deflection Measurements," California Test 356, Department of Transportation, Division of Construction, Office of Transportation Laboratory, Sacramento, CA, 1978.

[Carey 60] Carey, W. N., and P. E. Irick, "The Pavement Serviceability—Performance Concept," Highway Research Bulletin 250, 1960.

[CEBTP] Centre Experimental de Recherches et d'Etudes du Batinent et des Travaux Publics, "The CEBTP Curviameter, Product Bulletin," France.

[CGRA 59] Canadian Good Roads Association, "Manual on Pavement Investigations," CGRA Tech Publ. No. 11, 1959.

[CGRA 65] Canadian Good Roads Association, "A Guide to the Structural Design of Flexible and Rigid Pavements in Canada," CRGA, 1965.

[CGRA 67] Canadian Good Roads Association, "Pavement Evaluation Studies in Canada," Proc., First International Conference on Structural Design of Asphalt Pavements, University of Michigan, 1967.

[CGRA 71] Canadian Good Roads Association, "Field Performance Studies of Flexible Pavements in Canada," Proc., Second International Conference on Structural Design of Asphalt Pavements, University of Michigan, 1971.

[Chua 85] Chua, K. M., and R. L. Lytton, "Load Rating of Light Pavement Structures," Transportation Research Board, Research Record 1043, Washington, DC, 1985.

[Dehlin 62] Dehlin, G. L., "A Simple Instrument for Measuring the Curvature Induced in a Road Surfacing by a Wheel Load," Transactions, South African Institute of Civil Engineers, 1962.

[DeWilder 85] DeWilder, M., "GERPHO and APL" Road Evaluation Workshop, Proceedings, Report No. FHWA-TS-85-210, 1985.

[Dickerson 76] Dickerson, R. S., and D. G. W. Mace, "A High-Speed Profilometer Preliminary Description," Department of the Environment Supplementary Report 182, Crowthorne, England, 1976.

[Donnelly 88] Donnelly, D. E., W. Hutter, and J. P. Kiljan, "Pavement Profile Measurement, Seminar Proceedings, Volume II, Data Collection Equipment," Demonstration Project No. 172, Federal Highway Administration, Report FHWA DP-88-072-004, Washington, DC, 1988.

[Drake 59] Drake, W. B., and J. H. Havens, "Reevaluation of Kentucky Flexible Pavement Design Criteria," Highway Research Board, Bulletin 233, 1959.

# References to Part Two

| | |
|---|---|
| [Dynatest] | Dynatest Consulting, Inc., "Dynatest 8000 FWD Test System Owner Manual," Ojai, CA. |
| [Eaton 87] | Eaton, R. A., S. Gerard, and D. W. Cate, "Rating Unsurfaced Roads," Special Report 87-15, US Army Corps of Engineers, Cold Regions Research and Engineering Laboratory, Vermont, 1987. |
| [ERES 86] | ERES Consultants, "MTC Pavement Management System User's Guide," Metropolitan Transportation Commission, Oakland, CA, 1986. |
| [FDI] | Foundation Dynamics Inc., "Road Rater Operations Manual," El Segundo, CA. |
| [Fernando 83] | Fernando, E. G., and W. R. Hudson, "Development of a Prioritization Procedure for the Network Level Pavement Management System," Research Report 307-2, Center for Transportation Research, University of Texas at Austin, Austin, TX, 1983. |
| [FHWA 90] | Federal Highway Administration, "An Advanced Course in Pavement Management Systems," Course Text, 1990. |
| [Geolog] | Geo-Log, Inc., "Operations Manual for Dynaflect Dynamic Deflection Determination System," Granbury, TX. |
| [Gillespie 80] | Gillespie, T. D., M. W. Sayers, and L. Segal, "Calibration of Response-Type Road Roughness Measuring Systems," NCHRP Report 228, 1980. |
| [Haas 71] | Haas, R. C. G., and W. R. Hudson, "The Importance of Rational and Compatible Pavement Performance Evaluation," Highway Research Board Special Report 116, 1971. |
| [Haas 91] | Haas, Ralph, "Generically Based Data Needs and Priorities for Pavement Management," ASTM STP 1121, 1991. |
| [Horne 70] | Horne, W. B., and H. C. Sparks, "New Methods for Rating, Predicting and Alleviating the Slipperiness of Airport Runways," Paper Presented to National Air Transportation Meeting, New York, 1970. |
| [HPI 85] | Highway Products International Inc., "Automatic Road Analyzer, Mobile Data Acquisition Vehicle," Product Bulletin, Paris, Ontario, Canada, 1985. |
| [HRB 52] | Highway Research Board, "Maryland Road Test One MD: Final Report, Effect of Controlled Axle Loadings on Concrete Pavement," Highway Research Board Special Report 4, 1952. |
| [HRB 55] | Highway Research Board, "The WASHO Road Test, Part 2: Test Data Analysis and Findings," Highway Research Board Special Report 22, 1955. |
| [HRB 57] | Highway Research Board, "Pavement Condition Surveys—Suggested Criteria," HRB Special Report 30, 1957. |
| [HRB 62] | Highway Research Board, "The AASHO Road Test: Report 5—Pavement Research," Highway Research Board Special Report 61-E, 1962. |
| [HRB 72] | Highway Research Board, "Skid Resistance," NCHRP Synthesis of Highway Practice 14, 1972. |

| | |
|---|---|
| [Hudson 68] | Hudson, W. R., W. E. Teske, K. H. Dunn, and E. B. Spangler, "State of the Art of Pavement Condition Evaluation," Subcommittee Report to Pavement Condition Evaluation Committee of Highway Research Board, Special Report 95, 1968. |
| [Hudson 71] | Hudson, W. R., "Pavement Serviceability: The Surface Dynamics Profilometer Applied to Airport Pavements," Paper presented to Airline Pilots Association Meeting, Dallas, TX, 1971. |
| [Hudson 74] | Hudson, W. R., and F. N. Finn, "A General Framework for Pavement Rehabilitation," Proceedings, Pavement Rehabilitation Workshop, Transportation Research Board Report No. DOT-05-40022, Task Order 1, 1974. |
| [Hudson 84] | Hudson, W. R., D. S. Halbach, L. O. Moser, and J. P. Zaniewski, "Pavement Performance Model Development, Volume 4, Roughness Measurement & Calibration Guidelines," Report No. FHWA/RD-84/106, Federal Highway Administration, Washington, DC, 1984. |
| [Hudson 87a] | Hudson, W. R., G. E. Elkins, W. Uddin, and K. T. Reilley, "Evaluation of Pavement Deflection Measuring Equipment," Report No. FHWA-TS-87-208, Federal Highway Administration, Washington, DC, 1987. |
| [Hudson 87b] | Hudson, W. R., G. E. Elkins, W. Uddin, and K. T. Reilley, "Improved Methods and Equipment to Conduct Pavement Distress Surveys," Report No. FHWA-TS-87-213, Federal Highway Administration, Washington, DC, 1987. |
| [Hutchinson 64] | Hutchinson, B. G., "Principles of Subjective Rating Scale Construction," Highway Research Board, Research Record 46, 1964. |
| [Hutchinson 65] | Hutchinson, B. G., "Analysis of Road Roughness Records by Power Spectual Density Techniques," Dept. of Highways of Ontario Research Report 101, Jan. 1965. |
| [Hveem 60] | Hveem, F. N., "Devices for Recording and Evaluating Pavement Roughness," Highway Research Board, Research Record 264, 1960. |
| [Janoff 85] | Janoff, M. S., J. B. Nick, P. F. Davit, and G. F. Hayhoe, "Pavement Roughness and Rideability," NCHRP Report No. 275, Washington, DC, 1985. |
| [Janoff 88] | Janoff, M. S., "Pavement Roughness and Rideability Field Evaluation," NCHRP, Report 308, Washington, DC, 1988. |
| [Joseph 83] | Joseph, Ponniah, et al., "Universal Standard for Correlation of Road Roughness Measurements," Proceedings, Road and Transportation Association of Canada, Edmonton, Sept. 1983. |
| [Karan 77] | Karan, M. A., and R. C. G. Haas, "Urban Pavement Management on a Coordinated Network-Project Basis," Proceedings, Vol. II, Fourth International Conference on Structural Design of Asphalt Pavements, Univ. of Michigan, Ann Arbor, MI, 1977. |
| [Karan 83] | Karan, M. A., T. J. Christison, A Cheetham, and G. Burdahl, "Development and Implementation of Alberta's Pavement Information and Needs System," Paper presented to the Transportation Research Board, Washington, DC, Jan. 1983. |

# References to Part Two

| | |
|---|---|
| [KUAB] | KUAB Konsult and Utreckling AB, "KUAB50 Shop Repair Manual," Rattvik, Sweden. |
| [LCPC] | Laboratoire Central des Ponts et Chaussess, "Lacroix Deflectographs, Product Bulletin," France. |
| [LeClerc 71] | LeClerc, R. V., and T. R. Marshall, "Washington Pavement Rating System: Procedures and Applications," Highway Research Board Special Report 116, 1971. |
| [LeClerc 73] | LeClerc, R. V., T. R. Marshall, and K. W. Anderson, "Use of the PCA Road Meter in the Washington Pavement Condition Survey System," Highway Research Board Special Report 133, 1973. |
| [Lee 68] | Lee, H. R., and J. L. Sheffel, "Runway Roughness Effects on New Aircraft Types," J. Aerospace Transportation Division, ASCE, 1968. |
| [MAP 86] | MAP Inc., "Pavement Condition Monitoring Methods and Equipment, Final Report on GERPHO Survey for ARE Inc., Austin, TX," Federal Highway Administration, Strategic Highway Research Program, 1986. |
| [MAP 90] | MAP, Longitudinal Profile Analyzer (APL 25), Manufacture Literature, MAP S.A.R.L., Hegenheim, France, 1990. |
| [McComb 74] | McComb, R. A., and J. J. Labra, "A Review of Structural Evaluation and Overlay Design for Highway Pavements," Proceedings, Pavement Rehabilitation Workshop, Transportation Research Board Report No. DOT-05-40022, Task Order 1, 1974. |
| [McKenzie 78] | McKenzie, D. W., and M. Srinawarat, "Root Mean Square Vertical Acceleration (RMSVA) as a Basis for Mays Meter Calibration," Brazil Project Technical Memorandum BR-23, Center for Transportation Research, The University of Texas at Austin, Feb. 1978. |
| [McKenzie 82] | McKenzie, D. W., and W. R. Hudson, "Road Profile Evaluation for Compatible Pavement Evaluation," Presented at the 61st Annual Meeting of the Transportation Research Board, Washington, DC, 1982. |
| [Meyer 90] | Meyer, W. E., and J. Reichert (eds.), "Surface Characteristics of Roadways: International Research and Technologies," ASTM Special Technical Publication 1031, 1990. |
| [Moore 76] | Moore, W. M., J. W. Hall, Jr., and D. I. Hanson, "State of the Art on Nondestructive Structural Evaluation of Pavements," Paper presented to the 55th Annual Meeting of the Transportation Research Board, Washington, DC, 1976. |
| [MTCO 82] | Chong, C. J., W. A. Phang, and G. A. Wrong, "Manual for Condition Rating of Rigid Pavement—Distress Identification," Ministry of Transportation and Communications of Ontario, 1982. |
| [MTCO 89a] | Chong, C. J., W. A. Phang, and G. A. Wrong, "Manual for Condition Rating of Flexible Pavement—Distress Manifestations," Report SP-024, Ministry of Transportation and Communications of Ontario, 1989. |
| [MTCO 89b] | Chong, C. J., W. A. Phang, and G. A. Wrong, "Manual for Condition Rating of Surface Treatment Pavement—Distress Manifestations," Report SP-022, Ministry of Transportation and Communications of Ontario, 1989. |

| | |
|---|---|
| [MTCO 89c] | Chong, C. J., W. A. Phang, and G. A. Wrong, "Flexible Pavement Condition Rating, Guidelines for Municipalities," Report SP-021, Ministry of Transportation and Communications of Ontario, 1989. |
| [MTCO 89d] | Chong, C. J., and G. A. Wrong, "Manual for Condition Rating of Gravel Surface Roads," Report SP-025, Ministry of Transportation and Communications of Ontario, 1989. |
| [Nazarian 86] | Nazarian, S., and K. H. Skokie, "Evaluation of the Sensitivity of the SASW Method in Determining Thicknesses of Layers in Pavement Systems," Center for Transportation Research, Technical Memorandum, The University of Texas, Austin, TX, 1986. |
| [NC DOT 88] | Division of Highways, North Carolina Dept. of Transportation, "Geographic Information System Task Force Feasibility Report," May 1988. |
| [Novak 85] | Novak, R. L., "Swedish Laser Road Surface Tester," Road Evaluation Workshop, Proceedings, Report No. FHWA-TS-85-210, 1985. |
| [Paquet 77] | Paquet, J., "The CEBTP Curviameter—A New Instrument for Measuring Highway Pavement Deflections," 1977. |
| [PASCO 85] | "Report on Pavement Monitoring Methods and Equipment," PASCO USA Inc., Lincoln Park, NJ, 1985. |
| [PASCO 87] | "1 for 3 PASCO Road Survey System (PRS System) from Theory to Implementation," PASCO Corp., Mitsubishi International Co., NY, 1987. |
| [PCS 86] | Pavement Consultancy Services, Inc., "Phonix ML 10000 Falling Weight Deflectometer—Instrumentation Manual," College Park, MD, 1986. |
| [Philips 69] | Philips, M. B., and G. Swift, "A Comparison of Four Roughness Measuring Systems," Highway Research Board, Research Record 291, 1969. |
| [Potter 78] | Potter, D. W., "Measurement of Road Roughness in Australia," Transportation Research Board, Record 665, 1978. |
| [Pottinger 86] | Pottinger, M. G., and T. J. Yager, "The Tire Pavement Interface: A Symposium," ASTM Special Technical Publication 929, 1986. |
| [PTA 84] | Partner Technic As, "Falling Weight Deflectometer ML10001," Product Bulletin, Sept. 1984. |
| [PTI 85] | Pavement Technologies Inc., "PaveTech Failing Weight Deflectometer," Product Bulletin, 1985. |
| [Queiroz 81] | Queiroz, C. A. V. "A Procedure for Obtaining a Stable Roughness Scale From Rod and Level Profiles," Working Document 22, GEIPOT (Brazil), Sept. 1981. |
| [Roberts 70] | Roberts, F. L., and W. R. Hudson, "Pavement Serviceability Equations Using the Surface Dynamics Profilometer," Research Report 73-3, Center for Transportation Research, University of Texas, Austin, TX, 1970. |
| [RTAC 72] | Pavement Management Committee, Roads and Transportation Association of Canada, "Output Measurements for Pavement Management Studies in Canada," Proceedings of the Third International Conference on the Structural Design of Asphalt Concrete Pavements, University of Michigan, MI, 1972. |

# References to Part Two

| | |
|---|---|
| [RTAC 77] | Roads and Transportation Association of Canada, "Pavement Management Guide," 1977. |
| [Sayers 86] | Sayers, M. W., T. D. Gillespie, and C. A. V. Queiroz, "Establishing a Correlation and a Standard for Measurement," World Bank, Technical Report 45, Washington, DC, 1986. |
| [Schonfeld 70] | Schonfeld, R., "Photo-Interpretation of Skid Resistance," Highway Research Board, Research Record No. 311, 1970. |
| [Scrivner 66] | Scrivner, F. H., G. Swift, and W. M. Moore, "A New Research Tool for Measuring Pavement Deflection," Highway Research Record 129, Highway Research Board, Washington, DC, 1966. |
| [Shahin 79] | Shahin, M. Y., and S. D. Kohn, "Development of Pavement Condition Rating Procedures for Roads, Streets and Parking Lots—Volume I Condition Rating Procedure," Technical Report M-268, Construction Engineering Research Laboratory, United States Corps of Engineers, 1979. |
| [SHRP 89] | SHRP, "Distress Identification Manual for the Long-Term Pavement Performance Studies," The Strategic Highway Research Program, National Academy of Science, Washington, DC, 1989. |
| [Singleton 67] | Singleton, R. C., "On Computing Fast Fourier Transforms," Comm. ACM 10(10), 1967. |
| [Smith 80] | Smith, R. E., M. I. Darter, and S. M. Herrin, "Highway Pavement Distress Identification Manual for Highway Conditions and Quality of Highway Construction Survey," Federal Highway Administration, 1980. |
| [Spangler 62] | Spangler, E. B., and W. J. Kelley, "Servo-Seismic Method of Measuring Road Profile," Highway Research Board, Bulletin 328, 1962. |
| [Spangler 64] | Spangler, E. B., and W. J. Kelly, "GMR Road Profilometer: A Method for Measuring Road Profile," General Motors Corp., Warren Michigan, MI, Dec. 1964. |
| [Srinarawat 82] | Srinarawat, M., "A Method to Calibrate and Correlate Road Roughness Devices Using Road Profiles," Ph.D. Thesis, University of Texas, Austin, TX, 1982. |
| [Steitle 72] | Steitle, D. C., "Development of Criteria for Airport Runway Roughness Evaluation," Master's Thesis, University of Texas, Austin, TX, 1972. |
| [Stevens 59] | Stevens, S. S., "Measurement, Psychophysics and Utility," In "Measurement Definitions and Theories," edited by Churchman and Ratoosh, Wiley, 1959. |
| [Walker 87] | Walker, R., "A Self-Calibrating Roughness Measuring Process," Report No. 279-1, Texas Department of Highways and Public Transportation, Austin, TX, 1987. |
| [Wambold] | Wambold, J., "Pavement Texture Measurement," Federal Highway Administration, Final Report, Washington, DC. |
| [Way 78] | Way, G. B., "Arizona Pavement Management System—Pavement Monitoring Summary," Arizona Department of Transportation, Phoenix, AZ, 1978. |

| | |
|---|---|
| [WDM 72] | W.D.M. Limited, "SCRIM, Information Literature," W.D.M. Ltd., Western Works, Staple Hill, Bristol BS16, 4NX, Great Britain, 1972. |
| [Wilkins 68] | Wilkins, E. B., "Outline of a Proposed Management System for the CRGA Pavement Design and Evaluation Committee," Proceedings of the Canadian Good Roads Association, 1968. |
| [Yang 72] | Yang, N. C., *Design of Functional Pavements*, McGraw-Hill, 1972. |
| [Zaniewski 85] | Zaniewski, J. P., W. R. Hudson, R. High, and S. W. Hudson, "Pavement Rating Procedures," Federal Highway Administration Final Report, Contract DTFH61-83-C-00153, Washington, DC, 1985. |
| [Zaniewski 86] | Zaniewski, J. P., and W. R. Hudson, "Procedures for the Analysis of Pavement Condition Data," Report No. FH-13, Vol. I, ARE Inc., Austin, TX, March, 1986. |

# Part Three

# Determining Present and Future Needs, and Priority Programming of Rehabilitation and Maintenance

Chapter 15

# Establishing Criteria

## 15.1 REASONS FOR ESTABLISHING CRITERIA

A criterion is a specified limit for some measure of pavement behavior, response, performance, deterioration, or operating characteristic against which comparisons of actual measurements or estimates can be made. If the measurement or estimate exceeds the limit, then a deficiency or need exists. For example, say that a limit of PSI = 2.0 has been set as the minimum acceptable level for a collector highway pavement, then any sections with values equal to or less than 2.0 would represent a current deficiency or "now need."

The basic reasons for establishing criteria at the network level, therefore, include the following: (a) to provide an objective basis for identifying current needs and estimating future needs, (b) to provide consistency between sections and between classes of highway or street, and (c) to have a means for effectively portraying current and future backlogs of work or deficient mileage.

At the project level, criteria have usually been in terms of specifications. For example, a highway agency may set a limit on maximum, as-built roughness. The basic reasons for criteria in this instance include quality assurance and, in the case of some agencies, a means for assessing contractor performance.

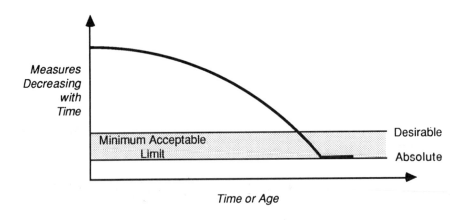

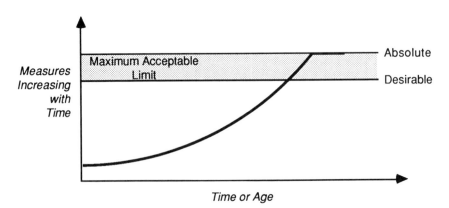

**Figure 15.1** Concept of limits or criteria for various measures of deterioration, behavior, response, or operating characteristics.

## 15.2 MEASURES OR CHARACTERISTICS TO WHICH CRITERIA CAN BE APPLIED

The measures described in Part Two are applicable for criteria to determine needs, in addition to some other characteristics such as traffic delays due to maintenance interruptions. Figure 15.1 illustrates schematically the concept of limits or criteria on the following:

1. Roughness, or some index such as Present Serviceability Index (PSI), International Roughness Index (IRI), or Riding Comfort Index (RCI)
2. Surface distress, or some index such as Surface Distress Index (SDI) or Pavement Condition Index (PCI)
3. Deflection, or some index such as Structural Adequacy Index (SAI)
4. Surface friction or skid resistance

**Establishing Criteria**                                                                              **189**

5. Combined or composite index, such as Pavement Quality Index (PQI), which incorporates RCI, SDI, SAI
6. Traffic delays due to maintenance interruptions
7. Vehicle operating costs

Some of the foregoing would obviously have limits in terms of maximum acceptable, such as roughness, while others would have a minimum acceptable limit such as surface friction. This is shown in Figure 15.1 in terms of measures increasing or decreasing with time. It also shows that rather than establish a single value for a limit it may have a lower and upper bound. For example, a desirable minimum acceptable PSI for a collector highway may be set at 2.0, and an absolute minimum set at 1.5 which could be kept at that level by maintenance even if rehabilitation were deferred for some period of time.

It would be unlikely that a road authority would set limits on all of the measures characteristics listed. Typically, there might be a "trigger value" for skid resistance (i.e., if a section falls to or below the specified limit, rehabilitation or maintenance to improve the surface friction is scheduled immediately for safety reasons), and a minimum acceptable for PSI, or RCI or PQI, plus in some cases minimum acceptable values for SDI and SAI.

## 15.3 FACTORS AFFECTING LIMITS, AND SOME EXAMPLES

The following factors can affect the limits by a road agency [Haas 89a]:

1. Type and functional class of facility (rural or urban, freeway, arterial, collector or local)
2. Size of pavement network and type of agency (federal, state or provincial, local)
3. Resources, budget, and policies of the agency

Usually the type and class of facility dominate the other factors. A minimum acceptable PSI, for example, would likely be affected far more by whether the facility were a freeway compared to a local highway than by the policy of the agency.

Example criteria are given in Table 15.1. These are typical of those used by various agencies and should be considered as such; they are not meant as recommendations, but may be used for comparison or as a reference in establishing limits for any particular situation or set of conditions. In a number of cases, such as for traffic delays during rehabilitation or maintenance, limits remain to be established.

## 15.4 EFFECTS OF CHANGING CRITERIA

The basic effect of changing any particular criterion is to advance or delay the needs years; conversely to decrease or increase the amount of deficient mileage. For example, if a minimum desirable or acceptable value of PSI = 2.0 were established for collector highways, then all sections at or below this level would be current needs. But if this

**Table 15.1 Example Limits for Various Measures or Characteristics**

Minimum or Maximum (Desirable) Acceptable Value

| Measure or Characteristic | Freeway | Arterial | Collector | Local | Remarks |
|---|---|---|---|---|---|
| 1. Roughness | | | | | |
|   a) PSI | Variable | Variable | Variable | Variable | Depends on how measured |
|   b) IRI | 3.0 | 2.5 | 1.5 | 1.5 | Remains to be established by IRI |
|   c) RCI | * | * | * | * | |
| | 6.0 | 5.0 | 4.0 | 3.0 | |
| 2. Surface Distress | | | | | |
|   a) SDI (scale of 0 to 10) | Variable | Variable | Variable | Variable | Depends on distress type |
|   b) PCI (scale of 0 to 10) | 6.0 | 5.0 | 4.0 | 3.0 | |
| | 60 | 50 | 40 | 30 | |
| 3. Deflection | | | | | |
|   a) SAI (scale of 0 to 10) | Variable | Variable | Variable | Variable | Depends on how measured |
| | 7.0 | 6.0 | 5.0 | 4.0 | |
| 4. Surface friction | Variable | Variable | Variable | Variable | Depends on how measured |
|   a) Skid number (ASTM) | * | * | * | * | Not specified by hwy agencies |
| 5. Combined index | | | | | |
|   a) PQI (scale of 0 to 10) | 6.0 | 5.0 | 4.0 | 3.0 | |
| 6. Traffic delays (Veh. hours) | * | * | * | * | Remains to be established |
| 7. Vehicle operating costs | * | * | * | * | Remains to be established |

# Establishing Criteria

were increased to PSI = 2.5, then it is likely that more sections in the network would become needs, thereby increasing the amount of deficient mileage.

Altering the criteria can also affect the list of feasible rehabilitation actions. For example, increasing the minimum acceptable PSI would likely make thin overlays more cost-effective than reconstruction. Conversely, if the minimum acceptable PSI were dropped to a very low level, then the more feasible alternatives would likely include thick overlays and full or partial reconstruction.

Another effect of lowering the minimum acceptable values for PSI and/or SDI would be to increase the amount and cost of corrective maintenance required (e.g., patching), traffic interruptions, and vehicle operating costs.

## REVIEW QUESTIONS

1. Why should an agency consider establishing both desirable and absolute criteria?
2. Is it necessary for an agency to establish criteria for all the characteristics that can be measured for a pavement?
3. Identity factors that can affect limits or criteria used by an agency.
4. Describe the relationship between the budgeting process and criteria levels.

Chapter 16

# Prediction Models for Pavement Deterioration

## 16.1 CLARIFICATION OF PERFORMANCE AND DETERIORATION PREDICTION

The serviceability-performance concept, as described in Chapter 8, has been a valuable and important part of pavement technology since the 1960s. Development of good models for predicting performance, in terms of PSI or RCI versus age or accumulated axle load applications, has been a major challenge for pavement engineers.

Despite the fact that performance has had a precise definition since the Carey-Irick development of the serviceability-performance concept [Carey 60], the term *performance* is often used in a loose way by people in the pavement field. A major reason may be that it has a general meaning in everyday life. Consequently, it has become common practice among practitioners and researchers to use alternate terms such as deterioration or damage.

## 16.2 PARAMETERS OR MEASURES TO BE PREDICTED, AND THE REQUIREMENTS

In order to estimate needs years for the sections in a pavement network, it is necessary to predict the rate of change of those measures for which criteria have been established.

# Prediction Models for Pavement Deterioration

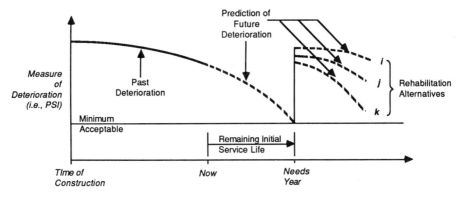

**Figure 16.1** Illustration of how a deterioration model is used to predict future deterioration of an existing pavement, and rehabilitation alternatives constructed in the needs year.

As well, it may be desirable to predict the rate of change of some of the components of a measure, such as the cracking component of surface distress, to estimate maintenance requirements.

Figure 16.1 is a schematic illustration of how deterioration prediction would be applied to an existing pavement section to estimate the rate of future deterioration, and the needs year. As well, it illustrates the application of the deterioration model to rehabilitation alternatives applied in the needs year. The basic requirements for any prediction model have been described by [Darter 80] and they include the following:

1. An adequate data base
2. Inclusion of all significant variables affecting deterioration
3. Careful selection of the functional form of the model to represent the physical, real-world situation
4. Criteria to assess the precision of the model

## 16.3 BASIC TYPES OF PREDICTION MODELS AND EXAMPLES

A classification of prediction models has been suggested [Mahoney 90], based on earlier work [Lytton 87], as summarized in Table 16.1. It considers the network and project levels of pavement management and two basic types or classes of models: deterministic and probabilistic. These are further broken down into primary response, structural, functional, and damage for the first type, and survivor curve and transition for the second type.

A convenient way of aggregating the breakdown of Table 16.1 into four basic types, for operational purposes, would be as follows:

1. Purely mechanistic, based on some primary response (behavior) parameter such as stress, strain, or deflection

Table 16.1 Classification of Prediction Models [Mahoney 90]

| | Types of Models | | | | | | |
|---|---|---|---|---|---|---|---|
| | Deterministic | | | | Probabilistic | | |
| | | | | | | Transition Process Models | |
| Levels of Pavement Management | Primary Response | Structural | Functional | Damage | Survivor Curves | Markov | Semi-Markov |
| | • Deflection<br>• Stress<br>• Strain<br>• etc. | • Distress<br>• Pavement Condition | • PSI<br>• Safety<br>• etc. | • Load Equiv. | | | |
| National Network<br>State Network<br>District Network<br>Project | • | •<br>•<br>• | •<br>•<br>• | •<br>•<br>•<br>• | •<br>•<br>• | •<br>•<br>• | •<br>•<br>• |

## Prediction Models for Pavement Deterioration

2. Mechanistic-empirical, where a response parameter is related to measured structural or functional deterioration, such as distress or roughness, through regression equations
3. Regression, where the dependent variable of observed or measured structural or functional deterioration is related to one or more independent variables like subgrade strength, axle load applications, pavement layer thicknesses and properties, environmental factors, and their interactions
4. Subjective, where experience is "captured" in a formalized or structured way, using transition process models, for example, to develop deterioration prediction models

The first type, purely mechanistic, has not yet been developed because pavement engineers do not use primary or fundamental response parameters as ends in themselves. Rather, they are only useful if they can be related to pavement distress, or to pavement properties used in other models such as for overlay design. Consequently, the mechanistic-empirical type of deterioration modeling approach has been developed. A good example is provided by Queiroz where linear elasticity was used as the basic constitutive relationship for pavement materials in a study of 63 flexible pavement test sections [Queiroz 83]. Calculated responses included surface deflection, horizontal tensile stress, strain and strain energy at the bottom of the asphalt layer, and vertical compressive stress and strain at the top of the subgrade. Various attempts were made to relate these responses to observed roughness and cracking through regression analysis. For example, the following predictive equation for roughness was developed (see Figure 16.2 for a plot of the equation):

$$\text{Log(QI)} = 1.297 + 9.22 \cdot 10^{-3} \text{AGE} + 9.08 \cdot 10^{-2} \text{ST} \\ - 7.03 \cdot 10^{-2} \text{RH} + 5.57 \cdot 10^{-4} \text{SEN1 Log } N \quad (16.1)$$

where

- QI = roughness (quarter-car index, in counts/km)
- AGE = pavement age in years
- ST = surface type dummy variable (0 for as constructed and 1 for overlayed)
- RH = state of rehabilitation indicator (0 for as constructed and 1 for overlayed)
- SEN1 = strain energy at bottom of asphalt layer ($10^{-4}$ kgf cm)
- $N$ = cumulative equivalent single axle loads (ESAL)

The squared correlation coefficient ($R^2$) for equation 16.1 is 0.52 and the standard error for the residuals is 0.11.

Another predictive equation from Queiroz involved cracking, as follows [Queiroz 83] (see also Figure 16.3):

$$\text{CR} = -8.70 + 0.258 \text{ HST} \cdot \text{Log } N + 1.006 \cdot 10^{-7} \text{ HST1} \cdot N \quad (16.2)$$

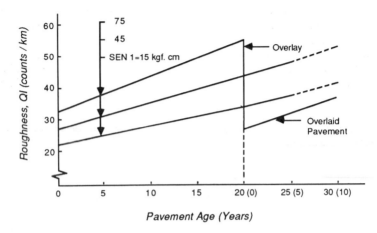

**Figure 16.2** Roughness prediction from mechanistic-empirical model [Queiroz 83].

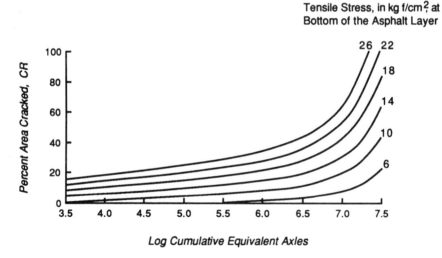

**Figure 16.3** Cracking prediction regression model [Queiroz 83].

where

$CR$ = percent of pavement area cracked
$HST$ = horizontal tensile stress at the bottom of the asphalt layer (kgf/sq cm)

The squared correlation coefficient ($R^2$) for equation 16.2 is 0.54 and the standard error for the residuals is 15.40.

The third type, direct regression, is particularly applicable where a long-term data base has been acquired. For example, up to 25 years of data on roughness, surface

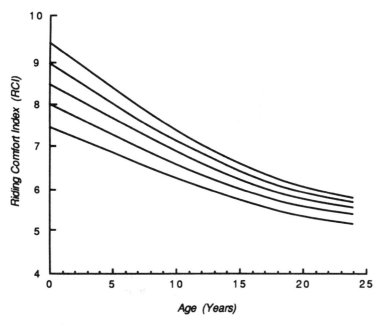

**Figure 16.4** Performance prediction from equation 16.3.

distress, traffic, deflection, and other factors were used in Alberta [Karan 83] to develop deterioration models, such as the following for conventional granular base pavements (see Figure 16.4 for a plot of the equation):

$$\begin{aligned}\text{RCI} = &-5.998 + 6.870 \cdot \text{LOG}_e (\text{RCI}_B) - 0.162 \cdot \text{LOG}_e (\text{AGE}^2 + 1) \\ &+ 0.185 \cdot \text{AGE} - 0.084 \cdot \text{AGE} \cdot \text{LOG}_e(\text{RCI}_B) - 0.093 \cdot \Delta\text{AGE}\end{aligned} \quad (16.3)$$

where

$\text{RCI}$ = Riding Comfort Index (scale of 0 to 10) at any AGE
$\text{RCI}_B$ = previous RCI
$\text{AGE}$ = age in years
$\Delta\text{AGE}$ = 4 years (for equation 16.3, but it can be 1, 2, 3, 4—etc. years)

The squared correlation coefficient ($R^2$) for equation 16.3 is 0.84 and the standard error of estimate is 0.38. While a number of variables were considered, such as traffic in terms of ESALs, climate zone, subgrade soil type and others, only pavement age, $\Delta$ AGE, and $\text{RCI}_B$ were found to be significant variables in equation 16.3, which is a recursive regression type of model. A possible reason is that the pavements were primarily designed in the first place for environmental deterioration, with structural sections significantly thicker than required by traffic alone.

Similarly, the state of Washington [Jackson 90] has developed a set of regression equations, based on a long-term pavement performance data base, of the form:

$$PCR = C - mA^P \tag{16.4}$$

where

PCR = pavement condition rating, scale of 0 to 100
$C$ = 100
$m$ = slope coefficient
$A$ = age of the pavement, years
$P$ = constant which controls the shape of the curve

Table 16.2 provides an example listing of the standard or default performance curves, for western Washington, using equation 16.4 for different pavement designs or types.

Capturing the subjective experience of local engineers and experts, the fourth basic type of approach, is an excellent way to create performance models for the development of a pavement management system. The one-step transition probability technique based on the Markov process is well suited for this purpose [Finn 74, Smith 74, Karan 76]. In the Markov process, the future state of the model element, such as a pavement, is estimated solely from the current state of the element.

The state of the element is defined by condition measures. For pavements, condition measures include roughness or serviceability, pavement condition index or percent surface cracking, and skid number. For modeling purposes, the state is defined with respect to a range of the condition measures. Table 16.3 demonstrates that nine condition states would result from using two condition measures at three levels each. The numbers within the table represent the various condition states; for example, condition state 5 corresponds to the condition state with medium cracking and roughness levels.

A transition probability matrix defines the probability that a pavement in an initial condition state will be in some future condition state. Table 16.4 is an example transition probability matrix for the nine condition states defined in Table 16.3. As shown in Table 16.4, the majority of pavements will stay in the same condition state from one year to the next, e.g., according to Table 16.4, a pavement in condition state 4 has a 92% probability of remaining in condition state 4 after one year. One of the requirements of the transition probability matrix is the sum of each row must equal 1.0. Formal interview methods, such as the Delphi method, can be used for developing the transition probability matrices. The experts are asked to determine to the best of their ability the probability a pavement in one of the condition states will go to each of the future condition states in one time period.

Because the Markov process estimates the future condition state solely from the current condition state, other factors that affect pavement behavior are handled by defining a transition matrix for each combination of factors that affect pavement performance, such as:

1. Pavement type
2. Pavement thickness
3. Traffic volumes or loads

**Table 16.2 Standard Performance Equations, of Equation 16.4 Type, in Washington State's Pavement Management System**

| Location | Type of Construction/ Pavement Surfacing | Number of Analysis Units | Performance Equation | Age to PCR = 40 |
|---|---|---|---|---|
| Western Washington | New or Reconstructed/ Bituminous Surface Treatment | 2 | $PCR = 100 - 0.086\,(AGE)^{2.50}$ | 13.7 |
| | New or Reconstructed/Asphalt Concrete | 26 | $PCR = 100 - 0.22\,(AGE)^{2.00}$ | 16.5 |
| | New or Reconstructed/Portland Cement Concrete | 19 | $PCR = 100 - 0.85\,(AGE)^{1.25}$ | 30.1 |
| | Resurfacing/BST over AC | 5 | $PCR = 100 - 8.50\,(AGE)^{1.25}$ | 4.8 |
| | Resurfacing/BST over BST | 6 | $PCR = 100 - 3.42\,(AGE)^{1.50}$ | 6.8 |
| | Resurfacing/AC Overlay (under 1.2 inches) | 75 | $PCR = 100 - 0.58\,(AGE)^{2.00}$ | 10.2 |
| | Resurfacing/AC Overlay (1.2 inches to 2.4 inches) | 126 | $PCR = 100 - 0.76\,(AGE)^{1.75}$ | 12.1 |
| | Resurfacing/AC Overlay (over 2.4 inches) | 19 | $PCR = 100 - 0.54\,(AGE)^{1.75}$ | 14.8 |

**Table 16.3 Example Condition State for a Markov Process Model**

| Pavement Roughness | Surface Distress, percent area cracked | | |
|---|---|---|---|
| | 0–3 | 3–7 | >7 |
| 0–40 | 1 | 4 | 7 |
| 41–90 | 2 | 5 | 8 |
| >90 | 3 | 6 | 9 |

**Table 16.4 Example Transition Probability Matrix for a Markov Process Model**

| Initial Condition State | Future Condition State | | | | | | | | |
|---|---|---|---|---|---|---|---|---|---|
| | 1 | 2 | 3 | 4 | 5 | 6 | 7 | 8 | 9 |
| 1 | 0.90 | 0.04 | 0.02 | 0.03 | 0.01 | 0 | 0 | 0 | 0 |
| 2 | 0.01 | 0.90 | 0.03 | 0 | 0.05 | 0.01 | 0 | 0 | 0 |
| 3 | 0 | 0.01 | 0.92 | 0 | 0.01 | 0.03 | 0 | 0.01 | 0.02 |
| 4 | 0 | 0 | 0 | 0.92 | 0.05 | 0.02 | 0 | 0.01 | 0 |
| 5 | 0 | 0 | 0 | 0.01 | 0.94 | 0.03 | 0.01 | 0.01 | 0 |
| 6 | 0 | 0 | 0 | 0 | 0.01 | 0.94 | 0 | 0.01 | 0.04 |
| 7 | 0 | 0 | 0 | 0 | 0.02 | 0 | 0.95 | 0.02 | 0.01 |
| 8 | 0 | 0 | 0 | 0 | 0 | 0 | 0.01 | 0.96 | 0.03 |
| 9 | 0 | 0 | 0 | 0 | 0 | 0.01 | 0 | 0.01 | 0.98 |

4. Subgrade type or strength
5. Environmental and regional effects

For example, one application of the Markov process uses two levels for surface thickness, three levels of traffic, and two levels of subgrade strength for a total of 2 × 3 × 2 or 12 combinations [Turay 91]. Thus, 12 transition probability matrices are required for this application, which is limited to one pavement type and environment. By multiplication of the transition probabilities together for several time steps, performance curves can be developed as shown in Figure 16.5 [Turay 91].

There are several advantages to the Markov process model, including direct local calibration by capturing, in a formalized way, the experience of local engineers and the ability to develop performance curves without any historical data base. In addition, after the pavement management system has been in place for several years it is possible to calibrate the models with field data [Wang 92]. Disadvantages of the approach include the need for developing a transition probability matrix for each combination of factors that affect the pavement performance. In addition, pavement history is difficult to include in the Markov model since the estimate of the future state of the

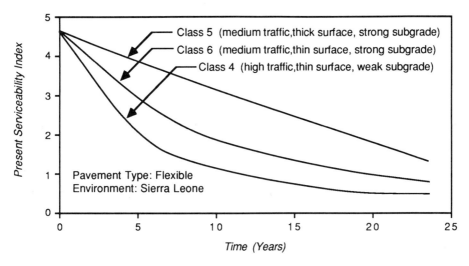

**Figure 16.5** Example performance curves from transition probability matrices [Turay 91].

pavement is based only on the current state. The Arizona Department of Transportation uses rate of crack change as one of the condition state factors, thereby allowing pavements with a deterioration history to have a greater probability of a pavement transitioning to a lesser condition state than pavements that do not have a history of rapid crack development.

## REVIEW QUESTIONS

1. Why are prediction models necessary for pavement management at both the network and project levels?
2. Examine Table 16.1 and describe why some models are only appropriate at particular levels of pavement management.
3. Make a table of the advantages and disadvantages of mechanistic, mechanistic-empirical, regression, and subjective models.
4. Define factors that could be used to define pavement classes in a Markov process model for an airport pavement network.

Chapter 17

# Determining Needs

## 17.1  NEEDS YEARS AND ACTION YEARS

The year in which a pavement section deteriorates to the minimum acceptable level would also be the action year for rehabilitation if sufficient funds are available. That is the case illustrated in Figure 16.1. However, under conditions of limited resources, the action year may have to be deferred, particularly if other sections in the network carry a higher priority. Alternatively, the action year can be advanced, which may be desirable for certain sections such as those carrying high traffic volumes, where significant economic benefits can accrue. The concept of action years varying from needs years is schematically illustrated in Figure 17.1.

Another way of varying the needs year, and in turn the feasible action years is to change the minimum acceptable level for the measure(s) of pavement deterioration. There are practical and economic limits to the range of action years. For example, continuing to advance the action year reaches some point where the level of deterioration may warrant, at best, only preventive maintenance such as crack sealing. On the other hand, deferring the action year excessively beyond the needs year will likely require increasingly extensive and costly corrective maintenance, as well as restrict the feasible rehabilitation alternatives to thick overlays or reconstruction.

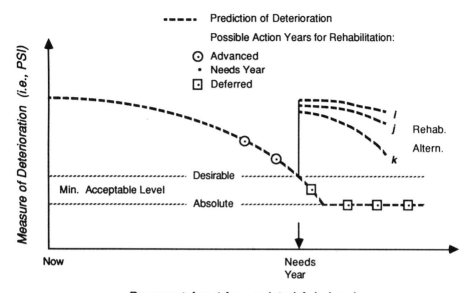

**Figure 17.1** Needs year and possible action years for rehabilitation.

## 17.2 EFFECT OF PREDICTION MODEL ERRORS

If the level of deterioration for a section is near the minimum acceptable level now, then the application of a deterioration model even with a high standard error would likely estimate say next year or the year after as the needs year. In effect, low reliability of the model is not important. However, if there are some years of remaining initial service life, as in the schematic example of Figure 17.1, then the needs year may be over or under predicted by, say, 4 or 5 years.

Consequently, deterioration predictions should be periodically updated, and the period over which needs years are estimated should be restricted to where some reasonable degree of reliability still exists (i.e., 10 years and certainly no more than 20 years), at least with current models available. Moreover, there should be consistency in relation to the high errors associated with such other variables as traffic.

## 17.3 A NEED VERSUS TYPE OF ACTION THAT SHOULD BE TAKEN

The establishment of a needs year for a section, and other possible action years, does not by itself indicate what rehabilitation alternatives should be considered and which is preferable or most cost-effective.

It is possible to select good alternatives, based on experience plus judgment, for the needs sections without comparative calculations. This may in fact be quite satisfactory for small networks where the familiarity and experience of the local engineer

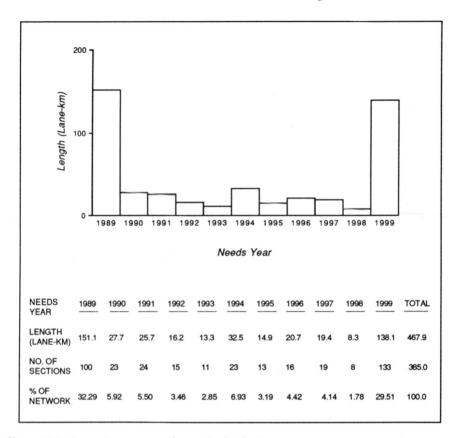

**Figure 17.2** Example summary of a needs distribution.

are quite extensive. However, where larger networks are concerned, the extensive number of possible combinations of sections versus action years versus feasible rehabilitation alternatives makes it almost if not impossible to select the best set of combinations for any given year based only on judgment. That requires a priority programming approach, as subsequently described in Chapter 19.

## 17.4 GRAPHICAL AND TABULAR PORTRAYAL OF NEEDS

The portrayal of needs within a highway authority should be tabular, for detailed review purposes, and graphical for summary purposes. In the first case, the listing would likely be section by section needs (with identifiers and descriptions such as reference number, from and to, length, year constructed, etc.) for each year in the program period. As well, sections estimated to be needs beyond the program period (i.e., 10 years +) might also be listed.

The graphical portrayal of needs is conveniently done by a histogram. For example, Figure 17.2 shows such a histogram for the arterial and collector street network,

280.8 miles (467.9 km), of a medium size city [Haas 89b]. It is based on a minimum acceptable Pavement Quality Index (PQI) level of 4.5 (where PQI is on a scale of 0 to 10) and a deterioration prediction model for PQI [Marcotte 87]. The high amount of needs for the first year of the programming period is typical of the backlog that exists for many agencies. If the backlog cannot all be financed in that first year, which is the usual case, then those needs remaining will be deferred to the second year in Figure 17.2, and so on.

## REVIEW QUESTIONS

1. What is the effect of model errors on determining the needs year?
2. What is the effect of model errors on the selection of an action?

Chapter 18

# Rehabilitation and Maintenance Strategies

## 8.1 IDENTIFICATION OF ALTERNATIVES

The alternatives considered by an agency for rehabilitation and for maintenance, both preventive and corrective, usually represent current practice. However, they almost invariably continue to change as new technologies become available, the experience of other agencies is reviewed, and long-term performance and cost-effectiveness is assessed. In any given situation, though, the feasible set of alternatives may be much smaller than the total available set because of costs, physical restraints, condition of the existing pavement, and the like.

Figure 18.1 provides an example list of rehabilitation and maintenance alternatives currently being used by the Ministry of Transportation of Ontario [MTO 90]. A number of subdivisions exist for most of the rehabilitation alternatives (i.e., different thickness for hot-mix resurfacing). The ministry has estimates of service life for each rehabilitation alternative, but actual deterioration models remain to be developed for most of the alternatives. That is a situation common to many other road authorities. Another example list of the 34 current rehabilitation alternatives used by Minnesota [Hill 91] is shown in Table 18.1.

What is important is that all the available rehabilitation and maintenance alter-

# Rehabilitation and Maintenance Strategies

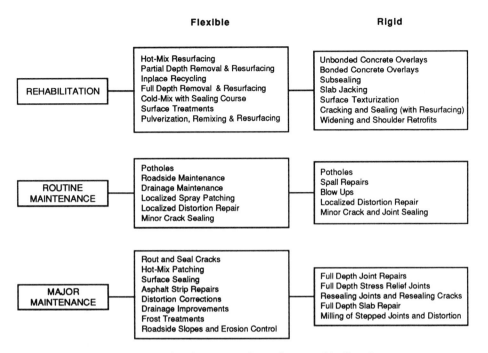

**Figure 18.1** Rehabilitation and maintenance alternatives used in Ontario.

native strategies are clearly identified by the agency, that provision exists for adding to or deleting existing strategies, and that a procedure exists for deciding which ones are feasible for given situations so that the priority programming process can be efficiently carried out.

Similar requirements hold for identifying maintenance treatment alternatives. These can generally be categorized as routine and major (Figure 18.1) or commonly as:

1. Preventive (such as chip seals, slurry seals, crack sealing, thin overlays), usually at levels of pavement deterioration considerably above the acceptable limits
2. Corrective (such as hot- and cold-mix patching), usually at levels of deterioration near or even below the acceptable limit(s)

## 18.2 DECISION PROCESSES AND EXPERT SYSTEMS APPROACHES TO IDENTIFYING FEASIBLE ALTERNATIVES

The process used to select feasible rehabilitation alternatives from a set of available alternatives can range from simple judgment to a decision tree of expert systems approach which considers various combinations of factors. The latter should represent

**Table 18.1 Rehabilitation Alternatives Used in Minnesota [Hill 91]**

| No. | Alternative Description | | Cost $/mi | No. | Alternative Description | | Cost $/mi |
|---|---|---|---|---|---|---|---|
| 1 | Overlay 1" | (BOB) | 25000 | 19 | Plane & Reseal | (CON) | 60000 |
| 2 | Overlay 1.5" | " | 36000 | 20 | Reconstruct | " | 500000 |
| 3 | Overlay 1" | " | 33000 | 21 | Plane, Joint Rep/Res | " | 195000 |
| 4 | Overlay 3" | " | 75000 | 22 | Seal Coat | (BOC) | 6000 |
| 5 | Overlay 5" | " | 150000 | 23 | Overlay 1" | " | 45000 |
| 6 | Overlay 6" | " | 180000 | 24 | Overlay 1" | " | 35000 |
| 7 | Mill 2" Overlay 3" | " | 82000 | 25 | Overlay 3" | " | 90000 |
| 8 | Mill 3" Overlay 4" | " | 97000 | 26 | Mill 4" Overlay 5" | " | 179245 |
| 9 | Seal Coat | (BIT) | 6000 | 27 | Mill 2" Overlay 3" | " | 100000 |
| 10 | Partial Reconst. | " | 350000 | 28 | Rehab Crack/O.L 1.5" | " | 81000 |
| 11 | Reconstruct | " | 500000 | 29 | Reconstruct | " | 500000 |
| 12 | Reconstruct | " | 750000 | 30 | Partial Recon/Reseal | " | 300000 |
| 13 | Minor Joint Reseal | (CON) | 30000 | 31 | Mill 3" Overlay 4" | " | 165000 |
| 14 | Major Joint Rep/Res | " | 125000 | 32 | Mill 5" Overlay 6" | " | 207165 |
| 15 | Overlay 3" | (BOC) | 100000 | 33 | Mill 6" Overlay 7" | " | 221455 |
| 16 | Overlay 1.5" | " | 60000 | 34 | Mill 7" Overlay 8" | " | 235045 |
| 17 | Overlay 5" | " | 180000 | | | | |
| 18 | Unbonded | (CON) | 332000 | | | | |

BOB = Bituminous over Bituminous  
BIT = Bituminous  
CON = Concrete  
BOC = Bituminous over concrete

## Rehabilitation and Maintenance Strategies

existing local practice. An example decision tree for a small city is given in matrix form in Figure 18.2 [Smeaton 85].

In this example, the city has a total of 12 available alternatives but the maximum feasible set for a given combination of conditions is 5, in three cases, while the minimum feasible set is 2, in one case. As well, this particular city subjects each feasible alternative to a series of screens (including structural adequacy, pavement type, surface thickness, curb height, section length, functional class, and curb and gutter condition), which can further reduce the number of feasible actions for a given section.

Because the number of available maintenance treatment alternatives is usually more limited, the feasible sets for preventive or corrective maintenance can be identified by applying screens. For example, crack sealing, slurry seals, or chip seals are not likely feasible alternative if both the level of severity and the density or extent of cracking are high. In this case, patching might be the only feasible alternative, provided the minimum acceptable level of deterioration has not yet been reached to warrant rehabilitation. Even if rehabilitation is selected, patching repairs should be carried out as a pre-rehabilitation maintenance treatment.

The general role of expert systems technology in pavement management has been described [Hendrickson 87, Ritchie 87], and it has been shown [Hall 87] how the technology can be applied to first identifying a main feasible rehabilitation approach and then developing detailed, feasible alternatives within the approach chosen. Figure 18.3 is an example of selecting such a main approach. It uses a variety of physical, traffic, cost, climate, and other data. Then the detailed set of alternatives are chosen through another set of decision trees and may include such items as subdrainage, full-depth repairs, etc.

A generic type of decision tree approach, involving over 1,000 branches, has been described [Haas 90]. This expert system uses pavement surface distress, ride quality, loading response, traffic, geometric, and structural data to assign one or more feasible rehabilitation strategies from over 500 contributions of 96 alternatives to each section. It is designed such that a tree applicable to any agency's set of conditions and requirements can be generated with a Knowledge Based Expert System (KBES) development tool or "shell," of which there are many commercially available. The field of KBES technology is rapidly evolving, and new and more powerful shells become available quite frequently. It is imperative that users match the capabilities of the shell to their individual requirements.

The application of Knowledge Based Expert Systems technology to selecting maintenance treatments includes the following:

1. PRESERVER, which utilized the OPS5 representation language and considers single and combined distresses, plus their severities and densities, in selecting the most cost-effective maintenance treatment [Haas 87]. It utilizes Ontario's Pavement Maintenance Guidelines.
2. ROSE, which has been developed for recommending routing and sealing of cracks in flexible pavements, for cold areas [Hajek 87]. The factors considered are crack severity and extent, pavement structure, serviceability, and the

| DISTRESS PRESENCE | COMBINATIONS OF DISTRESS (READ VERTICALLY) | | | | | | | | | | | | | | | | | |
|---|---|---|---|---|---|---|---|---|---|---|---|---|---|---|---|---|---|---|
| PSI < 4.0 | NO | NO | NO | NO | NO | NO | NO | NO | YES | YES | YES | YES | YES | YES | YES | YES | YES | YES |
| Cracking Major | NO | NO | NO | NO | NO | NO | YES | YES | | | | | | | | | | |
| Rutting > 30% | YES | NO | NO | NO | NO | YES | YES | YES | | | | | | | | | | |
| Ravelling > 30% | | YES | NO | NO | | | | | | | | | | | | | | |
| Bleeding > 30% | | | YES | NO | | | | | | | | | | | | | | |
| Alligator Crack > 30% | | | | | NO | NO | NO | YES | | | | | | | | | | |
| Edge Crack > 30% | | | | | NO | NO | YES | | | | | | | | | | | |
| Longitudinal Crack. > 30% | | | | | NO | YES | | | | | | | | | | | | |
| Excess Crown Major | | | | | | | | | YES | NO | NO | | | | | | | |
| AADT > 5000 | | | | | | | | | | NO | YES | NO | YES | NO | YES | | NO | YES |
| Alligator Crack Major | | | | | | | | | | | | NO | NO | YES | YES | | | |
| Rutting Major | | | | | | | | | | | | | | | | NO | YES | YES |
| FEASIBLE REHABILITATION ACTIONS | 3 4 6 11 | 1 5 7 12 | 1 8 12 | 2 4 5 | 2 5 7 | 3 4 6 9 10 | 2 6 9 11 | 3 6 11 | 4 10 | 1 4 10 | 2 9 11 | 2 4 5 9 10 | 3 4 9 11 | 2 4 6 10 | 3 9 11 | 2 4 5 8 | 3 4 5 8 10 | 3 6 9 11 |

REHABILITATION REHABILITATION CODES

1. 1 in. Overlay
2. 2 in. Overlay
3. 3 in. Overlay
4. Mill 1 in. + Chipseal
5. Recycle 1 in. + 1 in. Overlay
6. Recycle 1 in. + 2 in. Overlay
7. Heater Plane 1 in. + 1 in. Overlay
8. Heater Plane 1 in. + 2 in. Overlay
9. Heater Plane 1 in. + 3 in. Overlay
10. Reconstruct 2 in. AC/ 4 in. ABC
11. Reconstruct 2 in. AC/ 6 in. ABC
12. Chipseal

**Figure 18.2** Example decision tree (matrix form) for identifying feasible rehabilitation alternatives.

# Rehabilitation and Maintenance Strategies

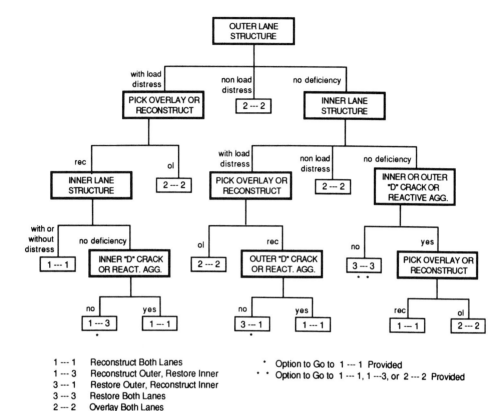

**Figure 18.3** Decision tree for selection of main rehabilitation approach [Hall 87].

existence of other distresses. It utilizes the EXSYS development tool and can be operated in either interactive or automatic mode.

3. ERASME, which was developed in France to assist the user in selecting maintenance and rehabilitation strategies [Allez 88]. It has a diagnostic component for assessing pavement condition and a selection component which is linked to the diagnosis. The EXSYS development tool is also used in this system.

4. PMAS, which is a Pavement Maintenance Advisory System developed for selecting the appropriate maintenance treatment in cold inland or coastal regions [Hanna 92]. In addition to climate, it considers type, severity, and density of distress, traffic volume, and Riding Comfort Index. A unique feature of this system is the capability of the user to interact with distress photographs. The system utilizes the EXSYS development tool for DOS computers and INSTANT EXPERT for the Apple Macintosh. An example maintenance treatment selection, expected life, and illustration of how the rules are extracted from the knowledge base are given in Figure 18.4.

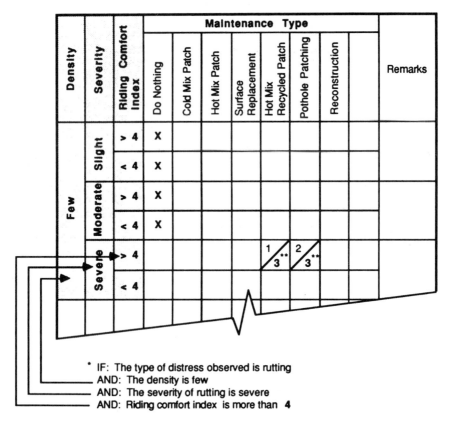

**Figure 18.4** Example maintenance treatment selection and rule extraction [Hanna 92].

## 18.3 DETERIORATION MODELING OF REHABILITATION AND MAINTENANCE ALTERNATIVES

The types of deterioration models described in Section 16.3 are entirely applicable to modeling the deterioration of rehabilitation alternatives. However, because rehabilitation only started to become the major emphasis area for pavement expenditures in the 1980s, particularly for North American, the degree of performance data available is significantly less than for original pavement designs. Consequently, while overlay deterioration models have been developed by a number of agencies, there is also considerable ongoing work.

An example of a mechanistically based overlay design procedure, originally developed for the United States Federal Highway Administration and in which deterio-

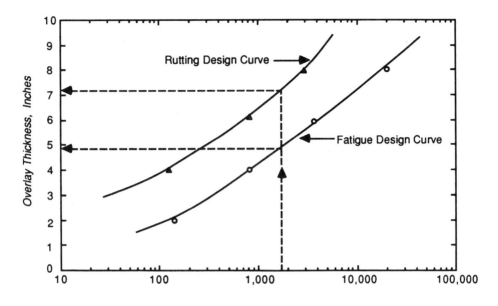

**Figure 18.5** Overlay design curves for an example problem [Treybig 77].

ration is modeled in terms of fatigue and rutting damage, has been described [Treybig 77]. The procedure includes a deflection and cracking survey based evaluation of the existing pavement; estimation of subgrade support from the deflections; laboratory testing of the surface, base, and subbase materials to determine their moduli; calculation of various stresses and strains using elastic layer analysis; calculation of fatigue and rutting damage as a function of equivalent axle load applications using AASHO Road Test criteria; and a "remaining life" analysis as the difference between the design fatigue life and the fatigue life used. Then, overlay thickness is determined as that which satisfies the more critical of the fatigue and rutting criteria. For example, Figure 18.5 shows overlay design curves for a sample situation [Treybig 77]. For the design lane traffic of $2 \times 10^6$ Equivalent Single Axle Load Applications (ESAL), the rutting criterion requires a 7.3-inch overlay, whereas fatigue only requires 4.9 inches. Thus, rutting controls in this situation and the design would be 7.3 inches. It should be noted, however, that in other situations the fatigue criterion can govern, and in still others the curves can cross so that for a low number of load applications one criterion will govern but then for high numbers the other criterion will govern.

One of the most useful approaches to developing rehabilitation deterioration models is through the application of Markov chain procedures, as described in Section 16.3. This approach has been used by the Province of Prince Edward Island in Canada, first to develop deterioration models for its existing pavements in the late 1970s and then to develop similar models for rehabilitation in the late 1980s [Haas 90a]. Similarly, extensive use of the Markov process has been made in developing performance curves

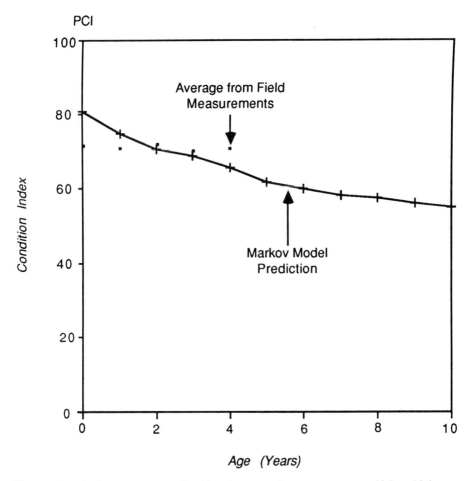

**Figure 18.6** Performance curve for bituminous surface treatment on Yukon highways [McLeod 89].

for bituminous surface treatments (BST) for Yukon highways [McLeod 89]. Figure 18.6 provides an example performance curve for their class 3 highways (fully designed subbase and base layers, where BST substitutes for hot-mix asphalt concrete). It also shows a comparison with the average performance curve from 4 years of actual measurements.

## 18.4 COSTS, BENEFITS, AND COST-EFFECTIVENESS CALCULATIONS

The calculations for costs, cost-effectiveness, or benefits for rehabilitation alternatives, described in the following paragraphs, are needed for priority programming (see Chap-

# Rehabilitation and Maintenance Strategies

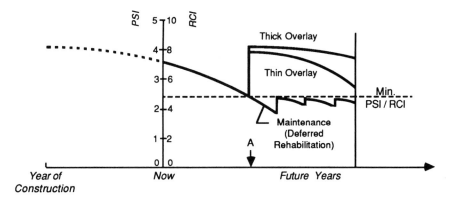

**Figure 18.7** Typical serviceability versus age relationship for a life cycle with rehabilitation at age *A*, and with only maintenance.

ter 19). Details of the economic evaluation principles underlying these calculations are subsequently provided in Chapter 29.

Rehabilitation costs include:

1. Cost of the actual work (materials, transport, placement, etc.), which are often on an in-place basis (volume, area, mass) such as dollars per lane-mile, or lane-km
2. Vehicle operating costs
3. Cost of user delays, including extra vehicle operating costs, during rehabilitation
4. Accident costs due to traffic hazards or interruptions associated with the rehabilitation
5. Environmental damage (air or water pollution, noise, etc.)

The latter two costs are very difficult to capture in numerical terms. While they can be real, they are usually omitted in cost calculations.

The procedures for calculating costs of the actual work require data on unit costs plus volume or mass or area involved and require relatively straightforward multiplications. Many agencies have manuals for carrying out the necessary calculations, with detailed examples, such as that used by the Ontario Ministry of Transportation [MTO 90]. As well, organizations such as The Asphalt Institute [TAI 87] and others have published documents for this purpose.

If the rehabilitation alternative is to be implemented in some future year, such as "A" in Figure 18.7, then the costs can be discounted back to the present, or now, as follows:

$$\text{Present Cost} = \text{Future Cost} \times \text{PWF} \quad (18.1)$$

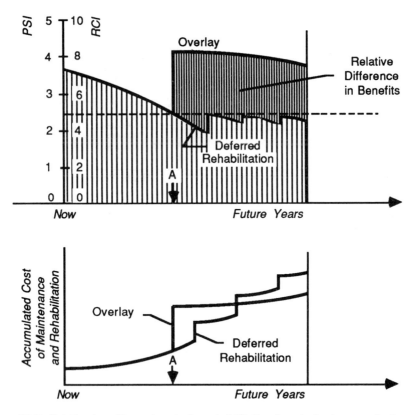

**Figure 18.8** Relative benefits and costs for rehabilitation (overlay) when required versus deferred rehabilitation.

where

    PWF = present worth factor
            = $1/(1 + i)^n$
      $n$ = number of years to the rehabilitation implementation (which is $A$ years in Figure 18.7)
      $i$ = discount rate

For example, if the cost of the thick overlay at time $A$ in Figure 18.7 were estimated to be $100,000 per lane-mile, if $A$ were 3 years, and if $i$ were chosen as 5%, then the present cost would be $100,000 \times [1/(1 + 0.05)^3] = \$80,600$ per lane-mile.

Vehicle operating costs (VOC) vary directly with level of serviceability. The difference in VOC occurring between a rehabilitation action and no rehabilitation can be considered a measure of the relative difference in benefits. Figure 18.8 illustrates the concept. It also illustrates the relative accumulation of costs.

VOC relationships are available for quantifying the relative difference in benefits

# Rehabilitation and Maintenance Strategies

**Table 18.2  Extra User (Vehicle Operation) Costs Per Mile of Deferring the Decision to Rehabilitate for the Example of Figure 18.7**

| Years of Deferral | Difference in PSI | AADT | Annual Extra Veh. Operating Costs $1,000 | Accum. Extra Veh. Operating Costs (P.W. Basis $1,000) |
|---|---|---|---|---|
| | −1.5 | 5000 | 27 | 26 |
| 2 | −1.8 | " | 47 | 66 |
| 3 | −2.1 | " | 66 | 118 |
| 4 | −2.4 | " | 89 | 184 |
| 1 | −1.5 | 10000 | 55 | 51 |
| 2 | −1.8 | " | 95 | 132 |
| 3 | −2.1 | " | 131 | 236 |
| 4 | −2.4 | " | 179 | 368 |

Assumptions: 1. Difference in PSI (column 2) is between PSI = 4.0 (for rehabilitation) and PSI for deferral decision (no rehabilitation)
2. Unit costs for calculating column 3 are from [Zaniewski 82] for small autos, 0% grade, 80 km/h operating speed.
3. Discount rate of 8% was used for col. 4; costs are discounted back from end of year of deferral.

in Figure 18.8 [Zaniewski 82]. An illustration is provided in Table 18.2 for two levels of traffic: 5,000 and 10,000 vehicles per day. It incorporates certain assumptions (i.e., the difference in PSI between the rehabilitated and non-rehabilitated pavement increases at 0.3 unit per year), as listed at the bottom of the table, but certainly the present worth of extra user costs, or relative difference in benefits, is quite large. For example, this difference for rehabilitation delayed 4 years, at an AADT of 5,000, would accumulate to about $184,000 per mile (or about $110,000/km) in present worth terms. Since the unit costs for the VOC relationships in [Zaniewski 82] change with time, the numbers in Table 18.2 should be seen as illustrative only of the magnitude of benefits achievable through rehabilitation.

User delay costs during rehabilitation were first developed in Texas [McFarland 72]. Subsequently, a user delay cost model based on queuing theory and capable of incorporating a variety of traffic handling methods plus such factors as type of facility, traffic volume, length of rehabilitation zone, and time of day was developed [Karan 74]. Table 18.3 is an example summary table from this work (for four-lane highways and the different traffic handling methods). While the numbers are based on mid 1970s costs, which can easily be factored up to current values using change in consumer price index, they do illustrate how extremely large user delay costs can become.

Very few agencies consider user delay costs directly, Minnesota being an exception where the calculations can be done as an option in its pavement management system [Hill 91]. One of the reasons advanced for not considering user delay costs is that these are not a part of the agency's budget. However, a total cost analysis should include them, particularly since it is the user who pays the total cost of rehabilitation construction, maintenance, and delay.

Calculations for cost-effectiveness are based on the areas under the performance or deterioration curve. A schematic illustration for a rehabilitation alternative deferred

**TABLE 18.3 EXAMPLE USER DELAY COSTS FOR REHABILITATION OF A FOUR-LANE DIVIDED HIGHWAY, USING DIFFERENT TRAFFIC HANDLING METHODS [Karan 74].**

### Four-Lane Divided highways - Method A

| AADT | User Delay Cost ($/day) |
|---|---|
| <10000 | Insignificant |
| 10000-15000 | 125 |
| 16000-20000 | 350 |
| 21000-23000 | 600 |
| 24000-25000 | 1100 |
| 26000 | 1950 |
| 27000 | 3300 |
| 28000 | 5950 |
| 29000 | 10650 |
| 30000 | 19500 |
| 31000 | 34800 |
| 32000 | 57000 |
| 33000 | 88150 |
| 34000 | 130850 |
| 35000 | 180150 |
| 36000 | 238125 |
| 37000 | 307650 |
| 38000 | 388000 |
| 39000 | 483500 |
| 40000 | 609500 |
| > 40000 | 700000 |

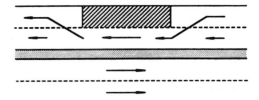

Assumptions:

Lane width = 12 feet
Shoulder width = 6 or more feet
Speed limit = 70 mph
Traffic composition:
  80% passenger cars
  15% transport trucks
  5% single unit trucks
Directional distribution factor = 0.50
Length of restricted area = 1 mile
Construction period = 7 a.m. - 6 p.m.

### Four-Lane Divided highways - Method B

| AADT | User Delay Cost ($/day) |
|---|---|
| <8000 | Insignificant |
| 8000-10000 | 150 |
| 11000-12000 | 350 |
| 13000-15000 | 750 |
| 16000 | 1050 |
| 17000 | 1250 |
| 18000 | 1500 |
| 19000 | 2400 |
| 20000 | 3150 |
| 21000 | 5150 |
| 22000 | 10250 |
| 23000 | 20500 |
| 24000 | 41500 |
| 25000 | 79800 |
| 26000 | 135000 |
| 27000 | 211000 |
| 28000 | 305500 |
| 29000 | 418500 |
| 30000 | 552000 |
| 31000 | 726250 |
| 32000 | 943600 |
| 33000 | 1188000 |
| > 34000 | 2000000 |

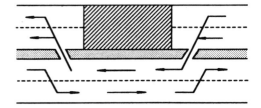

Assumptions:

Lane width = 12 feet
Shoulder width = 6 or more feet
Speed limit = 70 mph
Traffic composition:
  80% passenger cars
  15% transport trucks
  5% single unit trucks
Directional distribution factor = 0.50
Length of restricted area = 2 mile
Construction period = 7 a.m. - 6 p.m.

# Rehabilitation and Maintenance Strategies

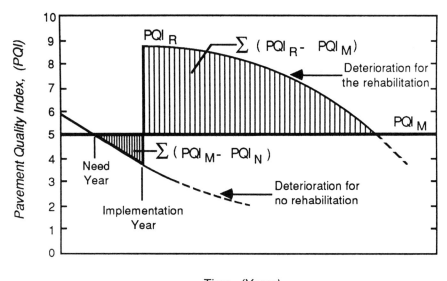

**Figure 18.9** Schematic illustration of effectiveness for a rehabilitation alternative.

from its needs year is given in Figure 18.9. Effectiveness is the net area under the rehabilitation deterioration curve (the shaded area to the right of the rehabilitation implementation year minus the shaded area to the left) multiplied by length of section and volume of traffic; that is,

$$\text{Effectiveness} = \left[ \sum_{\substack{\text{Rehab Year}}}^{\text{PQI}_R \geq \text{PQI}_M} (\text{PQI}_R - \text{PQI}_M) - \left( \sum_{\text{PQI}_N \geq \text{PQI}_M}^{\text{Rehab Year}} (\text{PQI}_M - \text{PQI}_N) \right) \right] \cdot [\text{ADT}] \cdot \begin{bmatrix} \text{Length of} \\ \text{Section} \end{bmatrix} \quad (18.2)$$

where

$\text{PQI}_R$ = Pavement Quality Index (PQI) after rehabilitation (i.e., for the implementation year) and for each year until $\text{PQI}_M$ is reached
$\text{PQI}_M$ = minimum acceptable level of PQI
$\text{PQI}_N$ = yearly PQI from the needs year to the implementation year

The calculation of cost-effectiveness, CE, would then be a simple ratio of effectiveness divided by cost. This ratio has no physical or economic meaning per se, but is valuable in the relative comparison of alternatives and in carrying out priority programming as described in the next chapter. It may be noted that while the terminology is cost-effectiveness, the actual calculation is the reverse (i.e., effectiveness/cost). The reason is that higher values represent more cost-effective alternatives; thus, it conforms with normal expectations such as with increasing benefit/cost ratio representing more economically attractive alternatives.

A simple approach to calculating the cost-effectiveness of maintenance treatment alternatives has been described [Chong 89]. It involves the following analysis for each feasible maintenance alternative:

$$\text{Unit Cost} = \frac{\text{Cost of (Manpower + Equipment + Materials)}}{\text{Accomplishment or Production per Day}} \quad (18.3)$$

and

$$\text{Average Annual Cost} = \frac{\text{Unit Cost}}{\text{Expected Life (Years) of the Treatment Alternative}}. \quad (18.4)$$

The treatment alternative with the lowest average annual cost would represent the most cost-effective solution. A step-by-step example summarized from [Chong 89] illustrates the procedure:

1. Alligator cracking of moderate severity and local density is involved. Table 18.4 tells us we should consider spray patching, cold-mix patching, and hot-mix patching.
2. For hot-mix patching, the performance standard 1002 (see Table 18.5) says that we need a 9-person crew and 6 different pieces of equipment.
3. Say that the labor rate for this location is $8.05 per hour; therefore the daily cost for the 9-person crew is $579.60.
4. Similarly, say that the cost of 6 pieces of equipment works out to $329.40 per day.
5. The accomplishment or production per day (see Table 18.5) would be an average of 70 tonnes of hot mix.
6. The cost of hot mix in this location is, say, $20.04 per tonne; thus 70 tonnes will cost $1,402.80.
7. The total cost per day for manpower, equipment, and materials will then be $579.60 + $329.40 + $1,402.80 = $2,311.80.
8. The unit cost is $2,311.80/70 = $33.03 per tonne.
9. The average annual cost (using an expected life of 4 years, from Table 18.4) is $33.03/4 = $8.26 per tonne per year.
10. Say that we want the patching to be 25mm in thickness. The MTO Estimating Manual says that this requires 0.0570 tonne per square meter. The equivalent annual cost on this basis would be $8.26 × 0.0570 = $0.47/m$^2$/year.
11. Similar calculations for this example gave equivalent annual costs of:
    a. spray patching = $1.93/m$^2$/year
    b. cold-mix patching = $2.02/m$^2$/year (for a 25 mm thickness)
12. The hot-mix patching is considerably more cost-effective in this example and would be the preferable maintenance treatment.

# Rehabilitation and Maintenance Strategies

**Table 18.4  Maintenance Treatment Alternatives for Alligator Type Cracking [Chong 89]**

| Evaluation | | Recommended Maintenance Treatment Alternatives | Maintenance Function Classification | | Expected Effective Life in Years |
|---|---|---|---|---|---|
| Severity | Density | | Routine patrol | Non-patrol | |
| Slight | Local | No action | | | |
| | General | No action but monitor closely for future development | | | |
| Moderate | Local | Spray Patch | 1004 | | 1 |
| | | Cold-Mix Patch | 1001 | | 1 |
| | | Hot-Mix Patch | 1001 1002 | 1002 | 4 |
| | | Hot-Mix Patch for multi-lanes | | 1002 | 4 |
| | General | Same as above but notify District Office of situation and maintain close monitoring | | | |
| Severe | Local | Cold-Mix Patch | 1001 | | 0.5 |
| | | Hot-Mix Patch | 1001 1002 | 1002 | 3 |
| | | Excavate, Granular and Hot-Mix Patch | 1002 | 1002 | 7 |
| | | Improve Drainage (additional) | * | * | |
| | General | Hot-Mix Patch for highways with AADT < 2000 and notify District Office | 1002 | | 2 |
| | | Mulching for AADT < 2000 | | 1014 | 3 |
| | | Granular Lift and Surface Treatment for highways with AADT < 2000 | | 1017 | 4 |
| | | Hot-Mix Patch over selected areas and notify District Office for further action for highways with AADT > 2000 | 1002 | 1002 | 3 |

*Contract only

**Table 18.5 Maintenance Performance Standard for Machine Patching, Placing, and Spreading [Chong 89]**

| DESCRIPTION  The machine placing and spreading of premixed asphaltic materials (hot or cold mix) to repair major surface defects such as depressions, bumps and other pavement defects. Includes preparation of patching of patching area and compaction. | | | |
|---|---|---|---|
| **ROAD TYPE**  **HOT MIX**  **MULTI-LANE** | **HOT MIX**  **2–3 LANE** | **MULCH**  **2-LANE** | **ALTERNATIVE METHOD FOR ALL ROAD TYPES** |
| CREW SIZE  1—Grader Opr.  1—Roller Opr.  2—Shovellers/ Rakers  1 to 5—Drivers (Asphalt Supply)  1 to 2—Drivers (Traffic Cont.) | 1—Grader Opr.  1—Roller Opr.  2—Shovellers/ Rakers  1 to 5—Drivers (Asphalt Supply)  1 to 3—Flaggers (Traffic Cont.) | 1—Grader Opr.  1—Roller Opr.  2—Shovellers/ Rakers  1 to 5—Drivers (Asphalt Supply)  1 to 3—Flaggers (Traffic Cont.) | 2—Screed Oprs.  1—Roller Opr.  2—Shovellers/ Rakers  1 to 5—Drivers (Asphalt Supply)  1 to 2—Drivers or 1 to 3—Flagmen (Traffic Cont.) |
| EQUIPMENT  1—Crew Carrier  1—Grader  1—Roller  1 to 5—Dump Trucks  1 or 2—Sign Trucks or Sign Trailers | 1—Crew Carrier  1—Grader  1—Roller  1 to 5—Dump Trucks | 1—Crew Carrier  1—Grader  1—Roller  1 to 5—Dump Trucks | 1—Crew Carrier  1—Tailgate Spreader  1—Roller  1 to 5—Dump Trucks  1 to 2 sign Trucks/ Sign Trailers |
| MATERIALS  Hot or Cold Mix | Hot or Cold Mix | Cold Mix | Hot Mix |
| ACCOMPLISHMENT  Tonnes | Tonnes | Tonnes | Tonnes |
| MAN-HOURS PER ACCOMPLISHMENT  0.7–2 | 0.7–35 | 0.7–2 | |
| ACCOMPLISHMENT PER DAY  105–35 | 105–35 | 105–35 | |

# Rehabilitation and Maintenance Strategies

## REVIEW QUESTIONS

1. Contact a local highway agency and determine the types of pavement rehabilitation and maintenance practices that are followed in your area. If you have access to a state, city, and county agency, compare the activities used by each. Try to determine what criteria are used for the selection of each type of action.
2. Based on local experience, rework Table 18.2, paying particular attention to the types of distress that trigger the different types of rehabilitation actions.
3. Draw Table 18.2 as a traditional decision tree for the selection of pavement rehabilitation actions.
4. For discount rates of 5%, 7%, and 9%, compute the present worth factor for 20, 25, 30, 35, and 40 years and discuss the results.

Chapter 19

# Priority Programming of Rehabilitation and Maintenance

## 19.1 BASIC APPROACHES TO ESTABLISHING ALTERNATIVES AND POLICIES

There are several basic approaches used by road agencies to establish alternatives and policies for their rehabilitation and maintenance programs, including the following:

1. Strategic approach where certain targets are set for the future, such as:
   a. maximum amount or percentage (say 10%) of the network mileage below the minimum desired level of serviceability
   b. average weighted serviceability of the network
   c. timing all rehabilitation for an optimum level of serviceability, and/or maximizing cost-effectiveness
2. Defining a set of "approved" rehabilitation and maintenance alternatives. This approach is complementary to or can be a part of the strategic approach in 1.
3. Specifying a limited number or type of alternatives, because of a political or social objective, such as only thin overlays or surface treatments to obtain maximum coverage for the available funds.

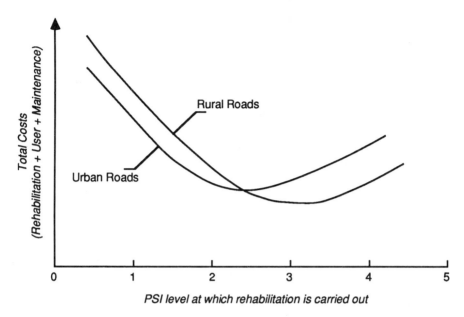

**Figure 19.1** Variation of total costs of rehabilitation with PSI level.

The strategic approach, combined with an approved set of rehabilitation and maintenance alternatives, is incorporated in the priority programming procedures subsequently described. Under conditions of sufficient available funds, only approach 1c will result in the realization of maximum benefits or minimum total costs. For urban road networks, it has been shown [Karan 77] that the optimum PSI level for minimum total costs of rehabilitation would lie between about 2.0 to 2.5. Because lower speeds accrue in urban areas, it has been suggested that the optimum PSI level for rural road or highway networks would be about 3.0 [Haas 85a]. Both situations are shown in Figure 19.1. They show rapidly rising costs for lower levels of PSI, as might be expected, and somewhat higher costs at levels higher than 3.0. This is because the extra construction costs of rehabilitation would not be offset by savings in vehicle operating costs.

The use of a specified or approved set of rehabilitation and maintenance alternatives should make provision for periodic updating. Such updating can include dropping alternatives which are clearly less cost-effective and adding new ones based on the experience of others and on research results.

The policy-oriented approach of specifying alternatives for social or political purposes is certainly practiced by some road agencies. Commonly, this would be only a certain type of pavement, or an alternative strategy like thin overlays or surface treatments to cover a maximum amount of mileage. The limitation of this approach is that it may involve considerably less than optimal use of funds from an economic standpoint.

## 19.2 SELECTING A LENGTH OF PROGRAM PERIOD

The length of program period for rehabilitation improvements is not necessarily the same as that for life-cycle economic evaluation. In fact, it would usually be less because the latter commonly covers 20 or more years. Certainly, the length of program period for maintenance would generally only be 1 to 5 years.

So the question becomes one of what is both a reasonable and useful program period for rehabilitation. It can range from single year by single year [Gunderson 90] to multiyear programs [Hill 90]. The former is convenient, but cannot consider alternative action years for rehabilitation (see Section 17.1). Thus, timing is excluded from the optimization.

The following guidelines for rehabilitation program periods are considered reasonable and have found substantial acceptance in practice:

1. Five years for networks or portions of networks where there is considerable uncertainty for future funding and/or no useful purpose is seen in developing even tentative priority programs beyond that time
2. Ten years for other network situations, particularly where it is desired to evaluate the longer term, strategic implications of greater or lesser levels of funding

A very practical approach is to develop a 5- or 10-year program, but considering the uncertainties in the future years, only fix the first 2 or 3 years of the program, and carry out annual or biannual updates. In this way, the flexibility exists for some projects to be advanced from the second year if bid prices come in below estimates; alternatively, higher than expected bids may cause some first-year program projects to be deferred to the second year.

## 19.3 BASIC FUNCTIONS OF PRIORITY PROGRAMMING

One of the key components of pavement management is to compare investment alternatives at both the network and project levels, within some funding or budget constraint. The result of the comparisons should be a network level priority program of new pavement construction, rehabilitation and reconstruction, and maintenance, and the most cost-effective alternative for each project or maintenance section within the program.

A number of road agencies in the United States and other countries have already incorporated some form of priority programming procedure in their pavement management systems. This practice will undoubtedly accelerate in the United States under the Federal Highway Administration's (FHWA's) policy requiring each state to establish a pavement management system.

The FHWA policy allows for considerable flexibility on the part of individual agencies. Moreover, this flexibility extends to prioritization in that a particular methodology is not specified.

The types of priority assessment methods vary from simple subjective ranking

# Priority Programming of Rehabilitation and Maintenance

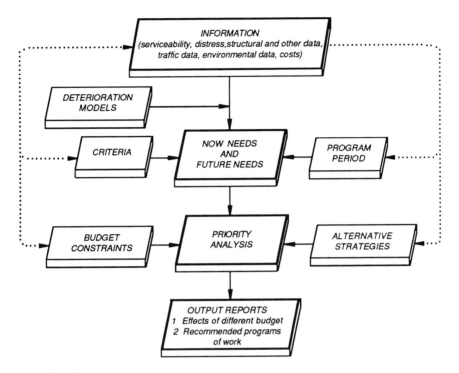

**Figure 19.2** Major steps in priority programming.

to sophisticated mathematical programming. Each has specific features in terms of the pavement rating parameters, type of economic analysis applied, and so forth. However, every priority programming method developed to date for network level pavement management basically incorporates four major central steps: information, identification of needs, priority analysis, and output reports, as shown in Figure 19.2. Simpler methods would involve simpler procedures within the steps and activities of Figure 19.2. Whatever type of priority programming method is used, it should be directed to answering the following questions:

1. What projects (or sections) should be rehabilitated or maintained (selection of candidates)?
2. How can they be built or maintained (selection of within-project alternatives or maintenance treatment alternatives)?
3. When should they be built or maintained (selection of timings)?

A model which can deal with these questions simultaneously (i.e., consider and evaluate all the possible combinations and trade-offs) will not be simple. Consequently, most agencies have models which can only deal with the first, or the first two, questions. However, if the investment is to be truly optimized, all possible combi-

**Table 19.1 Different Classes of Priority Programming Methods [Haas 85a]**

| CLASS OF METHOD | ADVANTAGES AND DISADVANTAGES |
|---|---|
| Simple subjective ranking of projects based on judgment | Quick, simple; subject to bias and inconsistency; may be far from optimal |
| Ranking based on parameters, such as serviceability, deflection, etc. | Simple and easy to use; may be far from optimal |
| Ranking based on parameters with economic analysis | Reasonably simple; should be closer to optimal |
| Optimization by mathematical programming model for year-by-year basis | Less simple; may be close to optimal effects of timing not considered |
| Near optimization using heuristics and marginal cost-effectiveness | Reasonably simple; can be used in a microcomputer environment; close to optimal results |
| Comprehensive optimization by mathematical programming model taking into account the effects | Most complex; can give optimal program (max. of benefits) |

nations of the three questions must be evaluated. Thus prioritization and optimization are not necessarily the same thing.

## 19.4 PRIORITY PROGRAMMING METHODS

Priorities can be determined by many methods, ranging from simple subjective ranking to the true optimization noted in the preceding section. Table 19.1 summarizes the various classes of methods, plus their advantages and disadvantages [Haas 85a].

Subjective ranking defeats many of the advantages of a pavement management system and should only be used on an interim basis, pending the development of one of the other methods.

Many agencies use parameter based ranking methods. The Rational Factorial Rating Method is one example [Fernando 83]. In this method, expert opinion was used to develop a priority index. The procedure is to first define a number of pavement sections with a wide range of conditions. Because the number of combinations of condition factors and levels can be very large, only a portion of the combinations were used on the questionnaire form. Statistical techniques for the design of experiments were used for selecting the fractional factorial. Then a panel of engineers rate the sections on the basis of the selected attributes. The rating obtained are analyzed statistically to develop an equation for prioritization. One of the regression equations obtained from the analysis [Fernando 83] is

$$Y = 5.4 - 0.0263 \cdot X1 - 0.0132 \cdot X2 - 0.4 \cdot \log(X3) + 0.749 \cdot X4 + 1.66 \cdot X5 \tag{19.1}$$

where $Y$ is the priority index in which a value of 1 represents very poor condition and a value of 10 represents excellent condition. Thus, a low value of $Y$ indicates the pavement should have a high priority for treatment. $X1$ to $X5$ are, respectively, rainfall (5 inches to 40 inches per year), freeze and thaw (0 to 60 cycle/year), traffic (100 to 100,000 ADT), PSI (4.0 to 2.0), and distress ($+1.0$ to $-1.0$).

Near optimization, based on a heuristic, marginal cost-effectiveness method, is the basis for the priority programming method adopted by Idaho, Minnesota, and South Carolina (in the United States), and by Alberta, Prince Edward Island, and Newfoundland in Canada [Haas 85a]. It is also the basis for a comprehensive, integrated system developed for cities in Alberta [Marcotte 87, Haas 89].

The near optimization method used by these agencies proceeds as follows:

1. Consider each combination of section, treatment alternative, and year in the program period.
2. Calculate the effectiveness, $E$, of each combination (which is the area under the performance curve, multiplied by ADT and section length—see Figure 18.9).
3. Calculate the cost, $C$, in net present value terms, of each treatment alternative in each combination.
4. Calculate the cost-effectiveness, CE, of each combination as the ratio of $E/C$ (i.e., a ratio inverse in concept to the benefit-cost ratio method).
5. Select the combination of treatment alternative and year for each section which has the best CE, until the budget is exhausted.
6. Calculate the marginal cost-effectiveness, MCE, of all other strategies (i.e., treatment alternative when) for each section, as follows:

$$\text{MCE} = (E_s - E_r)/(C_s - C_r)$$

where

$E_s$ = effectiveness of the combination selected in Step 5
$E_r$ = effectiveness of the combination for comparison
$C_s$ = cost of the combination selected in Step 5
$C_r$ = cost of the combination for comparison

7. If MCE is negative, or if $E_r < E_s$, then the comparative strategy is eliminated from further consideration; if not, it replaces the combination selected in Step 5.
8. The process is repeated until no further selections can be made in any year of the program period (i.e., when the budget is exhausted).

## 19.5 MATHEMATICAL PROGRAMMING (OPTIMIZATION METHOD)

Mathematical programming models can be applied to either single year or multiyear prioritization. They do not indicate priorities, per se; rather, alternatives are selected to satisfy a specific objective function (i.e., where some value is maximized, such as effectiveness, or minimized, such as costs). Formulations used in these models include several variations of linear programming and dynamic programming [Lytton 85]. Linear programming (LP) is a useful formulation for multiyear prioritization because it can be set up to model the trade-offs between project timing and benefit losses. Each implementation or action year (see Figure 17.1) can be treated as an independent alternative, along with the within-project alternatives and the projects or sections in the network. Thus, all possible combinations are considered and compared.

The formulation of an LP model for maximizing the total present value of $m$ pavement improvement projects, each with $k$ within-project alternatives, for a $T$ years programming period can be done in the following way:

$$\text{Maximize} \quad \sum_{i=1}^{m} \sum_{j=1}^{k} \sum_{t=1}^{T} X_{ijt} \cdot B_{ijt} \qquad (19.3)$$

$$\text{Subject to} \quad \sum_{t=1}^{T} \sum_{j=1}^{k} X_{ijt} \leq 1 \quad \text{for } i = 1, 2, \ldots m , \qquad (19.4)$$

$$\sum_{i=1}^{m} \sum_{j=1}^{k} X_{ijt} \cdot D_{ijtt'} \leq B_t \quad \text{for } t = 1, 2 \ldots T , \qquad (19.5)$$

and

$$X_{ijt} \leq 0 \qquad (19.6)$$

where

$X_{ijt}$ = section $i$ (of $m$ total sections) with alternative $j$ (of $k$ total treatment alternatives) in year $t$ (of the $T$ years in the program period)

$B_{ijt}$ = present value of annual benefits (including salvage value) of section $i$, with alternative $j$, built in year $t$, all discounted to base year at a discount rate of $R$

$D_{ijtt'}$ = the actual construction and/or maintenance cost of section $i$, with alternative $j$, built in year $t$, incurred in year $t'$

$B_t$ = budget for year $t$

$X_{ijt}$ is a discrete variable equal to 1 if a project is selected for section $i$ in year $t$ and equal to zero otherwise.

Equation 19.3 is the objective function for maximization of benefits. Maximization of net benefits (benefits minus costs) is not necessary because construction and maintenance costs are specifically dealt with in the annual budget constraints. Thus, there is no need for discounting costs back to a base year to calculate the net present value of the projects.

Equation 19.4 states that an $X_{ijt}$ combination is unique; that is, a section cannot be selected twice during the analysis period. Either it is or is not built. If a project is committed (by judgment, political intervention, etc.), then the constraint becomes one of equality (i.e., $= 1$).

Equation 19.5 represents the budget constraint each year in the programming

## Priority Programming of Rehabilitation and Maintenance

period which cannot be exceeded. It states that the summation over $m$ total sections and $k$ total alternative treatments of all the $X_{ijt}$ costs must be less than or equal to the budget, $B_t$.

Equation 19.6 is a common constraint of all linear programming problems which states that it is not possible to recapture construction costs by not constructing the project.

Figure 19.3 shows an example LP formulation for $m$ total sections, $k$ (within project) treatment alternatives, and $t$ years program period.

Figure 19.4 (top) shows the calculation of benefits for a section or project which is expected to reach its "trigger level" or needs year in year 6 of a 10 year program period. If the project is delayed to year 7 (middle of Figure 19.4), or to year 8 (bottom of Figure 19.4), and so on, the benefits are recalculated. The delay of a project obviously shifts the benefit stream to the right and changes the present value of benefits in the base year.

The costs are similarly affected by the delay, as shown in Figure 19.5. The upper portion of the diagram assumes that the pavement is constructed in year 6. There is a construction cost in that year followed by routine maintenance costs in the following years. A one-year delay results in a higher extra maintenance cost in year 6. This maintenance cost is spent to keep the pavement at its "trigger level" during the year in which the project is delayed. Then, the same costs (i.e., construction cost, routine maintenance cost) occur in the same order starting from year 7. A 2-year delay means additional maintenance costs in years 6 and 7, followed by the same sequence of costs.

It should be noted that costs are considered directly in the LP model up to the end of the programming period. After that, costs cannot be taken into account because of the absence of budget constraints. It is for this reason that costs incurred in each year after the programming period are considered in the analysis indirectly by subtracting them from the respective annual benefits.

## 19.6 EXAMPLES AND COMPARISONS

There are many descriptions of priority analysis methods and results in the literature, but few comparisons between methods for actual networks. However, one study compared marginal analysis (near optimization) and multiyear linear programming for several cities [Haas 85a]. The near optimization technique, which is much simpler, flexible, and operationally efficient, provided results close enough to actual optimization for all practical purposes.

Another comparison, between the Rational Factorial Rating Method (ranking) and Linear Programming (optimization), was carried out [Kikukawa 84] using an earlier set of data from [Karan 81]. The analysis was for a small network of approximately 72 miles (120 km), broken into 25 homogeneous sections, varying in length between 0.5 mile (0.9 km) and 7.0 miles (11.6 km) on the basis of traffic volume, pavement type and thickness, and geometry. A sectional summary of the Structural Adequacy Index (SAI), Riding Comfort Index (RCI), Surface Distress Index (SDI),

| Project \ Alternative | YEAR 1 | | | | YEAR 2 | | | | YEAR t | | | |
|---|---|---|---|---|---|---|---|---|---|---|---|---|
| | 1 | 2 | 3 ... | k | 1 | 2 | 3 ... | k | 1 | 2 | 3 ... | k |
| 1 | $X_{111}$ | $X_{121}$ | $X_{131}$ | $X_{1k1}$ | $X_{112}$ | $X_{122}$ | $X_{132}$ | $X_{1k2}$ | $X_{11t}$ | $X_{12t}$ | $X_{13t}$ | $X_{1kt}$ |
| 2 | $X_{211}$ | $X_{221}$ | $X_{231}$ | $X_{2k1}$ | $X_{212}$ | $X_{222}$ | $X_{232}$ | $X_{2k2}$ | $X_{21t}$ | $X_{22t}$ | $X_{23t}$ | $X_{2kt}$ |
| 3 | | | | | | | | | | | | |
| m | $X_{m11}$ | $X_{m21}$ | $X_{m31}$ | $X_{mk1}$ | $X_{m12}$ | $X_{m22}$ | $X_{m32}$ | $X_{mk2}$ | $X_{m1t}$ | $X_{m2t}$ | $X_{m3t}$ | $X_{mkt}$ |

**Figure 19.3** Linear programming formulation.

# Priority Programming of Rehabilitation and Maintenance

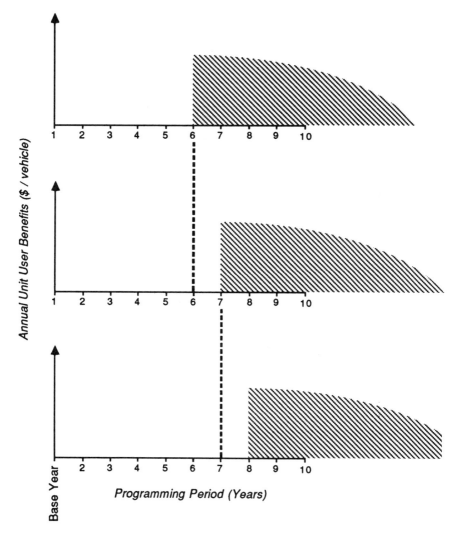

**Figure 19.4** Effect of project delays on user benefits.

and composite, Pavement Quality Index (PQI), each on a scale of 0 to 10, is provided in Table 19.2.

In order to conduct the analysis, the following assumptions were used:

1. A minimum acceptable PQI level of 5.0 was used in the analysis. Thus, 11 sections out of the 25 in Table 19.2 are now or current needs.
2. A programming period of 5 years was selected. For the purpose of performance predictions, curves originally developed using a Markov model were used [Karan, 81].

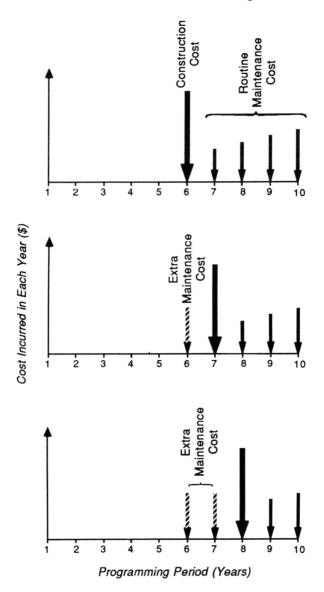

**Figure 19.5** Effect of project delays on costs.

3. The following three rehabilitation alternatives for each road section were analyzed by the priority program:
   a. Surface treatment: A double 9.5 mm (3/8 inch) chip seal at $2.40/m² (1.92/yd²)
   b. Overlay: An 89 mm (3 1/2 inch) asphalt concrete overlay at $7.65/m² (6.14/yd²)
   c. Reconstruction: A new pavement with 152 mm (6 inch) asphalt concrete

**Table 19.2 Sectional Data Summary [Kikukawa 84]**

| Section Number | SAI | RCI (2xPSI) | SDI | PQI | AADT | Section Length (m) |
|---|---|---|---|---|---|---|
| 50 | 7.5 | 7.5 | 8.0 | 7.5 | 1300 | 11600 |
| 51 | 8.0 | 7.5 | 8.4 | 7.8 | 1300 | 13350 |
| 52 | 5.8 | 8.2 | 8.2 | 5.8 | 1500 | 7900 |
| 53 | 4.3 | 7.7 | 8.5 | 5.7 | 1950 | 4650 |
| 54 | 4.3 | 5.1 | 4.3 | 4.8 | 1950 | 5400 |
| 55 | 2.0 | 5.0 | 10.0 | 3.8 | 3750 | 4800 |
| 56 | 6.3 | 6.0 | 6.3 | 6.1 | 3750 | 11500 |
| 57 | 5.0 | 5.0 | 4.1 | 5.0 | 10100 | 5450 |
| 58 | 6.3 | 5.5 | 8.2 | 5.8 | 7450 | 2150 |
| 59 | 6.3 | 7.8 | 10.0 | 6.9 | 4500 | 1650 |
| 60 | 7.0 | 7.7 | 10.0 | 7.3 | 3500 | 4750 |
| 61 | 6.3 | 7.8 | 10.0 | 6.9 | 3500 | 3250 |
| 62 | 5.3 | 8.4 | 9.9 | 5.3 | 3500 | 3300 |
| 63 | 5.0 | 2.9 | 2.8 | 2.9 | 3500 | 2100 |
| 64 | 5.0 | 8.6 | 10.0 | 5.0 | 3500 | 3050 |
| 65 | 1.5 | 7.7 | 10.0 | 4.0 | 3500 | 1000 |
| 66 | 7.8 | 8.1 | 10.0 | 7.8 | 3500 | 5850 |
| 67 | 4.8 | 8.1 | 10.0 | 4.8 | 3500 | 2000 |
| 68 | 5.5 | 5.4 | 10.0 | 5.4 | 2700 | 6500 |
| 69 | 4.8 | 5.3 | 10.0 | 5.1 | 2700 | 4750 |
| 71 | 4.3 | 6.5 | 6.2 | 5.2 | 3200 | 6200 |
| 701 | 1.0 | 3.0 | 5.2 | 3.0 | 2700 | 1000 |
| 702 | 3.5 | 1.9 | 4.5 | 1.9 | 2700 | 900 |
| 703 | 8.0 | 2.4 | 7.1 | 2.4 | 2700 | 1300 |
| 704 | 8.5 | 3.1 | 5.7 | 3.1 | 2700 | 1450 |

surface and 254 mm to 305 mm (10 to 12 inch) base at $22.50/m² (18.00/yd²).

4. Three levels of expected annual rehabilitation budget, $2,000,000, $1,000,000, and $500,000 were considered in the study. They were assumed to remain constant over the 5-year programming period.

5. Economic analyses, including direct agency costs (rehabilitation and routine maintenance) and user benefits (savings in vehicle operating costs due to improved pavement conditions) were conducted for each implementation year and for each combination of project and rehabilitation strategies.

The LP analysis used a package known as LP1, from Cyberware Computer Systems Ltd.

The ranking model of equation 19.1 was used without any modifications. It should be noted though that the equation may not be very applicable to regions other than for which it was developed. The RCI and SDI values were converted into the scale of the $X4$ and $X5$ variables of equation 19.1. Also, the variables on amount of rainfall and freeze-thaw cycles were considered to be constant throughout the area where the network exists.

Table 19.3 gives the priority rankings from calculation of the prioritization indexes, and the priority programs from the LP analysis for the three budget levels. In the first case, projects would be selected in the order of the indexes (starting with number 63, then 702, 67, etc.) until the available funds are spent. In the latter case, because of the limited number of sections, the LP analysis does result in some "discontinuities" (i.e., for year 2 at the lowest budget level). Nevertheless, a comparison of the results shows some major differences. For example, section 57 would be done in year 1 for all three budget levels, but it is fourth in the priority rankings. Also, section 55 appears in the first or third year for all three budget levels but is not within the top 10 by the ranking method.

Thus, while the ranking model approach is simple, it may result in a program which is significantly different from an optimal one.

## 19.7 BUDGET LEVEL EVALUATION

Road agencies often wish to assess the effects of different budget or funding levels on the average serviceability of their pavement networks. This may range to the extreme of a zero capital budget (i.e., no rehabilitation, only routine maintenance).

Figure 19.6 shows such a budget plot for the network example of the preceding section for four different budget levels (including zero capital funds). It appears that budget level 2 ($1,000,000 per year) would keep the network at approximately its current PQI level, while budget levels 3 and 4 ($500,000 and $0 per year, respectively) would result in significant deterioration. Budget level 1 would improve the average PQI somewhat, but at twice the cost of budget level 2.

It may also be desirable to show the accumulation of deficient mileage (i.e., at or below a minimum desirable level) for each budget level. Examples of this are shown in the working system descriptions of Part Six.

## 19.8 FUNDING LEVEL REQUIREMENTS FOR SPECIFIED STANDARDS

An alternative approach to priority programming is to specify a standard or requirement and then determine the funding required to meet this standard. Such standards could be a minimum average serviceability for the network and/or a maximum amount of deficient mileage, say no more than 10% of the total. The optimization involved for this approach would carry out a cost minimization, as compared to an effectiveness or benefit maximization for a specified budget level.

Several agencies, such as in Minnesota [Hill 91], have the capability in their pavement management systems of developing priority programs in either the cost minimization or effectiveness maximization mode. The first approach is useful for policy development.

## Table 19.3 Comparison of Results from Two Prioritization Methods [Kikukawa 84]

**(a) Pavement Rehabilitation Priorities by Sections Using the Rational Factorial Rating Method for Ranking**

| Ranking | Section Number | Prioritization Index |
|---------|----------------|----------------------|
| 1       | 63             | 3.34                 |
| 2       | 702            | 3.80                 |
| 3       | 67             | 3.85                 |
| 4       | 57             | 3.90                 |
| 5       | 701            | 4.20                 |
| 6       | 54             | 4.27                 |
| 7       | 704            | 4.38                 |
| 8       | 703            | 4.48                 |
| 9       | 56             | 4.95                 |
| 10      | 71             | 5.02                 |
| ⋮       | ⋮              | ⋮                    |
| 25      | 64             | 6.58                 |

**(b) Pavement Rehabilitation Priorities by Sections Using the Optimization (LP Formulation) Method**

| Year 1 | Year 2 | Year 3 | Year 4 | Year 5 |
|--------|--------|--------|--------|--------|
| (a) Budget level—$0.5 million/year | | | | |
| 57—R   |        | 55—O   | 64—O   | 54—O   |
|        |        | 63—R   | 701—O  | 64—R   |
|        |        | 67—R   | 702—O  |        |
|        |        |        | 703—O  |        |
|        |        |        | 704—O  |        |
| (b) Budget level—$1.0 million/year | | | | |
| 57—R   | 63—O   | 55—R   | 54—R   |        |
|        | 64—O   | 67—R   | 704—R  |        |
|        | 65—O   |        |        |        |
|        | 701—O  |        |        |        |
|        | 702—O  |        |        |        |
|        | 703—O  |        |        |        |
| (c) Budget level—$2.0 million/year | | | | |
| 55—R   | 63—R   | 54—R   |        |        |
| 57—R   | 64—R   | 703—R  |        |        |
|        | 65—R   |        |        |        |
|        | 67—R   |        |        |        |
|        | 701—R  |        |        |        |
|        | 702—R  |        |        |        |
|        | 704—R  |        |        |        |

O: Overlay
R: Reconstruction

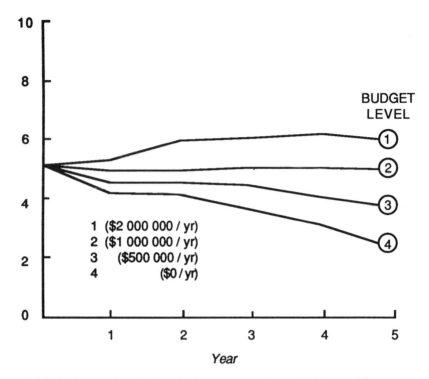

**Figure 19.6** Budget level evaluation for the example network [Kikukawa 84].

## 19.9 FINAL PROGRAM SELECTION

The techniques for developing either a priority or optimal program of pavement preservation as presented in this chapter will assist the pavement manager with the task of selecting sections, treatments, and timing for projects. However, the programs developed from these methods will only be as good as the model used for the analysis and the data that goes into the model. Frequently, data is collected well in advance of the program development and conditions may have changed between the time when the data was collected and the program is formulated. In addition, the priority or optimization model may not capture all of the factors that need to be considered for each individual section of pavement in the network. Frequently, in large highway organizations, the pavement management programming will be performed by the central office and it will be necessary to present the priority program to district engineers to reach a consensus of project selection. Hence, there may be a need for intervention by the engineer to assure the integrity of the program. As pointed out in Part One, the pavement management system provides decision support; it does not make the decisions. The output of the priority programming process, however, should present a useful guide for the engineer to follow.

## REVIEW QUESTIONS

1. What is the minimum programming period for effective pavement management?
2. What is the basic function of priority programming?
3. How can an agency develop a ranking based priority index?
4. Write an objective function for an optimization analysis.

Chapter 20

# Developing Combined Programs of Maintenance and Rehabilitation

## 20.1  DESIRABILITY OF COMBINED PROGRAMS AND BARRIERS

The question of what is an optimal allocation of total available funds to maintenance and to rehabilitation is difficult to answer. For example, can a road agency's policy to have no more than 10 percent of its network at or below the minimum desirable level for serviceability (see Section 19.7) best be achieved by doing: (a) cost minimization priority programming only for rehabilitation, or (b) by doing this in conjunction with specified pre-rehabilitation (corrective) and post-rehabilitation (preventive) maintenance, or (c) by doing a combined optimization for preventive maintenance and rehabilitation plus specified pre- and post-rehabilitation maintenance?

The third approach is desirable in that it should result in the best overall use of available resources. However, few systems have yet been developed to do this, for reasons that include the following: (a) provision would have to exist for integrating maintenance and rehabilitation budgets, (b) deterioration models for preventive maintenance have not yet been well developed, and (c) it has been common practice to separate routine maintenance from the rehabilitation programming.

The institutional barrier represented by the third reason is prevalent for larger road agencies. Maintenance is carried out on an ongoing basis by the agency's own

**Developing Combined Programs of Maintenance and Rehabilitation**     **241**

forces, under a separate division, while rehabilitation programming is determined by a planning division and implemented by the construction division.

Another barrier is the tendency to view rehabilitation only as something that is programmed when a pavement reaches a minimum acceptable level of serviceability, rather than to view maintenance throughout the life of a pavement as something that can have a major influence on the timing and type of rehabilitation.

The following sections illustrate the development of a combined program using pre- and post-rehabilitation maintenance strategies.

## 20.2 PRE-REHABILITATION MAINTENANCE STRATEGIES

A method for identifying pre-rehabilitation maintenance strategies has been described [Smeaton 85]. It uses local practice plus judgment as illustrated in Figure 20.1. This example also uses the decision tree in Figure 18.2 for identifying feasible rehabilitation alternatives.

Pre-rehabilitation maintenance starts between 1 and 10 years before rehabilitation, depending on the implementation year for the rehabilitation. For example, any sections scheduled for rehabilitation in year 3 would have been 60 percent crack sealed three years prior, 2 inches of cold mix patching on 5 percent of the section two years prior, and 40 percent crack sealed one year prior to the rehabilitation, according to Figure 20.1. With unit costs assigned to the maintenance treatments listed in Figure 20.1, it is easy to calculate the maintenance cost portion for the life-cycle economic analysis.

## 20.3 POST-REHABILITATION MAINTENANCE STRATEGIES

The methodology described by Smeaton also makes provision for post-rehabilitation maintenance strategy selection, using a decision tree approach [Smeaton 85]. Figure 20.2 provides an example, which again goes with the rehabilitation alternatives of Figure 18.2.

Each rehabilitation alternative has a particular post-rehabilitation set of maintenance treatments associated with it throughout the program period. The earliest of these starts 3 years after rehabilitation (for alternatives 1, 4, and 12), while for rehabilitation alternatives 10 and 11, maintenance does not begin until year 6.

For example, Figure 20.2 indicates that if rehabilitation alternative 10 were implemented (reconstruct 2 in. AC/4 in. ABC), 6 years later it would require maintenance treatment 1 (crack sealing) over 10% of the area and 8 years later it would also require maintenance treatment 1 over 15% of the area.

Again, the methodology of Figure 20.2, together with having unit cost data, makes it easy to calculate the maintenance cost portion for life-cycle economic analysis.

## 20.4 EXAMPLE RESULT OF A COMBINED PROGRAM

The methodology referred to in the preceding sections has been used to develop combined network programs for maintenance and rehabilitation, again for the example

| IMPLEMENTATION YEAR | MAINTENANCE STRATEGY & PERCENT OF SECTION FOR YEARS PRIOR TO IMPLEMENTATION | | | | | | | | | | NO. OF MAINT. YRS FOR REMOVAL (1) | NO. OF MAINT. YRS FOR RECONSTR (2) |
|---|---|---|---|---|---|---|---|---|---|---|---|---|
| | 1 | 2 | 3 | 4 | 5 | 6 | 7 | 8 | 9 | 10 | | |
| 1 | 1-80 | | | | | | | | | | 1 | 1 |
| 2 | 5-20 | 1-70 | | | | | | | | | 2 | 2 |
| 3 | 1-40 | 3-5 | 1-60 | | | | | | | | 2 | 2 |
| 4 | 2-7 | 3-5 | 5-20 | 1-50 | | | | | | | 2 | 3 |
| 5 | 1-60 | 2-8 | 3-5 | 5-20 | 1-45 | | | | | | 2 | 3 |
| 6 | 1-50 | 3-3 | 2-8 | 3-7 | 5-20 | 1-40 | | | | | 2 | 3 |
| 7 | 1-25 | 1-25 | 4-20 | 2-7 | 3-8 | 5-25 | 1-40 | | | | 2 | 4 |
| 8 | 1-10 | 1-25 | 4-20 | 4-20 | 5-20 | 3-10 | 5-20 | 1-40 | | | 2 | 4 |
| 9 | 1-5 | 1-15 | 5-15 | 2-5 | 4-20 | 2-7 | 5-20 | 3-15 | 1-40 | | 3 | 5 |
| 10 | 1-5 | 1-15 | 5-15 | 2-3 | 5-15 | 4-20 | 5-15 | 2-5 | 1-30 | 5-30 | 0 | 0 |

(1) If the implementation alternative is a removal, such as recycling or milling, maintenance is eliminated for the number of years specified immediately prior to the implementation year.

(2) If the implementation alternative is a reconstruction, maintenance is eliminated in the number of years specified immediately prior to the implementation year.

Maintenance Strategies
1. Crack seal
2. Hot mix patch, 2 in.
3. Cold mix patch, 2 in.
4. Slurry seal
5. Fog seal
6. Fabric + Thin overlay

**Figure 20.1** Example of pre-rehabilitation maintenance strategies [Smeaton 85].

| REHAB. (1) OPTION NO. | MAINTENANCE STRATEGY (2) & PERCENT OF SECTION FOR YEARS AFTER IMPLEMENTATION |||||||||
|---|---|---|---|---|---|---|---|---|---|
| | 1 | 2 | 3 | 4 | 5 | 6 | 7 | 8 | 9 |
| 1 | | | 3 - 5 | 1 - 20 | 5 - 40 | | 2 - 20 | | 6 - 25 |
| 2 | | | | 1 - 15 | 3 - 5 | 5 - 35 | 2 - 17 | | 6 - 20 |
| 3 | | | | 1 - 10 | 3 - 5 | 5 - 30 | | 2 - 13 | |
| 4 | | | 3 - 8 | | 1 - 20 | 5 - 40 | | 6 - 25 | 2 - 10 |
| 5 | | | | | 1 - 15 | 5 - 35 | 6 - 20 | | 2 - 10 |
| 6 | | | | | 1 - 10 | 5 - 30 | 6 - 15 | | 2 - 7 |
| 7 | | | | | 1 - 15 | 5 - 35 | 6 - 20 | | 2 - 10 |
| 8 | | | | | 1 - 10 | 5 - 35 | 6 - 20 | | 2 - 8 |
| 9 | | | | | 1 - 5 | 1 - 20 | 5 - 25 | 6 - 15 | 2 - 5 |
| 10 | | | | | | 1 - 10 | | 1 - 15 | |
| 11 | | | | | | 1 - 10 | | 1 - 15 | |
| 12 | | | 2 - 3 | 4 - 30 | | 3 - 8 | 2 - 5 | 6 - 20 | 6 - 25 |

(1) See Figure 18.2

(2) See Figure 20.1

**Figure 20.2** Example of post-rehabilitation maintenance strategies [Smeaton 85].

Table 20.1 EXAMPLE OF DETAILED COMBINED MAINTENANCE AND REHABILITATION REQUIREMENTS FOR AN INDIVIDUAL SECTION OF A NETWORK [Smeaton 85]

CLIENT:
DATE: FEBRUARY 1984
STRATEGY SUMMARY REPORT: 3. FINAL SUMMARY OF ANNUAL COSTS AND BUDGET REQUIREMENTS

| DESCRIPTION | SUMMARY ITEM | 1984 | 1985 | 1986 | 1987 | 1988 | 1989 | 1990 | 1991 | 1992 | 1993 | TOTAL |
|---|---|---|---|---|---|---|---|---|---|---|---|---|
| ROUTINE MAINTENANCE | LENGTH | 0.00 | 2.70 | 10.71 | 0.00 | 3.79 | 0.00 | 5.41 | 0.00 | 8.66 | 0.00 | 31.27 |
|  | COST | 0 | 900 | 10675 | 0 | 1260 | 0 | 1799 | 0 | 2879 | 0 | 17513 |
| TOTAL REHABILITATION | LENGTH | 27.73 | 3.77 | 10.80 | 10.93 | 5.68 | 8.86 | 2.19 | 2.93 | 4.15 | 6.82 | 83.L99 |
|  | COST | 187788 | 16220 | 38307 | 31627 | 14540 | 35059 | 4785 | 16862 | 17931 | 43629 | 406748 |
| TOTAL MAINTENANCE | LENGTH | 18.07 | 19.05 | 28.75 | 8.81 | 8.25 | 7.94 | 12.72 | 15.20 | 16.67 | 29.58 | 165.04 |
|  | COST | 8452 | 11115 | 26544 | 3342 | 2992 | 4350 | 5259 | 11194 | 5841 | 38076 | 117165 |
| BUDGET REQUIREMENTS | LENGTH | 45.80 | 22.83 | 39.56 | 19.74 | 13.93 | 16.80 | 14.91 | 18.13 | 20.82 | 36.40 | 248.92 |
|  | COST | 196239 | 27336 | 64851 | 34969 | 17531 | 39109 | 10044 | 28056 | 23772 | 81705 | 523913 |

NOTE: 1. LENGTH IN LANE-MI
2. COSTS IN 10'S OF 1984 DOLLARS
3. BUDGET REQUIREMENTS = TOTAL REHAB + TOTAL MAINTENANCE (INCLUDES ROUTINE MAINTENANCE)

# Developing Combined Programs of Maintenance and Rehabilitation

used in Figures 18.2, 20.1, and 20.2. Detailed requirements for an individual example section of the network are illustrated in Table 20.1. It shows that the most cost-effective combined program for this section would involve crack sealing in year 1 (maintenance treatment M1), rehabilitation consisting of 1 in. recycle and 1 in. overlay in year 3 (alternative R6), and so on.

A summary of costs and budget requirements for the network as a whole is provided in Table 20.2. It shows, for example, that rehabilitation will require about $188,000 in year 1 and maintenance will only require about $8,000. For subsequent years, these relative magnitudes of costs vary substantially. Consequently, to achieve the optimum represented in Table 20.2 necessitates complete flexibility between maintenance and rehabilitation budget allocations, which can be achieved with a small agency as described in the example.

This example demonstrates the feasibility of developing a cohesive pavement preservation strategy at the network level. A preservation *strategy* is a combination of specific pavement maintenance and rehabilitation actions which are performed over the life of the pavement. While the example was developed for a small agency, the concepts are equally applicable to larger agencies. However, implementation of these concepts requires overcoming institutional barriers that have historically been a major deterrent to the development of a total pavement management system.

## REVIEW QUESTIONS

1. Use a spread sheet program to develop an example pavement network, include 50 sections in a range of conditions as identified by the rational factorial method. Compute the priority index for these pavements and select a timing for rehabilitation projects. Assume the budget is only adequate to do five projects per year. Use Figures 20.1 and 20.2 to develop pre- and post-maintenance strategies for these pavements.
2. Contact a local highway agency and arrange to visit a highway maintenance site. Observe the crew, equipment, and materials being used on the job to develop an estimate of the resources required to perform the activity. Use a standard pricing guide, such as Best Construction Estimating Guide, and develop a cost for the activity. Use this as a basis for estimating costs for the pavement preservation program developed for question 1.

**Table 20.2  Example of Final Summary of Costs and Budget Requirements for a Combined Maintenance and Rehabilitation Program [Smeaton 85]**

CLIENT:
REPORT: DETAILED SECTIONAL STRATEGY EVALUATION
DATE: FEBRUARY 1984

SECTION: 0707
LOCATION: ARBOL COURT FROM PIEDRA LOOP TO NORTH END > 0721

AREA: 1253 SQ. YD.

| 1984 | 1985 | 1986 | 1987 | 1988 | 1989 | 1990 | 1991 | 1992 | 1993 | Cost-Effect | 1984 $/sq yd | Costs Total |
|---|---|---|---|---|---|---|---|---|---|---|---|---|
| 0.38M1 | 0.00 | 3.00R6 | 0.00 | 1.62M3 | 0.00 | 0.00 | 0.07M1 | 0.00 | 0.00 | 0.3481 | 5.07 | 6356 |
| 0.38M1 | 0.00 | 16.00RG | 0.00 | 1.62M3 | 0.00 | 0.00 | 0.00 | 0.07M1 | 0.00 | 0.3070 | 18.07 | 22646 |
| 0.38M1 | 0.00 | 4.48RD | 0.00 | 0.00 | 0.05M1 | 0.00 | 0.09M1 | 0.00 | 0.14M1 | 0.2942 | 5.14 | 6442 |
| 0.38M1 | 0.00 | 3.05R1 | 0.00 | 0.00 | 0.00 | 0.09M1 | 0.00 | 0.14M1 | 0.00 | 0.2554 | 3.66 | 4591 |
| 0.38M1 | 0.00 | 5.35R2 | 0.00 | 1.62M3 | 0.00 | 0.07M1 | 0.00 | 0.00 | 0.12M1 | 0.1806 | 7.54 | 9449 |

| Code | Rehab. Strategy | Code | Maint. Strategy |
|---|---|---|---|
| R1 | 1" Overlay | M1 | Crack Seal |
| R2 | 2" Overlay | M2 | Rejuvenator |
| R3 | SAMI + PMSC | M3 | SEAL COAT |
| R4 | SAMI + 1" OVERLAY | M4 | SPARE |
| R5 | SAMI + 2" OVERLAY | M5 | SPARE |
| R6 | RECYCLE 1" + 1" O/L | M6 | SPARE |
| R7 | RECYCLE 1.5" + 1" O/L | M7 | SPARE |
| R8 | MILLOCURB + 1" O/L | M8 | SPARE |

| Code | Rehab. Strategy | Code | Maint. Strategy |
|---|---|---|---|
| R9 | MILL 2" + SAMI + PMSC | M9 | SPARE |
| RA | MILL + 2" SAMI + 1" O/L | MA | SPARE |
| RB | MILL 2" + 1" O/L | MB | SPARE |
| RC | MILL 2" + 2" O/L | MC | SPARE |
| RD | MTR PLANE + PMSC | MD | SPARE |
| RE | MTR PLANE + SAMI + PMSC | ME | SPARE |
| RF | MTR PLANE + SAMI + 1" O/L | MF | SPARE |
| RG | RECON 3" SURF + 6" BASE | MG | SPARE |
| RH | RECON 5" SURF + 6" BASE | MH | SPARE |
| RI | MILLOCURB + 2" O/L | MI | SPARE |
| RJ | MILLOCURB + SAMIPMSC | MJ | SPARE |
| RK | MILLOCURB + SAMI + 1" O/L | MK | SPARE |
| | ROUTINE MAINTENANCE | | |

NOTE:
THE ANNUAL COSTS (IN 1984 $'S) FOR EACH STRATEGY CODE REPRESENT TO TOTAL SECTIONAL COST DIVIDED BY THE TOTAL SECTIONAL AREA. THE STRATEGY COSES ARE EXPLAINED IN THE TABLE AT LEFT

EXAMPLE: ALTERNATIVE 01 YEAR 1986 IS: 3.00R6 WHICH MEANS THAT RECYCLE 1" + 1" O/L WILL COST $3.00 PER SQ. YD.

# References to Part Three

[Allez 88]      Allez, F., M. Dauzats, P. Joubert, G. P. Labat, and M. Puzelli, "ER-ASME: An Expert System for Pavement Maintenance," Transportation Research Record 1205, 1988.

[Carey 60]      Carey, W. N., and P. E. Irick, "The Pavement Serviceability-Performance Concept," Highway Research Board Bulletin 250, 1960.

[Chong 89]      Chong, G. J., F. W. Sewer, and K. Macey, "Pavement Maintenance Guidelines: Distresses, Maintenance Alternatives and Performance Standards," Ministry of Transportation of Ontario, Report SP-001, Aug. 1989.

[Darter 80]     Darter, M. I., "Requirements for Reliable Predictive Pavement Models," Transportation Research Board Research Record 766, 1980.

[Fernando 83]   Fernando, E. G., and W. R. Hudson, "Development of a Prioritization Procedure for the Network Level Pavement Management System," Center for Transp. Res., The Univ. of Texas at Austin, 1983.

[Finn 74]       Finn, F., R. Kulkarni, and K. Nair, "Pavement Management Feasibility Study," Final Report Prepared for Washington State Hwy. Commission, Olympia, Wash., Aug. 1974.

[Gunderson 90]  Gunderson, B., and R. Kher, "Single Year Prioritization," Module C of FHWA Course Notes on Pavement Management Systems, Washington, DC, Aug. 1990.

# References to Part Three

[Haas 85a] Haas, R., M. A. Karan, A. Cheetham, and S. Khalil, "Pavement Rehabilitation Programming: A Range of Options," Proc., First North American Pavement Management Conf., Toronto, March 1985.

[Haas 85b] Haas, R., "The Long Term Effects of Spending Decisions on Paved Roads," Proc., European Asphalt Paving Assoc. Congress, Berlin, May 1985.

[Haas 87] Haas, C., and H. Shen, "Preserver: A Knowledge Based Pavement Maintenance Consulting Programme," Proc., Vol. 2, Second North American Conf. on Managing Pavements, Toronto, Nov. 1987.

[Haas 89a] Haas, R., W. R. Hudson, and C. A. V. Queiroz, "A Strategic Approach to Pavement Rehabilitation Data Needs," Proc., 2nd Int. Symp. on Pavement Evaluation and Overlay Design, Rio de Janiero, Sept. 1989.

[Haas 89b] Haas, R., M. A. Karan, and F. Meyer, "Municipal Pavement Management Tailored to Varying Needs and Resources," Proc. Roads and Trans. Assoc. of Canada, Calgary, Sept. 1989.

[Haas 90a] Haas, R., "A Study to Establish Future Road Construction and Rehabilitation Practices in Prince Edward Island," Report prepared for the Province of Prince Edward Island by Pavement Management Systems Ltd., May 1990.

[Haas 90b] Haas, R., T. Triffo, and M. A. Karen, "The Use of Expert Systems in Network Level Pavement Management," Proc., OECD Workshop on Knowledge-Based Expert Systems in Transportation, Espoo, Finland, June 1990.

[Hajek 87] Hajek, J. J., G. J. Chong, R. C. G. Haas, and W. A. Phang, "ROSE: A Knowledge Based Expert System for Routing and Sealing," Proc., Vol. 2, Second North American Conf. on Managing Pavements, Toronto, Nov. 1987.

[Hall 87] Hall, K. T., J. M. Conner, M. I. Darter, and S. H. Carpenter, "Development of an Expert System for Concrete Pavement Evaluation and Rehabilitation," Proc., Second North American Conf. on Managing Pavements, Toronto, Nov. 1987.

[Hanna 92] Hanna, P. B., T. Papagiannakis, and A. S. Hanna, "Knowledge Based Advisory System for Flexible Pavement Maintenance," Paper presented to Transp. Assoc. of Canada Annual Conf., Quebec, Sept. 1992.

[Hendrickson 87] Hendrickson, C. T., and B. N. Janson, "Expert Systems and Pavement Management," Proc, Second North American Conf. on Managing Pavements, Toronto, Nov. 1987.

[Hill 90] Hill, L., and R. Haas, "Multi Year Prioritization," Module E of FHWA Course Notes on Pavement Management Systems, Washington, DC, Aug. 1990.

[Hill 91] Hill, L., A. Cheetham, and R. Haas, "Development and Implementation of Pavement Management System for Minnesota," Paper presented to Annual Meeting, Transp. Res. Board, Washington, DC, Jan. 1991.

| | |
|---|---|
| [Jackson 90] | Jackson, N., and J. Mahoney, "Washington State Pavement Management System," Federal Hwy. Admin. Text for Advanced Course on Pavement Management, Nov. 1990. |
| [Karan 74] | Karan, M. A., and R. Haas, "User Delay Cost Model for Highway Rehabilitation," Final Rept. on Project W-30, Ministry of Transportation and Communications of Ontario, 1974. |
| [Karan 76] | Karan, M. A., and R. Haas, "Determining Investment Priorities for Urban Pavement Improvements," J. of Assoc. of Asphalt Paving Technology, Vol. 45, 1976. |
| [Karan 77] | Karan, M. A., C. Bauman, and R. Haas, "An Inventory and Priority Programming System for Municipal Pavement Improvements," Proc. Roads and Transp. Assoc. of Canada, 1977. |
| [Karan 81] | Karan, M. A., R. C. G. Haas, and T. Walker, "Case Illustration of Pavement Management: From Data Inventory to Priority Analysis," Transportation Research Board Research Record 814, 1981. |
| [Karan 83] | Karan, M. A., T. S. Christison, A. Cheetham, and G. Berdahl, "Development and Implementation of Alberta's Pavement Information and Needs System," Transportation Research Board Research Record 938, 1983. |
| [Kikukawa 84] | Kikukawa, S., and R. Haas, "Priority Programming for Network Level Pavement Management," Proc., Workshop on Paving in Cold Areas, Tsukuba, Japan, Oct. 1984. |
| [Lytton 85] | Lytton, R. L., "From Ranking to True Optimization," Moderators Report, Proc., First North American Pavement Management Conf., Toronto, March 1985. |
| [Lytton 87] | Lytton, R. L., "Concepts of Pavement Performance Prediction and Modeling," Proc., Second North American Conf. on Managing Pavements, Vol. 2, Toronto, Nov. 1987. |
| [McFarland 72] | McFarland, W. F., "Benefit Analysis for Pavement Design Systems," Res. Rept. 123-13, Texas Transp. Institute of Texas A&M University, 1972. |
| [McLeod 89] | McLeod, D. R., "A BST Management Study for Yukon Highways," Report prepared for the Government of the Yukon by Public Works Canada, Oct. 1989. |
| [Mahoney 90] | Mahoney, J., "Introduction to Prediction Models and Performance Curves," Course Text, FHWA Advanced Course on Pavement Management, Nov. 1990. |
| [Marcotte 87] | Marcotte, B., A. Cheetham, M. A. Karan, and R. Haas, "Alberta's Municipal Pavement Management System," Proc., Second North American Conf. on Managing Pavement, Toronto, Nov. 1987. |
| [MTO 90] | Ministry of Transportation of Ontario, "Pavement Design and Rehabilitation Manual," SDO-90-01, 1990. |
| [Queiroz 83] | Queiroz, C., "A Mechanistic Analysis of Asphalt Pavement Performance in Brasil," J. of Assoc. of Asphalt Paving Technology, Vol. 52, 1983. |

[Ritchie 87]   Ritchie, S., "Applications of Expert Systems for Managing Pavements," Proc, Second North American Conf. on Managing Pavements, Toronto, Nov. 1987.

[Smeaton 85]   Smeaton, W. K., M. A. Karan, and R. Haas, "Determining the Most Cost-Effective Combination of Pavement Maintenance and Rehabilitation for Road and Street Networks," Proc., First North American Pavement Management Conf., Toronto, March 1985.

[Smith 74]   Smith, W. S., "A Flexible Pavement Maintenance Management System," Ph.D. Dissertation, University of California, Berkeley, 1974.

[TAI 87]   The Asphalt Institute, "Calculating Pavement Costs," Information Series No. 174, Nov. 1987.

[Treybig 77]   Treybig, H. J., B. F. McCullough, F. N. Finn, R. McComb, and W. R. Hudson, "Design of Asphalt Concrete Overlays Using Layer Theory," Proc., Int. Conf. on Structural Design of Asphalt Pavements, Vol. 1, Univ. of Michigan, 1977.

[Turay 91]   Turay, S., and R. Haas, "A Road Network Investment System (RONIS) for Developing Countries," Transportation Research Board Research Record 1291, Vol. 1, 1991.

[Wang 92]   Wang, J. T., and J. P. Zaniewski, "Revisions for the Arizona Department of Transportation Pavement Management System," Paper submitted to the 1993 Annual Meeting of the Transportation Research Board, August 1992.

[Zaniewski 82]   Zaniewski, J. P., et al., "Vehicle Operating Cost, Fuel Consumption and Pavement Type and Condition Factors," Report Prepared for U.S. Federal Highway Admin. by Texas Res. and Development Foundation, June 1982.

# Part Four

# Project Level Design: Structural and Economic Analysis

Chapter 21

# A Framework for Pavement Design

## 21.1 INTRODUCTION

Many procedures are available for designing pavements. Many agencies have their own particular design techniques, ranging from very simple in concept to highly sophisticated. Whatever the particular method is, and to whatever degree it differs from others, there are some common features to all methods. In other words, there is an identifiable framework that characterizes most pavement design methods.

This chapter describes the framework and its major components in general terms. The context of the design framework subsystem within the overall pavement management system has previously been identified in Chapter 4. This chapter presents a suggested classification of the structural analysis methods and describes the application of these models for flexible pavements. In addition, design objectives and constraints are described which define the role of pavement design within the general framework of pavement management.

## 21.2 EVOLUTION OF PAVEMENT DESIGN TECHNOLOGY

The framework for pavement design presented in this chapter is general and quite comprehensive. Many existing pavement design methods do not include all the com-

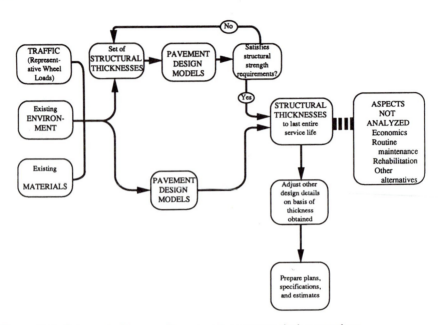

**Figure 21.1** Schematic diagram of most past pavement design practices.

ponents defined in the general framework. Our method, however, has been developed largely as a logical extension of past practices. Therefore, it is useful to review briefly the evolution of such past and existing practices, with particular relevance to flexible pavements.

Past design practices have been concerned primarily with layer thickness selection or structural analysis. This can be represented schematically by Figure 21.1. More specifically, past practice can be divided into a relatively small number of identifiable classes as shown on Figure 21.2 for flexible highway pavements [Hutchinson 68]. To indicate the evolutionary nature of the development of design approaches, an approximate time of initial development is shown for each class of pavement design methods. These classifications are highly simplified and should not be taken to represent a sharp distinction between the approaches. In fact, any particular flexible pavement design method may incorporate features of several classes of the approaches shown.

The general approach used by most of the agencies up to World War II, and still used by many today, consists essentially of designating standard sections for the various ranges of traffic and environmental conditions. This is especially true for rigid pavements. The inherent implication in such methods is that performance will be "satisfactory" with standard sections and "unsatisfactory" for anything less and "uneconomical" for anything greater.

The approaches for flexible pavement design using simple strength tests, soil formulas, triaxial tests, and inplace strength tests were initiated in the 1930s and early

| Classification (Approximate time of initiation) | Features | Limitations |
|---|---|---|
| 1. Methods Based on Experience (i.e. Standard Sections) Examples: Most urban areas in North America | Attempt to provide a pavement for some standard service life, without excessive distress; simple in concept and low design costs; can be relatively reliable for particular jurisdictions. | Poor basis for economic comparison of alternatives; no recognition of varying serviceability with age; materials properties only subjectively tied to layer thicknesses; section selected is only for "average" expected traffic and environment. |
| 2. Methods Based on "Soil Formula" Examples: Group Index Methods | Attempt to provide sufficient pavement thickness to prevent excessive distress before end of service life; simple soil classification tests for index-based assignment of subgrade strength; empirical correlations with pavement thickness. | Traffic and environmental effects only subjectively reflected in design curves; no recognition of varying serviceability with age; no recognition of variations of pavement layer material properties. |
| 3. Methods Based on Simple Strength Tests Examples: CBR Methods (i.e., US Corps of Engineers) | Attempt to provide pavement thickness and materials that will not undergo excessive distress before end of service life; simple procedures for measuring material properties; empirical correlations between properties and pavement thicknesses. | No recognition of varying serviceability with age; repeated traffic load effects only subjectively included; recognition only of "weakest" condition of materials due to moisture effects. |
| 4. Methods Based on Field or Laboratory Strength Tests Examples: Canadian DOT Plate Bearing Method, Triaxial Test based methods | Attempt to provide sufficient layer thicknesses to limit deflection or ensure stability; full-scale testing of subgrade and pavement reaction to load in field tests; moduli of layers obtained for use in theoretically based structural analysis. | Testing is time consuming and expensive; varying effects of environment difficult to measure; only limited number of static load repetitions in place bearing and only single dynamic load in field vibratory tests. |
| 5. Methods Based on Structural Analysis of Layered Systems Examples: Shell n-layer Method | Attempt to provide pavement thickness and materials such that certain critical stresses, strains, or deflections are not exceeded; direct means of considering distress mechanisms. | Require extensive testing for material properties; difficult to account for environmental effects; subjective links between calculated structural response and performance, methodology for permanent deformation distresses not yet well developed. |
| 6. Methods Based on Statistical Evaluation of Pavement Performance Examples: AASHTO Guide for the Design of Pavement Structures, Canadian Good Roads Association Design Guide | Predictions made of serviceability age history of alternative designs, using models developed from evaluation of performance of in-service pavement; realistic basis for economic comparison of alternatives. | Models require additional long term and regional verification; relationships between distress and performance not yet adequately established; considerably more work needed on quantifying probabilistic considerations in methods |

Approximate time of initiation: 1900, 1910, 1920, 1930, 1940, 1950, 1960, 1970

**Figure 21.2 A classification of flexible pavement design approaches.**

1940s. They represented advancements from the earlier approach because they used material properties in selecting layer thicknesses. The design curves also allowed for flexibility in considering thickness alternatives for different materials. These approaches are simple and relatively reliable in regions where they have been widely used with substantial experience. However, performance analysis with such methods is implicitly either "satisfactory" or "unsatisfactory."

Theoretical techniques for analyzing response to load were initiated in the late 1930s and early 1940s but received extensive attention only with the advent of large computers in the 1960s. With the development of powerful and low cost microcomputers, these methods are moving to the forefront of pavement design. Methods based on structural analysis have an inherent appeal to many engineers because of the fundamental nature of the procedures. However, the methodology for relating computed structural response to pavement performance needs further development.

In the late 1950s and early 1960s, design methods were developed based on extensive data from the AASHO Road Test [AASHO 61] and Canadian studies [CGRA 65]. The methods developed in these studies explicitly considered the serviceability-age history of a pavement structure. Many people feel this is the best basis for structural analysis because the performance relationship must be known to conduct an economic analysis. To improve the art of pavement design and analysis, the Strategic Highway Research Program (SHRP) was initiated in 1987. In this program, pavement performance data throughout the United States and several other countries is being collected. The resulting data base should provide a valuable tool to relate structural analysis and pavement performance.

## 21.3 PAVEMENT DESIGN FRAMEWORK AND COMPONENTS

In a pavement management system, the pavement design phase requires several activities. These may be broadly classified as:

1. Information needs relating to inputs, objectives, constraints, etc.
2. The generation of alternative design strategies
3. The structural analysis, economic evaluation, and optimization of these strategies

Figure 21.3 is a diagram of the major components of a project level pavement management system. The general nature of this diagram applies to both flexible and rigid pavements. Another version of the design phase [Hudson 73] is shown on Figure 21.4. Although the organization of this diagram is somewhat different from Figure 21.3, both contain the same basic activities and represent the same basic philosophy. Figure 21.4 is an updated version of Figure 3.1. The reader should compare these two figures to note the subtle refinements that were made to the pavement management framework as it evolved.

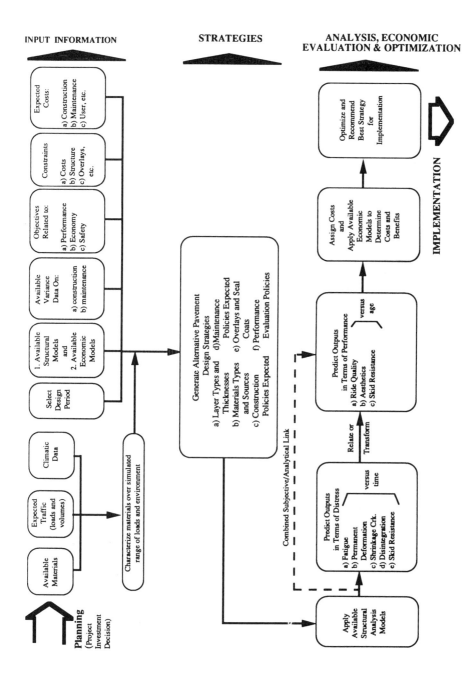

**Figure 21.3** Major pavement design components.

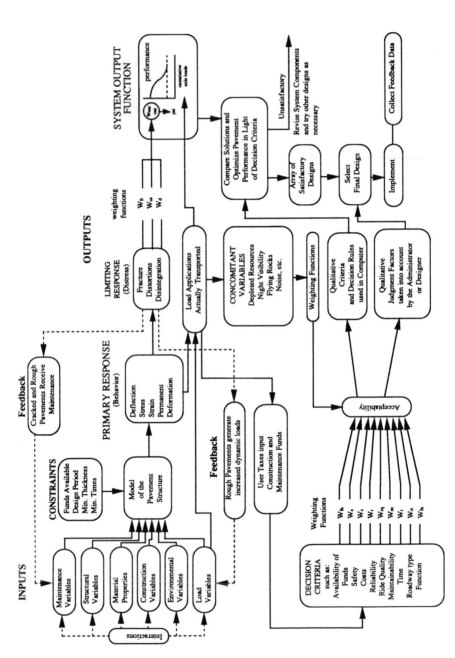

Figure 21.4 Block diagram of pavement design system [Hudson 73].

# A Framework for Pavement Design

## 21.3.1 Information Needs

The top row of Figure 21.3 represents the information and analysis methods that should be available or acquired before generating alternative design strategies. Data on available materials, expected traffic, and climatic factors are often the first information items acquired. Design methods that include materials characterization use this data for establishing a range of load and environment for testing purposes. It might also use the data in both generating and analyzing design strategies.

Depending on the design method, the structural model(s) can range from simple to complex. For example, methods based on empirically based limiting strength values are simple in concept and use; alternatively, a linear-viscoelastic analysis is relatively complex both in terms of material characterization and mathematical analysis. This also demonstrates an interaction between the material characterization and the structural analysis processes. Prior to testing for material characteristics, one must know the type of analysis model that will be used in the design process.

The selection of a design period (analysis period, planning horizon, etc.) is only implicitly included in some methods. These involve the situation where design charts are used to provide a particular pavement structure to satisfy the particular input requirements. The charts would then inherently be based on a certain service life (e.g., 20 years) being achieved by the pavement structure. Other design methods might explicitly select a design period, such as 25 years, over which the alternatives are compared. Without a consistent analysis period, detailed economic analysis comparison of alternatives cannot be properly performed.

The economic analysis models also range in complexity from one design method to another. A straightforward estimate of the initial capital costs of construction may be used in one method, whereas another method might use a net present value model that incorporates present and future costs and benefits over the design period.

Available construction and maintenance variance data is listed along the top row of Figure 21.3, but most design methods use such information only in a subjective manner. The 1986 AASHTO Guide for the Design of Pavement Structures incorporated a measure of the observed variance in pavement performance into the design equation. This represents a significant improvement in the state of the art in pavement design. However, it also presents a significant need to quantify the true construction and maintenance variability.

The objectives established for design should be primarily related to performance, safety, and economy. Performance measures, in terms of pavement outputs, were described in Chapter 6. The objectives would define certain minimums and maximums for these outputs. For example, minimum levels of serviceability and skid resistance, and a maximum level of structural capacity, may be defined. Many existing design methods, however, incorporate objectives only on an implicit basis.

Constraints on a design method, per se, or on the designs produced by that method, are usually stated more explicitly. For example, there should be a limit on expenditures, a minimum time to the first overlay, a minimum thickness of pavement, etc.

Expected costs are a vital part of the information needed for design. Among the major cost categories, both present and future, are materials, construction, maintenance, and user costs.

### 21.3.2 Generating Alternative Pavement Design Strategies

The generation of alternative pavement design strategies is shown along the middle row of Figure 21.3. The term *strategies* emphasizes that a design alternative should consist not only of pavement layer types and thicknesses, including future rehabilitation action such as overlays and seal coats, but also material types and sources. Additionally, it is desirable to include the expected construction, maintenance, and performance evaluation policies.

The need for a design alternative to specify material types and layer thicknesses is apparent. However, without including the construction and maintenance policies, there may be appreciable error in the predicted performance outputs for that pavement structure.

Overlays, seal coats, or other rehabilitation alternatives also become part of any design strategy it the serviceability of the pavement drops to the minimum acceptable or terminal level before the end of the design period. An exception would be a maintenance policy that kept the pavement at or about the minimum acceptable level of serviceability to the end of the design period. Alternatively, if financial constraints prevailed, maintenance at this level might continue only until funds were available for rehabilitation.

In the formulation of rehabilitation alternatives, there are two major, interrelated aspects that should be considered:

1. Structural aspects, with respect to dealing with excessive distress, lack of adequate serviceability, lack of adequate safety, etc.
2. Policy aspects, with respect to traffic handling procedures and time of day and season of the rehabilitation

The method of handling traffic can be a very important consideration in certain situations, primarily in terms of the user delay costs.

The specification of a policy for performance evaluation of a pavement throughout the design period, as part of a design strategy, might be considered unusual in conventional pavement design methodology. An example can illustrate the concept involved. Suppose a particular strategy included an overlay of a certain thickness and materials at 10 years. Some error is, of course, associated with any such prediction. Consequently, it is desirable to have some means for periodically monitoring the performance and structural capacity of the pavement as it is in service so that the design strategy may be updated.

### 21.3.3 Analysis, Economic Evaluation, and Optimization

The bottom row on Figure 21.3 lists the main component activities that would ideally be involved in the analysis, economic evaluation, and optimization of the various alternative design strategies. Most design methods do not include all of these activities;

however, design methodology in general appears to be moving toward this idealized form.

The first major step in the analysis of any pavement design alternative is the application of structural models. If they are sufficiently comprehensive, they are used first to predict the outputs of that alternative in physical terms, for example, the distress that is expected to occur over the design period. The major distress modes are fatigue cracking, permanent deformation (primarily rutting of flexible pavements and faulting at joints of rigid pavements, and distortions associated with foundation and environmental influences), shrinkage cracking, disintegration, and loss of skid resistance. Fatigue and usually permanent deformation are associated with repeated loads, whereas shrinkage cracking is associated with temperature change. Disintegration could be associated with both traffic and environment, and the loss of skid resistance is usually associated mainly with traffic effects.

Current technology cannot predict adequately the type and degree of all these distress modes as a function of time and traffic. Consequently, several structural models used today attempt to make a direct prediction of serviceability (or ride quality) versus age. This approach is represented by the dashed line of Figure 21.3, noted as a "combined subjective/analytical link." These terms are used to indicate that some methods might make only a subjective estimate of serviceability versus age, or perhaps only of service life, based on experience. Alternatively, they might involve some combination of such subjective estimates with calculated estimates of service life. The other two outputs, aesthetics versus age and skid resistance versus age, are difficult to predict with current technology. In fact, no acceptable measure of aesthetics or appearance has been developed, probably because most agencies feel that this is a relatively minor component of performance.

Research on pavement performance has failed to produce a definitive relationship of distress outputs and pavement performance. The need for these relationships was defined as a first priority research need in 1970 [HRB 71] and remains an elusive relationship.

The economic evaluation of a pavement design alternative, as shown on Figure 21.3, should involve the assignment of costs and benefits to the predicted outputs, in terms of the user, plus the costs of providing that alternative to the agency in terms of materials, construction, maintenance, etc. These are then incorporated into an economic model to determine the total costs and benefits of the strategy. It is desirable to include benefits in the economic evaluation, but this is a relatively undeveloped aspect of pavement design. One reason is that it is difficult to separate pavement benefits from overall highway benefits. Another is that the differences in user costs between pavement design alternatives, which can be considered as benefits, are difficult to determine. Research by the World Bank and subsequently by the Federal Highway Administration has successfully related vehicle operating costs to pavement condition [Zaniewski 82, Chesher 85]. Thus, the information is available for including user cost analysis in the economic analysis of a pavement strategy.

When all the alternative design strategies have been analyzed and evaluated, optimization should be used to select the best strategy. Frequently, this is accomplished by selecting the alternative with the least discounted total cost. The design is then completed by recommending this strategy for implementation.

## 21.4 DESIGN OBJECTIVES AND CONSTRAINTS

Pavement management helps the engineer focus on the basic function and constraints of the design process. This avoids the rote application of the design procedure while ignoring other available alternatives. As indicated in Chapter 2, one of the first activities that should be accomplished under the systematic approach is to carefully define the objectives and constraints of the problem. Objectives and constraints apply to the pavement and the design process or the designer's activities. Some apply to both.

### 21.4.1 Pavement Objectives

The principal objectives to be fulfilled by a pavement, both during construction and in service, are economic and social in nature. They include the following:

1. Maximum or reasonable economy, in terms of both agency and user costs
2. Maximum or adequate safety
3. Maximum or reasonable pavement serviceability over the design period
4. Maximum or adequate load-carrying capacity, for both the magnitude and number of repetitions of the loads
5. Minimum or limited physical deterioration due to environmental and traffic influences
6. Minimum or limited noise and air pollution during construction
7. Minimum or limited disruption of adjoining land use
8. Maximum or good aesthetics

These objectives pose several major contradictions, which occur with any complex system that attempts to fulfill social and economic needs. Consequently, each pavement constructed represents some compromise or trade-off between competing objectives. The relative influence of the objectives in a particular design situation varies with such factors as rural or urban area and traffic volumes.

The first three objectives listed are very important. Economy is a prime requisite, because most public agencies feel that they have insufficient funds available for all the desired pavement investments. Safety, as achieved through sufficient skid resistance, proper lane demarcation, light reflectivity during night driving, etc., is another major requirement the pavement must fulfill. Serviceability should also be sufficiently high over the design period to ensure the desired level of quality to the traveling public in terms of speeds, vehicle operating costs, user delays, and comfort.

The fourth objective, maximizing load-carrying capacity, is applicable to both the pavement and to the design process. This objective competes directly with the objective of maximizing economy.

The fifth objective, minimizing physical deterioration, is related to serviceability in that deterioration is a major factor in causing serviceability loss. However, it is

# A Framework for Pavement Design

identified as a separate objective because it also competes with the economic objective in terms of maintenance and rehabilitation costs.

The last three objectives are usually of lesser importance. Aesthetics or appearance can be a significant objective for certain special applications (e.g., pedestrian walkways) or in those cases where patching and crack filling are extensive and appear ugly to motorists or adjacent property owners.

## 21.4.2 Pavement Design Objectives

In one sense, the objective of the pavement design process is to design a pavement structure that, when constructed, will accomplish the functional objectives defined above. However, the specific objectives of the design process can be stated in specific technical, economic, and social terms:

1. Develop a design strategy of maximum (or reasonable) economy, safety, and serviceability.
2. Consider all possible design strategy alternatives.
3. Recognize the variability in the design factors.
4. Maximize the accuracy of predictions of serviceability, safely, and physical deterioration for the alternatives considered.
5. Maximize the accuracy of estimating cost and benefits.
6. Minimize the costs of design, including labor, testing, and computer time.
7. Maximize information transfer and exchange between construction and maintenance personnel.
8. Maximize use of local materials and labor in the design strategies considered.

Some of the foregoing objectives correspond with the objectives for the pavement itself, whereas others are peculiar to the design process. Most of the objectives compete with other objectives and require trade-offs.

The first objective simply reflects the need for the design to recognize the pavement objectives of economy, serviceability, and safety. In order to accomplish this, it is necessary, as stated in the second objective, to consider many possible design strategy alternatives. Because these can amount to several hundred combinations for any given situation, the use of computer analysis is necessary for accomplishing this objective.

The third objective—recognition of the stochastic nature of pavement design, construction, and maintenance factors—is accepted as an important criteria for a pavement design method. Early contributions to the inclusion of variability in the pavement design process were presented by researchers at the University of Texas in the early 1970s [Darter 71, 72, 73a, 73b]. Variability concepts have been incorporated into the 1986 version of the AASHTO Guide for the Design of Pavement Structures.

The fourth objective of maximizing the accuracy of the design predictions is related to the third objective and also to the quality of the input data. Decreased uncertainty in design predictions is important to planning activities in that investment programming can be accomplished more reliably.

The accuracy of the cost and benefit prediction, the fifth objective, is directly related to the accuracy of the design predictions as well as the ability to estimate the construction, maintenance, and user costs.

Minimizing the costs of the design process in terms of materials testing, personnel time, preparation of drawings, etc., the sixth objective, is important but obviously competes with the other objectives. If the cost of the design process is underfunded, the result can be an incomplete, inadequate, or excessively costly design strategy.

The seventh objective of information exchange between those responsible for construction and maintenance has been neglected to a large extent in design practice. There have been instances in North America of maintenance crews unfamiliar with continuously reinforced concrete pavements trying to fill with asphalt the numerous hairline cracks that are a design feature of this pavement type. The resulting mess might be more readily ascribed to the designers' failure to communicate with the maintenance people than to the maintenance people themselves. It is extremely important that such communication occurs, because it is construction and maintenance policies that "carry" the design through its design period.

The last objective, involving the use of local materials and labor, is perhaps a self-evident desirability for most situations. This objective may complement the first objective of maximum economy, but in some cases local materials may require stabilization or treatment.

### 21.4.3 Design Constraints

The major design constraints are economic and physical or technical in nature and include the following:

1. Availability of time and funds for conducting the design and the construction
2. Minimum level of serviceability allowed for the pavement before rehabilitation
3. Availability of materials
4. Minimum and/or maximum layer thicknesses
5. Minimum time between overlays or seal coats
6. Capabilities of construction and maintenance personnel and equipment
7. Testing capabilities
8. Capabilities of the structural and economic models available
9. Quality and extent of the design information available

The first constraint might involve not only a limit on the funds that can be applied to a particular project, but also the opportunity costs of those funds. All the other constraints, which are essentially physical or technological, are relatively self-explanatory. Each has economic implications and is therefore related to the first constraint.

## REVIEW QUESTIONS

1. Pavement design technology has evolved from the early 1900s. Original methodologies were based on a general understanding of soil properties. Outline the first major breakthrough in soil testing that helped to codify flexible pavement design. Who developed the method? When was it developed?

# A Framework for Pavement Design

2. Outline at least four major objectives of a good pavement.
3. Pavement design objectives differ from pavement objectives. Name and outline at least four pavement design objectives.
4. All designs are constrained by various factors. There are at least nine constraints associated with pavement design. Name and outline the details of at least four of these design constraints.

Chapter 22

# Characterization of Physical Design Inputs

## 22.1 INTRODUCTION

The physical inputs to the pavement design process are material properties, traffic characteristics, and environmental factors. The purpose of this chapter is to describe in a general manner the nature of these inputs with respect to the pavement design process. The focus of this chapter with respect to material characterization is on the fundamental evaluation of the material properties rather than on the traditional empirical testing methods.

## 22.2 MATERIAL PROPERTIES

The primary materials used for pavement construction are the asphalt concrete or portland cement concrete surface, granular bases and subbases, asphalt, cement, and lime stabilized bases, and steel used for reinforcement and dowel bars in concrete pavements. In addition, due to the importance of the subgrade in the performance of pavements, any discussion of pavement materials must address subgrade soil properties.

### 22.2.1 Fundamental Material Characterization

Many pavement design methods use empirical tests of material quality as opposed to methods that quantify the engineering properties of the materials. Examples of empirical tests include the California Bearing Ratio (CBR) and the Marshall stability test. Empirical material characterization has many drawbacks, the primary one being the inability to extrapolate historical knowledge to changing conditions. Since pavement design is a dynamic process with new materials being introduced and traffic loads changing, there is increasing interest in the use of mechanistic analysis procedures. Mechanistic evaluation requires testing procedures that determine the load-response properties of the materials, usually under test conditions that simulate field conditions.

Mechanistic models of a structural response are derived based on assumptions about the response of the material to loads. The terminology used to described the load-response characteristics of materials is defined in many textbooks on strength of materials and will not be reviewed here. Materials can display a wide range of response to load ranging from elastic to plastic and viscous; the properties can be dependent on the rate of loading and temperature during testing. The simplest type of behavior is where the deformation of the material is proportional to the stress level and the material returns to its original shape when the load is removed. This proportional response would be independent of other factors that can affect material response such as the rate of loading, the temperature of the test, and the age of the material. While this type of behavior is commonly assumed for model simplification, it rarely occurs in nature. Thus, the mechanistic evaluation of materials always requires compromises between the complexity of the testing and analytical procedures and the ability of engineers to perform the tests and the calculations in a cost-effective manner; hence, the need to perform tests under simulated field conditions. However, in-service conditions can never be fully duplicated in the lab [Sentler 71]:

> The strength characteristic of materials is to a large extent based on the results obtained in standardized tests. Such information is valuable because it is often the only information available. But very few structural members, if any, fail in a way which resembles a standardized test. Instead other types of failures like fatigue play a much more important role in practice. It is also obvious that the environmental influence has to be considered in a more appropriate manner.

Deacon, in the conclusion to an indepth literature survey of pavement material properties, states [Deacon 71]:

> This discussion indicates that one may characterize the behavior of pavement materials in numerous ways depending in part on the nature of the problem and in part on personal preferences. It must be emphasized, however, that in most cases pavement materials do not possess idealized properties and that the measured properties are often significantly influenced by the test procedures and equipment. It is important, therefore, for laboratory procedures to simulate to as great a degree as possible actual field loading conditions.

Test procedures that result in nearly homogeneous stress and strain states are necessary to investigate the properties of a small volume element.

As the characteristics of materials are reviewed in the following sections, it is important to realize that these discussions are based on the state-of-the-art methods of analysis for pavement design and do not necessarily represent the "true" response of the material that would be observed by a materials specialist. It is also important to recognize that the state of the art in pavement design is limited by the development of modeling assumptions rather than representing the ultimate development in pavement analysis techniques. In other words, application of current mechanistic testing and analysis techniques represents an improvement over empirical methods but there remains a need to further refine our analytical capability.

### 22.2.2 Steel

Steel is used for pavement reinforcement and load transfer devices. Compared to other materials used in pavements, the properties of steel are relatively easy to quantify. For the conditions encountered in pavement performance, steel is a linear elastic material. Structural design of the steel in pavements is relatively straightforward as Hooke's law applies and the modulus of elasticity and Poisson's ratio are well defined.

Generally, the corrosion is the limiting factor in the performance of the steel used for reinforcing and load transfer devices in pavements. Epoxy coatings can be used to limit the corrosion of reinforcing steel. Load transfer devices can either be epoxy coated or made with stainless steel.

### 22.2.3 Portland Cement Concrete

Portland cement concrete is a combination of cement, aggregates, and water. The chemical reaction between the cement and water forms the bonds in the concrete that determine the properties of the hardened material.

The failure mode of concrete depends on the stress state of the material. Uniaxial compressive tests are generally used in concrete design and quality control tests to determine the ultimate compressive strength. However, concrete is much weaker in tension than in compression. Since pavements carry traffic loads in a flexural mode, the tensile-flexural test is usually recommended for the design and quality control of concrete used in pavements.

In addition to mechanical failures, many concrete pavements fail due to deterioration of the concrete as a result of environmental and chemical attack. Freezing water in the pores and cavities of the concrete expands about 9 percent, producing hydrostatic pressure [Derucher 88]. Generally an air-entraining admixture is added to the concrete to provide protection against freeze-thaw deterioration. Deicing salts and chemicals can increase the water retention of the concrete and contribute to recrystallization and weathering. These effects will cause the properties of the concrete to vary with time.

The stress-strain behavior of concrete is nonlinear. Generally the modulus of elasticity that is used for design is determined as a secant modulus with the end points of the cord being the origin and the point on the curve where the stress equals one half of the compressive strength.

### 22.2.4 Asphalt Concrete

Asphalt concrete mix design consists of selecting the aggregate gradation and asphalt content required to meet design criteria, such as the Marshall or Hveem stability and flow, and void content. In the past little attention has been paid to the relationship between the asphalt concrete quality and the thickness of the pavement. Based on the Marshall or Hveem mix design method, the asphalt concrete was either acceptable or not. A new asphalt concrete mix design procedure was proposed based on evaluation of asphalt mixtures using engineering measures of the properties [Monismith 85]. One of the unique features of this method is simultaneous consideration of the mix design and pavement design. In other words, greater consideration of the application of the material needs to be included in the mix design process. The mixture properties that should be considered during mix design include:

1. Mixture stiffness
2. Resistance to permanent deformation
3. Durability
4. Fatigue resistance
5. Low temperature response (including stiffness at long loading times and fracture characteristics)
6. Permeability

With the exception of durability and permeability, it is recommended that the properties of the mixes are measured in a form which permits mechanistic analysis [Monismith 85]. The purpose of the proposed mechanistic approach is to improve the predictive capability of the design models, particularly with respect to changes in materials and traffic loading conditions.

### 22.2.5 Stabilized Materials

Granular materials can be stabilized with either portland cement or asphalt cement. Generally the amount of cement used for stabilization and the specifications for the aggregate gradation are lower for the production of the stabilized materials than for concrete. Therefore, the quality of the stabilized materials will not be equal to asphalt concrete or portland cement concrete. However, the nature of the load response curves are similar, for example, cement stabilized bases are nonlinear elastic and asphalt-stabilized bases are viscoelastic nonlinear.

### 22.2.6 Soils and Granular Materials

The properties of the granular materials depend on gradation, moisture content, density, stress state, and the aggregate shape and texture. Depending on the amount and type of fine material in the aggregates, they may be classified as either cohesive or cohesionless. From a general mechanistic response viewpoint, soils and aggregate materials can be considered within these two classifications. However, the properties of unstabilized aggregate materials are discussed separately due to the extensive use of these materials in bases and subbases.

The characteristics of aggregates and soils are strongly influenced by the moisture content. During construction, care is taken to compact these materials at the optimum moisture content. However, during the life of the pavement, various mechanisms such as percolation of water through cracks in the pavement surface and capillary action tend to allow water to enter the base and the subgrade. As a result, many pavement foundations, even in arid regions, are saturated for a major portion of the time. In essence, this means the strength of the foundation in field conditions is less than the optimum strength.

Cedergren developed a damage factor for comparing the damage to pavements with saturated bases and subgrades to well-drained pavements. This damage factor ranged from 10 to 70,000 demonstrating the importance of subgrade moisture on the strength of the pavement [Cedergren 88]. Cedergren also demonstrated that pavements can remain saturated for up to 22 days following a rainstorm.

### 22.2.6.1  Cohesive Soils

Most investigators describe the behavior of cohesive soils as highly nonlinear [Deacon 71]. Clays show immediate and time dependent recoverable and permanent strains, the immediate strains being predominant under short duration loads and the permanent strain per cycle decreasing to an insignificant amount after many cycles of stress. Stress history may have a significant effect on response. The nonlinear response to load varies in two ways:

1. The stiffness of these materials is dependent on the initial stress state and increases as the effective mean principal stress increases.
2. The stiffness decreases with an increase in the incremental stress amplitude (deviator stress in triaxial tests).

The effects of load, mixture features, and environment on the stiffness of cohesive soils are summarized in Table 22.1 [Deacon 71]. As this table demonstrates, the load deformation characteristics, or stiffness, of cohesive soils are very complex. In addition, cohesive soils are cross-isotropic with the horizontal stiffness exceeding the vertical stiffness.

### 22.2.6.2  Cohesionless Soils

The stiffness of cohesionless soils, such as sand, is affected by many of the same factors as cohesive soils. However, the response to an increase in the number of cycles differs for cohesive and cohesionless soils. Table 22.2 is a summary of how the various factors affect the stiffness of cohesionless soils [Deacon 71]. Many investigators relate the stiffness, $S$, of cohesionless soils to the mean effective stress, $\sigma_0$, as

$$S = K\sigma_0^n \tag{22.1}$$

where $K$ and $n$ are experimentally determined constants. Cohesionless soils by definition do not have tensile strength. However, they are probably more isotropic than other paving materials [Deacon 71].

# Characterization of Physical Design Inputs

**Table 22.1 Effect of Variables on the Stiffness of Cohesive Soils [Deacon 71]**

| Variable That Is Increased | Effect on Stiffness | Remarks |
|---|---|---|
| **Loading** | | |
| Number of cycles | Decrease | |
| | Minimum | Minimum at 1 to 5000 cycles |
| Incremental strain amplitude | Decrease | Rate of decrease depends on maximum stiffness and shear stress |
| Effective mean initial principal stress | Increase | Effect depends on stress or strain amplitude |
| Transverse stress | | No effect |
| Initial octahedral shear stress | | Effect negligible after 10 cycles |
| Frequency of loading | Increase | Effect minor above 10 cps |
| Strain rate | Increase | |
| Over consolidation ratio | Increase | Any effect can be explained on basis of effective pressure and void ratio |
| Stress path | | Large dependency |
| **Mixture** | | |
| Soil disturbance | Decrease | |
| Void ratio | Decrease | Maximum effect at low confining pressure |
| Dispersion | Decrease | At small strains |
| Structure | | Little effect on max. shear modulus |
| Degree of saturation at compaction | Decrease | Modulus of resilient deformation |
| Plasticity | Decrease | |
| Compaction Energy | Maximum | Impact compaction |
| **Environmental** | | |
| Aging | Increase | |
| Degree of saturation | Decrease | |
| Time (thixotropy) | Increase | Recovery after high amplitude cyclic loading or many load cycles |
| Densification | Increase | |
| Time (during secondary compression) | Increase | Bentonite |

## 22.2.6.3 Untreated Granular Materials

The effects of various factors on the stiffness of granular materials are summarized in Table 22.3 [Deacon 71]. The major effect is the initial confining pressure on the sample. The modulus of resilient deformation, $M_R$, to the initial stress state is

$$M_R = K\sigma_3^n \tag{22.2}$$

and

**Table 22.2 Effect of Variables on the Stiffness of Cohesionless Soils [Deacon 71]**

| Variable That Is Increased | Effect on Stiffness | Remarks |
|---|---|---|
| Loading | | |
|   Number of cycles | Increase | Approaches a maximum |
|   Incremental strain amplitude | Decrease | Rapid decrease |
|   Incremental stress amplitude | Decrease | |
|   Load duration | Decrease | Pulsating loads |
|   Load rate or frequency | Constant | No effect after the first few cycles |
|   Initial effective mean principle stress | Increase | |
|   Initial octahedral shear stress | Decrease | Very small effect after 10 load cycles |
| Mixture | | |
|   Void ratio | Decrease | |
| Environmental | | |
|   Degree of saturation | Constant | Effective stresses must be used |

**Table 22.3 Effect ov Variables on the Stiffness of Untreated Granular Aggregates [Deacon 71]**

| Variable That Is Increased | Effect on Stiffness | Remarks |
|---|---|---|
| Loading | | |
|   Number of cycles | Constant | After 50 to 100 cycles |
|   Initial confining pressure | Increase | Triaxial compression |
|   Initial effective mean principal stress | Increase | |
|   Incremental stress level | Constant to increase | Differences in literature, large effect if shear failure |
|   Load duration | Constant | 0.1 to 0.25 s |
|   Load rate or frequency | Increase | Small increase |
|   Drainage | Constant | |
| Mixture | | |
|   Void ratio | Decrease | At low moisture content |
| | Increase | At high moisture content |
|   Angularity and surface roughness | Increase | |
|   Fines | Decrease | Minor effect |
|   Compaction water content | Decrease | |
| Environmental | | |
|   Degree of saturation | Decrease | |

# Characterization of Physical Design Inputs

$$M_R = K\sigma_0^n. \tag{22.3}$$

These equations are similar to Equation 22.1 for the cohesionless soil, except $\sigma_3$ is the initial confining pressure in a triaxial test.

## 22.3 TRAFFIC CHARACTERISTICS

Traffic loading and variation comprises one of the most difficult classes of variables confronting the pavement designer. Actual values can vary markedly from design estimates and thus result in actual performance significantly different from original predictions. There are several important traffic load variables, including:

1. Total vehicle weight
2. Distribution of the load to the wheel assemblies
3. Geometry of the wheel assemblies and the distribution of the loads to the individual wheels
4. Characteristics of the tire, including inflation pressure
5. Lateral distribution of the load across the pavement structure
6. Duration of the load
7. Dynamic nature of the wheel loads
8. Number of repetitions

The first four parameters are defined by the design characteristics of the vehicles. The other parameters are defined by an interaction between the design features of the vehicles and the type of pavement facility. It should be noted that these are generic features of the vehicle-pavement system and are equally applicable to highway and airport pavements.

The traffic-loading characteristics that need to be considered in design are primary differences between the designs of highway and airport pavements. In general, airfields are designed for greater load magnitudes but fewer load repetitions than highway pavements.

### 22.3.1 Highway Traffic Characteristics

Highway traffic loads, as far as pavement design is concerned, are usually considered in terms of axle loads and the number of repetitions of the axle loads that the pavement must carry. Truck axle types and terminology are shown on Figure 22.1. Conventional single tires are used on the steering axle. Super-single tires are used in Europe as the main load carrying tire and are being introduced to the United States and Canada. The super-single replaces the dual tire and improves the fuel economy of the truck and therefore, its use will probably increase. Dual and dual-tandem axles are currently the most common main load carrying units in North America. Triple axles with dual tires are used in many parts of the world.

The legal limits for axle loads in most states in the United States and provinces in Canada range between 18,000 to 22,000 pounds for single axle-dual tires. The legal limit for super-singles ranges between 60 to 100 percent of the allowable load for dual tires. The legal axle limit for tandem axles ranges from 32,000 to 36,000 pounds. In

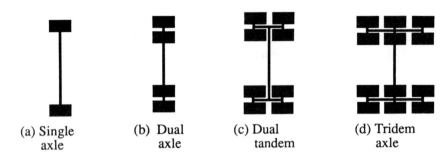

**Figure 22.1** Truck axle configurations.

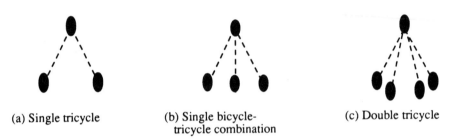

**Figure 22.2** Types of aircraft gear assemblies [TAI 91].

addition, there are legal limits on the spacing between the axles and on the gross vehicle weight. Legal axle load limits in Europe are generally greater than in North America and can be as high as 13 tonnes (28,600 pounds). There are currently no restrictions on tire pressures and there has been a general trend toward higher tire pressures to improve fuel economy. Traditionally truck tire pressure was assumed to be approximately 75 psi; however, recent studies have demonstrated average tire pressures of 110 psi [Papagiannakis 87].

There is considerable variation among the axle loads that actually occur on highways from unloaded trucks to fully loaded and overloaded trucks. Compliance to weight laws is limited by the degree of enforcement. The development of reliable weigh-in-motion scales should improve the understanding of the true nature of the axle load distributions of trucks.

### 22.3.2 Airport Traffic Characteristics

Aircraft characteristics are available in several references [Sargious 75, TAI 91] and are only briefly summarized here. As shown in Figure 22.2, there are three basic types of gear assemblies for civilian aircraft: single tricycle, single bicycle-tricycle combination, and double tricycle. The Boeing 727, McDonnell-Douglas DC-10, and the Boeing 747 are respectively examples of each of these types of assemblies. The types of tire assemblies are shown on Figure 22.3. Nose gear assemblies are predominantly twin tire assemblies. The main truck assemblies of the heavier aircraft are predomi-

**Characterization of Physical Design Inputs**

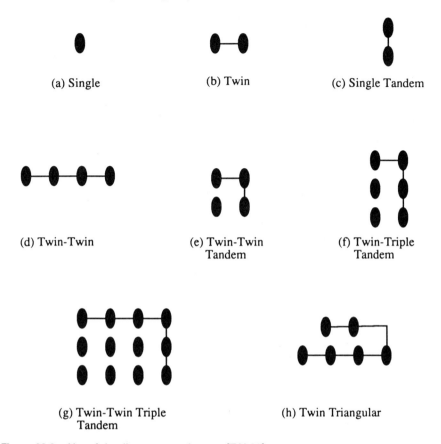

**Figure 22.3** Aircraft landing gear-truck types [TAI 91].

nantly twin-twin tandems, for example, B-747, B-707, DC-8, and some models of the L-1011 (model L). The DC-10 has a twin-twin assembly under the wings and a twin gear in the center. The main truck assemblies of the lighter commercial jets, B-727, B-737 and DC-9, are twin tire assemblies.

The introduction of the wide-body aircraft greatly increased the maximum gross weight of aircraft during the 1970s. The Boeing 747-F is the heaviest civilian aircraft at 778,000 pounds. Boeing is in the process of developing a model of the B-747 estimated to weigh 987,000 pounds.

Due to the load-spreading effect of multiple gears and tires, the weight on the truck assembly is of greater concern than the weight on the individual tires. The DC-10, DC-30, and DC-20 models have the highest maximum weight per tire of any civilian aircraft. These tires have a spacing of 54 inches. By comparison the B-727 also has a high tire weight but the spacing between the tires is 34 inches. This narrower spacing can result in a greater concentration of the stresses and in some cases, the B-727 can actually cause greater damage to the pavement than the heavier aircraft.

## 22.4 ENVIRONMENTAL EFFECTS

As noted in Section 22.2 on material properties, environmental conditions have a major influence on the characteristics of materials and the performance of pavements. The purpose of this section is to summarize the direct influences of the environment on pavements. The environmental properties that influence the pavement are rainfall or moisture, temperature, and solar radiation.

Moisture in the pavement structure has the following effects:

1. Concrete swell
2. Transport of contaminants into cracks and joints
3. Reduction of the strength and stability of base, subbase and subgrade
4. Corrosion of reinforcing steel
5. Stripping of asphalt concrete

Both low and high temperatures can have detrimental effects on the performance of pavements. High temperature produces the following effects:

1. Expansion of concrete producing high compressive stresses at the joints and curling and warping stresses in the slab
2. Excessively rapid curing of concrete
3. Softening of asphalt cement, reduced stiffness
4. Reduction in the viscosity of asphalt cement

Low temperatures promote the following effects:

1. Widening of joints and cracks in concrete pavements
2. Thermal contraction stresses in both asphalt and concrete pavement
3. Increased modulus of asphalt concrete resulting in loss of flexibility
4. Expansion of frozen moisture in pavement layers generating internal hydrostatic stresses in the pavement

One of the most critical conditions that can develop in a pavement structure is the spring thaw process. Low temperatures freeze the moisture in the pavement structure. As the temperature warms, the pavement thaws from the top down. The ice in the lower areas of the pavement traps the water in the pavement structure, greatly reducing its strength. Traffic on the pavement will then result in an excess amount of destruction to the pavement. The freeze-thaw problem is frequently the greatest problem in moderate temperature areas where several cycles can occur annually.

The primary effect of solar radiation on pavement materials is the hardening of asphalt concrete surfaces. This is the result of volatilization of the light molecular weights portion of the asphalt cement, and reduces the flexibility.

## 22.5 INTERACTION EFFECTS

Obviously the performance of pavements is a complex process affected by the interactions of the material properties, traffic loadings, and environment. Static loadings on pavements softened by high temperatures can result in shoving and permanent deformation of the surface. Wheels crossing joints opened by low temperatures and subgrades saturated by moisture can cause a thousand times the damage produced under ideal conditions. Development of a uniform model of the performance of pavements must incorporate the combined effects of traffic, environment, and materials and their variations over the life of the pavement.

## REVIEW QUESTIONS

1. There are several types of physical design inputs in the pavement process. The major set of inputs involve material properties. For an asphalt concrete pavement, outline the types of properties that must be defined and give at least one method for defining each. How do they relate to each other?
2. List and discuss four variables associated with highway traffic characteristics for pavement design and management. Which of these is the more important and why?
3. What environmental inputs are most often used in pavement design and management?
4. What is meant by interaction effects for environment and traffic?

Chapter 23

# Basic Structural Response Models

## 23.1 INTRODUCTION

Structural models used for the analysis of pavement response to traffic and environmental loads range in complexity from simple empirical models for limiting strength to sophisticated models that attempt to realistically describe the behavior of the materials. The selection of the type of model depends on the ability of the designer to quantify the required material inputs and interpret the results of the models. This chapter describes some of the structural response models available for the mechanistic analysis of a pavement as an engineering structure.

The central feature of a mechanistic design method is the structural analysis model used for computing the response of the pavement with respect to the load and environmental inputs or stimulants. There are several excellent reviews of the various structural analysis models [Yoder 75]. The review of structural models poses a dilemma as to whether to strive for completeness of the theoretical aspects of the models or to address in broad terms the features and abilities of the models. The latter approach was selected for this chapter.

The formulation of mechanistic models of pavement response involves idealizing the real physical problem and casting it into mathematical form [Nair 71]. The general mathematical form of pavement response models consists of a set of partial differential equations subjected to various initial and boundary conditions. There are two basic approaches to the solution of the boundary value problems: (1) analytical or classical methods and (2) numerical or approximation techniques. Elastic layer theory [Burmister 45] and thin plate solutions [Westergaard 27] are examples of the analytical approach. Finite differences and finite element models are examples of the numerical approach.

The theories discussed in this chapter are for predicting the primary response of the pavements, such as stresses, strains, and displacements. Concepts for the analysis of the limiting response of the pavements, for example, cracking, are discussed in the following chapter.

## 23.2 ELASTIC LAYER THEORY

Burmister's solution of the elastic two-layer problem laid the foundation for the extension of the theory to multiple layers. The equations for the two layer case are relatively simple and can be solved on a pocket calculator. However, the extension of the theory to multiple layers greatly complicates the problem and practical application of the theory requires computer analysis. Fortunately, several computer programs are available for performing this analysis.

The general concept of elastic layer theory (ELT) is shown on Figure 23.1. The assumptions used for model development are [Yoder 75]

1. Homogeneous material properties.
2. Finite layer thickness except for the bottom layer which is assumed to be infinite.
3. Isotropic material properties.
4. Full friction between the layers.
5. No shear stresses at the surface.
6. Materials are linear-elastic and obey Hooke's law, i.e., the constitutive behavior of the material is defined by the elastic modulus and Poisson's ratio.

In addition, the load is assumed to be static and uniformly distributed over a circular area. Models of the response of a pavement to dynamic load are available; however, the complexity of these models has limited their utility. The characteristics of dynamic models are not addressed.

Strictly speaking, elastic layer theory is not a good model of a pavement structure. Comparison of the material characteristics, pavement geometry, and traffic loading conditions indicates that real pavement structures violate virtually every assumption specified for the theory development. Yet when the theory was introduced, engineers recognized the potential of the model, if properly applied, to improve the state of the art in pavement design. Numerous researchers investigated the effects of the differences between the theory and reality on the utility of elastic layer theory for pavement analysis. The basic conclusion is that elastic layer theory is a useful model for the analysis of pavements provided the input data is properly formulated and the output

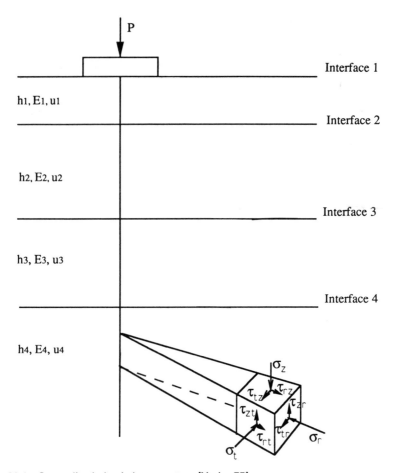

**Figure 23.1** Generalized elastic layer system [Yoder 75].

is properly interpreted [Avramesco 67, Brown 67, Lister 67, McCullough 69, Ahlvin 73, Saraf 87].

The assumption of infinite horizontal dimensions of the layers is a major drawback to the use of the model for pavement structures. Edges, joints, and cracks in pavements increase the stresses generated by wheel loads. Because the interior loading condition is a more realistic assumption for flexible pavements than for concrete pavements, ELT has been more widely applied to flexible pavements.

There are several computer codes for the solution of elastic layer theory equations. Three widely used programs in the United States are:

1. CHEV5L or LAYER developed by California Research Company, a division of Chevron Oil [Chevron 63]
2. BISAR developed by Shell Oil Co. [De Jong 73]
3. ELSYM5 developed at the University of California [Ahlborn 72]

## Basic Structural Response Models

The CHEV5L and ELSYM5 programs are widely available in the public domain. The capabilities of the CHEV5L program have been expanded by several researchers. For example, an iteration method was developed for the analysis of stress sensitivity [Shahin 71]. Other modifications permit the analysis of up to 15 layers. The Federal Highway Administration sponsored the modification of ELSYM5 to operate on a microcomputer with a full screen editor for inputting the data. This program is available from the McTRANS Center at the University of Florida.

BISAR is the most powerful of the ELT programs. It can handle up to 10 layers and 10 different loads. Burmister's theory has been modified in this program to permit the analysis of shear loads at the pavement surface and varying interface continuity between the pavement layers ranging from full continuity to full slip. The mathematical techniques used in the BISAR programs are reported to be more sophisticated than the CHEV5L program [De Jong 73].

### 23.3 VISCOELASTIC LAYER ANALYSIS

Several theories have been proposed for the analysis of pavement structures. [Aston 67]. The only working viscoelastic model for pavement analysis is VESYS developed for the Federal Highway Administration [Aston 67, Moavenzadeh 72]. VESYS starts with the Burmister formulation of the primary response of the pavement system. Modifications are then introduced to model a limited time duration of the load and the viscoelastic form of the constitutive equation for the material characteristics.

The inputs to the VESYS IIIA model are [Khosla 87]:

1. Geometry of the Pavement System: the geometry requirement for the first $(N - 1)$ layers with the thickness of the $N^{th}$ layer being infinite.
2. Traffic Loadings: number of 18,000 pound equivalent axle loads per day, intensity, and duration of loads.
3. Temperature: average seasonal temperature and winter design temperature.
4. Material Response Properties: modulus of resilience and Poisson's ratio of every layer for every season; these properties are needed in order to calculate the stresses, strains, and deflection response in a pavement system under the application of external loadings.
5. Material Damage Properties: fatigue coefficients of the surface layer and permanent deformation parameters of every layer; these coefficients and parameters together with the stress, strain, and deflection response in the pavement system are used to estimate the pavement damage in terms of cracking, rutting, and roughness under various stages of its life.

The input requirements demonstrate that VESYS considers fatigue cracking as well as the rutting of the pavement from viscoelastic strains. The fatigue model will be addressed later.

Khosla concluded the structural subsystem of VESYS IIIA predicts pavement performance accurately and triaxial compression tests can be used for measuring the material properties. However, Beckedahl criticized the ability of the VESYS model, especially with respect to characterization of the material properties. Beckedahl proposed several improvements to the model including the development of procedures for capturing the fluctuations in the material properties over the life of the pavement [Beckedahl 87].

## 23.4 THIN PLATE THEORY

As opposed to flexible pavements, which spread the load downward through the pavement structure, rigid pavement slabs act as structural elements. Since the deflection of rigid pavements is small relative to their thickness, they can be analyzed as thin plates [Sargious 75]. The following approximations are required for the development of the thin plate models:

1. There is no deformation in the middle plane of the slab; this plane remains neutral.
2. The planes in the slab initially normal to the middle plane of the slab remain normal after bending.
3. The normal stresses in the direction transverse to the plane of the slab can be ignored.

Sargious reports that the differential equation describing the deflected surface of a slab subjected to a uniform load was developed by LaGrange in 1811 and identifies Westergaard as the first to develop a theoretical solution for rigid pavement design. Assumptions used in the development of the Westergaard equations are [Sargious 75]:

1. The concrete slab acts as a homogeneous elastic solid in equilibrium.
2. The reaction of the subgrade is solely vertical, and proportional to the deflection of the slab.
3. The reaction of the subgrade per unit area at any given point is equal to a constant, $K$ (modulus of subgrade reaction), multiplied by the deflection at that point.
4. The thickness of the slab is uniform.
5. The load at the interior and at the corner of the slab is distributed uniformly over a circular contact area; for the corner loading the circumference of this circular area is tangential to the edge of the slab.
6. The load at the interior edge of the slab is distributed uniformly over a semicircular contact area, the diameter of the semicircle being along the edge of the slab.

Although not commonly stated, it should be noted that Westergaard also assumed a static load condition. Based on these assumptions Westergaard developed equations for computing the stresses in the slab for the following cases:

1. Maximum tensile stress at the bottom of the slab due to central loading
2. Maximum tensile stress at the bottom of the slab for an interior edge loading in a direction parallel to the edge

## Basic Structural Response Models

3. Maximum tensile stress at the top of the slab in a direction parallel to the bisector of the corner angle for corner loading

Subsequently, Westergaard modified the equations specifically for the analysis of airport pavements assuming elliptical load areas, and load transfer across the joint or edge of the pavement. In 1951, Picket presented equations for "protected" and "unprotected" corners. The Westergaard equations are widely used for the design of concrete pavements. Influence charts were developed for the solution of these equations and Packard incorporated them into a generalized program for the design of portland cement concrete pavements [Picket 51, Packard 67].

Westergaard's work on the development of equations for the prediction of stresses in slabs was an extremely important contribution to pavement engineering. However, it has been recognized that some of the development was in error for the edge stresses. In addition, some of the equations used for pavement analysis that are attributed to Westergaard have errors or misapplications [Ioannides 85].

### 23.5 NUMERICAL METHODS

There are two basic numerical techniques that can be applied to the analysis of pavement structures: finite differences and finite element. Finite element methods (FEM) for the analysis of rigid pavements were developed at the University of Texas in the 1960s and several successful computer programs were produced [Stelzer 67]. The SAP program, developed at the University of California, has been used for pavement analysis [Duncan 68]. Compared to elastic layer theory, FEM estimates were almost identical for stresses and strains, but there were slight differences in the computed deflections attributed to the differences in the boundary conditions [Pichumani 72]. This application did not offer any advantage over elastic layer theory. However, it demonstrated the applicability of this approach to the analysis of pavements. Subsequent applications in the use of FEM for the analysis of pavements have permitted modeling of more complex material characteristics and pavement geometries than is possible with the analytical solution methods developed by Burmister and Westergaard.

ILLI-PAVE has been widely used for the analysis of flexible pavements [Raad 80]. The characteristics of this model are [Gomez-Achecar 86]:

1. An axisymetric solid of revolution, e.g., Figure 23.2
2. Nonlinear, stress-dependent resilient modulus of layer materials
3. Limits on the principal stresses in granular and fine-grain soils so that they do not exceed the Mohr-Coulomb theory of failure

Gomez-Achecar and Thompson report that several authors have tested the validity of the program with favorable results.

### 23.6 ENVIRONMENTAL MODELS

Several environmental variables affect pavement behavior and performance, including:

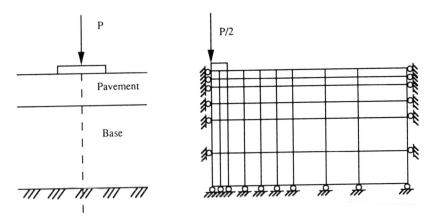

**Figure 23.2** Axisymetric representation of two layer pavement using finite element [Raad 80].

1. Moisture
2. Temperature
3. Solar and atmospheric conditions
4. Site geological conditions

The first two classes of variables receive primary consideration in many pavement design procedures.

Moisture variations have two general classifications:

1. External to the pavement, i.e., rainfall amount, intensity and duration; snowfall amount, intensity and duration; site drainage conditions; water table
2. Within the pavement and subgrade, i.e., variations with time, depth, laterally, longitudinally, type and depth of pavement component layers, type of subgrade material

Moisture variations are difficult to measure and characterize. Although they may be as important as temperature effects, they receive relatively little attention. Frequently, pavement drainage is considered as a separate issue from the pavement design process, with the assumption in the pavement design that the drainage will be adequate to control the moisture in the materials to a proper level.

All materials shrink as their temperature drops. Surface layers resting on a base are constrained by their weight; thus, any attempt to move causes forces at the interface. When the temperature of the pavement drops, the constraint to movement causes the development of stresses in the surface. When these stresses exceed the strength of the material, transverse cracks develop. This basic mechanism is the reason conventional concrete pavements are constructed with joints. The need for design models for

## Basic Structural Response Models

the selection of joint spacings in concrete pavements has led to the development of models of the environmental stresses in concrete slabs.

### 23.6.1 Environmental Effects on Flexible Pavements

Figure 23.3 summarizes the four classes of environmental variables and identifies the potential causes for their variations. Temperature conditions affecting pavement behavior and performance can be further subdivided into three ranges according to the type of distress that might occur:

1. High temperature range during mixing and placing, which can affect the durability
2. Medium temperature range for the pavement in service, which can affect stability, permanent deformation susceptibility under repeated loads, and fatigue cracking susceptibility under repeated loads
3. Low temperature range for the pavement in service, which can affect the thermal shrinkage cracking susceptibility

The higher temperature range is controlled and the medium temperature range applies to the response under traffic loads previously discussed in this chapter. The low temperature range can result in significant stresses and strains in the pavement that produce distress.

Although low temperatures can result in stresses high enough to cause cracking under certain conditions, it should be noted that such cracking might occur as a result of freezing shrinkage and cracking of the subgrade, or as a result of temperature-induced stresses in the bituminous surface combined with traffic-imposed stresses.

Generally, however, the temperature-induced stresses are the primary cause of low-temperature cracking in pavements [CGRA 70]. A complete listing of the factors of possible significance in low-temperature cracking is shown in Figure 23.4 [Haas 69].

Temperature can vary markedly through the depth of a pavement, depending on the pavement component layer thicknesses and properties, air temperature variances, wind conditions, and the solar conditions. These vertical gradients of temperature are important and are considered in many design procedures. There can be lateral and longitudinal variations of temperatures, but they are generally considered less important.

The calculation of temperature induced stresses in flexible pavements has been formulated by several investigators [Humphreys 63, Monismith 65, Hills 66, Lamb 66, Haas 69, Christison 72]. Monismith et al. used a stress equation developed by Humphreys and Martin for an infinite slab composed of a linear viscoelastic material, subjected to a time-dependent temperature field. The equation provides for the calculation of the horizontal tensile stress field as a function of depth, time, and temperature. However, the stresses computed by this method appear to be unrealistically high because of the assumption of infinite extent of the pavement slab in the lateral direction.

A simpler model that considers the pavement surface as an elastic, infinitely long

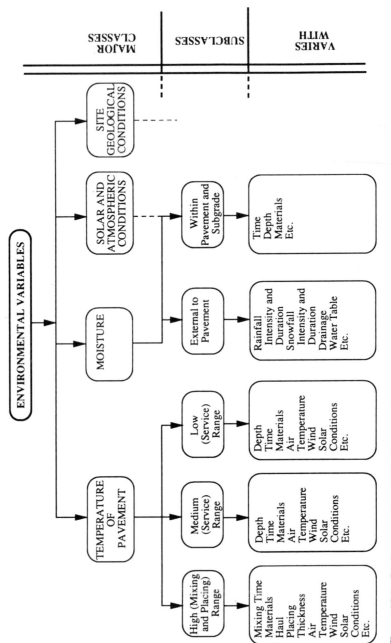

**Figure 23.3** Environmental variables that can affect flexible pavement behavior and performance.

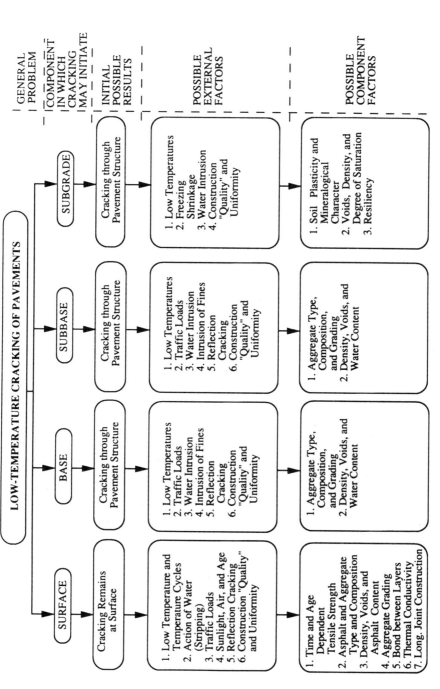

**Figure 23.4** Factors of possible significance in low-temperature cracking of flexible pavements [Haas 69].

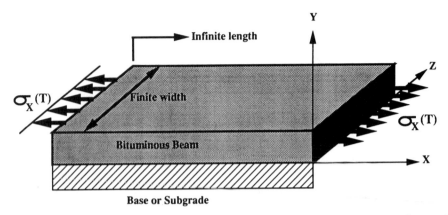

**Figure 23.5** Representation of a bituminous layer under shrinkage stresses as a beam of infinite length and finite width [Lamb 66].

beam of finite width is shown on Figure 23.5 [Lamb 66]. For this model, the restraint is considered in two directions and no temperature gradient exists through the depth of the layer. The unit tensile stress in the longitudinal direction is given by:

$$\sigma_x(T) = E\alpha(\Delta T) + \mu f L \tag{23.1}$$

where

- $E$ = Young's modulus
- $\alpha$ = average coefficient of thermal contraction, over the temperature drop, $\Delta T$
- $\mu$ = Poisson's ratio for the material
- $f$ = coefficient of friction between the bituminous layer and the base
- $L$ = one half the width of the pavement section

Because the modulus of asphalt concrete is sensitive to temperature, Young's modulus is replaced by the stiffness modulus $S(T, t)$, which varies with both temperature and load duration. Lamb and his co-workers evaluated this modulus for the midpoint of their most extreme daily temperature drop range in the winter and computed tensile stresses as high as 110 psi. The contribution of lateral restraint to this stress was found to be relatively minor.

The method used by Hills and Brien has extended the concept of stiffness modulus to the entire temperature range and used experimentally determined values of the modulus at the midpoint of small discrete temperature intervals [Hills 66]. The equation for longitudinal stress for the same infinitely long beam of Figure 23.5 (neglecting lateral restraint) then becomes:

$$\sigma_x = \alpha \sum_{T=T_0}^{T_f} (S\,\Delta T) \tag{23.2}$$

where

$S$ = stiffness modulus, determined experimentally, at the midpoint of each $\Delta T$

$\Delta T$ = temperature interval of which a finite number are taken between $T_0$ and $T_f$

Hills and Brien used this equation to calculate thermally induced stresses in a beam of pure asphalt where the stiffness modulus was obtained at the midpoint of small temperature intervals, using a load duration corresponding to the cooling time for the temperature interval. They contend that any error involved in this use of corresponding times is likely to be small. The stress values obtained were then compared to tensile strengths and probable fracture, by direct measurement, and were found to be in relatively good agreement for a limited range of experimental work and no temperature gradients.

Haas and Topper extended the work of Lamb et al. and Hills and Brien to account for temperature gradients through the bituminous layer [Hills 66, Lamb 66, Haas 69]. In addition, Haas and Topper incorporated a "stiffness gradient" due to the greater hardening of the binder at the surface of the pavement. This method was presented in terms of several possible situations of cases for calculating stresses at the top and bottom of the bituminous layer. Intermediate stresses can be determined if the temperature and stiffness gradients are known.

Temperature-induced stresses in asphalt pavements were also modeled as a "psuedo-elastic" beam analysis that considered the influence of mix stiffness, binder stiffness, and volume concentration of aggregates in the mix to determine the temperature-induced stresses [Christison 72].

### 23.6.2 Environmental Effects on Rigid Pavements

Temperature variations cause rigid pavements to shrink and expand and mobilize friction at the bottom of the slab. The model developed for computing the stresses in a concrete slab due to friction is shown in Figure 23.6. Balancing the forces defined in the figure yields [Yoder 75]:

$$s_c = WLf/(24h) \qquad (23.3)$$

where

$s_c$ = friction stress
$W$ = weigth of the slab (psf)
$L$ = length of slab (ft.)
$f$ = average coefficient of subgrade resistance
$h$ = depth of the slab (in.)

Here, $f$ is sometimes called the friction between the slab and the subgrade. However, this is incorrect terminology since the contraction of the slab results in shear forces that are transmitted into the subgrade and dissipate [Yoder 75]. For jointed concrete pavements, the stress due to friction can be computed and compared to the

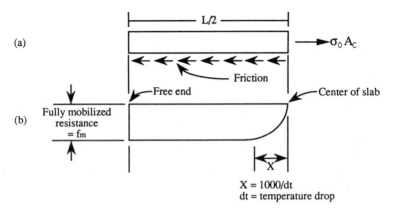

**Figure 23.6** Stresses resulting from contraction. (a) Forces acting to contract slab; (b) variation of subgrade resistance with length [Yoder 75].

concrete strength. If the stress is excessive, the length of the slab is reduced. In the design of continuously reinforced concrete pavements, the slab length is very long or the distance between the cracks is unknown. Numerical techniques are available for the analysis of crack spacing in reinforced concrete pavements [McCullough 75]. This model estimates the crack spacing as a function of temperature drop, drying shrinkage, moisture change, wheel load, and coefficient of subgrade resistance.

Warping of a concrete slab is developed by a thermal gradient in the slab. Yoder and Witczak presented equations formulated by Westergaard along with the solution developed by Bradbury as the edge stress:

$$s = CE\alpha t/2 \quad (23.4)$$

and the interior stress:

$$s = (E\alpha t/2)((C_1 + \mu C_2)/(1 - \mu^2)) \quad (23.5)$$

where

$C_1, C_2$ = coefficients defined as functions of the relative stiffness of the slab
$\alpha$ = coefficient of thermal expansion of concrete
$\mu$ = Poisson's ratio
$E$ = elastic modulus
$t$ = temperature differential between the top and bottom of the slab

Traditionally the principal of superposition is used to add the warping, friction, and traffic stresses to obtain the total stress in the pavement.

## 23.7 CUMULATIVE DAMAGE THEORY

Traffic loads on both highways and airfields are a mixture of different wheel weights, tire pressures, and axle or truck load configurations. In addition, the critical strain

## Basic Structural Response Models

or stress in the pavement will vary throughout the life of the pavement as the materials respond to fluctuations in environmental conditions. Thus, application of mechanistic analysis for pavements requires a method for determining the cumulative effects of the different strain and stress levels.

Linear summation of relative damage, Minor's rule, can be used for determining the relative damage of different strain levels according to the equation [Monismith 73]:

$$D = \sum (n_i/N_i) \qquad (23.6)$$

where

$D$ = total damage caused by strain repetitions
$n_i$ = number of repetitions of strain level $i$ applied to the pavement
$N_i$ = theoretical number of applications the pavement can withstand, strain or stress level $i$.

The summation in equation 23.6 is performed over the number of different strain conditions the pavement will experience. Failure due to fatigue is indicated by a damage ratio of 1.0 or greater.

## REVIEW QUESTIONS

1. Outline the basic structural response models used in pavements. Discuss them briefly.
2. What is meant by cumulative damage theory?

Chapter 24

# Variability, Reliability, and Risk in Pavement Management

## 24.1 DEFINITIONS AND BASIC CONCEPTS

Properly programmed computer codes such as those described in the preceding chapters will precisely compute pavement response to loads. However, that does not mean the resulting calculations are accurate predictors of pavement performance. Many factors affect the analyst's ability to accurately predict pavement performance. These are the result of our inability at this time to accurately model pavement behavior and the inherent variability of the factors that affect pavement performance. As noted in the previous chapter, progress is being made on the development of increasingly sophisticated analytical tools for modeling pavement behavior. The purpose of this chapter is to address the issue of variability in the pavement design, construction, use, and performance process.

### 24.1.1 Sources of Variation

Just as people vary in height and weight, materials vary with respect to any measured parameter, such as CBR, modulus of elasticity, modulus of subgrade reaction, and flexural strength. In addition, it is impossible for contractors to build pavements exactly as specified in the construction contract. Construction specifications recognize this by allowing a variance in the thickness of each pavement layer. Due to variability, it

### Variability, Reliability, and Risk in Pavement Management

is impossible to know the exact material properties and layer thicknesses at any specific point along the pavement.

The second major source of variation in pavement life prediction rests with the ability to estimate the amount and type of traffic that will use a facility. This problem is common to the design of both highway and airfield pavements. Historically, traffic quantities have been underestimated for both types of facilities.

The third area where variability enters the pavement design and analysis process is in the lack of fit error of pavement performance models, or transfer functions that relate pavement response to pavement life. Almost all performance models are developed from empirical evidence. These models generally have the conceptual form:

$$N = f(D_i, R_i, E_i, \ldots) \pm e \qquad (24.1)$$

where

$N$ = the number of loads the pavement can carry
$D_i$ = design variables
$R_i$ = response variables
$E_i$ = environmental variables
$e$ = error term describing a lack of fit between the predicted and the observed data

The error term is an expression of the variability in the data that is not explained or captured by the design (independent) variables in the equation. Statisticians use the term *error* to describe a lack of fit in a regression equation. Lack of fit is the result of unexplained variances and does not indicate a blunder or mistake in the data collection or modeling process.

#### 24.1.2 Variance in Pavement Design

In pavement design, the variability in material characterization combines with error in the performance model to produce uncertainty in the ability to predict the life of the pavement, in terms of number of axle loads, stress or strain repetitions, or aircraft loadings. However, most agencies design a pavement to last a certain number of years. Hence, inaccurate traffic estimates directly affect the designer's ability to predict how long a pavement will last in terms of years.

Several design methods recognize that variability in materials will affect pavement performance. Selection of the 85$^{th}$ percentile value for the subgrade strength or flexural strength of concrete is common to some pavement design methods. During the 1970s researchers investigated the application of reliability concepts to the pavement design process [Darter 71, 72, 73a, 73b, Moavenzadeh 72, Kher 73]. However, the AASHTO Guide for Design of Pavement Structures was the first pavement design method to fully recognize the effect of several variance terms on pavement design [AASHTO 86]. This design method is presented in a subsequent chapter; however, the statistical concepts discussed in the guide are so important and fundamentally sound as to warrant special attention in this chapter. Details of the development of the AASHTO Guide are presented in Appendix EE of the guide [AASHTO 85]. A summary of these concepts, presented by Irick, Hudson, and McCullough, forms the basis for this chapter [Irick 87].

A consequence of the variation of material properties, layer thicknesses, traffic estimates, etc., combined with the model error term, is that two pavement sections can exhibit different lives, even when they were designed and constructed "identically." Improving the pavement design process requires quantification of variations in pavement design, construction, and performance and the use of statistically sound analyses to assure the design-performance process has a known and specified degree of reliability. The 1986 AASHTO Guide defines reliability as the probability that pavement sections will withstand the actual number of equivalent axle loadings and environmental conditions that will be applied over the design life of the pavement.

## 24.2  FORMULATION OF PAVEMENT RELIABILITY

The pavement design process essentially consists of selecting a pavement structure such that the structural capacity of the pavement, expressed in terms of the number of axle applications that can be carried, is equal to the number of axles that will traverse the pavement during the design life or:

$$N_t = n_T \tag{24.2}$$

where

$N_t$ = the actual performance of the pavement in terms of equivalent single axle loads, ESAL, the pavement carried between construction and failure
$n_T$ = the actual number of ESAL during the design period

If $N_t$ was less than $n_T$ the pavement was underdesigned and did not last the design life. If $N_t$ was greater than $n_T$ the pavement was overdesigned in the sense that it carried more traffic than required by the design problem. Note that $N_t$ and $n_T$ refer to *actual* performance of a pavement and number of ESALs during the design life. Therefore, these are observed variables and not subject to variation. However, they can only be determined at the end of the pavement's life.

Due to the magnitude of the number of axle loads pavements carry, and statistical evidence related to the variance of pavement performance, highway pavement design equations are generally expressed in terms of the log of ESAL. Thus, the *process deviation* between the design period ESALs and the pavement performance is:

$$\partial_{nN} = \log(N_t) - \log(n_T) . \tag{24.3}$$

Equation 24.3 demonstrates that pavement design requires the prediction of two load variables: the structural capacity of the pavement and the traffic to be carried. The *predicted* variables for pavement structural capacity and design period traffic are defined as $W_t$ and $w_T$ respectively. Because these are predicted variables (unobserved) they are subject to variability. It must be remembered that a performance prediction is not for a single pavement section but is the prediction for the mean performance of possibly hundreds of replicate sections. Replicate sections are designed to be iden-

# Variability, Reliability, and Risk in Pavement Management

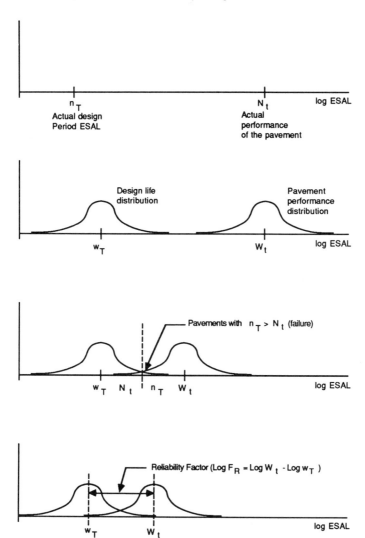

**Figure 24.1** Conceptual relationship between actual and predicted traffic loads and pavement performance.

tical with respect to all variables that affect performance. Differences between the predicted and actual performance are expressed as:

$$\partial_{wn} = \log(w_T) - \log(n_T) \tag{24.4}$$

and

$$\partial_{WN} = \log(W_t) - \log(N_t) . \tag{24.5}$$

Figure 24.1 demonstrates the relationships between actual and predicted performance and design life ESALs. Note that $w_T$ and $W_t$ are mean values for distributions of several pavements whereas $n_T$ and $N_t$ are section specific. The overlap between the pavement performance and design life distributions results in cases where $n_T$ exceeds $N_t$, or failure of the pavement. The amount of overlap between the two distributions is a function of the differences between the means and the spread (standard deviation) of the distributions. If the spread of the distributions cannot be controlled, then the only option available to the designer is to increase the distance between the means of the distributions. Reducing the overlap of the distributions reduces the incidents of failure and hence the designs are more reliable. Thus, a *reliability* term, $F_R$, may be defined as

$$\log(F_R) = \log(W_t) - \log(w_T) . \quad (24.6)$$

Substitution of equations 24.4, 24.5, and 24.6 into equation 24.3 and reducing yields:

$$\partial_{nN} = \partial_{wn} + \partial_{wN} + \log F_R . \quad (24.7)$$

Since $w_T$ and $W_t$ are random variables, the sign resulting from equations 24.4 and 24.5 can be either positive or negative. Consideration of this is necessary for the derivation of equation 24.7.

Equation 24.7 is composed of a random or "chance" part ($\partial_{nw} + \partial_{NW}$) and a fixed part ($\log F_R$). The fixed part of the equation can be selected in the design process to define reliability. The random parts of the equation must be defined empirically. Using equation 24.7 as a model of the chance deviations in the design and performance process, the task now required is to define each of the components of chance deviations in the pavement performance and the design life ESALs predictions. The AASHTO Guide contains models for the individual components of variation. These are accumulated into an overall variance term, $S_0^2$, that represents the total variance in the pavement design process.

The reliability for a design is then related to the standard deviation of pavement design process as

$$\log(F_R) = -z_R S_0 . \quad (24.8)$$

The term $z_R S_0$ has been added to the AASHTO Guide design equations for both flexible and rigid pavements. Selected values for the reliability factor for different reliability levels are given in Table 24.1.

## 24.3 INFLUENCE OF VARIABILITY ON PAVEMENT MANAGEMENT

Due to the inherent variations in the pavement construction, maintenance, and use factors, it will never be possible to predict the performance of a specific section of pavement with 100 percent certainty. Inclusion of the reliability concept in the pavement design process will result in more rational expectation of the performance of pavements. This can lead directly to improved models for the economic analysis of pavements and more effective life-cycle cost analysis.

Reliability calculations in the design process require an estimate of the overall

**Table 24.1 Reliability Factors for Selected Reliability Levels and Process Standard Deviation [Irick 87]**

| Reliability Levels $R(\%)$ | Normal Curve $Z_R$ | Process Standard Deviation | | |
|---|---|---|---|---|
| | | 0.25 | 0.35 | 0.45 |
| 50 | 0.00 | 1.00 | 1.00 | 1.00 |
| 60 | −0.25 | 1.16 | 1.23 | 1.30 |
| 70 | −0.52 | 1.35 | 1.53 | 1.72 |
| 80 | −0.84 | 1.62 | 1.87 | 2.39 |
| 90 | −1.28 | 2.09 | 2.81 | 3.77 |
| 95 | −1.64 | 2.58 | 3.76 | 5.50 |
| 99 | −2.32 | 3.82 | 6.52 | 11.1 |

process standard deviation, $S_0$. A complete description of the steps required to quantify this value is found in Appendix EE of the AASHTO Guide [AASHTO 85]. However, there is a lack of data available to support the analysis [Irick 87]. According to Irick:

> Available data for the forgoing estimates are indeed sparse. A few studies have been made of traffic prediction variance; the AASHO Road Test is virtually the only performance study that has produced comprehensive field data on performance prediction variance.

Based on an analysis of this data, values of $S_0$ of 0.35 for rigid pavements and 0.45 for flexible pavements have been included in the AASHTO Guide. Verification of these variances should be a high priority research task.

## REVIEW QUESTIONS

1. Define variability in general terms.
2. Define variance in terms of the factors related to pavement materials and pavement design.
3. Define pavement reliability.
4. What is the effect of variability on the results of pavement management?

Chapter 25

# Generating Alternative Design Strategies

## 25.1 INTRODUCTION

Common pavement design practice considers only structural thickness selection. Although this is in many cases the most important aspect of design, it is an incomplete formulation of the problem. Chapter 21 (see Figure 21.1) suggests that a more comprehensive concept is required, which additionally includes consideration of material types, expected policies of construction, maintenance, rehabilitation, and performance evaluation. This more comprehensive concept is emphasized by using the term *design strategy* to describe a pavement design alternative.

The term *strategy* implies that design should be concerned with the pavement throughout the entire design or analysis period. Predictions of performance and costs and benefits need to be made over the design period. In order to do this adequately, all components affecting performance, such as material types, must be included in the analysis.

These components are represented conceptually in Figure 25.1. The diagram contains only the general order involved in generating alternative strategies, but it does illustrate that there can be a large number of possible alternatives to be considered in any particular design situation.

# Generating Alternative Design Strategies

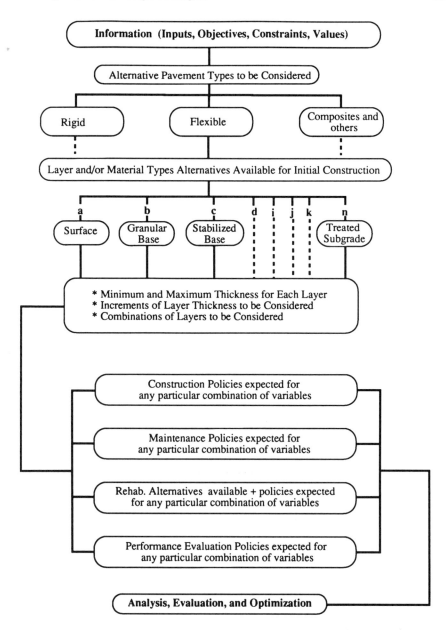

**Figure 25.1** Key components of generating alternative pavement design strategies.

## 25.2 NEW PAVEMENT ALTERNATIVES

The designer of a new pavement structure has a great deal of flexibility with respect to the structural cross section selected for the pavement type. The cross section can consist of a variety of material types and thicknesses. In addition to cross-section selection, the designer should consider construction methods which affect pavement performance. Selection of the most economical pavement structure should consider the life-cycle costs of the structure. Thus, the pavement designer should be aware of the pavement evaluation methods and their influence on the selection and applications of maintenance treatments. The interaction of these factors on the pavement design process is shown on Figure 25.1.

### 25.2.1 Structural Section Alternatives

The pavement types considered for any design situation should include both rigid and flexible. Additionally, composite pavements and flexible or rigid with stabilized layers might be considered. Usually, however, existing design methods consider only one type. The reasons for this include:

1. Personal preference of the designer
2. Prior successful experience with one particular pavement type for the design situation under consideration
3. Lack of adequate means for comparison

No methodology for comparing pavement types on a completely objective basis has yet been developed. Although there is a great deal of comparative evidence available, no conclusive agreement has been reached.

The following paragraphs use flexible pavements as an example to illustrate the aspects of generating alternative pavement design strategies except for pavement type. Many of the considerations involved would also be applicable to the rigid pavement alternatives.

The structural sections which combine layer types and thicknesses that might be considered in a design problem present the designer with a large array of alternatives. An example will be used to illustrate the points involved.

Suppose that a design for a secondary highway is to consider a flexible pavement type, with four possible layer types: (a) an asphalt concrete surface, (b) an unbound granular base, (c) an asphalt treated base, and (d) an unbound granular subbase, as shown on Figure 25.2. The minimum and maximum layer thicknesses must be chosen and are listed in the diagram. For practical reasons layer types (b) and (c) are mutually exclusive, that is, if one is used, the other will be omitted.

Figure 25.2 shows, using 1/2-inch increments, some of the alternative initial structural sections to be considered within the thickness range constraints. It begins by increasing the surface layer from 2 inches to 4 inches, in 1/2-inch increments, while keeping the underlying layers at a constant thickness. Next, as shown for alternatives $k$ and $l$, a change is made to layer type (c) and the surface layer is again

| Layer Type | Is layer to be included in all alternatives | Thickness Range min max | 1 | 2 | i | k | l | n | p | q |
|---|---|---|---|---|---|---|---|---|---|---|
| (a) Asphalt Concrete Surface | yes | 2" 4" | 2" | 2½" | 4" | 2" | 2½" | 4" | 2" | 2" |
| (b) Unbound Granular Base or | yes unless (c) is used | 4" 8" | 4" | 4" | 4" | 0 | 0 | 0 | 4½" | 5" |
| (c) Asphalt Treated Base | yes unless (b) is used | 2" 4" | 0 | 0 | 0 | 2" | 2" | 0 | 0 | 0 |
| (d) Unbound Granular Subbase | yes unless (c) is used | 4" 8" | 4" | 4" | 4" | 4" | 4" | 4" | 4" | 4" |
| **Total Pavement Thickness** | | | 10" | 10½" | 12" | 8" | 8½" | 10" | 10½" | 11" |

**Figure 25.2** Example representations of initial possible structural section alternatives for a pavement design problem.

incremented through the range. There are several hundred such possible section alternatives.

It is impractical for the designer to generate, analyze, and evaluate all of these alternatives manually. Consequently, a computerized design system must be used to combine the generation of all possible solutions with analysis and economic evaluation (see Figures 3.1, 3.2, and 21.4). The designer in effect generates structural section solutions by specifying layer types to be considered, any mutual exclusivities of layer types, and minimum and maximum layer thicknesses, as shown on Figure 25.2. Other components of alternative pavement design strategies are similarly combined in a computerized system. Nevertheless, it is useful to consider them separately for conceptual purposes.

### 25.2.2  Material Type Alternatives

Several material type alternatives may be available in a design problem. This should not, however, be confused with layer types. For example, such layer types as asphalt concrete surface, asphalt-treated base, cement-treated base, unbound granular base or subbase, crushed stone base or subbase, and lime-treated subgrade might all be considered within one design problem, although some types would be mutually exclusive with respect to the others.

A variety of materials may then be used for any one of these layers. For example, in asphalt concrete, a manufactured lightweight aggregate (perhaps for skid-resistance properties) may be compared to a locally available crushed stone. Also, more than one type of asphalt binder may be considered. As another example, a granular material may be stabilized with an emulsified asphalt, or with an asphalt cement to produce a "black base."

If there are several material types available, then they are combined with the incremental layer type and thickness alternatives illustrated in Figure 25.2, generating a large number of possible design strategies for any problem. This number is increased by the alternative construction, maintenance, rehabilitation, and performance evaluation policies that the designer might expect or specify. The success of a design strategy in performing properly is largely dependent on the construction policies used to implement the strategy (i.e., actually to build the pavement). Yet there traditionally is a distinct and often uncoordinated division between design and construction. All too often, design is treated as an end within itself, as is construction. Designers often do not define the construction policies they want to be applied.

A major reason for this incomplete approach to design perhaps lies with the traditional administrative structure of highway agencies. Construction personnel do not want to be told by designers how to build pavements; and maintenance personnel similarly do not want to be told how to perform maintenance.

However, a properly developed design system need not create the impression of dictating to the other related divisions, such as construction and maintenance. Rather, the design should indicate what construction and maintenance policies were expected or assumed. In this way, it is much easier to evaluate any subsequent deviations from expected behavior or performance.

The construction policies that might be considered in design include:

1. Variations or control limits allowed in:
   a. Layer thicknesses
   b. Material properties such as density, voids, moisture content, gradation, etc.
   c. Asphalt or portland cement concrete mix design
   d. Initial pavement roughness
   e. Temperature or moisture extremes during construction operations
2. Traffic-handling method
3. Time of day and season for construction operations
4. Material sources

The first policy aspect is relatively self-evident in its importance. Although the specific traffic-handling method may not be self-evident, it is very important in that it can significantly affect costs and benefits. It may be even more important for pavement rehabilitation than for initial construction, as is subsequently described in more detail. The time of day and season of construction may not only affect the "quality" of the finished product but also the costs involved in handling traffic, and perhaps the traffic-handling method that is used.

There are also a variety of construction intangibles that can be important but are difficult for the designer to consider in anything but a subjective manner. These can include such factors as the characteristics and conscientiousness of the contractor and the experience and idiosyncrasies of the resident engineer on the job. Most highway agencies can readily point to examples where the only apparent difference between one or more years of extra service life between comparative projects was some such intangible factor.

During actual construction operations, or perhaps even before, deviations from those policies expected during the design phase can often occur. These can include the use of a new source of materials, increased costs, or unexpected variations in subgrade support conditions requiring a change in quantities or layer thicknesses. Such deviations during construction should be noted in a proper feedback data system. This can be immediately important to designers in the sense that they may wish to update their performance predictions. They may then also wish to update or modify parts of the design strategy, such as the future rehabilitation schedule.

It is also possible for construction policies to vary with the layer and/or material type used in the design strategy. For example, alternatives $i$ and $k$ of Figure 25.2 might involve different traffic-handling methods and different sources of materials.

### 25.2.3 Maintenance Policy Alternatives

The consideration of maintenance policies in generating alternative design strategies is similar in many ways to the consideration of construction policies. In addition, it might include a consideration of the level or degree of maintenance expected. For example, one extreme would be to do absolutely nothing about observed distress until the pavement requires an overlay or other major rehabilitation. At the other extreme, every crack could be cleaned and filled as soon as it occurs.

Maintenance policies can vary with type of facility, traffic volumes, available budget, complaints from the public, nature of the maintenance work crews, maintenance district, etc. Also actual practice, even for a seemingly well-defined policy, can vary and thereby affect performance. For example, crack filling with excess asphalt, without sand blotting, can result in unsightly tracking by traffic.

The degree and type of shoulder maintenance practiced is another policy aspect that can significantly affect pavement performance, especially where the shoulders are unpaved.

Still another aspect of maintenance policy that can be important is the accounting procedures used to distinguish between capital and maintenance expenditures. For example, many agencies will place a limit on the length or amount of material to be used in a full-width patch, which is really a short overlay. Depending on this limit, the policy for patching can range from spot repairs to thin overlays of substantial length.

A number of other aspects of maintenance policies might need to be considered in design. However, on the basis of current technology, maintenance policies usually can be handled in only a relatively subjective manner. Methods for quantitatively relating level of maintenance to serviceability loss have not yet been developed. Thus, it is not yet possible to consider adequately alternative levels of maintenance in terms of their cost and benefit effects on a design strategy. Nevertheless, the designer should indicate what maintenance policies and costs were expected when documenting the recommended strategy.

### 25.2.4 Overlay and Seal Coat Alternatives

The most common types of pavement rehabilitation are overlays and seal coats. Either type, or both, may be applied up to several times during the design period. It has been pointed out in Chapter 21 that rehabilitation alternatives include two major components:

1. Structural aspects, i.e., correcting excessive distress, inadequate serviceability, inadequate safety, or inadequate strength
2. Policy aspects, i.e., time of day and season for construction, traffic control, etc.

The first component is related directly to the initial structural section alternative (Figure 25.2). For example, assume alternatives 1, 2, . . . , $i$, . . . $n$ were estimated to have the initial service life curves shown by the heavy lines of Figure 25.3. Strategy 2 has an initial service life, for example, of 10 years. At that time, a number of overlay alternatives may be considered. Overlay alternative $j$ requires a subsequent overlay at about 16 years, whereas $k$ and $l$ go past the end of the design period. Obviously, a large number of combinations of initial structural sections and overlay strategy alternatives exist.

Although Figure 25.3 is an example of rehabilitation required by minimum acceptable serviceability considerations, it should be recognized that other factors such as safety or distress might require an overlay or seal coat before this minimum serviceability is reached.

# Generating Alternative Design Strategies

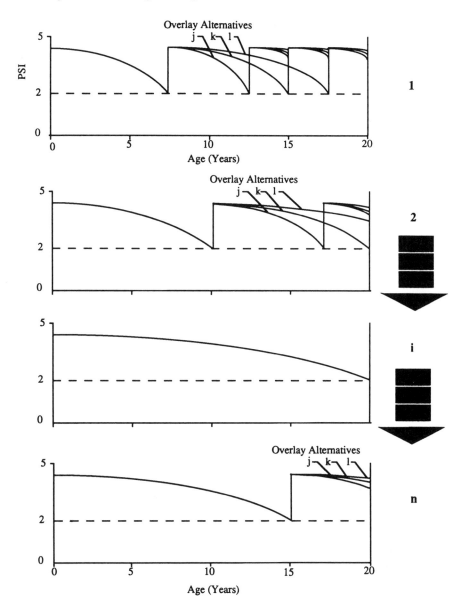

**Figure 25.3** Performance curves for several initial pavement sections and subsequent overlay strategy alternatives.

The rehabilitation policy can be an important consideration, because it can markedly affect the extra user costs associated with such rehabilitation. For example, if traffic volumes were relatively high in the previous example, then overlay alternative $j$ in strategy 2 could be more costly because it results in still another overlay being required before the end of the design period. On the other hand, for low traffic volumes, strategy 2 could represent a lower total cost than either $k$ or $l$. The economic analysis procedures for such situations are presented in Chapter 29.

Several alternative methods of handling traffic have been defined in the FPS and RPS design systems developed in Texas [Scrivner 68, Hudson 70, Kher 70]. These are illustrated in Figure 25.4. The first two methods consider two-lane, undivided facilities, whereas the last three methods consider divided facilities with four or more traffic lanes. Length of the overlay operation is designated as LO in the diagrams, and LSO and LSN designate the distances in the overlay and nonoverlay directions, respectively, over which vehicles travel at a reduced speed.

The designer may have several options in generating alternative traffic-handling methods. For example, consider Method III, Figure 25.4(c), for a six-lane facility. The alternatives can be to overlay one lane at a time, two lanes at a time (i.e., with a double width paver), to do either of these in normal working hours, or to pave only during certain nighttime hours. If heavy traffic volumes are involved, there can be substantial user cost differences among these alternatives.

### 25.2.5 Pavement Evaluation Alternatives

Designers should have a keen interest in the procedures used for pavement evaluation because such feedback information is directed primarily toward investment programming and design needs. It has been previously pointed out that designers should be concerned with the performance of the pavement throughout the design period, not just with the original structure. Consequently, designers need to be concerned with the information available to them throughout this period so that they may update their design predictions. In addition they may use the information to develop better design models.

For example, suppose that a highway agency conducts roughness measurements every 3 years on its secondary road network. The data, along with other periodic evaluation information, is stored in a data bank. Thus, designers can expect that they will have certain information available on the behavior and performance of any particular project, and that they can use this to monitor their design predictions. Although they may not be able to influence these evaluation policies, it is important they at least be able to specify as a part of the design strategy the expected evaluation policies.

The expected evaluation policies should be communicated to those responsible for actually conducting the evaluation throughout the analysis period. In this way, any changes in policies can be communicated back to the designers. The responsibility for such evaluation often lies with the materials and testing division of the highway agency, and often pavement design is also conducted within the division. In these cases, designers are probably quite aware of the evaluation policies. However, in some agencies, especially larger ones, design responsibility may be administratively separated from subsequent evaluation responsibility. In such cases, designers' expectations (or recommendations) of evaluation policies should be documented as a part of the design strategy and communicated to those responsible for actually carrying out the evaluation.

## Generating Alternative Design Strategies

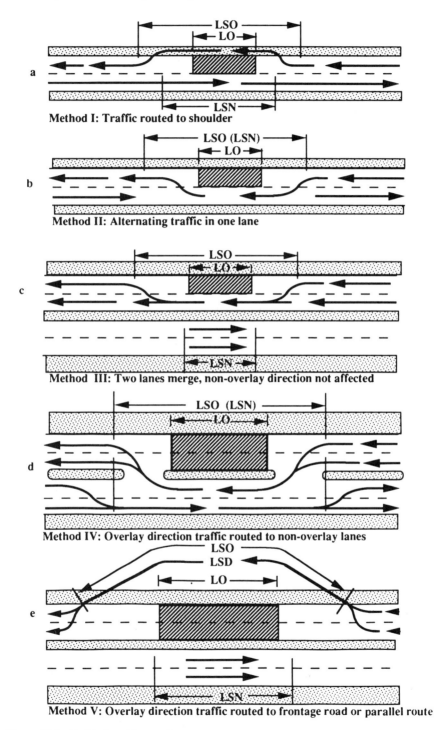

**Figure 25.4** Traffic handling methods.

### 25.2.6 Automation in Generating Alternative Design Strategies

This chapter has pointed out that a design strategy includes not only the layer thicknesses but also the material types, and the expected construction, maintenance, and evaluation policies. When all the alternatives for these design strategy components have been generated, they can be analyzed and evaluated economically to select the best one. For most design situations it is not possible to generate all the alternatives manually. Even if they could be so generated, it would probably constitute a considerable waste of time for two reasons: (1) computers can be used to generate, analyze, and evaluate the alternatives economically; and (2) many alternatives are essentially the same in economic terms, but there is no way of knowing which ones are the same until the economic evaluation has been completed.

Consequently, the generation of alternative design strategies can be combined, on a computerized basis, as in such working design systems described by various authors [Kher 70, Lytton 74, Kher 77]. In these systems, the designers simply include the following input specifications:

1. Layer materials to be considered
2. Minimum and maximum layer thicknesses that are to be used
3. Minimum time to first overlay and between overlays

With this approach, using automation, the designers can be certain that all the possible alternatives have been generated and that they will not miss the potential best one.

## 25.3 REHABILITATION AND RECONSTRUCTION ALTERNATIVES

The preceding section addressed the importance of consideration of maintenance and rehabilitation as part of the initial pavement design process to permit the evaluation of the life-cycle costs. However, as highway networks throughout the world mature, the design of rehabilitation and reconstruction projects is gaining importance. While the systematic technique described for new pavement design applies to reconstruction and rehabilitation projects, the designer is faced with additional constraints:

1. Traffic management during construction
2. Structural capacity of the existing pavement
3. Extent of distress of the existing pavement
4. Layer thickness constraints due to existing structures, e.g., need to maintain curb height
5. Construction time

In addition to these constraints, the designer is presented with site specific information that can assist with the design, such as the actual traffic on the section and the performance of the original pavement.

For highways, the alternatives available to the design engineer are generally

**Generating Alternative Design Strategies**

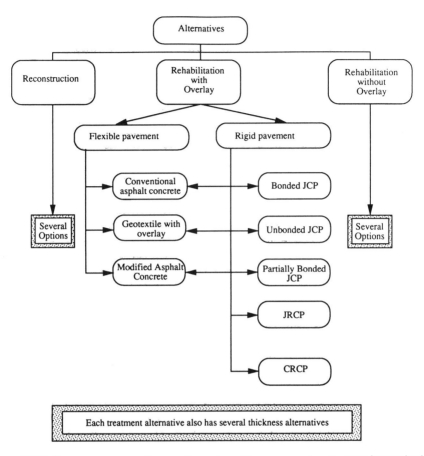

**Figure 25.5** Key components of generating alternative pavement reconstruction and rehabilitation alternatives.

referred to as 4R: reconstruction, rehabilitation, restoration, and resurfacing. The 1986 AASHTO Guide for the Design of Pavement Structures defines major rehabilitation as any activity undertaken to significantly extend the service life of an existing pavement. Reconstruction requires complete removal and replacement of the entire existing pavement structure. Once the reconstruction alternative has been selected, a new pavement design procedure can be used to design the replacement structure. Generally, complete reconstruction is an expensive alternative that is selected when the alignment of the facility must be altered. The guide classifies rehabilitation, restoration, and resurfacing as "rehabilitation other than overlay and rehabilitation with overlay."

Figure 25.5 summarizes the rehabilitation alternatives in a manner analogous to the new pavement design alternatives, Figure 25.1. Using available information on the pavement structure, the type of pavement rehabilitation method would be considered. For each rehabilitation method, the designer is faced with a range of options concerning both the types of materials and thicknesses of the layers. The ultimate decision is then constrained by the features of the design problem.

## REVIEW QUESTIONS

1. Why is it necessary to generate multiple alternative design strategies in pavement management?
2. How are various alternative designs generated?

Chapter 26

# Structural Analysis and Design of Asphalt Concrete Pavements

## 26.1 INTRODUCTION

Figure 21.4 defines a framework for a pavement design system. Chapter 22 discussed the physical characteristics of the pavement structure required for this design process and Chapter 23 described available models for the analysis of pavement structures. These models produce estimates of the primary response of the pavement structure in terms of stresses, strains, and displacements (both elastic and permanent) in the pavement structure. The purpose of this chapter is to present an overview of the models for the estimating limiting responses of the pavement structure. Detailed information on these models can be reviewed in the cited literature.

The models presented in this chapter are different from those presented in Chapter 16 with respect to the depth of analysis and amount of data required to use the model. The reason for these differences is related to the way the models are applied. The models presented in Chapter 16 are suitable for network level analysis whereas the models presented in this chapter are for application at the project level which requires greater detail.

Although there are many types of pavement distress, as discussed in Chapter 10, the state of the art in asphalt pavement design is limited to the prediction of distress types:

1. Load associated cracking (fatigue)
2. Load-associated permanent deformation (rutting)
3. Low-temperature (shrinkage) cracking

In addition, the empirical model for predicting performance, developed from the AASHO Road Test is described in this chapter.

## 26.2 PREDICTING FATIGUE CRACKING

Fatigue cracking has been reported to be the most prevalent form of structural distress of asphalt pavements in the United States [HRB 73]. Considerable research has been conducted on developing design methods to minimize fatigue cracking. Two basic approaches have been developed for the prediction of fatigue cracking: the phenomenological approach and the fracture mechanics approach.

### 26.2.1 Phenomenological Approach

In this approach, the critical tensile strains in the pavement structure are used to estimate the fatigue life of the pavement with an equation of the form:

$$Nf = a(1/\varepsilon)^b \qquad (26.1)$$

where

$Nf$ = number of load applications to initiate cracking
$\varepsilon$ = critical tensile strain
$a, b$ = constants relating to the nature of the mixture, determined from laboratory testing

Some authors have included a term in the equation for the stiffness of the asphalt concrete mixture.

Numerous researchers have investigated the fatigue properties of asphalt concrete using a variety of test methods, as shown on Figures 26.1, 26.2, and 26.3. Figure 26.1 demonstrates the effect of the stiffness of the asphalt concrete on the measured fatigue properties [Finn 73a, 73b]. The Kingham equations demonstrated on this graph indicate that stiffness affects the $a$ coefficient of equation 26.1, whereas the University of California data indicates that stiffness affects the slope of the relationship (the $b$ coefficient of a log-log plot). Figure 26.2 [Bergan 73] supports the University of California approach since the slope changes for different temperatures. Figure 26.3 [Rauhut 82] shows the state of the art in fatigue analysis based on the work of several researchers using a variety of test methods. The $b$ coefficients for the curves shown on Figure 26.3 range from 2.70 to 5.51 with a mean of 3.84.

Two of the curves on Figure 26.3 were developed through an analysis of the AASHO Road Test data. Generally, curves developed from field data indicate a longer service life than laboratory test results. Reasons why fatigue life estimates using laboratory test values fall short of inservice fatigue lives include [Rauhut 75]:

# Structural Analysis and Design of Asphalt Concrete Pavements

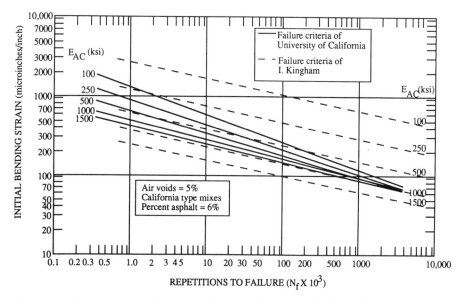

**Figure 26.1** Fatigue curves for some California mixes using different failure criteria [Finn 73].

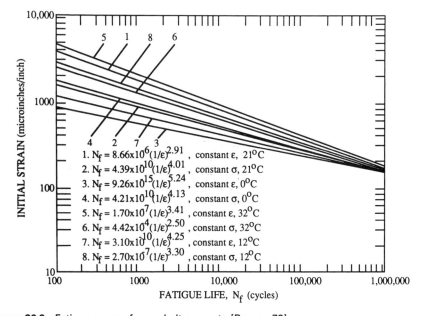

**Figure 26.2** Fatigue curves for asphalt concrete [Bergan 73].

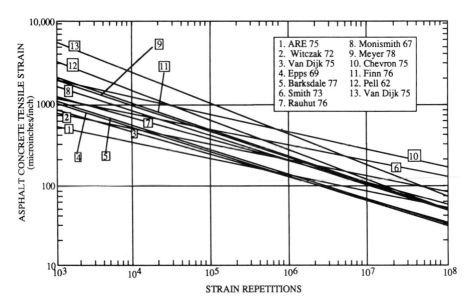

**Figure 26.3** Fatigue relationships, including laboratory testing, wheel-track testing, and efforts to represent field conditions [Rauhut 82].

1. Failure in laboratory tests is relatively sudden, soon after cracks initiate, whereas some maximum acceptable amount of surface cracks may occur in the field long after initial cracking at the bottom of the asphalt layer.
2. Actual stress and strain responses of a pavement surface layer reticulated by cracks at the bottom are considerably different from those estimated with elastic layer theory.
3. Minor material or test variations in the exponent coefficient $b$ can cause major variations in the fatigue life because of the exponential nature of the relationship.

Additional factors include healing of the asphalt between loads, relaxation of residual stresses, and variability of the traffic in the wheel path [Hoyt 87].

Considerable work is required to obtain field verification of fatigue relationships and to develop predictive methods for estimating the amount of cracking. For example, one researcher determined that the coefficient $a$ should be multiplied by a factor of 13 to adjust for the difference between laboratory and fatigue behavior of asphalt concrete [Finn 77].

With this approach to the fatigue analysis, Minor's rule can also be used for the development of equivalency factors for different traffic loadings. Research has demonstrated that load equivalency factors can be computed as the ratio of the strain caused by a particular load to the strain caused by the reference load raised to the power of the fatigue equation [Deacon 69, Zaniewski 78]. Using an equation developed through an analysis of the AASHO Road Test data, Zaniewski demonstrated that the load equivalency factors in the AASHTO Interim Guide for the Design of Pavement

Structures [AASHTO 72] could reasonably be applied when computing the fatigue damage of a pavement. However, concluding the AASHTO equivalency factors can be used in a mechanistic analysis is only valid when the fatigue behavior of the surface material is similar to the performance of the materials at the AASHO Road Test. If a new material type is used for a pavement structure, then new load equivalency factors should be developed.

### 26.2.2 Fracture Mechanics Approach

The range of fatigue equations available in the literature and variability of fatigue testing demonstrates a need for an improved method for fatigue analysis. Several researchers have questioned the validity of the phenomenological approach to fatigue analysis. For example, the coefficients $a$ and $b$ in equation 26.1 are influenced by the type of load applied, dimensions of the test sample, loading rate, test type, temperature and properties of the mix, including air voids, aggregate gradation, type and texture, and asphalt content and viscosity [Hoyt 87]. Thus, $a$ and $b$ are not material constants. Therefore, one would anticipate the need to establish a new fatigue equation for each design situation. However, since fatigue testing is costly and time consuming, this is not practical for routine pavement design.

Crack propagation can be estimated using Paris's law and fracture mechanics [Paris 63, Majidzadeh 71, 73, Jayawickrama 87]. The basic Paris's law for crack propagation is:

$$dc/dN = A(\Delta K)^n \qquad (26.2)$$

where

$c$ = crack length
$N$ = the number of load cycles
$dc/dN$ = rate of crack growth, or the change in crack length for one load cycle
$\Delta K$ = the change in the stress intensity factor during the application of the load
$A$ = fracture coefficient
$n$ = fracture exponent

The stress intensity factor is calculated with finite element analysis. When applied to pavement analysis, equation 26.2 is integrated with respect to the thickness of the pavement surface. The primary advantage of the fracture mechanics approach over the phenomenological approach is that the coefficients $A$ and $n$ in equation 26.2 are material constants, rather than empirical observations dependent on testing configuration. Theoretical analysis has demonstrated that the fracture coefficient and exponent of a viscoelastic material are material constants [Schapery 81, Hoyt 87]. The fracture coefficient is a function of the tensile strength and creep compliance of an asphalt sample tested in tension. The fracture exponent depends solely on the slope of the creep compliance curve. The theoretical derivations have been verified with empirical observations of asphalt concrete behavior using the overlay tester at Texas A&M University.

Although the fracture mechanics approach appears to be fundamentally more sound than the phenomenological approach to fatigue analysis the method has not been widely applied to the analysis and design of pavement structures. This could be due to a lack of standard test methods for determining the creep compliance properties of asphalt materials. It is anticipated that the Strategic Highway Research Program will provide the tools required to apply the fracture mechanics approach to the analysis of asphalt pavements.

## 26.3 PREDICTING PERMANENT DEFORMATION

Excessive rutting or permanent deformation in highway or airport pavements can accelerate other forms of structural deterioration and can create a safety hazard. With increasing magnitudes and repetitions of loads and increased tire pressures, the rutting problem has become severe in many highway pavements. Rutting appears as deformation of the transverse profile of the pavement surface. However, studies of trenches at the AASHO Road Test demonstrated deformations can occur in all layers of the pavement structure and not just in the surface of the pavement. These studies showed the rutting occurred as a result of lateral deformations of the pavement materials rather than densification.

Design approaches for the control of permanent deformation can be described as either indirect-prescriptive methods or predictive methods. The indirect methods require (1) providing minimum layer thicknesses and component material strengths, stability, or density or (2) limiting the vertical compressive strain at the top of the subgrade to an acceptable level. Four predictive methods are in use or under development:

1. Relating permanent strain to calculated elastic strains or stresses
2. Calculating permanent strains by use of viscoelastic layer theory
3. Relating permanent strains to creep test results
4. Developing predictive equations or rutting indicators from repeated load tests

Only the first two methods are presented in this text.

### 26.3.1 Limiting Permanent Strain by Indirect Means

The approach of limiting permanent deformations through the use of minimum layer thicknesses and minimum component material properties is inherently contained within many structural design methods. The California Bearing Ratio, CBR, method was originally developed to prevent excessive permanent deformation, with thickness selection based on field experience correlated with strength indexes. The U.S. Corps of Engineers CBR design method provides thicknesses based on the CBR of the subgrade and specifies minimum CBR values for the untreated pavement materials and minimum Marshall stability values for the asphalt concrete [Yoder 75].

One approach to preventing excessive permanent deformation is to limit the vertical compressive strain on the subgrade, as calculated with elastic layer theory [Dorman 62, Klomp 64]. The Dorman and Klomp criteria shown in Figure 26.4 were

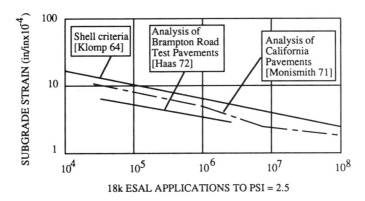

**Figure 26.4** Relationship between vertical compressive strain on subgrade and axle load applications.

developed based on designing pavements according to the CBR procedure and an analysis of the AASHO Road Test data for a limiting rut depth of 3/4 inch. Further criteria using the limiting subgrade strain approach were developed based on an analysis of the Brampton Road Test [Haas 72] and for California [Monismith 71]. Both of these criteria are more conservative than the Shell criteria.

### 26.3.2 Use of Viscoelastic Layer Theory

Several investigators have used a direct approach to prediction of permanent deformation through representing the pavement as a linear viscoelastic layer system [Barksdale 67, Moavenzadeh 72, 74a, 74b, Soussou 74]. There are several basic forms of equations that can be used to capture the time dependent deformation characteristics of materials. The most common form uses the hereditary integral. Time dependency is evaluated as a creep compliance function of the form [Moavenzadeh 72]:

$$Dj = \sum_{i=1}^{n} (G_i e^{-t\partial i}) \quad j = 1, 2, 3 \ldots \quad (26.3)$$

where

$Dj$ = creep compliance function
$t$ = time interval
$G_i$ = coefficient of the Dirichlet series
$\partial i$ = exponent for the coefficients for the Dirichlet series
$j$ = number of intervals used to evaluate the Dirichlet series

This formulation of the creep compliance function is used in the permanent deformation prediction subroutine of the U.S. Federal Highway Administration's VESYS III design system. A sensitivity analysis of VESYS indicated the predictions of permanent deformation are quite sensitive to the $Gi$ and $\partial i$ parameters of the Dirichlet series [Rauhut 75].

VESYS requires 47 input variables to describe the pavement layer thicknesses,

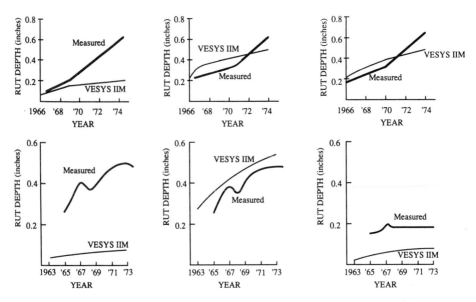

**Figure 26.5** Comparison of measured and predicted rut depths for several Brampton and AASHO Road Test sections [Rauhut 77].

traffic and load data, materials characteristics, and control variables. The materials are characterized with respect to the viscoelastic response by creep compliance curves. For the surface layer, a master curve is used along with a time-temperature shift function describing the behavior of the material at different load durations and temperatures. Long-term repetitive load tests are used to define the permanent deformation potential for each pavement material in terms of the coefficients ALPHA and GNU. Figure 26.5 compares measured versus predicted rut depths for the AASHO Road Test data [Rauhut 77]. There is reasonable agreement for some of the sections but several of the sections show appreciable divergence.

## 26.4  LOW-TEMPERATURE SHRINKAGE CRACKING

Low-temperature shrinkage cracking of asphalt pavements is a serious problem in many areas of the United States and Canada. A considerable amount of research has been directed toward developing a design technology for the problem. The basic design approaches can be categorized as:

1. Selection of an appropriate grade of asphalt cement for the design temperature conditions, within certain specification limits
2. Designation of a limiting stiffness for the asphalt, or the mix, for the design temperature conditions
3. Prediction of cracking temperature using estimated or measured stiffness values of the mix and expected field temperature conditions

4. Prediction of frequency of cracking at various ages for the design under consideration, based on empirical relationships

The first two methods are prescriptive in nature, that is, they are based on the assumption that if the designer specifies asphalts, or an asphalt mix design, with the desired characteristic then distresses in the pavement will not exceed "acceptable" limits. Such methods are useful as interim measures for the practitioner but they suffer from a lack on knowledge of the mechanisms that precipitate the distress.

### 26.4.1 Fracture Temperature

The design approach based on an estimation of fracture temperature for expected inservice temperature conditions is a variation on the limiting stiffness approach [Haas 68, Peutz 68]. Fracture temperature is estimated by comparing the tensile strength of the mix with the development of thermal shrinkage stresses in a restrained layer. If the calculated fracture temperature is higher than the design temperature, cracking is expected to occur, and the design being checked should be rejected or modified if possible. The stiffness of the mix must be reduced.

This approach has given results that compare quite well with field observations. However, it is not able to estimate the amount of cracking in those cases where the expected design temperature is lower than the calculated fracture temperature. Figure 26.6 is an example from one of the sections of the Ste. Anne Test Road [CGRA 70]. The estimated cracking temperature for this situation would be about $-35°C$.

### 26.4.2 Estimating Cracking Frequency

Figure 26.7 presents an approach for estimating the cracking frequency due to low temperatures [Haas 73]. The method uses easily available design inputs, including asphalt stiffness, winter design temperature, thickness of the asphalt layer, and subgrade soil type. McLeod's method is used for estimating asphalt stiffness [McLeod 72]. An empirical relationship is used to estimate the cracking index. Criteria developed from field experience are used to determine if the mix is acceptable or should be modified to meet the design conditions.

## 26.5 PERFORMANCE PREDICTION

Although design to limit distresses as described in the preceding sections is important, the primary operating characteristic of a pavement is the serviceability-age history, or performance. The major function of a design method, therefore, is the ability to predict the performance of alternative design strategies. The performance prediction needs to cover the entire design period so that a complete economic analysis can be conducted. Moreover, unless the actual serviceability-age relationship is predicted, the effects of varying serviceability on user costs and the effects of changing the minimum acceptable level of serviceability cannot be evaluated completely. In some design methods, however, only the initial service life is predicted, in effect, the age at which the serviceability reaches a minimum acceptable level.

If the design period extends beyond the initial service life of the alternative being

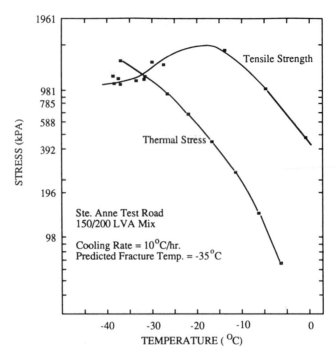

**Figure 26.6** Example of thermal stress calculation and prediction of fracture temperature for a restrained strip of asphalt concrete [CGRA 70].

considered, then the prediction of the serviceability-age relationship must also include the effects of any overlay or other major rehabilitation.

A functional form of the serviceability of a pavement at any age can be expressed as:

$$PA_i = f(PA_0, Ai, C, hp, Sp, Li, W, Md, I) \tag{26.4}$$

where

- $PA_i$ = serviceability (i.e., in terms of Present Serviceability Index),
- $PA_0$ = serviceability at time 0
- $Ai$ = pavement age at time interval $i$
- $C$ = construction effects
- $hp$ = thickness of the pavement layers
- $Sp$ = stiffness of the pavement layers and subgrade
- $Li$ = traffic loads carried by the pavement to time $Ai$
- $W$ = set of climatic or environmental variables
- $Md$ = type and degree of maintenance
- $I$ = possible interactions of the above factors

The variables in equation 26.4 are really classes of factors, except for the serviceability and age terms. Each class can be further subdivided into a set of individual

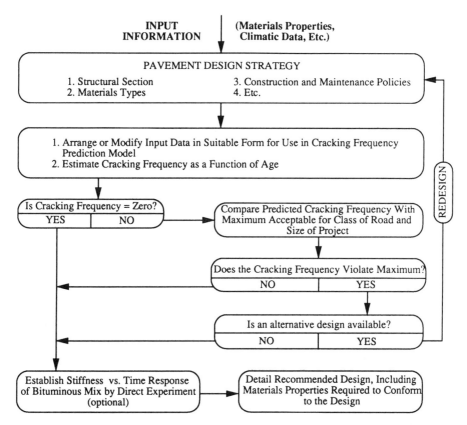

**Figure 26.7** Basic steps in checking a design for low-temperature shrinkage cracking frequency [Haas 73].

variables. The interaction class, $I$, can contain a large number of terms. Thus, a truly comprehensive, reliable, and universal performance prediction model represented by this equation would be extremely complex. This is why previous efforts at developing models of this sort, such as those of the Canadian Good Roads Association in the late 1950s and early 1960s contained sizeable errors of estimate in predicting serviceability [CGRA 65].

If distress could be related directly, as a predictor of performance, and if fundamental, reliable, and universally applicable methods of predicting distress could be developed, then the need for separate performance prediction models might not be so critical. The current state of technology is, in effect, the reverse; distress predictions are used only as a check on those design alternatives that have been predicted to have an acceptable performance history.

### 26.5.1 Performance Prediction Models

A true performance prediction model is one that can calculate the expected serviceability-age, or serviceability-traffic, relationship over the entire design period. This

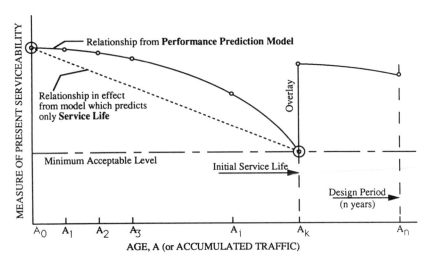

**Figure 26.8** Illustration of the differences between the application to a pavement design alternative of a true performance prediction model and a service life prediction model.

definition can be extended, for the purposes of the scope of this chapter, to include those models that predict only the service life. Figure 26.8 illustrates the distinction. Whereas a performance prediction model produces the solid-line relationship of Figure 26.8, for the design alternative under consideration, a service life estimation model actually yields only the point $A_k$. However, a dashed line has been shown in the diagram to illustrate the relationship that is produced by the service life estimate.

It should be noted that a service life estimate also corresponds to a single fixed minimum acceptable level of serviceability, as shown on Figure 26.8. With a performance prediction model, this level can be allowed to vary, and consequently the effects of such variations can be evaluated. The particular design alternative of Figure 26.8 includes an overlay at age $A_k$. A comprehensive performance prediction model should also be able to predict that portion of the relationship between $A_k$ and $A_n$ as a function of the type and thickness of the overlay. If a service life estimation model has been used to obtain $A_k$, then it is usual practice to have a separate service life estimation model for the overlay itself. Alternatively, some design methods have one performance prediction model for the initial pavement structure, that is, for $A_0$ to $A_k$, and another performance prediction model for the overlaid pavement, $A_k$ to $A_n$.

Still other methods apply the performance prediction model used for initial or new pavement structures to the overlaid pavement and simply consider the overlaid pavement to be a new structure. They usually do this by adding some portion of the existing thickness to the overlay thickness to obtain an effective thickness of "new" pavement for purposes of performance prediction.

### 26.5.2 AASHTO Guide Performance Models

The original AASHO Interim Guide for the Design of Pavement Structures was published in 1961 and revised in 1972. Extensive modifications were performed and published in 1986 as the AASHTO Guide for the Design of Pavement Structures. While the 1986 Guide introduced several new concepts, the underlying models for the prediction of pavement performance are similar to the original interim guide and were derived from the AASHO Road Test. The basic performance equation in the guide is:

$$G_t = \beta(\log W_t - \log \rho) \tag{26.5}$$

where

$G_t$ = the log of the ratio of the serviceability loss at time $t$ to the potential loss to a serviceability index level of 1.5

$$G_t = \log((p_o - p_t)/(p_o - p_T)) \tag{26.6}$$

where

$p_o$ = the initial serviceability of the pavement
$p_t$ = the serviceability at time $t$
$p_T$ = the terminal serviceability; 1.5 was used for the development of the guide
$\beta$ = a function of design and load variables that influence the shape of the performance curve
$\rho$ = a function of design and load variables denoting the expected number of axle load applications to the terminal serviceability level

For the conditions at the AASHO Road Test, regression equations were developed for $\beta$ and $\rho$ for flexible pavements:

$$\beta = 0.40 + 1094/(SN + 1)^{5.19} \tag{26.7}$$

and

$$\rho = 9.36 \log(SN + 1) - 0.20 \tag{26.8}$$

where

$SN$ = the structural number for the pavement section

Combining equations 26.5 through 26.8 and rearranging to solve for the number of axle loads that can be carried yields:

$$\log W_{18} = \rho + G_t/\beta$$

or

$$\log W_{18} = 9.36 \log(SN + 1) - 0.20 + \frac{\log((p_o - p_t)/(p_o - p_T))}{0.40 + 1094/((SN + 1)^{5.19})} \tag{26.9}$$

For the AASHO Road Test, $p_o$ was equal to 4.2.

Equation 26.9 can be used for design; however, the results are applicable only to the AASHO Road Test conditions. For application to other regions, the equation must be modified. In the earlier versions of the guide, the modifications took the form of a regional factor and a subgrade support modification. The 1986 version of the guide uses a significantly different approach to the modification of equation 26.9 for pavement design.

**Table 26.1  Suggested Levels of Reliability for Various Functional Classifications [AASHTO 86]**

| Functional Classification | Recommended Level of Reliability | |
|---|---|---|
| | Urban | Rural |
| Interstate and other freeways | 85–99.9 | 80–99.9 |
| Principal arterials | 80–99 | 75–95 |
| Collectors | 80–95 | 75–95 |
| Local | 50–80 | 50–80 |

The regional and subgrade support values of the previous guide have been replaced with the "effective roadbed soil resilient modulus," $M_R$. This is a weighted value for the resilient modulus based on the relative damage caused to the pavement during periods of different subgrade strength. The subgrade strength can vary throughout the year due to fluctuations in the moisture content of the subgrade and in periods of frozen subgrade conditions. The guide suggests the seasonal subgrade modulus can be estimated using a laboratory established relationship between moisture content and the resilient modulus, and estimates of the moisture content of the subgrade throughout the year. Alternatively, deflection measurements can be collected throughout the year and the modulus back-calculated. Using either method, the relative damage factor, $u_f$, is computed for each period of differing subgrade modulus using:

$$u_f = 1.18 \times 10^8 \times (M_R)^{-2.32}. \qquad (26.10)$$

The average relative damage factor is computed and equation 26.10 is used to determine the effective roadbed soil resilient modulus.

For design, the relative damage term is used to modify the predicted allowable traffic. Since the design equation is expressed in terms of the log of traffic, the log of $u_f$ is added to the design equation.

The second major modification to the pavement performance equation 26.9 was the inclusion of the reliability concept into the equation as described in Chapter 24. This requires determining the overall standard deviation of the pavement performance process and selecting a reliability factor. Currently the only source of information for the standard deviation of pavement performance is the AASHO Road Test, modified for variations in traffic estimates. The AASHTO Guide recommends a value of 0.45 for the standard deviation for flexible pavement design. The reliability factor is a function of the "importance" of the road. Pavements that serve a high level of traffic are more critical than pavements for low volume roads and therefore should be designed to a higher level of reliability. Suggested reliability factors from the guide are given in Table 26.1. For use in the design equation, the reliability factor is converted to the standard normal deviate, $Z_R$, using Table 26.1.

The effective roadbed soil resilient modulus and the reliability factor were incorporated into the performance equation to define the design equation. With these modifications, the design equation becomes:

$$\log(W'_{18}) = Z_R \cdot S_0 + \log(W_{18}) + \log(u_f) \qquad (26.11)$$

## 26.6 USE OF STRUCTURAL MODELS FOR PROJECT LEVEL PAVEMENT MANAGEMENT OF ASPHALT PAVEMENTS

As pointed out in Chapter 21, predictive models lie at the heart of project level pavement management. Inputs are transferred into a model. Models are used to predict behavior, and/or distress, and/or performance. This chapter summarizes the available models for use in project level pavement management or what is sometimes called "the design function" for project level pavement management. Specific versions of these models have been used in computerized project level pavement management systems FPS, SAMP, and OPAC [Hudson 73, Finn 77, Kher 77]. To examine the implementation of these models, please refer to Part Six, in particular Chapter 37 which summarizes the working details of the SAMP program.

### REVIEW QUESTIONS

1. What are the primary factors used to assess the structural integrity of asphalt concrete pavements?
2. What are the basic factors related to low-temperature shrinkage cracking?

Chapter 27

# Structural Analysis and Design of Portland Cement Concrete Pavements

## 27.1 INTRODUCTION

Portland cement concrete pavements must be designed for both traffic and environmental stresses. However, unlike asphalt pavements, the portland cement concrete slab carries load through bending action. The traffic load stresses greatly spread out before being transmitted to the subgrade soil. The concrete slab acts as a rigid structural member hence concrete pavements are frequently referred to as rigid pavements. Because the slab is a structural member, cracking and distress of the pavement surface have a considerably different influence on the performance of the pavement than is the case for asphalt pavements. Although both types of pavements are susceptible to transverse cracking, the consequences of the transverse crack in the load-carrying concrete slab are more important than in the flexible pavement. Hence, joints and/or reinforcement are designed to control the longitudinal movement of the concrete pavements. Pavements without reinforcement are called jointed plain concrete pavements (JPCP), pavements with both joints and reinforcement are jointed reinforced concrete pavements (JRCP), and pavements with considerable reinforcement but no joints are continuously reinforced concrete pavements (CRCP).

Both mechanistic and empirical models are used for the analysis and design of concrete pavements. In a Federal Highway Administration course on pavement design

[FHWA 87a] the discussion of mechanistic procedures for concrete pavement distress is limited to fatigue analysis. Empirical models are presented for slab cracking, transverse joint faulting, joint deterioration, pumping, and serviceability loss. Westergaard, Spangler, Hudson, and many others have treated these subjects in detail.

## 27.2 STRUCTURAL DESIGN OF THE PAVEMENT SLAB

Rigid pavement design requires determination of not only the slab thickness, but also the size of the slabs, the type of load transfer, and reinforcement of the slab. These can be described as independent topics, but it should be realized the final design requires consideration of the interaction between the design factors.

### 27.2.1 Mechanistic Analysis for Fatigue Cracking

Rigid pavement cracking is generally a result of a combination of load stresses and temperature stresses. Design is usually based upon an analysis of the load stresses combined with an estimate of temperature or moisture warping stresses.

Historically, most pavement design methods have been based upon the prediction of stresses as compared to a working strength of the pavement concrete which is usually the breaking tensile strength divided by a safety factor.

Conceptual models of fatigue behavior of concrete have been studied and used by some. The basic model is:

$$\log(N) = a - b(\sigma/f_r)$$

or

$$N = a(f_r/\sigma)^b \tag{27.1}$$

where

$N$ = number of applications of stress to fail the concrete
$f_r$ = modulus of rupture of the concrete
$\sigma$ = critical stress in the concrete
$a, b$ = fatigue coefficients

Many people have carried out fatigue tests of concrete samples; however, since the stress ratio of concrete is a function of the logarithm of the number of applications, long-duration tests are required at low stress ratios. Very little testing has been carried out beyond 10 million applications. This is illustrated in Figure 27.1 where it is shown that there are a number of specimens at 10 million applications which did not fail. In the very early days, it was assumed that concrete structures, including pavements, were designed such that if the stress to strength ratio was always below 0.50 the material would never crack. This fallacy grew out of the fact that up to the 1950s, fatigue testing was almost never carried out beyond a million load applications. Subsequent testing has shown that there is no such thing as a fatigue limit of 0.50. There is merely a limit on the amount of testing that can be carried out.

Figure 27.1 demonstrates the results of flexural fatigue testing of concrete samples

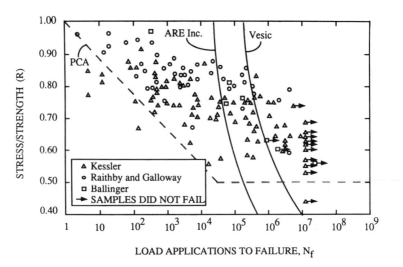

**Figure 27.1** Typical fatigue data and relationships for portland cement concrete [FHWA 87].

is highly variable. The PCA model shown on Figure 27.1 typifies the fatigue limit model previously discussed. As pointed out, this is not a valid methodology.

Fatigue models developed from an analysis of the AASHO Road Test data are also shown on Figure 27.1 [Vesic 69, Treybig 77]. Stresses for the Treybig analysis were computed using ELSYM5 so an interior load condition was assumed and failure was defined based on the extent of cracking of the sections. Vesic and Saxena used a Westergaard analysis and a failure criteria of a serviceability level of 2.5. The Treybig and Vesic equations are, respectively:

$$N = 23440(f_r/\sigma)^{3.21} \tag{27.2}$$

and

$$N = 225000(f_r/\sigma)^4 \tag{27.3}$$

As shown on Figure 27.1, these two analyses of the fatigue behavior of the AASHO Road Test sections demonstrate a considerably different behavior compared to either the PCA model or the laboratory data. Basically, the differences between the fatigue failure based on load stresses and observation at the Road Test relate to the combined forces of environment and load as well as the environmental effects on the strength of concrete itself. Hudson carried out significant work in this area after the Road Test [Hudson 62, 63a, 63b, 64a, 64b, 66a, 66b, 66c, 67].

### 27.2.2 Use of Theory to Modify the Road Test Equations

The most valuable use of theoretical analysis of portland cement concrete pavement slabs was to extend the results of the AASHO Road Test. The Road Test concrete strength was fixed to a single value and there were only two levels of subgrade support. While the Road Test equations themselves clearly show the field performance of concrete pavements, it was desirable to add additional design variables such as concrete strength, modulus of subgrade reaction ($K$), and type of joint factor. As a part of his work for the Texas Highway Department, and as a senior staff member of the AASHTO Pavement Design Task Force in 1962 and 1963, Hudson contributed to the development of the modifications of the Road Test equations [Hudson 63a, 67].

It is important to note that all modern rigid pavement design methods for highways, including the AASHTO Design Guide and the RPS Rigid Pavement Management System use the modified AASHO Road Test equations, which combine practice and theory as outlined here.

### 27.2.3 Empirical Models of Rigid Pavement Damage

Given the complexity of the analytical situation, it is understandable that much of the research on pavement behavior has extended empirical data. In addition to the AASHO Road Test, there are several data bases of the performance of concrete pavements developed through Federal Highway Administration and National Cooperative Highway Research Program projects. In addition, the Long Term Pavement Performance project under the Strategic Highway Research Program may provide further insight into the performance of concrete pavements. While empirical methods can produce predictive models that carry forward the state of the art in pavement engineering, it should be remembered that these models are only as good as the data base used for the model development. Frequently pavement engineers are forced to extrapolate the application of empirical models beyond the inference space of the data base. This can produce a flawed analysis.

Among other relationships, equations for estimating the cracking of jointed reinforced concrete slabs were developed at the AASHO Road Test for dual and tandem axles:

Single axle:
$$\log(W) = 4.95 + 0.5 \cdot \log(C) - 2.30 \cdot \log(L_1) + 3.57 \cdot \log(D) \quad (27.4)$$

Tandem axle:
$$\log(W) = 6.37 + 0.5 \cdot \log(C) - 3.13 \cdot \log(L_1) + 3.96 \cdot \log(D) \quad (27.5)$$

where

$W$ = predicted number of axle applications of magnitude $L_1$
$C$ = cracking index, feet cracking/1000 ft$^2$
$D$ = slab thickness, in.
$L_1$ = magnitude of the load, kips

These equations are not convenient for pavement design since the number of applications of specific load magnitudes are estimated as a function of an allowable cracking index. One could solve the equations for $C$ as a function of $W$, and use the resulting equation for estimating the amount of cracking that a certain traffic mix would cause.

However, this would violate the way the equation was developed and could result in incorrect answers.

An equation for estimating joint faulting in plain concrete pavements is [Packard 77]:

$$F = 1.29 + 48.95 \cdot \left[\frac{(T \cdot A^2)}{D^{3.9}}\right] \cdot S^{0.610} \cdot (J - 13.5)^b \qquad (27.6)$$

where

$F$ = average joint fault in 32nds of an inch
$T$ = average number of tractor semi-trailer combination trucks in one direction per day over the analysis period
$A$ = analysis period, years
$S$ = subgrade drainage factor, 1 for good and 2 for poor
$J$ = joint spacing, ft
$b$ = subbase factor, 0.241 for granular bases and 0.037 for stabilized bases

This equation can be used for evaluating the economics of stabilizing the subbase and for determining an optimum joint spacing. However, it has yet to be implemented into a working pavement design method.

A number of equations have been developed based on the FHWA COPES data base [Darter 85]. It is vital to note that none of these equations have a statistical validity which makes them practical for use in routine pavement design. This is in no way the fault of the authors, but is because the data base is too incomplete to support the development of reliable models.

It is likely that the SHRP data base will suffer from this same problem since all rigid pavements are built in accordance with the general understanding of load-thickness-strength relationships. This fact forces an autocorrelation between the factors. This confounding of the factors makes it impossible to separate out the effects of the individual factors with a statistical analysis. The AASHO Road Test is the only source of pavement performance data with a significantly complete factorial arrangement. If the special pavement studies of the LTPP portion of SHRP are completed, then data will be developed in the next 10 to 15 years which can be used for the development of new equations. Failing this, all highway pavement design in current use relies either on pure theory, generally based on Westergaard's model or some modification thereof, or on modified AASHO Road Test equations. In the balance of this chapter, we will address the modified AASHTO design equations.

### 27.2.4 AASHTO Model of Rigid Pavement Performance

The basic model for rigid pavements in the AASHTO Guide for Design of Pavement Structures is similar to the equation for asphalt pavements. However, the parameters in the equation are different to reflect the differences between asphalt and concrete.

# Structural Analysis and Design of Portland Cement Concrete Pavements

The performance equation for the design of rigid pavements can be expressed as:

$$\log(W_{18}) = R + \rho + G_t/\beta + c \tag{27.7}$$

where

$R$ = reliability, $Z_R \cdot S_o$
$\beta = 1 + 1.624 \cdot 10^7/(D + 1)^{8.46}$
$\rho = 7.35 \cdot \log(D + 1) - 0.06$
$c$ = factor that accounts for corner stresses, load transfer, loss of subgrade support, and drainage

$$c = (4.22 - 0.32 p_t) \times \log\left[\frac{S_c' \times C_d(D^{0.75} - 1.132)}{215.63 \times J[D^{0.75} - {}^{18.42}/(E_c/k)^{0.25}]}\right]$$

where

$p_t$ = terminal serviceability
$S_c'$ = modulus of rupture of the concrete
$C_d$ = drainage coefficient
$J$ = load transfer coefficient
$D$ = slab thickness
$E_c$ = modulus of elasticity of the concrete
$k$ = modulus of subgrade reaction

Application of this model for pavement design requires iterating the slab thickness until the minimum thickness required to carry the design traffic is determined.

The factors $S_c$, $J$, $E_c$, and $k$ all result from the modified Westergaard equations developed by Hudson and subsequently modified by Hudson and McCullough [Hudson 64a]. These are the equations used in the only known rigid pavement management system at the project level, RPS [Kher 70].

## 27.3 JOINT DESIGN

Joints in concrete pavements are often used to accommodate the volume changes of the material due to temperature. However, all joints represent a discontinuity in the pavement. The joint also presents a tedious construction detail. As a result, many distress types in concrete pavements are associated with the behavior of the joints. Thus, the design of transverse and longitudinal joints is critical to the performance of rigid jointed pavements. The components of joint design that must be considered include [FHWA 87a]:

1. Load transfer at the transverse joints
2. Spacing of the transverse joints
3. Spacing between longitudinal joints

4. Joint sealant reservoir
5. Ties between longitudinal joints

### 27.3.1 Load Transfer

Load transfer refers to the ability of the pavement to distribute the deflections and loads from one side of a joint to the other side. Inadequate load transfer can result in faulting, pumping, and corner breaks. Load transfer is accomplished with dowel bars, keyed joints, and tie bars. Aggregate interlock can be effective at providing load transfer for narrow joints. However, slab contraction due to shrinkage and temperature changes increases the joint opening and reduces load transfer at such joints and cracks.

Mechanical load transfer devices are required for jointed reinforced concrete pavements due to the length of the slab. The need for mechanical load transfer devices in plain concrete pavements is a debated issue. Dowels are frequently not used in the western states; however, many of these pavements fault and get rough. In general, there is no evidence in the United States that jointed concrete pavements can successfully perform without load transfer devices. All test sections at the AASHO Road Test had positive mechanical load transfer, and their performance was excellent. Observations of rigid pavement performance supports the need to use mechanical load transfer devices in all jointed concrete pavements.

### 27.3.2 Joint Spacing

Longitudinal joint spacing is generally determined by construction and geometric considerations. Transverse joint spacing is generally determined as a function of slab thickness. The AASHTO Guide recommends that for JPCP, transverse joint spacing in feet be less than twice the slab depth in inches, for example, the maximum joint spacing for a 10-inch pavement should be less than 20 feet.

## 27.4 REINFORCEMENT DESIGN

Reinforcement is used to control slab movement due to temperature and drying shrinkage. In JRCP, this movement is manifested primarily at the cracks in the pavement between the transverse joints. Therefore the steel design is based on keeping these cracks tightly closed to maintain aggregate interlock. The AASHTO design equation is:

$$P_S = 100LF/2f_S \tag{27.8}$$

where

$P_S$ = percent steel reinforcement required
$L$ = joint spacing, ft
$F$ = friction factor between the base and slab
$f_S$ = working stress of the steel, usually 75 percent of the yield strength

Note that equation 27.8 is not sensitive to the strength of the concrete or the amount of temperature drop that generates the stresses the steel is designed to resist.

# Structural Analysis and Design of Portland Cement Concrete Pavements

The design of reinforcing steel for CRCP is more complicated because the distance between the cracks is unknown and must be determined. Factors considered in the design process are:

1. Concrete indirect tensile strength
2. Concrete shrinkage at 28 days
3. Concrete thermal coefficient
4. Reinforcing bar or wire diameter
5. Steel thermal coefficient
6. Design temperature drop
7. Wheel load tensile stresses developed during initial loading of the pavement either by construction equipment or truck traffic

The design procedure requires iterating on the amount of reinforcing steel to satisfy criteria on the stress in the steel, the crack spacing, and the crack opening.

## 27.5 OTHER CONCRETE PAVEMENT DESIGN CONSIDERATIONS

The design of concrete pavements is a complicated process and requires consideration of many factors that will affect the performance of the pavement. Included in this list are the drainage, erodibility of the base, and reliability. The effect of many of these factors on the performance of rigid pavements is difficult to quantify. Although these factors have not been presented due to space limitations, they are important and are generally considered in a comprehensive design procedure such as the AASHTO Guide and RPS.

## REVIEW QUESTION

1. What is the basis of structural design most commonly used with portland cement concrete?

Chapter 28

# Rehabilitation Design Procedures

## 28.1 TYPES OF REHABILITATION

With the maturation of the highway system in many countries, the emphasis in highway construction is shifting from new facilities to the maintenance and restoration of existing facilities. Thus, methods for the design of rehabilitation of existing pavements are now critical in pavement management.

As pointed out in previous chapters, pavement management covers the full life-cycle of the pavement, including the initial performance period following rehabilitation. The methods described in this chapter are subsystems of a project level pavement management program. Such rehabilitation subsystems are used in SAMP, RPS, and FPS among other project level pavement methodologies.

The life-cycle of a pavement involves several types of behavior and different levels. The initial stage is typified by good performance with low maintenance required. As the pavement ages, the need for maintenance increases as more of the pavement surface is affected by distress that affects the ride quality. At some point in the pavement's life, the condition of the surface deteriorates to the point where routine maintenance is no longer cost-effective and major treatment is needed to restore the quality of the pavement. The most economical rehabilitation treatment will depend on the condition of the pavement, anticipated traffic loads, and environmental con-

# Rehabilitation Design Procedures

ditions. Rehabilitation design is a complicated process because the engineer is faced with several options and sometimes with incomplete data on the pavement section. The AASHTO Guide for the Design of Pavement Structures states:

> Therefore, a considerable amount of both analysis and engineering judgement must be applied to each project. Due to state of the art limitations relative to the entire rehabilitation process, a definite need exists for continuous feedback from agencies on the performance of various rehabilitation methods.

Figure 28.1 shows the overall process for the selection of a rehabilitation strategy [AASHTO 86]. This is a three-step process: define the problem, examine potential solutions, and select the preferred solution. Similarities should be noted between these steps and the systems method as described in Chapter 2. The problem definition process essentially establishes the environment for solutions. Defining the problem requires data collection and analysis of the condition of the existing pavement. Generating potential solutions requires generating and analyzing alternative approaches to the problem of rehabilitating the pavement. Based on the results of the preliminary analysis, the detailed design is performed and the solution implemented.

The AASHTO Guide for the Design of Pavement Structures broadly classifies pavement rehabilitation into methods for overlay, non-overlay methods, and complete reconstruction. The design procedures for new pavements are used for complete reconstruction projects. Overlay design methods determine the amount of additional pavement structure required to serve the future traffic. The major non-overlay methods include:

1. Full depth repair
2. Partial depth patching
3. Joint and crack sealing
4. Subsealing and undersealing
5. Grinding and milling
6. Subdrainage
7. Pressure relief joints
8. Load transfer restoration
9. Surface treatments

The AASHTO Guide methodology is presented here because it is widely known and has been widely examined by state DOTs. This methodology will be replaced based on an NCHRP study to be completed in 1992. Thus, nothing herein is intended to support or deny the specifics of the AASHTO methodology. The chart and the method, however, do serve to illustrate the complete aspects of a rehabilitation method.

In considering the data factors, the economic analysis, and the life-cycle cost analysis to be used in rehabilitation, it is essential that the data and the methodology be compatible with the data methodology available in the project level pavement management system. It cannot be overemphasized that the rehabilitation design meth-

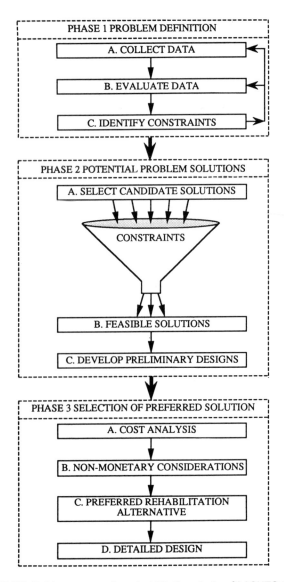

**Figure 28.1** AASHTO Guide structure for rehabilitation design [AASHTO 86].

odology used in project level pavement management must be compatible with the balance of the project level pavement management system.

## 28.2 DESIGN INPUTS AND FIELD DATA REQUIRED

The design of a pavement rehabilitation project starts with the collection of data about the condition of the project. If the agency has a network level pavement management system, a summary of required information may be available from the data base.

## Rehabilitation Design Procedures

However, it will always be necessary to supplement this information with data taken from the specific project under review.

The AASHTO Guide includes data collection guidelines for evaluating pavement rehabilitation projects [AASHTO 86]. In addition, the Federal Highway Administration developed a training course on pavement rehabilitation [FHWA 87b] that contains valuable information on pavement repair and rehabilitation methods.

The AASHTO Guide defines several elements of data collection for rehabilitation design:

1. Delineation of analysis units
2. Drainage evaluation
3. Pavement distress survey
4. Nondestructive testing
5. Field sampling and testing

As previously discussed, the data collection requirements for project level work are more intense than are required for network level analysis.

### 28.2.1 Delineation of Analysis Units

Frequently in large pavement rehabilitation projects, the properties of the pavement will vary along the length of the project. If the variations in these properties are not recognized in the selection of the rehabilitation alternative, then parts of the project will be over or under designed. Thus, it is important to isolate unique factors that influence potential pavement performance into separate sections called analysis units. The factors that can be used for delineation include:

1. Pavement type
2. Construction history, including rehabilitation and major maintenance
3. Pavement structural characteristics, layer material type, and thicknesses
4. Subgrade soil type
5. Traffic
6. Pavement condition

Figure 28.2 demonstrates how these factors can be used to separate a project into analysis units. Historical records can be used to assist in selecting analysis units. Generally, the subgrade support conditions are the most difficult factor to assess. Even in areas with uniform soil types, other factors such as drainage, compaction, cut and fill sections, and topography can cause variations in the subgrade soil. Usually the variability in subgrade support should be assessed with deflection measurements taken along the length of the pavement.

Deflection and other pavement response variables, such as skid resistance or distresses, can be used to define boundaries in the pavement condition using a variety of methods. The first step is to plot the response variable versus the distance along the pavement as shown in Figure 28.3. The analysis units can then be identified subjectively or with an analytical procedure such as the cumulative difference method recommended in the AASHTO Guide. In this method, the response variable is trans-

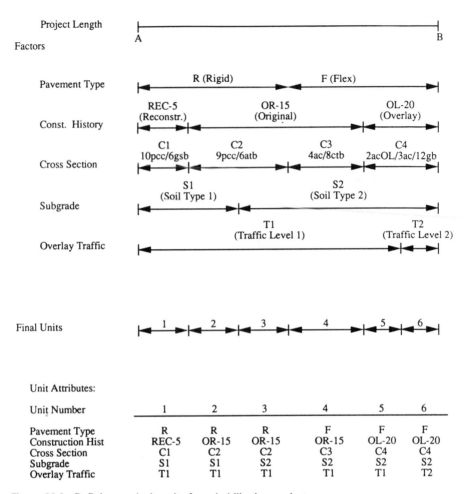

**Figure 28.2** Defining analysis units for rehabilitation projects.

formed or normalized into a new parameter, $Z_c$, defined as the difference between the area under the response curve at any distance and the total area developed from the overall project average response at the same distance. The transformed variable is plotted versus the distance along the project, and points where the slope of the resulting curve changes sign define the boundaries of the analysis units, as shown on Figure 28.3. Once the analysis units have been established through a consideration of the above factors, the engineer needs to review the location of the boundaries to ensure the sections are logical. Sections may need to be combined for practical construction considerations or economic reasons.

## 28.2.2 Drainage Evaluation

Moisture in a pavement structure is a major contributor to pavement deterioration. All rehabilitation projects should consider the drainage of the pavement structure.

# Rehabilitation Design Procedures

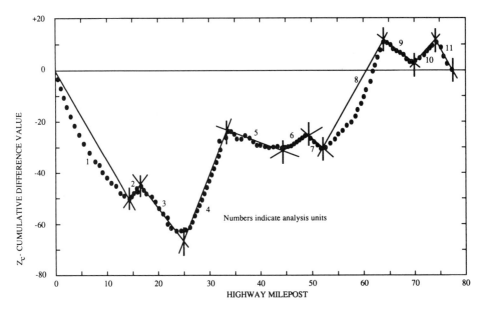

**Figure 28.3** Cumulative difference approach for establishing analysis units [AASHTO 86].

Distresses in flexible pavements that are caused or accelerated by moisture include stripping, rutting, depressions, fatigue cracking, and potholes. Distresses in rigid pavements that are moisture related include D-cracking, joint deterioration, faulting, and corner breaks. Freeze-thaw, differential frost heave, and spring breakup all indicate the pavement structure retains excessive moisture. Sometimes, the need for drainage improvements is related to excessive pavement distress. However, even in the absence of pavement distress, the drainage features of the existing pavement should be evaluated and any needed corrections identified.

The factors that should be evaluated in a drainage survey include the existing drainage features, topography, permeability of the pavement materials and subgrade soil, and the climatic zone. There are several excellent references on drainage surveys and drainage design [Cedergren 74, Moulton 80]. A summary of these methods is beyond the scope of this book. However, the pavement manager should appreciate the importance of drainage and the need to include drainage considerations in the design of a pavement rehabilitation project.

## 28.2.3 Detailed Distress Survey

Distress surveys have been previously defined as an important feedback data item related to pavement management. Once a project section for project level pavement management or rehabilitation is selected, it will be necessary to conduct a detailed distress survey. Chapter 10 describes a variety of pavement survey methods and should be referred to at this point. One of the more detailed methodologies should be adopted and a method of summarizing the data or evaluating possible design response to the observed distress types should be selected. More details related to rehabilitation selection based on the distress type can be obtained from various sources [Yoder 75, AASHTO 86].

### 28.2.4 Nondestructive Pavement Evaluation

As described in Chapter 9, nondestructive pavement evaluation through deflection measurements is an extremely valuable and cost-effective tool for determining the structural condition of an existing pavement. For the design of rehabilitation projects, deflection evaluations provide information on the structural capacity of the existing pavement for overlay design, the location of differences in the subgrade characteristics, and voids and load transfer problems in rigid pavements. As identified in Chapter 9, the sampling plan is an important feature of the nondestructive pavement evaluation. The engineer must make a tradeoff between the cost of the data collection and the number of test locations required for the design of the pavement rehabilitation strategy. The number of samples depends on the design situation. For flexible pavements on a relatively uniform subgrade soil, the Arizona Department of Transportation collects 5 deflection measurements per mile. Other agencies collect as many as 20 deflection measurements per mile. Agencies investigating void location and joint load transfer of rigid pavements may need to take 2 or more measurements per slab.

### 28.2.5 Field Sampling and Testing

Even with an effective nondestructive testing program, it will be necessary to collect some cores of the pavement structure. These are required for verifying the thicknesses of the pavement layers and for providing samples for laboratory analysis. The extent of the sampling program required depends on the condition of the existing pavement. If the pavement is in relatively sound condition, then the majority of the information required can be determined with historical records and nondestructive testing. However, if the pavement is displaying a materials related problem, such as stripping or major rutting, then a more thorough analysis of the materials may be required.

## 28.3 PERFORMANCE OR DETERIORATION MODELS

The ability to design a rehabilitation strategy depends on the availability of models for predicting the performance of the rehabilitation strategy. Without knowledge of the performance of the rehabilitation strategy, the life of the repair cannot be estimated and therefore, the cost-effectiveness cannot be compared to other alternatives. Unfortunately there is not a reliable set of performance models for rehabilitated pavements which is universally accepted. Whereas the AASHO Road Test provides us with a great deal of information on new pavements, there is no comparable data set for rehabilitated pavements. Thus, most of the models used in rehabilitation performance prediction are based on modified Road Test equations or on sketchy observations of small experiments in the field.

### 28.3.1 Restoration Performance Models

Little is known about the performance of pavements which have been restored. Restoration is generally a term associated with rigid pavements and the restoration process

# Rehabilitation Design Procedures

is one that has been developed by the industry which involves undersealing, joint repair, and replacement. While we call the reader's attention to some of these concepts, we will not present additional information at this point because there is no proven performance record for such pavements. Even where such pavements have performed well, there was not complete information on the "before restoration" condition. The AASHTO Guide presents one model for the prediction of distress of restored rigid pavements. The reader may refer to that reference for additional details.

### 28.3.2 Overlay Performance Models

Conceptually, the performance model for an overlay may be different than the model for a new pavement. The subgrade soil, subbase, and base have been subjected to repeated stress applications which should densify the material and reduce the potential for further deformation. On the other hand, if the existing pavement has cracked, the unbound layers may be saturated and therefore have a lower strength than when they were new. The bound pavement materials have been subjected to repeated traffic applications and therefore some of the fatigue resistance of these materials has been used up.

While the relative performance of a new pavement versus an overlaid pavement is subject to speculation, the pavement manager must have a performance model for use in the pavement management system. The basic overlay equation used in the AASHTO Guide is

$$SC_{OL}^n = SC_y^n - F_{RL}(SC_{xeff})^n \qquad (28.1)$$

where

$SC_{OL}$ = the structural capacity that the overlay must provide

$SC_y$ = the total structural capacity required to support the overlay traffic over the existing subgrade (foundation) conditions

$SC_{xeff}$ = the effective structural capacity of the existing pavement immediately prior to the time of the overlay, which reflects the damage to that point in time

$F_{RL}$ = the remaining life factor which accounts for the damage of the existing pavement as well as the desired degree of damage to the overlay at the end of the future traffic. It is always less than or equal to 1.0

$n$ = a coefficient to account for the bonding condition between an existing rigid pavement and a rigid overlay

The structural capacity is expressed either as structural number for flexible pavements or as surface thickness for rigid pavements. The required structural capacity is determined from the design procedures for a new pavement. The effective structural capacity and remaining life factor of the existing pavement are determined from an analysis of the pavement condition. Examination of equation 28.1 reveals that the performance of the overlay pavement is estimated based on the same performance equation as a new pavement. In other words, once the condition of the existing pavement is established, the rate of serviceability loss with traffic is equal to the serviceability loss for a new pavement.

## 28.4 SUMMARY OF REHABILITATION DESIGN

This chapter demonstrates the dearth of good performance information or models for rehabilitation techniques. This is an area of strong potential research for the future and pavement managers should keep this in mind. It is necessary, however, to have performance models for second and third analysis periods after a pavement is rehabilitated. At this point, it is best to look at the sections of this book that treat specific project level pavement management systems. In each of these a set of rehabilitation models have been selected based on the best available information and a cohesive project level pavement management approach has been adopted. The pavement manager is warned at this point not to adopt one set of models from one methodology and use it with another set of models from another methodology. The total project level pavement management system must be cohesive if the results of the pavement management process are to be effective.

## REVIEW QUESTIONS

1. How does rehabilitation differ from basic design considerations in pavements?
2. What two kinds of data are most important in the evaluation of existing pavements for rehabilitation?
3. How is it possible to predict the life of a restoration methodology?

Chapter 29

# Economic Evaluation of Alternative Pavement Design Strategies

## 29.1 INTRODUCTION

The application of principles of engineering economy to transport projects, including pavements, occurs basically at two levels. First, there are the management decisions required to determine the feasibility and timing of a project; second is the requirement to achieve the maximum economy for the project once it is selected.

Project feasibility is determined at the network level, by comparison with other potential projects, whereas within-project economy is achieved by considering a variety of alternatives capable of satisfying the overall project requirements.

The major difference in economic evaluation between these two levels of pavement management concerns the amount of detail or information required. Otherwise, the basic principles involved are the same. This chapter considers both these principles and their incorporation into models or methods of economic evaluation. Such models are a vital part of the pavement management system.

## 29.2 BASIC PRINCIPLES

Considerable literature is available on the principles of engineering economy and methods of economic evaluation. The principles applicable to the pavement field may be summarized as follows:

1. The management level at which the evaluation is to be made must be clearly identified; i.e., investment programming or network level; or project level optimization of a life-cycle strategy.
2. The economic analysis provides the basis for a management decision but does not by itself represent a decision.
3. Criteria, rules, or guides for such decisions must be separately formulated before the results of the economic evaluation are applied, even though such criteria may be straightforward and simple.
4. The economic evaluation itself has no relationship to the method or source of financing a project. Such financing considerations can either limit the number of feasible projects (on a network planning level basis) or limit the amount available for a particular project. They do not affect the methodology or principles controlling the economic evaluation per se.
5. An economic evaluation should consider all possible alternatives, within the constraints of time and other planning and design resources. This includes comparing alternatives with an existing situation and also with each other.
6. All alternatives should be compared over the same time period. (An alternative approach, favored by some economists, is to use a "floating" time period, which is determined as that point in the future where the costs and benefits, discounted to present-day terms, become negligible, i.e., they fall below some arbitrarily selected limit. The discount rate used is then the prime factor in determining the extent of this time period.) The time period should be chosen so that the factors involved in the evaluation can be forecast with some reasonable degree of reliability. Uncertainties can be considered in the decision made to choose the best alternative.
7. The economic evaluation of pavements should include agency costs and user costs and benefits if possible.

The seventh principle need not normally be stated for transport projects in general because it is an accepted requirement. However, in the pavement field the usual practice has been to consider only capital and maintenance costs, with the implied assumption that user costs do not vary with level of serviceability, condition, extent and time of rehabilitation, extent and timing of maintenance, and so on. This approach is inadequate because, as demonstrated by researchers in Texas, Canada, the Transport and Road Research Laboratory, and the World Bank, user costs can vary significantly with these factors. Benefits can then be considered as cost reductions [Winfrey 69].

## 29.3 PAVEMENT COST AND BENEFIT FACTORS

The major initial and recurring costs that a public agency may consider in the economic evaluation of alternative pavement strategies include the following:

1. Agency costs:
   a. Initial capital costs of construction
   b. Future capital costs of construction or rehabilitation (overlays, seal coats, reconstruction, etc.)
   c. Maintenance costs, recurring throughout the design period

d. Salvage return or residual value at the end of the design period (which may be a "negative cost")
   e. Engineering and administration
   f. Costs of investments
2. User costs:
   a. Travel time
   b. Vehicle operation
   c. Accidents
   d. Discomfort
   e. Time delay and extra vehicle operation costs during resurfacing or major maintenance
3. Nonuser costs [Friesz 74]:
   a. Air pollution
   b. Noise pollution
   c. Neighborhood disruption

### 29.3.1  Identification of Pavement Benefits

The benefits of a transport project can accrue from direct or indirect cost reductions, and from advantages of gains in business, land use and values, aesthetics, and community activities in general. Pavement benefits would accrue primarily from direct reductions in transportation costs of the user, as listed in the preceding section. It is also possible to consider benefits in terms of additional road user taxes generated by a project [Wohl 67]. Although this concept could be applicable to an overall highway project, it can also be questioned in terms of its extra energy consumption implications. It has several other deficiencies and is not recommended for pavements.

In order to measure or calculate pavement benefits, it is necessary to define those pavement characteristics that will affect the previously noted user costs of vehicle operation, travel time, accidents, and discomfort. These could include level of serviceability, slipperiness, appearance, color, light reflection characteristics, and so on. However, the first two factors of serviceability (as it affects vehicle operating costs, travel time costs, accident costs, and discomfort costs) and slipperiness (as it affects accident costs) would have a major influence.

Figure 29.1 is a schematic representation of the effects of different pavement design strategies on user costs. Considering only the variation in serviceability, for example, the diagram shows the following:

1. As serviceability decreases, travel time costs increase because average travel speed decreases (in a nonlinear manner).
2. When rehabilitation occurs (i.e., major maintenance, resurfacing, or reconstruction), high travel time costs occur because of traffic delays.

The other three components of user costs illustrate two major points:

1. As pavement serviceability approaches a terminal level, user costs increase at an increasing rate.

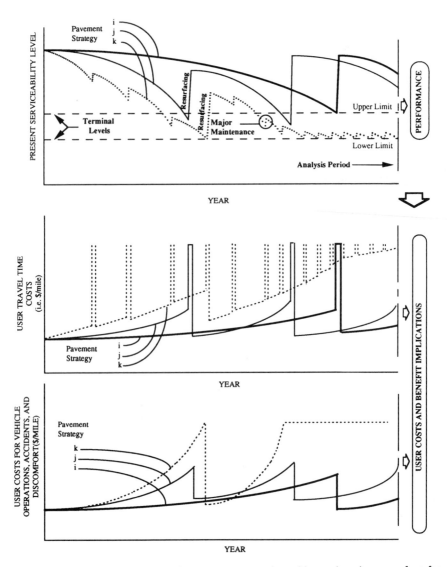

**Figure 29.1** Effects on user costs of pavement strategies with varying degrees of performance profiles.

2. Pavement strategies that use lower terminal serviceability levels prior to rehabilitation (i.e., strategy $k$ of Figure 29.1), result in higher user costs.

In assigning benefits for the purpose of project evaluation, the question of whether generated or transferred traffic should be included must be considered. Usually, this question arises with respect to the overall highway project, rather than with the pave-

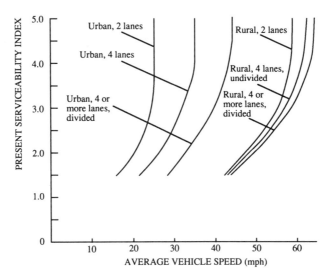

**Figure 29.2** Average vehicle speed related to present serviceability index by type of road [McFarland 72].

ment per se. Nevertheless, it is conceivable that a pavement improvement can by itself result in generated or transferred traffic.

This question can relate to the public policy viewpoint adopted [Winfrey 69]. One viewpoint involves a "sales concept" where all traffic is considered regardless of source; the other involves a concept of true savings where only existing plus normal traffic growth over the analysis period is considered. It is usual for highway agencies to use the first concept. Thus, the use of total expected traffic in calculating costs and benefits for pavement improvement projects should not be unreasonable.

### 29.3.2 Quantification of User Costs and Benefits for Pavement Projects

A considerable amount of reference material is available on user cost data for various highway types and design characteristics [AASHO 60, Winfrey 69, Claffey 71, SRI 76, Zaniewski 82]. In addition, there is a variety of material available on the costs of travel time and accidents. McFarland first quantified the effects of varying pavement serviceability on user costs, as illustrated in Figure 29.2 [McFarland 72]. This provided the information required to evaluate pavement benefits.

Another example of the quantification of user cost variation with pavement serviceability and speed is shown in Figure 29.3 [Kher 76]. The measure used for pavement ride quality in this example is the Canadian Riding Comfort Index (RCI). Based on experiments conducted in Texas, Zaniewski concluded that fuel consumption is not affected by roughness for the range of conditions encountered in the United States

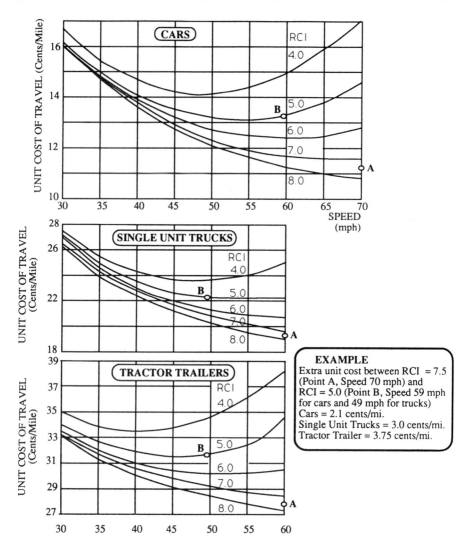

**Figure 29.3** Unit costs (vehicle operating and travel time) of speed reductions at different RCI levels on a rolling tangent in Ontario [Kher 76].

[Zaniewski 82]. However, roughness factors were developed for the other components of vehicle operating costs based on empirical data from Brazil [Harrison 80].

### 29.3.3 Design or Analysis Period

A general guideline for selecting the length of design or analysis period is that it should not extend beyond the period of reliable forecasts. For traffic, 20 years is often used as an upper limit. For other factors, 30 years may not be unreasonable; however, the present worth of costs or benefits at such future times may be insignificant, depending on the discount rate used. Most transport studies use a range of 20 to 30 years, and this seems reasonable for pavements. The particular period chosen is basically a policy decision for the agency concerned and can vary with a number of factors. An alternative approach of using a "floating" time period was mentioned in Section 29.2.

### 29.3.4 Discount Rate and Interest Rate

A discount rate is used to reduce future expected costs or benefits to present-day terms. It provides the means to compare alternative uses of funds, but it should not be confused with interest rate, which is associated with actually borrowing money although the terms are sometimes used interchangeably.

The rate to be used in the agency's calculations is basically a policy question. Also, this rate could vary with the facilities being evaluated to reflect the associated degree of uncertainly. Most agencies, however, use a single rate for all analyses. In the pavement field, discount rates between 4 and 10 percent have often been used. It should be emphasized that discount rate is a highly significant factor and can have a major influence on the results of an economic analysis.

### 29.3.5 Inflation

The question of how to take inflation into account in an economic evaluation is of concern to many engineers and administrators. Basically, the answer is that inflation is not used in the evaluation, except where substantial evidence exists that real prices will change (i.e., where "real price" is price in constant-value money).

Lee and Grant suggest that inflation should not be recognized when forecasting future prices, costs, and benefits, and that current prices should be used [Lee 65]. An exception exists when the price of one item in the analysis inflates at a rate that does not follow the general trend of inflation in the society, e.g., at times land values inflate more rapidly than the general inflation rate.

Some reasons for not using an inflation factor in economic evaluation of pavement projects are:

1. Inflation is difficult to forecast and merely introduces another uncertainty into the evaluation. In fact, if inflation were used in highway economy studies, it would tend to justify higher capital investments today. This could in turn contribute to more inflation, and for a public agency the morality of such actions could be questioned (i.e., governments are usually committed to fighting inflation).
2. If inflation were considered, benefits as well as costs would be increased so that their relative magnitude would still be the same.
3. The purpose of economic evaluation is to provide management with a basis for decision making. Inserting a factor for inflation is no guarantee that the results will be more reliable and lead to a different or better selection of alternatives.
4. The uninflated value of the alternatives in "real constant dollars" is a better tool for economic analysis than inflated values.

### 29.3.6 Discount Rate and Inflation

Some engineers believe it is not necessary to use a discount rate to capture the time value of money. These engineers claim that since money will not be extracted from the taxpayers until it is needed to fund construction, there is no opportunity to invest the money. This is a false argument in that taking the money from the taxpayers denies them the opportunity to use the funds for other purposes, opportunity cost of capital. Also, many projects are financed by bonds and thus have a true time value of money added.

On the other hand, some engineers argue that it is necessary to include inflation in an economic analysis since ignoring inflation leads to underestimating out-of-pocket costs and therefore budgets will be incorrect. This argument indicates a misunderstanding of the objective of an economic analysis which is to provide management with a tool for the selection of a specific option from a set of alternatives. Once the option is selected, a separate budget analysis is required to determine the cash flow requirements. The budget analysis generally includes inflation as well as actual sources of funds, such as taxes, federal grants, bonds, etc.

### 29.3.7 Salvage or Residual Value

Salvage or residual value is used by some agencies in economic evaluation. It can be significant in the case of pavements because it involves the value of reusable materials at the end of the design period. With depleting resources, such materials can become increasingly important in the future, especially when used in a new pavement by reworking or reprocessing.

Salvage value of a material depends on several factors such as volume and position of the material, contamination, age or durability, anticipated use at the end of the design period, and so on. It can be represented as a percentage of the original cost (i.e., that part of the unit price applicable to the material itself, not to the incremental unit price which results from placement and processing costs).

### 29.4 METHODS OF ECONOMIC EVALUATION

A number of methods of economic analysis are applicable to the evaluation of alternative pavement design strategies. They can be categorized as follows:

1. Equivalent uniform annual cost method, often simply termed the *annual cost method*
2. Present worth method for:
   a. Costs
   b. Benefits
   c. Benefits minus costs, usually termed the *net present worth* or *net present value* method
3. Rate-of-return method
4. Benefit-cost ratio method
5. Costs-effectiveness method

These methods have the common feature of being able to consider future streams of costs (i.e., methods 1, 2a, and 5), or of costs and benefits (i.e., methods 2b, 2c, and 4), so that alternative investments may be compared. Differences in the worth of money over time, as reflected in the compound interest equations used, provide the means for such comparison.

The following paragraphs briefly consider the essential characteristics of each method, their functional forms, and some of their limitations. The net present value method is given somewhat more attention than the others.

### 29.4.1 Basic Considerations in Selecting an Evaluation Method

There are several basic considerations in selecting the most appropriate (but not necessarily the best) method for economic evaluation of alternatives. They include the following:

1. How important is the initial capital expenditure in comparison to future expected expenditures? Often, public officials and private interests (especially, say, in the case of paving a large parking lot) are concerned primarily with initial costs. An economic analysis may indicate, for example, that a small capital expenditure today can result in excessive future costs for a particular alternative. Yet the low capital expenditure may be the most relevant consideration to decision makers especially in the face of limited funding. Such situations may not represent good economy but may be the harsh reality.
2. What method of analysis is most understandable to the decision maker? This consideration again represents reality. For example, consider an agency that has used a benefit-cost ratio method for several years, with a good understanding of the results of the analysis. It may well be that this is not the best overall method for its situation; however, changing to a better method could be quite difficult and lengthy.

    Another aspect is the level of decision making involved (i.e., at the network level or the project level). It is possible, for example, that a highway agency could use a rate-of-return method for analyzing its proposed investments over the network, whereas a net present value analysis is used by the pavement designer at the project level.
3. What method suits the requirements of the particular agency involved? Although the net present value method is preferable for public agencies providing pavements, as subsequently indicated, an annual cost method might be more suitable for a privately provided pavement (such as that for a large shopping complex).
4. Are benefits to be included in the analysis? Any method that does not consider the differences in benefits between pavement alternatives is basically incomplete for use by a public agency. However, for the private situation, an implicit assumption of equal benefits for various alternatives may be satisfactory.

### 29.4.2 Equivalent Uniform Annual Cost Method

The equivalent uniform annual cost method combines all initial capital costs and all recurring future expenses into equal annual payments over the analysis period. In equation form, this method may be expressed as follows:

$$AC_{x_1, n} = \text{crf}_{i, n}(ICC)_{x_1} + (AAMO)_{x_1} + (AAUC)_{x_1} - \text{sff}_{i, n}(SV)_{x_1, n} \quad (29.1)$$

where

$AC_{x_1, n}$ = equivalent uniform annual cost for alternative $x_1$, for a service life or analysis period of $n$ years

$\text{sff}_{i, n}$ = sinking fund factor for interest rate $i$ and $n$ years
  = $i(1 + i)^n/[(1 + i)^n - 1]$

$(ICC)_{x_1}$ = initial capital costs of construction (including actual construction costs, materials costs, engineering costs, etc.)

$(AAMO)_{x_1}$ = average annual maintenance plus operation costs for alternative $x_1$

$(AAUC)_{x_1}$ = average annual user costs for alternative $x_1$ (including vehicle operation, travel time, accidents, and discomfort if designated)

$(SV)_{x_1, n}$ = salvage value, if any, for alternative $x_1$ at the end of $n$ years

Equation 29.1 considers annual maintenance and operating costs, and user costs, on an average basis. This can be satisfactory for many purposes. Where such costs do not increase uniformly, however, an exponential growth factor can easily be applied.

The basic appeal of the equivalent uniform annual cost method is its simplicity and understanding for public officials. However, it cannot be used, except intuitively, to determine whether or not a project is economically justifiable because it does not include benefits in the evaluation. Consequently, comparisons between alternatives must be on the basis of costs alone, with the inherent assumption that they have equal benefits. However, especially where differences in vehicle operating costs are involved between such alternatives, this assumption is questionable.

The present worth of costs method is directly comparable to the equivalent uniform annual cost method.

### 29.4.3 Present Worth Method

The present worth method can consider either costs alone, benefits alone, or costs and benefits together. It involves the discounting of all future sums to the present, using an appropriate discount rate. The factor for discounting either costs or benefits is

$$\text{pwf}_{i, n} = 1/(1 + i)^n \quad (29.2)$$

where

$\text{pwf}_{i, n}$ = present worth factor for a particular $i$ and $n$
$i$ = discount rate
$n$ = number of years to when the sum will be expended, or saved

# Economic Evaluation of Alternative Pavement Design Strategies

The total present worth method for costs can be expressed in terms of the following equation:

$$\text{TPWC}_{x_1, n} = (\text{ICC})_{x_1} + \sum_{t=0}^{n} [\text{pwf}_{i, t}((\text{CC})_{x_1, t} + (\text{MO})_{x_1, t} + (\text{UC})_{x_1, t})]$$
$$- (\text{SV})_{x_1, n}\text{pwf}_{i, n} \qquad (29.3)$$

where

$\text{TPWC}_{x_1, n}$ = total present worth of costs for alternative $x_1$, for analysis period of $n$ years
$(\text{ICC})_{x_1}$ = initial capital costs of construction, etc., for alternative $x_1$
$(\text{CC})_{x_1, t}$ = capital costs of construction, etc., for alternative $x_1$, in year $t$, where $t < n$
$\text{pwf}_{i, t}$ = present worth factor for discount rate, $i$, for $t$ years
    $= 1/(1 + i)^t$
$(\text{MO})_{x_1, t}$ = maintenance plus operation costs for alternative $x_1$ in year $t$
$(\text{UC})_{x_1, t}$ = user costs (including vehicle operation, travel time, accidents, and discomfort if designated) for alternative $x_1$, in year $t$
$(\text{SV})_{x_1, n}$ = salvage value, if any, for alternative $x_1$, at the end of the design period, $n$ years

Although the present worth of costs method is directly comparable to the equivalent uniform annual cost method, it is only in recent years that it has begun to be applied to the pavement field.

The present worth of costs is used in the equivalent uniform annual cost method when additional capital expenditures occur before the end of the analysis period, that is, where the service life is less than the analysis period and future rehabilitation, such as overlays or seal coats, is needed. The equation for this situation, including user costs, is

$$\text{AC}_{x_1, n} = \text{crf}_{i, n}((\text{ICC})_{x_1} + R_1\text{pwf}_{i, a_1} + R_2\text{pwf}_{i, a_2} + \ldots + R_j\text{pwf}_{i, a_j}$$
$$- (1 - y/L)(R_1, R_2, \ldots, R_j)\text{pwf}_{i, a_1, a_2, \ldots, a_j}$$
$$+ (\text{AAUC})_{x_1} + (\text{AAMO})_{x_1} \qquad (29.4)$$

where

$\text{AC}_{x_1, n}$ = equivalent uniform annual cost for alternative $x_1$, for an analysis period of $n$ years
$R_1, R_2, \ldots, R_j$ = costs of first, second, $\ldots, j^{\text{th}}$ resurfacings
$a_1, a_2, \ldots, a_j$ = ages at which the first, second, $\ldots, j^{\text{th}}$ resurfacings occur, respectively
$y$ = number of years from time of last resurfacing to end of analysis period

$L$ = estimated life in years of last resurfacing, and $(ICC)_{x_1}$, $(AAUC)_{x_1}$, $(AAMO)_{x_1}$, are defined in equation 29.1.

The present worth of benefits can be calculated in the same manner as the present worth of costs, using the following equation:

$$TPWB_{x_1,n} = \text{sum}(pwf_{i,t}((DUB)_{x_1,t} + (IUB)_{x_1,t} + (NUB)_{x_1,t})) \qquad (29.5)$$

where

$TPWB_{x_1,n}$ = total present worth of benefits for alternative $x_1$ for an analysis period of $n$ years
$(DUB)_{x_1,t}$ = direct user benefits accruing from alternative $x_1$ in year $t$
$(IUB)_{x_1,t}$ = indirect user benefits accruing from alternative $x_1$ in year $t$
$(NUB)_{x_1,t}$ = nonuser benefits accruing from project $x_1$ in year $t$

It is questionable, for pavements, if nonuser benefits and indirect user benefits can be measured adequately. Consequently, it is perhaps reasonable to consider only direct user benefits until such time as the state of the art is sufficiently advanced to be able to measure other factors.

The net present value method follows from the foregoing methods, because it is simply the difference between the present worth of benefits and the present worth of costs. Obviously, benefits must exceed costs if a project is to be justified on economic grounds. The equation for net present value is as follows:

$$NPV_{x_1} = TPWB_{x_1,n} - TPWC_{x_1,n} \qquad (29.6)$$

where $NPV_{x_1}$ is the net present value of alternative $x_1$, and $TPWB_{x_1,n}$ and $TPWC_{x_1,n}$ are as defined by equations 29.5 and 29.3, respectively.

However, for a pavement project alternative, $x_1$, equation 29.6 is not applicable directly to $x_1$ itself but rather to the difference between it and some other suitable alternative, say $x_0$. Considering only direct user benefits $(DUB)_{x_1,t}$, these are then calculated as the user savings (resulting from lower vehicle operating costs, lower travel time costs, lower accident costs, and lower discomfort costs) realized by $x_1$ over $x_0$.

Thus, the net present value method can be applied to pavements only on the basis of project comparison, where the project alternatives are mutually exclusive. When a project alternative is evaluated, it needs to be compared not only with some standard or base alternative but also with all the other project alternatives. In the case of pavements, the base alternative may be that of no capital expenditures for improvements (where increased maintenance and operation costs are required to keep it in service). The equation form of the net present value method for pavements may then be expressed as follows:

$$NPV_{x_1} = TPWC_{x_0,n} - TPWC_{x_1,n} \qquad (29.7)$$

# Economic Evaluation of Alternative Pavement Design Strategies

where

$\text{NPV}_{x_1}$ = net present value of alternative $x_1$

$\text{TPWC}_{x_0, n}$ = total present worth of costs, for alternative $x_0$ (where $x_0$ can be the standard or base alternative, or any feasible mutually exclusive alternative $x_1, x_2, \ldots x_k$) for an analysis period of $n$ years

$\text{TPWC}_{x_1, n}$ = as defined in equation 29.3

Wohl and Martin state a preference for the net present value method for analyzing transport projects [Wohl 67]. Winfrey feels that it has no particular advantage in economic studies of highways [Winfrey 69]. Although there are certain limitations to the method, the advantages outweigh the disadvantages. Thus, it is the preferred approach for evaluating alternative pavement strategies when public investments are involved. Moreover, with increasing use of this approach in the overall transport planning field, its application to pavements will undoubtedly find much greater acceptance in the next decade.

Most agencies, however, use equation 29.3 without the user costs term. The comparison between alternatives is conducted in such cases on the basis of least total present worth of costs.

There are a number of advantages inherent in the net present value method that make it perhaps the most feasible for the highway field in comparison to the "traditional" annual cost and benefit-costs methods. These advantages include the following:

1. The benefits and costs of a project are related and expressed as a single value.
2. Projects of different service lives, and with stage development, are directly and easily comparable.
3. All monetary costs and benefits are expressed in present-day terms.
4. Nonmonetary benefits (or costs) can be evaluated subjectively and handled with a cost-effectiveness evaluation.
5. The answer is given as a total payoff for the project.
6. The method is computationally simple and straightforward.

There are several disadvantages to the net present value method, including the following:

1. The method cannot be applied to single alternatives where the benefits of those single alternatives cannot be estimated. In such cases, each alternative must be considered in comparison to the other alternatives, including the standard or base alternative.
2. The results, in terms of a lump sum, may not be as easily understandable to some people as a rate of return or annual cost. In fact, the summation of costs in this form can tend to act as a deterrent to investment in some cases.

Wohl and Marlin have extensively considered these advantages and disadvantages not only for the net present value method, but also for other methods of economic analysis [Wohl 67]. They conclude that the net present value method is the only one that will always give the correct answer. The other methods may, under certain situations, give incorrect or ambiguous answers.

### 29.4.4 Rate-of-Return Method

The rate-of-return method, which is used by a number of highway agencies, considers both costs and benefits and determines the discount rate at which the costs and benefits for a project are equal. It can be in terms of the rate at which the equivalent uniform annual cost is exactly equal to the equivalent uniform annual benefit, that is,

$$AC_{x_1, n} = AB_{x_1, n} \tag{29.8}$$

where

$AC_{x_1, n}$ = equivalent uniform annual cost for alternative $x_1$ for an analysis period of $n$ years
$AB_{x_1, n}$ = equivalent uniform annual benefit for alternative $x_1$ for an analysis period of $n$ years

Alternatively, the rate of return can be expressed in terms of the rate at which the present worth of costs is exactly equal to the present worth of benefits; that is, by setting equations 29.3 and 29.5 equal to each other,

$$TPWC_{x_1, n} = TPWB_{x_1, n}. \tag{29.9}$$

In applying the rate-of-return method, each alternative is first compared with the standard or base alternative to establish the difference in benefits. Then, using equation 29.9, which is preferable, or equation 29.8, the rate of return can be calculated for alternatives $x_1, x_2, \ldots, x_x$. However, this is only a comparison with the standard. It is also necessary to calculate the rate of return on comparisons between alternatives $x_1, x_2, \ldots, x_x$. This is done on the increase in costs between alternatives having successively higher first costs. Proceeding on the basis of such paired comparisons will eliminate all but one alternative, that having the highest rate of return.

This alternative may or may not be economically attractive, depending on the decision for a minimum attractive rate of return. For example, if it were decided that an investment must have a minimum rate of return of 10 percent to be economically viable, any alternative yielding a lesser return would be rejected.

The rate-of-return method has a major advantage in that the results are well understood by most people. It is easy to comprehend a return on investment because of the association and familiarity with normal business terms. Because of reasons such as this, and its applicability to evaluating network investments (i.e., those that are not mutually exclusive), it has been favored by a number of highway agencies. However, situations can arise where the rate-of-return method gives ambiguous answers [Wohl 67].

# Economic Evaluation of Alternative Pavement Design Strategies

## 29.4.5 Benefit-Cost Ratio Method

The benefit-cost ratio method has perhaps seen more widespread use in the highway field than any other method. It involves expressing the ratio of the present worth of benefits of an alternative to the present worth of costs, or the ratio of the equivalent uniform annual benefits to the equivalent uniform annual costs. The benefits are established by comparison of alternatives. Using the present worth formulation, which is preferred by most engineering economists, the benefit-cost ratio may be expressed in equation form as follows:

$$BCR_{x_j, x_k, n} = (TPWB_{x_j} - TPWB_{x_k})/(TPWC_{x_j} - TPWC_{x_k}) \quad (29.10)$$

where

$BCR_{x_j, x_k, n}$ = benefit-cost ratio of alternative $x_j$ compared to alternative $x_k$ (where $x_j$ yields the greater benefits, and represents the larger investment), over an analysis period of $n$ years

$TPWB_{x_j}, TPWC_{x_j}$ = total present worth of benefits, and of costs, respectively, for alternative $x_j$

$TPWB_{x_k}, TPWC_{x_k}$ = total present worth of benefits, and of costs, respectively, for alternative $x_k$

The calculation of benefit-cost ratios for a set of proposed alternatives is first done by comparison with the standard or base alternative, using equation 29.10.

Then, those alternatives exhibiting a ratio greater than 1.0 are compared on an incremental basis. This involves calculation of the benefit-cost ratio on the increments of expenditures for successively higher cost alternatives. Proceeding on the basis of such paired comparisons, using equation 29.10, will reveal the alternative of highest economic desirability.

The use of the benefit-cost ratio method in the highway field has been "promoted" by the American Association of State Highway and Transportation Officials through what is commonly known as the Red Book [AASHO 60]. It also has a certain appeal in public projects because of the emphasis on benefits [Taylor 64]. It has been suggested that the wording of federal laws in the United States has largely (and unintentionally) resulted in this popularity [Winfrey 69].

There are also some disadvantages to the benefit-cost ratio method. A major one is the abstract nature of the ratio, which is difficult to comprehend alone. Another disadvantage is the possible confusion over whether maintenance cost reductions should be in the numerator or the denominator, whether cost reductions are benefits or negative costs, and so on.

## 29.4.6 Cost-Effectiveness Method

The cost-effectiveness method can be used to compare alternatives where significant, nonmonetary outputs are involved [Barish 62]. It involves a determination of the advantages of benefits to be gained, in subjective terms, of additional expenditures.

This requires the establishment of subjective measures of effectiveness or benefit, such as some aesthetic rating, some comfort index, some serviceability index, and so on.

The expenditures in this method of analysis are usually stated in terms of present worth of costs. However, the effectiveness measures cannot be reduced to a present worth basis; therefore, they must be represented either by average values over some period of time, or by values at certain specific times. For example, consider pavement alternative design strategies $x$ and $y$ with present worths of costs of $200,000 and $250,000, respectively. Consider further that the measure of effectiveness to be used is the Present Serviceability Index (PSI) after 10 years. If this PSI were predicted to be, say, 3.8 for alternative $y$ and 3.0 for alternative $x$, then by the cost-effectiveness method a decision would have to be made as to whether the additional $50,000 for $y$ justified the higher PSI at 10 years.

Where more than one measure of effectiveness is used, weights must be assigned to each measure. This will require skilled judgment.

Comparing alternative design strategies on the basis of costs (i.e., present worth basis) is really a cost-effectiveness method [McFarland 72]. This is based on the equal effectiveness criterion in that some minimum serviceability index is specified by the designer and all alternative strategies must always exceed the minimum, over the analysis period. However, because user costs, and hence savings or benefits, vary with level of serviceability (see Figure 29.1), the contention that all alternatives have equal effectiveness if their serviceability stays above some minimum can be questioned. Moreover, the minimum level of serviceability may not be the same for all alternative strategies; in fact, as shown in Figure 29.1, it should be allowed to vary for a complete economic evaluation.

The cost-effectiveness method, by itself, may not be the method most applicable to the pavement field, but it can be used to supplement other methods. For example, it is known that pavement appearance has an effect on users. If some measure of aesthetics of appearance were developed, this could be used in a cost-effectiveness analysis along with the results of the net present value analysis. Such an approach might be especially useful where the net present value analysis reveals little difference among several alternatives.

## 29.5 ECONOMIC ANALYSIS EXAMPLE

Economic analysis should be used for the design of either new construction or rehabilitation projects. The following example is for a new pavement design. Simple pavement design models were used to generate the pavement alternatives. Only 2-inch overlays were considered, and it was assumed the overlay life would be equal to the life of the original pavement. It was also assumed a seal coat would be placed every six years, unless an overlay was planned within two years. The performance for five pavement designs are shown in Figure 29.4. The initial layer thicknesses, in inches, for the designs considered in this example were:

# Economic Evaluation of Alternative Pavement Design Strategies

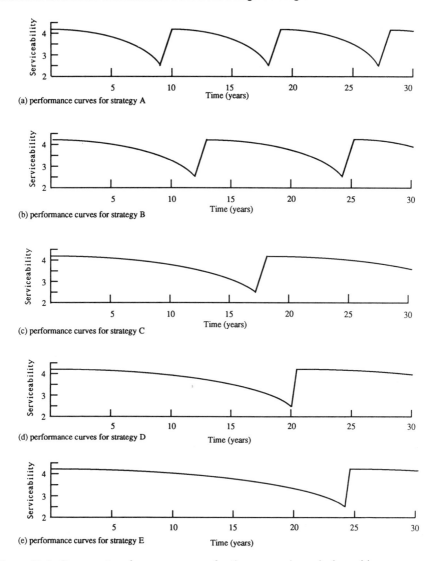

**Figure 29.4** Pavement performance curves for the economic analysis problem.

| Layer | Strategy | | | | |
|---|---|---|---|---|---|
| | A | B | C | D | E |
| HMAC | 2 | 2.5 | 3 | 3.5 | 4 |
| Aggregate Base | 8 | 8 | 8 | 8 | 8 |
| Select Subbase | 10 | 10 | 10 | 10 | 10 |

Agency costs are shown in Table 29.1 and present value are given in Table 29.2 for a 5 percent discount rate. These were based on the following unit prices:

| Material | Unit Price | Units |
|---|---|---|
| HMAC | 3.00 | $/sy/in |
| Aggregate Base | 1.00 | $/sy/in |
| Subbase | 0.50 | $/sy/in |
| Seal Coast | 0.80 | $/sy |

| Maintenance Costs SI range | Unit Price | Units |
|---|---|---|
| 5.0–3.51 | 0.015 | $/sy |
| 3.5–3.01 | 0.025 | $/sy |
| 3.0–2.51 | 0.030 | $/sy |
| 2.5–2.51 | 0.040 | $/sy |

The 2-inch overlays were assumed to have the same unit price as the hot mix asphalt concrete used for new construction. No adjustments were made for inflation over the analysis period. Based on this analysis, strategy C would produce the lowest total present worth of agency costs.

Next the user costs were computed following the method developed for the Federal Highway Administration [Zaniewski 82]. In this method, the component consumption is determined for fuel, oil, tires, maintenance and repair, and depreciation. The last three components are sensitive to the level of pavement roughness. The units of consumption are multiplied by unit prices to determine costs. Unit prices used in this analysis were:

| Vehicle Type | Fuel $/gal | Oil $/qt | Tires each | M & R $/1,000 mi. | Depreciation each |
|---|---|---|---|---|---|
| Medium auto | 0.75 | 2.50 | 70 | 60 | 10,000 |
| Pickup | 0.75 | 2.50 | 80 | 70 | 9,000 |
| 2 axle single unit truck | 0.75 | 1.00 | 220 | 120 | 16,000 |
| Semi truck | 0.80 | 1.00 | 500 | 150 | 55,000 |

Consumption is determined from a series of look up tables for each vehicle class. For this example, it was assumed traffic traveled at a constant 55 mph except when there was either an overlay or a seal coat. In years when a seal coat was applied, it was assumed 5 percent of all traffic slowed to a stop and idled for one minute. A vehicle-stop factor of 10 percent was assumed for overlays. User daily costs of $10 per vehicle per hour of delay were assumed. The vehicle operating costs for each option are given in Table 29.3 and the present worth of costs are given in Table 29.4 for a 5 percent discount rate.

As with the agency costs, strategy C has the lowest total present worth of costs for the user. In this case, strategy C should be selected since it has the lowest total present worth of costs as shown in Table 29.5.

**Table 29.1 AGENCY COSTS FOR ECONOMIC ANALYSIS EXAMPLE**

| Year | \multicolumn{5}{c}{DESIGN STRATEGIES} | | | | |
|---|---|---|---|---|---|
| | A | B | C | D | E |
| Initial Construction | 267,520 | 288,640 | 309,760 | 330,880 | 352,000 |
| 1 | 2,956 | 2,956 | 2,956 | 2,956 | 2,956 |
| 2 | 2,956 | 2,956 | 2,956 | 2,956 | 2,956 |
| 3 | 2,956 | 2,956 | 2,956 | 2,956 | 2,956 |
| 4 | 3,520 | 2,956 | 2,956 | 2,956 | 2,956 |
| 5 | 3,520 | 2,956 | 2,956 | 2,956 | 2,956 |
| 6 | 14,784 | 14,784 | 14,220 | 14,220 | 14,220 |
| 7 | 4,224 | 3,520 | 2,956 | 2,956 | 2,956 |
| 8 | 5,632 | 3,520 | 2,956 | 2,956 | 2,956 |
| 9 | 84,480 | 4,224 | 3,520 | 2,956 | 2,956 |
| 10 | 2,956 | 4,224 | 3,520 | 2,956 | 2,956 |
| 11 | 2,956 | 5,632 | 3,520 | 2,956 | 2,956 |
| 12 | 2,956 | 84,480 | 14,784 | 14,784 | 14,220 |
| 13 | 3,520 | 2,956 | 4,224 | 3,520 | 2,956 |
| 14 | 3,520 | 2,956 | 4,224 | 3,520 | 2,956 |
| 15 | 14,784 | 2,956 | 4,224 | 3,520 | 3,520 |
| 16 | 4,224 | 2,956 | 5,632 | 4,224 | 3,520 |
| 17 | 5,632 | 2,956 | 84,480 | 4,224 | 3,520 |
| 18 | 84,480 | 14,784 | 2,956 | 4,224 | 14,784 |
| 19 | 2,956 | 3,520 | 2,956 | 5,632 | 3,520 |
| 20 | 2,956 | 3,520 | 2,956 | 84,480 | 4,224 |
| 21 | 2,956 | 4,224 | 2,956 | 2,956 | 4,224 |
| 22 | 3,520 | 4,224 | 2,956 | 2,956 | 5,632 |
| 23 | 3,520 | 5,632 | 2,956 | 14,220 | 5,632 |
| 24 | 14,784 | 84,480 | 2,956 | 2,956 | 84,480 |
| 25 | 4,224 | 2,956 | 2,956 | 2,956 | 2,956 |
| 26 | 5,632 | 2,956 | 3,520 | 2,956 | 2,956 |
| 27 | 84,480 | 2,956 | 3,520 | 2,956 | 2,956 |
| 28 | 2,956 | 2,956 | 3,520 | 2,956 | 2,956 |
| 29 | 2,956 | 2,956 | 3,520 | 14,220 | 2,956 |
| 30 | 2,956 | 3,520 | 4,224 | 2,956 | 2,956 |
| Total | 651,481 | 577,280 | 514,764 | 554,892 | 566,720 |

## 29.6 LIMITATIONS OF ECONOMIC ANALYSIS

An economic evaluation of the type shown in the previous examples provides a basis for making a management decision to select the best strategy. However, factors in addition to economics may be used to make this judgment.

For example, not only the economically optimum strategy but also several nearly optimum strategies should be identified. These may differ so little in economic terms that the decision might be made on the basis of past experience with a particular alternative, preference of one material type to another, expected availability of initial capital funds and of future funds for overlap, and so on.

TABLE 29.2  PRESENT WORTH OF AGENCY COSTS FOR ECONOMIC ANALYSIS EXAMPLE

|  | DESIGN STRATEGIES | | | | |
|---|---|---|---|---|---|
| Year | A | B | C | D | E |
| Initial Construction | 267,520 | 288,640 | 309,760 | 330,880 | 352,000 |
| 1 | 2,816 | 2,816 | 2,816 | 2,816 | 2,816 |
| 2 | 2,681 | 2,681 | 2,681 | 2,681 | 2,681 |
| 3 | 2,554 | 2,554 | 2,554 | 2,554 | 2,554 |
| 4 | 2,895 | 2,432 | 2,432 | 2,432 | 2,432 |
| 5 | 2,758 | 2,316 | 2,316 | 2,316 | 2,316 |
| 6 | 11,032 | 11,032 | 10,611 | 10,611 | 10,611 |
| 7 | 3,001 | 2,501 | 2,101 | 2,101 | 2,101 |
| 8 | 3,811 | 2,382 | 2,001 | 2,001 | 2,001 |
| 9 | 54,456 | 2,722 | 2,269 | 1,905 | 1,905 |
| 10 | 1,815 | 2,593 | 2,160 | 1,815 | 1,815 |
| 11 | 1,728 | 3,292 | 2,058 | 1,728 | 1,728 |
| 12 | 1,646 | 47,041 | 8,232 | 8,232 | 7,918 |
| 13 | 1,866 | 1,568 | 2,240 | 1,866 | 1,568 |
| 14 | 1,777 | 1,493 | 2,133 | 1,777 | 1,493 |
| 15 | 7,111 | 1,422 | 2,031 | 1,693 | 1,693 |
| 16 | 1,935 | 1,354 | 2,580 | 1,935 | 1,612 |
| 17 | 2,457 | 1,290 | 36,858 | 1,842 | 1,535 |
| 18 | 35,103 | 6,143 | 1,228 | 1,755 | 6,143 |
| 19 | 1,170 | 1,392 | 1,170 | 2,228 | 1,392 |
| 20 | 1,114 | 1,326 | 1,114 | 31,839 | 1,591 |
| 21 | 1,061 | 1,516 | 1,061 | 1,061 | 1,516 |
| 22 | 1,203 | 1,443 | 1,010 | 1,010 | 1,925 |
| 23 | 1,146 | 1,833 | 962 | 4,629 | 1,833 |
| 24 | 4,584 | 26,194 | 916 | 916 | 26,194 |
| 25 | 1,247 | 873 | 873 | 873 | 873 |
| 26 | 1,583 | 831 | 989 | 831 | 831 |
| 27 | 22,627 | 791 | 942 | 791 | 791 |
| 28 | 754 | 754 | 897 | 754 | 754 |
| 29 | 718 | 718 | 855 | 3,454 | 718 |
| 30 | 684 | 814 | 977 | 684 | 684 |
| Total | 446,865 | 424,771 | 410,840 | 432,026 | 446,038 |

## REVIEW QUESTIONS

1. A 9.2-mi section of two-lane highway is currently at a PSI level of 3.2. It is expected to deteriorate at a uniform rate to a level of 2.0 in 3 years, at which time some resurfacing funds should be available. Traffic volume is currently 400 vehicles per day, expected to increase uniformly by 25 vehicles per day per year. The resurfacing, estimated to cost $100,000 per two-lane mile, is expected to raise the PSI level to 3.5, after which it is again expected to deteriorate uniformly to a level of 2.0 after 5 years. Maintenance costs are $300 per two-lane mile per year at a PSI level of 3.5, and they increase by $50 for each 0.1 drop in PSI level.

**TABLE 29.3 VEHICLE OPERATING COSTS FOR ECONOMIC ANALYSIS EXAMPLE**

|      | Traffic | DESIGN STRATEGIES | | | | |
| Year | AADT | A | B | C | D | E |
| --- | --- | --- | --- | --- | --- | --- |
| 1  | 1000 | 52,211 | 52,194 | 52,191 | 52,191 | 52,191 |
| 2  | 1050 | 54,922 | 54,829 | 54,806 | 54,802 | 54,801 |
| 3  | 1102 | 57,893 | 57,639 | 57,562 | 57,546 | 57,542 |
| 4  | 1157 | 61,208 | 60,656 | 60,474 | 60,435 | 60,422 |
| 5  | 1215 | 65,086 | 63,917 | 63,559 | 63,480 | 63,450 |
| 6  | 1276 | 69,499 | 67,502 | 66,838 | 66,694 | 66,635 |
| 7  | 1340 | 80,186 | 76,646 | 75,408 | 75,170 | 75,066 |
| 8  | 1407 | 82,169 | 76,100 | 74,067 | 73,697 | 73,524 |
| 9  | 1477 | 102,922 | 81,290 | 78,109 | 77,526 | 77,256 |
| 10 | 1551 | 80,997 | 87,844 | 82,570 | 81,604 | 81,199 |
| 11 | 1628 | 85,203 | 95,876 | 87,425 | 85,959 | 85,372 |
| 12 | 1710 | 89,812 | 119,146 | 99,207 | 97,231 | 96,277 |
| 13 | 1795 | 94,954 | 93,734 | 98,616 | 95,941 | 94,499 |
| 14 | 1885 | 100,970 | 98,466 | 105,759 | 101,574 | 99,504 |
| 15 | 1979 | 115,316 | 103,512 | 113,766 | 107,709 | 104,970 |
| 16 | 2078 | 116,519 | 108,930 | 123,768 | 114,677 | 110,857 |
| 17 | 2182 | 127,471 | 114,787 | 152,064 | 122,877 | 117,209 |
| 18 | 2292 | 159,667 | 129,907 | 119,624 | 132,039 | 132,765 |
| 19 | 2406 | 125,653 | 128,528 | 125,617 | 143,546 | 131,543 |
| 20 | 2526 | 132,178 | 136,665 | 131,933 | 176,033 | 140,290 |
| 21 | 2653 | 139,328 | 145,986 | 138,608 | 138,479 | 150,157 |
| 22 | 2785 | 147,306 | 157,756 | 145,679 | 145,406 | 161,122 |
| 23 | 2925 | 156,638 | 172,179 | 153,194 | 163,770 | 174,992 |
| 24 | 3071 | 178,894 | 213,969 | 161,202 | 160,353 | 213,969 |
| 25 | 3225 | 180,760 | 168,333 | 169,764 | 168,431 | 168,321 |
| 26 | 3386 | 197,749 | 176,831 | 179,027 | 176,959 | 176,738 |
| 27 | 3555 | 247,696 | 185,893 | 189,252 | 185,980 | 185,579 |
| 28 | 3733 | 194,930 | 195,624 | 200,380 | 195,541 | 194,867 |
| 29 | 3920 | 205,051 | 206,141 | 212,535 | 220,551 | 204,633 |
| 30 | 4116 | 216,144 | 217,702 | 226,031 | 216,520 | 214,906 |
| Total costs | | 3719,346 | 3648,598 | 3599,049 | 3612,733 | 3620,670 |

Another section of two-lane highway, 5.1 mi long, currently at a PSI level of 2.0 can be kept at that level for 3 years by $1,800 per two-lane mile per year of maintenance expenditures. At that time, a resurfacing, expected to cost $38,000 per two-lane mile, should raise the PSI level to 3.5. This resurfacing would also be expected to deteriorate uniformly to a PSI level of 2.0 in 5 years. Maintenance costs after resurfacing would be the same as for the first section. The traffic volume is currently 2,000 vehicles per day, and is expected to grow uniformly at 50 vehicles per day per year.

If sufficient funds to do only one project became available, which one would you recommend? Use a discount rate of 8 percent. Would your recommendation change for a discount rate of 4 percent? 12 percent?

2. Prepare a table of salvage values, as a percent of original materials costs, for your local

**TABLE 29.4 PRESENT WORTH OF VEHICLE OPERATING COSTS FOR ECONOMIC ANALYSIS EXAMPLE**

| Year | Traffic AADT | DESIGN STRATEGIES | | | | |
|---|---|---|---|---|---|---|
| | | A | B | C | D | E |
| 1 | 1000 | 49,725 | 49,709 | 49,706 | 49,706 | 49,705 |
| 2 | 1050 | 49,816 | 49,732 | 49,710 | 49,707 | 49,706 |
| 3 | 1102 | 50,010 | 49,791 | 49,724 | 49,711 | 49,707 |
| 4 | 1157 | 50,356 | 49,902 | 49,752 | 49,720 | 49,709 |
| 5 | 1215 | 50,996 | 50,081 | 49,800 | 49,738 | 49,714 |
| 6 | 1276 | 51,861 | 50,371 | 49,875 | 49,768 | 49,724 |
| 7 | 1340 | 56,986 | 54,470 | 53,591 | 53,422 | 53,348 |
| 8 | 1407 | 55,615 | 51,507 | 50,132 | 49,881 | 49,764 |
| 9 | 1477 | 66,345 | 52,400 | 50,349 | 49,974 | 49,799 |
| 10 | 1551 | 49,725 | 53,929 | 50,690 | 50,097 | 49,849 |
| 11 | 1628 | 49,816 | 56,056 | 51,115 | 50,258 | 49,915 |
| 12 | 1710 | 50,010 | 66,345 | 55,242 | 54,142 | 53,610 |
| 13 | 1795 | 50,356 | 49,709 | 52,298 | 50,879 | 50,115 |
| 14 | 1885 | 50,996 | 49,732 | 53,415 | 51,301 | 50,256 |
| 15 | 1979 | 55,469 | 49,791 | 54,723 | 51,810 | 50,492 |
| 16 | 2078 | 53,379 | 49,902 | 56,699 | 52,535 | 50,785 |
| 17 | 2182 | 55,615 | 50,081 | 66,345 | 53,611 | 51,138 |
| 18 | 2292 | 66,345 | 53,979 | 49,706 | 54,864 | 55,166 |
| 19 | 2406 | 49,725 | 50,863 | 49,710 | 56,806 | 52,056 |
| 20 | 2526 | 49,816 | 51,507 | 49,724 | 66,345 | 52,874 |
| 21 | 2653 | 50,010 | 52,400 | 49,752 | 49,706 | 53,897 |
| 22 | 2785 | 50,356 | 53,929 | 49,800 | 49,707 | 55,079 |
| 23 | 2925 | 50,996 | 56,056 | 49,875 | 53,318 | 56,972 |
| 24 | 3071 | 55,469 | 66,345 | 49,983 | 49,720 | 66,345 |
| 25 | 3225 | 53,379 | 49,709 | 50,132 | 49,738 | 49,705 |
| 26 | 3386 | 55,615 | 49,732 | 50,349 | 49,768 | 49,706 |
| 27 | 3555 | 66,345 | 49,791 | 50,690 | 49,814 | 49,707 |
| 28 | 3733 | 49,725 | 49,902 | 51,115 | 49,881 | 49,709 |
| 29 | 3920 | 49,816 | 50,081 | 51,634 | 53,582 | 49,714 |
| 30 | 4116 | 50,010 | 50,371 | 52,298 | 50,097 | 49,724 |
| Total costs | | 1594,698 | 1568,185 | 1547,952 | 1549,617 | 1548,005 |

conditions, as a function of class of road, type of material, type of pavement, PSI level at time of rehabilitation, and subgrade type.

3. For the data in Table 29.2, alternatives A and E are essentially the same on a basis of total present worth of costs for the discount rate used. If this rate were 0 percent, would they still be the same, or would you recommend one in preference to the other?

**Economic Evaluation of Alternative Pavement Design Strategies**

**TABLE 29.5 TOTAL PRESENT WORTH OF COSTS FOR ECONOMIC ANALYSIS EXAMPLE**

| Year | \multicolumn{5}{c}{DESIGN STRATEGIES} | | | | |
|---|---|---|---|---|---|
| | A | B | C | D | E |
| 1 | 304,506 | 324,604 | 344,715 | 364,829 | 384,944 |
| 2 | 52,498 | 52,414 | 52,392 | 52,389 | 52,388 |
| 3 | 52,565 | 52,345 | 52,278 | 52,265 | 52,261 |
| 4 | 52,789 | 52,335 | 52,184 | 52,152 | 52,142 |
| 5 | 53,754 | 52,398 | 52,117 | 52,054 | 52,031 |
| 6 | 54,488 | 52,577 | 52,082 | 51,974 | 51,930 |
| 7 | 67,493 | 64,977 | 63,698 | 63,528 | 63,454 |
| 8 | 58,474 | 53,890 | 52,133 | 51,882 | 51,765 |
| 9 | 69,975 | 54,669 | 52,255 | 51,880 | 51,705 |
| 10 | 101,588 | 56,522 | 52,851 | 51,913 | 51,664 |
| 11 | 51,545 | 58,526 | 53,173 | 51,987 | 51,644 |
| 12 | 51,657 | 69,481 | 57,202 | 55,788 | 55,257 |
| 13 | 51,924 | 94,510 | 60,138 | 58,720 | 57,656 |
| 14 | 52,774 | 51,225 | 55,549 | 53,079 | 51,749 |
| 15 | 57,162 | 51,213 | 56,755 | 53,503 | 51,914 |
| 16 | 60,151 | 51,256 | 58,634 | 54,147 | 52,397 |
| 17 | 57,458 | 51,371 | 68,802 | 55,454 | 52,673 |
| 18 | 68,685 | 55,207 | 84,809 | 56,620 | 56,629 |
| 19 | 83,157 | 56,713 | 50,881 | 58,477 | 57,906 |
| 20 | 50,930 | 52,834 | 50,838 | 68,467 | 54,200 |
| 21 | 51,072 | 53,664 | 50,813 | 80,029 | 55,414 |
| 22 | 51,367 | 55,373 | 50,811 | 50,717 | 56,523 |
| 23 | 52,142 | 57,432 | 50,838 | 54,281 | 58,806 |
| 24 | 56,560 | 68,091 | 50,900 | 54,129 | 68,091 |
| 25 | 57,744 | 74,656 | 51,005 | 50,611 | 74,653 |
| 26 | 56,803 | 50,563 | 51,181 | 50,599 | 50,537 |
| 27 | 67,853 | 50,583 | 51,633 | 50,606 | 50,499 |
| 28 | 71,275 | 50,656 | 52,013 | 50,635 | 50,463 |
| 29 | 50,534 | 50,799 | 52,489 | 54,300 | 50,433 |
| 30 | 50,695 | 51,055 | 53,112 | 53,388 | 50,408 |
| Total PWC | 2019,633 | 1971,953 | 1938,298 | 1960,419 | 1972,151 |

Chapter 30

# Selection of an Optimal Design Strategy

## 30.1 ROLE OF THE DECISION MAKER

A pavement management system does not make decisions. This is the role of the appropriate manager. Unfortunately, there still are people who erroneously feel that the use of a pavement management system inhibits the decision-making function of the true decision makers concerned.

On the contrary, the use of a pavement management system enhances the role of the decision maker. It does not make the decisions but rather it processes information or inputs in accordance with a preselected set of models. Thus, it provides an efficient tool to organize and process information which expands the scope and efficiency of the decision maker.

The role of the pavement manager is:

1. To help ensure that the pavement management system is properly structured for the conditions involved, and properly understood and used by agency personnel involved
2. To ensure that proper data is acquired and used correctly for each project under consideration

# Selection of an Optimal Design Strategy

The role of the decision maker on the other hand is:

1. To analyze and evaluate the output of the pavement management system and use the results in arriving at a final decision
2. To follow through and use the information in both updating estimates and making subsequent decisions including maintenance and rehabilitation

Thus, it is the responsibility of the decision maker to ensure that a valid pavement management system is used as a tool in the decision-making function. The major levels of decision making appropriate to pavement management have been shown in Figure 4.1 as network and project levels. Although an initial decision to select a project is developed using an estimated design at the network level, it is at the project level that the final design decisions must be made, including:

1. The detailed information to be acquired on project design factors such as traffic, material properties, climatic factors, and unit costs
2. Applicable constraints including minimum layer thicknesses, minimum time to first overlay and between overlays, minimum length of "homogeneous" sections, and maximum costs
3. The layer materials and thicknesses strategies for the project

The decisions on layer materials and thicknesses may rely largely on the results of economic evaluation of the various alternatives. However, additional information on previous experience with local or imported materials, previous experience with some of the design alternatives under consideration, likely construction effects on the various design alternatives, expected maintenance, reliability of the unit cost information, etc., may also be used in arriving at a final design decision.

Some of the foregoing types of decisions can be used directly or internally in a computerized design system. Other decisions are subjective in the way they are used, for example, previous experience with some of the alternatives under consideration. Such subjective information must then be used external to the program.

## 30.2 BASIS FOR OPTIMAL STRATEGY SELECTION

The final selection of a design strategy for implementation is partially subjective. Although the economic analysis must form a key basis for the decision, no firm decision rules exist that can be followed exactly.

In order to make a final selection, and to make maximum use of the economic analysis results, not only the economically optimal strategy but also a number of near optimal strategies should be listed. This is the case with such computerized design systems as SAMP6 and OPAC, where a listing of a number of near optimal strategies, usually by rank-order based on, for example, minimum total cost, is presented. For many design situations, experience has shown that the economically optimal strategy is only marginally more attractive than a number of other nearly equal alternatives.

**Table 30.1 Sample Description of Optimal Pavement Strategy for Communications to Construction, Maintenance, and Evaluation Personnel**

### Basic Identification Information

Rairon County
Highway 21 from Jct. Hwy. 102 to Jct. 156
Mile 20.22 to 30.42
Two lanes undivided, 12 ft. wide
Shoulders, gravel surfaced, 8 ft. wide

### Basic Design Input and Information

| | |
|---|---|
| Two way ADT at construction | 2,400vpd |
| Two way ADT, end of analysis | 6,000vpd |
| Amount of trucks in ADT | 8% |
| Initial serviceability index | 4.2 |
| Serviceability index after overlay | 3.8 |
| Time of initial construction | June 1992 |

### Recommended Material Types and Sources

Gravel from Markle pit
Crushed rock and screenings from Brinks quarry
Asphalt cement, AC 20

### Expected Maintenance Policies

Level of maintenance, normal procedures for crack sealing, patching and shoulder grading for secondary roads
Maintenance expenditures, $250/lane mile for first 2 years after construction and overlays, and increasing by $250/lane mile each year thereafter

### Basic Design Criteria Used

| | |
|---|---|
| Length of analysis period | 20 yrs |
| Minimum serviceability index allowed | 2.5 |
| Minimum time to first overlay | 6 yrs |
| Minimum time between overlays | 5 yrs |
| Maximum present worth of all costs | $13.00/sy |

### Recommended for Initial Thicknesses

| | |
|---|---|
| Prepared subgrade | 12 in. |
| Gravel subbase | 6 in. |
| Black base | 4.0 in. |
| Asphalt concrete | 3.0 in. |

### Expected Construction Practices

Variation in tolerances for materials within departmental specifications
Traffic handling method II for initial construction
Hours of construction 7 a.m. to 6 p.m.

### Rehabilitation Strategy

First overlay of 1.5 in. required at year 8
Second overlay of 1.5 in. required at year 14
Traffic handling method II for overlay construction
Hours of overlay construction 7 a.m. to 6 p.m.
Crushed rock screenings from Minton quarry
Asphalt cement, AC 20

### Expected Evaluation Policy

Roughness evaluation immediately after construction and overlays, otherwise every 3 years
Structural evaluation after initial construction and overlays, otherwise every 3 years with FWD
Skid resistance testing 4 years after construction and overlays, otherwise every 2 years

# Selection of an Optimal Design Strategy

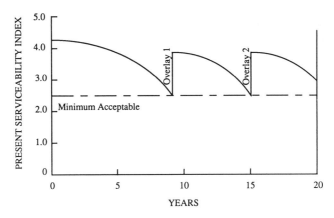

**Figure 30.1** Performance profits for sample optimal pavement design strategy.

Thus, there is considerable flexibility available and scope for judgment in choosing the "best" alternative.

A factor that might be considered in strategy selection is that of risk, that is, the probability that the desired performance will be achieved. Chapter 22 has shown that as the reliability of any particular design alternative increases, so do the costs to the agency, although user costs will probably decrease.

Another factor of potentially practical significance in strategy selection is the expected availability of funds. For example, consider that two alternatives are essentially equal in economic terms. The first has a relatively thick initial structure with no overlays during the design period, whereas the other has several intervening future overlays. As a consequence, the first alternative may be selected if adequate funds exist now. If current funds are short then a thinner initial design might be selected.

Finally, it should be emphasized that the decision making does not end with the selection of a design strategy. After the pavement is constructed, decisions will undoubtedly have to be periodically "upgraded" with respect to rehabilitation, in-service evaluation, etc.

## 30.3 COMMUNICATING RESULTS

The result of the project level pavement management systems analysis of a design project will be the selection of an optimum strategy for the pavement section considering both the quantifiable items that are input to the pavement management systems and the subjective factors that are evaluated by the pavement manager-engineer. In order for pavement management to be effective, the results of this analysis must be implemented by the agency. This includes not only the initial construction of the project, but the maintenance and rehabilitation schedules that were developed during the project level pavement management system analysis. The importance of communicating the design strategy, especially the predicted performance and future maintenance and rehabilitation requirements to construction, maintenance and performance evaluation personnel cannot be overemphasized. Table 30.1 and Figure 30.1 demonstrate the essential design strategy information that must be communicated to the construction, maintenance, and evaluation personnel. Communication of the design strategy in this manner can be most useful in improving coordination and eventually the efficiency of the pavement management system as a whole.

## REVIEW QUESTIONS

1. Explain how decisions are made in a pavement management system.
2. What is the basis for optimal strategy selection?
3. How are results of a pavement management evaluation communicated?

# References to Part Four

[AASHO 60] American Association of State Highway Officials, "Road User Benefit Analysis for Highway Improvements," AASHO, 1960.

[AASHO 61] American Association of State Highway Officials, "AASHO Interim Guide for the Structural Design of Pavements," 1961.

[AASHTO 72] American Association of State Highway and Transportation Officials, "AASHTO Interim Guide for Design of Pavement Structures," 1972.

[AASHTO 85] American Association of State Highway and Transportation Officials, "AASHTO Guide for the Design of Pavement Structures," Volume 2, Appendix EE, 1985.

[AASHTO 86] American Association of State Highway and Transportation Officials, "AASHTO Guide for the Design of Pavement Structures," 1986.

[Ahlborn 72] Ahlborn, G., "Elastic Layer System with Normal Loads," Institute for Transportation and Traffic Engineering, University of California, Berkeley, 1972.

[Ahlvin 73] Ahlvin, R. G., Y. T. Chou, and R. L. Hutchinson, "The Principle of Superposition in Pavement Analysis," Highway Research Record 466, Highway Research Board, 1973.

[Aston 67] Aston, J. E., and F. Moavenzadeh, "Analysis of Stresses and Displacements in a Three-Layered Viscoelastic System," Proc., Second

Intl. Conf. on the Structural Design of Asphalt Concrete Pavements, Univ. of Michigan, 1967.

[Avramesco 67] Avramesco, A., "Dynamic Phenomena in Pavements Considered as Elastic Layered Structures," Proc., Second Intl. Conf. on the Structural Design of Asphalt Pavements, Univ. of Michigan, 1967.

[Barish 62] Barish, N. N., *Economic Analysis for Engineering and Managerial Decision Making*, McGraw-Hill, 1962.

[Barksdale 67] Barksdale, R. D., and G. A. Leonards, 'Predicting Performance of Bituminous Surfaced Pavements," Proc., Second Intl. Conf. on Structural Design of Asphalt Pavements, Univ. of Michigan, 1967.

[Beckedahl 87] Beckedahl, H., A. Gerlach, H. Locke, and W. Schwaderer, "On Improvements of the Existing VESYS Concepts," Proc., Sixth International Conference Structural Design of Asphalt Pavements, Univ. of Michigan, 1987.

[Bergan 73] Bergan, A. T., and B.C. Pulles, "Fatigue Design Prodecures for Cold Climates," Proc., Can. Tech. Asphalt Assoc., 1973.

[Brown 67] Brown, S. F., and P. S. Pell, "An Experimental Investigation of the Stresses, Strains and Deflections in a Layered Pavement Subjected to Dynamic Loads," Proc., Second Intl. Conf. on the Structural Design of Asphalt Pavements, Univ. of Michigan, 1967.

[Burmister 45] Burmister, D. M., "The General Theory of Stresses and Displacement in Layered Systems," Vol 16(2), Vol 16(3), Vol 16(5), Journal of Applied Physics, 1945.

[Cedergren 74] Cedergren, H. R., *Drainage of Highway and Airfield Pavements*, John Wiley and Sons, New York, 1974.

[Cedergren 88] Cedergren, H. R., "Why All Important Pavements Should Be Well Drained," Transportation Research Record 1188, Transportation Research Board, 1988.

[CGRA 65] Canadian Good Roads Association, "A Guide to the Structural Design of Flexible and Rigid Pavements in Canada," 1965.

[CGRA 70] Canadian Good Roads Association, "Low-Temperature Pavement Cracking in Canada: The Problem and Its Treatment," Proc., Canadian Good Roads Association, 1970.

[Chesher 85] Chesher, A., and R. Harrison, "User Benefits from Highway Improvements, Evidence from Developing Countries," World Bank, 1985.

[Chevron 63] Chevron Oil Company, "Development of the Elastic Layer Theory Program," 1963.

[Christison 72a] Christison, J. T., D. W. Murray, and K. O. Anderson, "Stress Prediction and Low-Temperature Fracture Susceptibility of Asphalt Concrete Pavements," Proc., Association of Asphalt Paving Technology, 1972.

[Christison 72b] Christison, J. T., "The Response of Asphaltic Concrete Pavements to Low Temperatures," Ph.D. Dissertation, University of Alberta, 1972.

# References to Part Four

| | |
|---|---|
| [Christison 72c] | Christison, J. T., and K. O. Anderson, "The Response of Asphalt Pavements to Low-Temperature Climatic Environments," Proc., Third Intl. Conf. on Structural Design of Asphalt Pavements, London, England, 1972. |
| [Claffey 71] | Claffey, P. T., "Running Costs of Motor Vehicles as Affected by Road Design and Traffic," NCHRP Rept. 111, 1971. |
| [Darter 71] | Darter, M. I., "Uncertainty Associated with Predicting 18-kip Equivalent Single Axles for Texas Pavement Design Purposes," Center for Highway Research, The University of Texas at Austin, October 1971. |
| [Darter 72] | Darter, M. I., B. F. McCullough, and J. L. Brown, "Reliability Concepts Applied to the Texas Flexible Pavement Design System," Highway Research Record 406, Highway Research Board 1972. |
| [Darter 73a] | Darter, M. I., W. R. Hudson, and J. L. Brown, "Statistical Variations of Flexible Pavement Properties and Their Consideration in Design," Proc., Association of Asphalt Paving Technology, 1973. |
| [Darter 73b] | Darter, M. I., and W. R. Hudson, "Probabilistic Design Concepts Applied to Flexible Pavement System Design," Research Report 123-18, Center for Transportation Research, University of Texas, 1973. |
| [Darter 85] | Darter, M. I., J. M. Becker, M. B. Snyder, and R. E. Smith, "Portland Cement Concrete Pavement Evaluation System—COPES," NCHRP Report 277, Transportation Research Board, 1985. |
| [Deacon 69] | Deacon, J. A., "Load Equivalency in Flexible Pavements," Proceedings, Association of Asphalt Paving Technology, 1969. |
| [Deacon 71] | Deacon, J. A., "Material Characterization—Experimental Behavior," Special Report 126, Highway Research Board, 1971. |
| [De Jong 73] | De Jong, D. L., M. G. F. Peutz, and Korswagen, "Computer Program BISAR", External Report Koninklyke/Shell Laboratorium, Amsterdam, Holland, 1973. |
| [Derucher 88] | Derucher, K. N., and Korfiatis, G. P., *Materials for Civil and Highway Engineers*, 1988. |
| [Dorman 62] | Dorman, G. M., "The Extension to Practice of a Fundamental Procedure for the Design of Asphalt Pavements," Proc., First Intl. Conf. on Structural Design of Asphalt Pavements, Univ. of Michigan, 1962. |
| [Duncan 68] | Duncan, J. M., C. L. Monismith, and E. L. Wilson, "Finite Element Analysis of Pavements," Highway Research Record 228, Highway Research Board, 1968. |
| [FHWA 87a] | "Pavement Design, Principles and Practices," Federal Highway Administration short course, developed by ERES Consultants Inc., Champaign, Il., 1987. |
| [FHWA 87b] | "Techniques for Pavement Rehabilitation, A Training Course," Federal Highway Administration short course, third revision, developed by ERES Consultants Inc., Champaign, Il., 1987. |
| [Finn 73a] | Finn, F. N., "Relation Between Cracking and Performance," Special Report 140, Highway Research Board, 1973. |

[Finn 73b]       Finn, F. N., K. Nair, and J. Hilliard, "Minimizing Premature Cracking of Asphalt Concrete Pavements," Final Report, NCHRP Project 9-4, 1973.

[Finn 77]        Finn, F. N., C. Saraf, R. Kulkarni, K. Nair, W. Smith, and A. Abdullah, "The Use of Distress Prediction Subsystems for the Design of Pavement Structures," Proc., Fourth Intl. Conf. on the Structural Design of Asphalt Pavements, Volume I, Univ. of Michigan, 1977.

[Friesz 74]      Friesz, T. L., and J. M. Zuieback, "Discussion" of Paper "General Concepts of Systems Analysis as Applied to Pavements," by R. Haas, Transportation Research Record 512, Transportation Research Board, 1974.

[Gomez-Achecar 86]  Gomez-Achecar, M., and M. R. Thompson, "ILLI-PAVE-Based Response Algorithms for Full-Depth Asphalt Concrete Flexible Pavements," Transportation Research Record 1095, Transportation Research Board, 1986.

[Haas 68]        Haas, R. C. G., "The Performance and Behavior of Flexible Pavement Surfaces at Low-Temperatures," Proc., Can. Tech. Asphalt Assoc., 1968.

[Haas 69]        Haas, R. C. G., and T. H. Topper, "Thermal Fracture Phenomena in Bituminous Surfaces," Special Report 101, Highway Research Board, 1969.

[Haas 72]        Haas, R. C. G., N. I. Kamel, and J. Morris, "Brampton Test Road, An Application of Layer Analysis to Pavement Design," Ministry of Transp. and Communic. of Ontario, Rept. RR182, 1972.

[Haas 73]        Haas, R. C. G., "A Method for Designing Asphalt Pavements to Minimize Low-Temperature Shrinkage Cracking," Asphalt Institute, Research Report 73-1, 1973.

[Harrison 80]    Harrison, R., and J. D. Swait, "Relating Vehicle Use to Highway Characteristics: Evidence From Brazil," Transportation Research Record 747, Transportation Research Board, 1980.

[Hills 66]       Hills, J. F., and D. Brien, "The Fracture of Bitumens and Asphalt Mixes by Temperature Induced Stresses," Proc., Association of Asphalt Paving Technology, 1966.

[Hills 74]       Hills, J. F., D. Brien, and P. J. Van de Loo, "The Correlation of Rutting and Creep Tests on Asphalt Mixes," Paper IP 74-001, Institute of Petroleum, 1974.

[Hoyt 87]        Hoyt, D. M., R. L. Lytton, and F. R. Roberts, "Criteria for Asphalt-Rubber Concrete in Civil Airport Pavements, Volume II Evaluation of Asphalt-Rubber Concrete," Report No. DOT/FAA/PM-86/39, II, Federal Aviation Administration, 1987.

[HRB 71]         Highway Research Board, "Structural Design of Asphalt Concrete Pavement Systems," Special Report 126, 1971.

[HRB 73]         Highway Research Board, "Structural Design of Asphalt Concrete Pavements to Prevent Fatigue Cracking," Special Report 140, 1973.

# References to Part Four

[Hudson 62]      Hudson, W. R., and F. H. Scrivner, "AASHO Road Test Principal Relationships—Performance versus Stress, Rigid Pavements," Special Report 73, Highway Research Board, 1962.

[Hudson 63a]     Hudson, W. R., "Comparison of Concrete Pavement Load-Stresses at AASHO Road Test with Previous Work," Highway Research Record 42, Highway Research Board, 1963.

[Hudson 63b]     Hudson, W. R., "Comparison of Concrete Pavement Load-Stresses at AASHO Road Test with Previous Work," Departmental Research Record Report Number 62-2, Research Section, Highway Design Division, Texas Highway Department, 1963.

[Hudson 64a]     Hudson, W. R., and B. F. McCullough, "An Extension of Rigid Pavement Design Methods," Highway Research Record 60, Highway Research Board, 1964.

[Hudson 64b]     Hudson, W. R., and P. Irick, "Guidelines for Satellite Studies of Pavement Performance," NCHRP Report 2A, 1964.

[Hudson 66a]     Hudson, W. R., and H. Matlock, "Analysis of Discontinous Orthotropic Pavement Slabs Subjected to Combined Loads," Highway Research Record 131, Highway Research Board, 1966.

[Hudson 66b]     Hudson, W. R., and H. Matlock, "A Finite Element Analysis of Discontinous Orthotropic Plates," Abstract No. 413, Proc. of the Fifth U.S. National Congress of Applied Mechanics, 1966.

[Hudson 66c]     Hudson, W. R., and H. Matlock, "Discontinous Orthotropic Plates and Pavement Slabs," Research Report 56-6, Center for Highway Research, University of Texas, 1966.

[Hudson 67]      Hudson, W. R., and H. Matlock, "Cracked Pavement Slabs with Nonuniform Load and Support," Journal of the Highway Division, Vol. 93, No. HW1, Proc., American Society of Civil Engineers, 1967.

[Hudson 70]      Hudson, W. R., B. F. McCullough, F. H. Scrivner, and J. L. Brown, "A Systems Approach Applied to Pavement Design and Research," Research Report 123-1, Center for Highway Research, University of Texas, 1970.

[Hudson 73]      Hudson, W. R., and B. F. McCullough, "Flexible Pavement Design and Management: Systems Formulation," NCHRP Report 139, 1973.

[Humphreys 63]   Humphreys, J. S., and C. J. Martin, "Determination of Transient Thermal Stresses in a Slab with Temperature-Dependent Viscoelastic Properties," Trans., Society of Rheology, Vol. III, 1963.

[Hutchinson 68]  Hutchinson, B. G., and R. C. G. Haas, "A Systems Analysis of the Highway Pavement Design Process," Highway Research Record 239, Highway Research Board, 1968.

[Ioannides 85]   Ioannides, A. M., M. R. Thompson, and E. J. Barenberg, "The Westergaard Solutions Reconsidered," Presented at the Transportation Research Board Meeting, 1985.

[Irick 87]       Irick, P., W. R. Hudson, and B. F. McCullough, "Application of Reliability Concepts to Pavement Design", Proc., Sixth Intl. Conf. on the Structural Design of Asphalt Concrete Pavements, Univ. of Michigan, 1987.

[Jayawickrama 87]  Jayawickrama, P. W., and R. L. Lytton, "Methodology for Predicting Asphalt Concrete Overlay Life Against Reflection Cracking," Proc., Sixth Intl. Conf. on the Structural Design of Asphalt Concrete Pavements, Univ. of Michigan, 1987.

[Kher 70]  Kher, Ramesh K., W. R. Hudson, and B. F. McCullough, "A Systems Analysis of Rigid Pavement Design," Research Report 123-5, Center for Highway Research, University of Texas, 1970.

[Kher 73]  Kher, Ramesh K., and Michael I. Darter, "Probabilistic Concepts and Their Applications to the AASHO Interim Guide for Design of Rigid Pavements," Highway Research Record 466, Highway Research Board, 1973.

[Kher 76]  Kher, Ramesh, W. A. Phang, and R. C. G. Haas, "Economic Analysis Elements in Pavement Design," Transportation Research Record 572, Transportation Research Board, 1976.

[Kher 77]  Kher, Ramesh, and W. A. Phang, "OPAC Design System," Proc., Fourth Intl. Conf. on the Structural Design of Asphalt Concrete Pavements, Univ. of Michigan, 1977.

[Khosla 87]  Khosla, N. P., "A Field Verification of VESYS IIIA Structural Subsystems," Proc., Sixth Intl. Conf. on the Structural Design of Asphalt Concrete Pavements, Univ. of Michigan, 1987.

[Klomp 64]  Klomp, A. G. J., and G. M. Dorman, "Stress Distribution and Dynamic Testing in Relation to Road Design," Proc., Australian Road Reserch Board, 1964.

[Lamb 66]  Lamb et al., "Roadway Failure Study No. 1," Highway Engineering Research Publication H-14, Natural Resources Research Institute, University of Wyoming, August 1966.

[Lee 65]  Lee, R. T., and E. L. Grant, "Inflation and Highway Economy Studies," Highway Research Record 100, Highway Research Board, 1965.

[Lister 67]  Lister, N. W., and R. Jones, "The Behavior of Flexible Pavements Under Moving Wheel Loads," Proc., Second Intl. Conf. on the Structural Design of Asphalt Concrete Pavements, Univ. of Michigan, 1967.

[Lytton 74]  Lytton, R. L., and W. F. McFarland, "Systems Approach to Pavement Design-Implementation Phase," Final Report, NCHRP Project 1-10A, 1974.

[Majidzadeh 71]  Majidzadeh, K., E. M. Kauffmann, and D. J. Ramsamooj, "Application of Fracture Mechanics in the Analysis of Pavement Fatigue," Proc., Vol. 40., Association of Asphalt Paving Technology, 1971.

[Majidzadeh 73]  Majidzadeh, K., E. M. Kauffmann, and C. W. Chang, "Verification of Fracture Mechanics Concepts to Predict Cracking of Flexible Pavements," FHWA Final Report No. FHWA-RD-73-91, 1973.

[McCullough 69]  McCullough, B. F., "A Pavement Overlay Design System Considering Wheel Loads, Temperature Changes and Performance," Doctoral Dissertation, University of California, Berkeley, 1969.

# References to Part Four

[McCullough 75]   McCullough, B. F., A. Abou-Ayash, W. R. Hudson, and J. P. Randall, "Design of Continuously Reinforced Concrete Pavements for Highways," Final Report, NCHRP Project 1-15, 1975.

[McFarland 72]   McFarland, W. F., "Benefit Analysis for Pavement Design Systems," Res. Rept. 123-13, Texas Transportation Institute, Texas A & M Univ., 1972.

[McLeod 72]   McLeod, N. W., "A 4-Year Survey of Low-Temperature Transverse Pavement Cracking on Three Ontario Test Roads," Proc., Association of Asphalt Paving Technology, 1972.

[Moavenzadeh 72]   Moavenzadeh, F., and J. F. Elliott, "A Stochastic Approach to Analysis and Design of Highway Pavements," Proc., Third Intl. Conf. on the Structural Design of Asphalt Concrete Pavements, London, England, 1972.

[Moavenzadeh 74a]   Moavenzadeh, F., J. E. Soussou, and H. K. Findakly, "Synthesis for Rational Design of Flexible Pavements," Part I, Final Report for FHWA Contract 7776, 1974.

[Moavenzadeh 74b]   Moavenzadeh, F., J. E. Soussou, H. K. Findakly, and B. Brademeyer, "Synthesis for Rational Design of Flexible Pavements-Operating Instructions and Documentation," Vol. III, Final Report for FHWA Contract 7776, 1974.

[Monismith 65]   Monismith, C. L., G. A. Secor, and K. E. Secor, "Temperature Induced Stresses and Deformations in Asphalt Concrete," Proc., Association of Asphalt Paving Technology, 1965.

[Monismith 71]   Monismith, C. L., and D. B. McLean, "Design Considerations for Asphalt Pavements," Report No. TE 71-8, Institute of Transportation and Traffic Engineering, Univ. of Calif., 1971.

[Monismith 73]   Monismith, C. L., "Pavement Design, The Fatigue Subsystem," Special Rept. 140, Highway Research Board, 1973.

[Monismith 85]   Monismith, C., J. A. Epps, and F. N. Finn, "Improved Asphalt Concrete Mix Design," Proc., Vol. 54, Association of Asphalt Paving Technology, 1985.

[Moulton 80]   Moulton, L. K., "Highway Subdrainage Design", Report No. FHWA-TS-80-224, Federal Highway Administration, 1980.

[Nair 71]   Nair, K., "Solutions and Solution Techniques for Boundary Value Problems," Special Report 126, Highway Research Board, 1971.

[Packard 67]   Packard, R. G., "Computer Program for Airport Pavement Design," Portland Cement Association, SR029.02P Skokie, IL, 1967.

[Packard 77]   Packard, R. G., "Design Considerations for Control of Joint Faulting of Undoweled Pavements," Proc., Intl. Conf. on Concrete Pavement Design, Purdue University, 1977.

[Papagiannakis 87]   Papagiannakis, A. T., R. C. G. Haas, and W. A. Phang, "Wide Base Truck Tires: Industry Trends and State of Knowledge of Their Impact on Pavements," Proc., Roads and Transportation Association of Canada, Saskastoon, 1987.

[Paris 63] Paris, R. C., and F. Erdogan, "A Critical Analysis of Crack Propagation Laws," Journal of Basic Engineering, Series D, 85, No. 3, Transactions of the ASME, 1963.

[Peutz 68] Peutz, M. G. F., H. P. M. Van Kemper, and A. Jones, "Layered Systems Under Normal Surface Loads," Highway Research Record 228, Highway Research Board, 1968.

[Pichumani 72] Pichumani, R., "Application of Computer Codes to the Analysis of Flexible Pavement," Proc., Third Intl. Conf. on the Structural Design of Asphalt Pavements, London, England, 1972.

[Pickett 51] Pickett, G., and G. K. Ray, "Influence Charts for Concrete Pavements," Transactions, ASCE, 1951.

[Raad 80] Raad, L., and J. L. Figueroa, "Load Response of Transportation Support Systems," Vol. 106, No. TE1, Transportation Engineering Journal, ASCE, 1980.

[Rauhut 75] Rauhut, J. B., J. C. O'Quinn, and W. R. Hudson, "Sensitivity Analysis of FHWA Structural Model VESYSIIM, Prepatory and Related Studies," Volumes I and II, Report Prepared for Federal Highway Administration, ARE Inc., 1975.

[Rauhut 77] Rauhut, J., R. C. G. Haas, and T. W. Kennedy, "Comparison of VESYSIIM Predictions to Brampton and AASHO Performance Measurements," Proc., Vol. I, Fourth Intl. Conf. on the Structural Design of Asphalt Pavements, Univ. of Michigan, 1977.

[Rauhut 82] Rauhut, J. B., and T. W. Kennedy, "Characterizing Fatigue Life for Asphalt Concrete Pavements," Transportation Research Record 888, Transportation Research Board, 1982.

[Saraf 87] Saraf, C., K. Marshek, H. Chen, and W. R. Hudson, "The Effect of Truck Tire Contact Pressure Distribution of the Design of Flexible Pavements," Proc., Sixth Intl. Conf. on the Structural Design of Asphalt Pavements, Univ. of Michigan, 1987.

[Sargious 75] Sargious, M., *Pavements and Surfacings for Highway and Airports*, John Wiley and Sons, New York, 1975.

[Schapery 81] Schapery, R. A., "Non-Linear Fracture Analysis of Viscoelastic Composite Materials Based on a Generalized J Integral Theory," Proc., Japan-U.S.A. Conference on Composite Materials, Tokyo, 1981.

[Scrivner 68] Scrivner, F. H., W. M. Moore, W. F. McFarland, and G. R. Carey, "A Systems Approach to the Flexible Pavement Design Problem," Res. Rept. 32-11, Texas Transportation Institute, Texas A&M Univ., 1968.

[Sentler 71] Sentler, L., "A Strength Theory for Viscoelastic Materials," Document D9, Swedish Council for Building Research, 1971.

[Shahin 71] Shahin, M. Y., "A Computer Program for an Iterative Method of Pavement Structural Analysis," Unpublished technical memorandum, Center for Highway Research, University of Texas, Austin, 1971.

[Soussou 74] Soussou, J. E., F. Moavenzadeh, and H. K. Findakly, "Synthesis for Rational Design of Flexible Pavements," Vol. II, Final Report for FHWA Contract 7776, 1974.

# References to Part Four

[SRI 76] Stanford Research Institute, "User Benefit Analysis for Highway and Bus Transit Improvements," NCHRP Project 2-12, 1976.

[Stelzer 67] Stelzer, C. F., and W. R. Hudson, "A Direct Computer Solution for Plates and Pavement Slabs," Research Report 569, Center for Highway Research, University of Texas, Austin, 1967.

[TAI 91] The Asphalt Institute, "MS 11 Thickness Design—Asphalt Pavements for Air Carriers," Third Edition, 1991.

[Taylor 64] Taylor, G. A., *Managerial and Engineering Economy—Economic Decision Making*, Van Nostrand, 1964.

[Treybig 77] Treybig, H. J., B. F. McCullough, P. Smith, and H. Von Quintus, "Overlay Design and Reflection Cracking Analysis for Rigid Pavements, Vol. 1, Development of New Design Criteria," Report FHWA-RD-77-76, Federal Highway Administration, 1977.

[Vesic 69] Vesic, A. S., and S. K. Saxena, "Analysis of Structural Behavior of Road Test Rigid Pavements," Highway Research Record No. 291, Highway Research Board, 1969.

[Westergaard 27] Westergaard, H. M., "Theory of Concrete Pavement Design," Proc., Highway Research Board, 1927.

[Winfrey 69] Winfrey, R., *Economic Analysis for Highways*, International Textbook Company, 1969.

[Wohl 67] Wohl, M., and B. Martin, "Evaluation of Mutually Exclusive Design Projects," Special Rept. No. 92, Highway Research Board, 1967.

[Yoder 75] Yoder, E. J., and M. W. Witczak, *Principles of Pavement Design*, John Wiley & Sons, 1975.

[Zaniewski 78] Zaniewski, J. P., "Design Procedure for Asphalt Concrete Overlays of Flexible Pavements," Doctoral Dissertation, University of Texas, Austin, 1978.

[Zaniewski 82] Zaniewski, J. P., G. Elkins, B.C. Butler, M. Paggi, G. Cunningham, and R. Machemel, "Vehicle Operating Costs, Fuel Consumption and Pavement Type and Condition Factors," Federal Highway Administration, 1982.

Part Five

# Implementation

Chapter 31

# Implementing a Pavement Management System

## 31.1  INTRODUCTION

The best laid plans for a pavement management system are useless unless they are implemented, or put into practice. Likewise, all the work toward identifying an optimum pavement strategy is wasted unless the strategy is actually put into practice; i.e., the selected pavement must be constructed, used, and maintained.

Consequently, implementation can be defined in two ways:

1. Implementation; putting into practice a pavement management system itself, or certain selected components of a system
2. Actualization of the decisions made through the pavement management system; i.e., carrying out the selected construction and maintenance strategy

This chapter deals broadly with the whole process of implementing a pavement management system. Chapters 32 and 33 deal with the necessary and practical aspects of construction and maintenance. Chapter 34 on research management falls within the first category noted above, but it is presented after construction and maintenance because it relates not only to those two parts of pavement management, but also priority programming, pavement evaluation, and design.

Guidelines and procedures for implementation of pavement management systems were first set out in the Pavement Management Guide by the Roads and Transportation Association of Canada [RTAC 77]. These guidelines are directed toward agencies who want to: (1) develop and implement a pavement management system, (2) improve their existing system, or (3) adopt only certain practices and procedures. Both technical and administrative aspects have been considered, and the guidelines provide for flexibility and for partial or staged implementation.

Since the RTAC Guide was published, a considerable amount of implementation experience has been gained by federal, state/provincial and local agencies, and by consulting organizations. The following sections present a synthesis of that experience for both the implementation of a new system and for the implementation of improvements to an existing system.

## 31.2 MAJOR STEPS IN IMPLEMENTING A PAVEMENT MANAGEMENT SYSTEM

Previous chapters of this book have emphasized the many alternative methods, models, and procedures that can be used by individual agencies within a generic pavement management framework. For example, in the structural analysis or design subsystem, one agency might use a sophisticated layer analysis method whereas another might use a simplified empirical method. Thus, although there is a clearly identifiable framework which characterizes all pavement management systems, sufficient flexibility exists to accommodate the particular requirements and resources of each individual agency.

Similarly, implementation of a pavement management system may vary for different agencies or groups depending on their needs and resources. However, several major steps can be identified, as summarized in Figure 31.1:

1. Decision for implementation of a pavement management system, or implementation of improvements to an existing system
2. Formation of a steering committee to identify needs, develop basic scope and agenda for the system, and monitor implementation
3. Review of existing organization, data bases, methods, and procedures within a defined pavement management system framework
4. Development of an implementation plan with recommendations and schedule
5. Definition of selection procedures for network programs and individual project designs or maintenance treatments
6. Actual work activities for each implementation stage of the pavement management system, including plan and schedule updates, documentation, training, and software installation
7. Monitoring of the system and periodic improvements

Clearly, a large agency with several divisions and districts will need to develop and follow a pavement management system implementation plan as outlined in Figure 31.1. Good communication and recognition of organizational issues will determine to a large extent the success of the system. However, an implementation plan is also

# Implementing a Pavement Management System

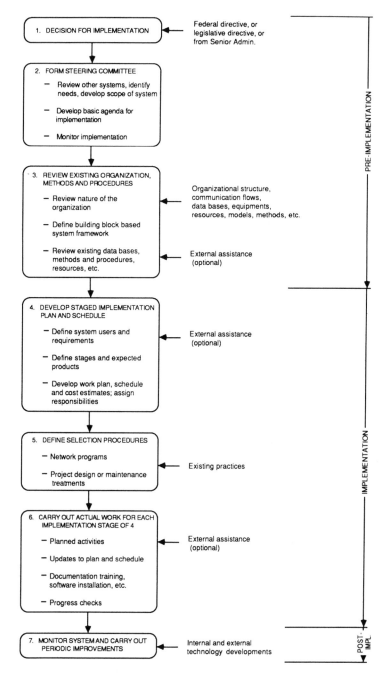

**Figure 31.1** Major steps in implementing a pavement management system

vital to the success in a small agency where communication and organizational issues are not as restrictive as in the large agencies. Even if a local agency has only one or two engineers, an implementation plan helps organize the steps needed for developing a successful system.

### 31.2.1 Decision for Implementation

A decision to implement must be the first major step, as shown in Figure 31.1, and it can be externally initiated, such as by a federal or legislative directive, or by senior administration within the agency. If the initiative derives from lower level staff, senior administrative support is still essential. A major part of the initial implementation decision may be the establishment of a pavement management office within the organization. This has been done by many states and provinces.

### 31.2.2 Steering Committee

A task force or steering committee (step 2) is required for any large scale pavement management implementation, particularly at the state/provincial level. Smaller local agencies may handle this, for example, through the director of roads and traffic, or some similar office. This steering committee has a number of responsibilities, including the following:

1. Review other pavement management systems.
2. Define the needs of the agency and develop the general scope of a system appropriate to these needs.
3. Develop a basic agenda for implementation, including commitments and financing.
4. Develop monitoring guidelines, provide advice to actual working groups, and monitor progress and effectiveness of the implementation.

The steering committee should include representation from the major divisions of the agency, such as planning, design, construction, maintenance, soils and materials, and research.

If outside consultants are retained for all or part of the implementation, then the necessity for a steering committee is obvious. However, even if all the work is done internally, such a steering committee is still essential not only for overall guidance and monitoring reasons but also because it must represent the range and needs of the users within the agency.

Many agencies, such as the Minnesota Department of Transportation [Maurer 87], have formed steering committees early in the pre-implementation stage. They have realized that successful development and implementation of a pavement management system require careful planning, staging, taking full advantage of their existing methods, procedures, and data, and recognizing the needs of the users within the agency.

Small agencies may find it advantageous to work in association with neighboring agencies to share experience and insights. Since many of the issues faced by the local

# Implementing a Pavement Management System

agencies in a geographic area or region are similar, formalization of this association into a type of users group can enhance efficiency within each agency.

### 31.2.3 Review Existing Organization, Methods, and Procedures

This review (step 3 in Figure 31.1) provides the basis for step 4, development of the staged implementation plan. In addition, the review can serve a valuable educational function within the agency itself [RTAC 77].

The activities in step 3 should include:

1. Review of the nature of the organization, i.e.,
   a. administrative structure and responsibilities
   b. communication flows
   c. program categories (such as major construction, reconstruction, rehabilitation, maintenance, and bridge rehabilitation)
2. Definition of a building block based system framework
3. Review of existing data bases, methods and procedures, resources, etc., within the framework of the building block based system

If a pavement management system is to be successfully implemented, it is important to know how the organization actually functions, which is not necessarily well represented by the usual flow charts of the administrative structure. Specific responsibilities, communication flows, and very importantly, the existing process for programming and executing projects within the program categories, need to be known because these represent the actual decision making. A pavement management system can significantly enhance the process but will be difficult to implement if it is configured as an abrupt replacement process.

The definition of a building block based systems framework, as detailed in step 4, can provide a valuable context for reviewing the agency's existing data bases, methods and procedures, resources, etc. It is desirable to make maximum use of such existing assets, and any gaps can then be identified by comparison with the requirements and expected products that are specified in step 4. An example building block framework for the network level of Minnesota's pavement management systems is given in Figure 31.2 [Maurer 87].

Supporting documentation in the form of manuals, guides, specifications, and reports should also be acquired for all existing practices. Even where existing practices are working well, this step can aid in identifying inadequate documentation of these practices. Frequently, changes in operating procedures are poorly documented; i.e., when a procedure was initially developed a manual was prepared, but after the procedure was in operation for some period of time it was modified to improve efficiency. However, the manual was not updated. An example of this sort demonstrates the need to include field personnel in the evaluation of existing practices. Adequate documentation is especially important for junior people in the agency and to provide continuity over time or staff turnover.

Deficiencies or roadblocks in existing practices are usually institutional, technical, financial, or political in nature [Clark 87, Jackson 87]. They can include the following:

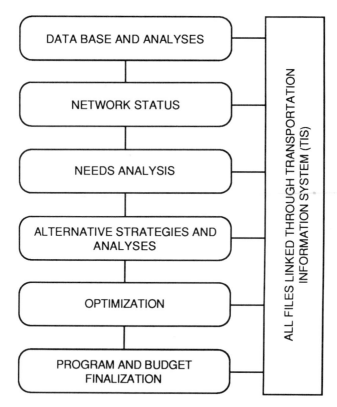

**Figure 31.2** Basic framework of building blocks defined for Minnesota's network level pavement management system [Maurer 87].

1. Institutional:
    a. Lack of sufficient communication, coordination, or support
    b. Personnel shortages and changes of personnel
    c. Perspective differences (view at top level pavement management function to provide program development versus view at operational level function of project design or selection)
2. Technical:
    a. Inadequate, outdated methods, equipment, or procedures
    b. Lack of sufficiently skilled people
    c. Lack of data or poor quality of data and/or resources for operating the system
    d. Inconsistencies (such as a sophisticated analysis procedure but crude or inadequate data)
3. Financial:
    a. Funding shortfalls
    b. Funding uncertainties
4. Political or policy-related:
    a. Perception of the role of the system and unrealistic expectations

b. Regional equity
c. Overrides of optimal network programs or project strategies

### 31.2.4 Development of Staged Implementation Plan and Schedule

Step 4, a staged implementation plan and schedule can include the following:

1. Pavement management system users within the organization, and their requirements
2. Specific stages for the implementation, and the expected products
3. Work plan, assignment of resources and responsibilities, schedule, and cost estimates

An example categorization of users and their requirements for the Minnesota Department of Transportation is shown in Figure 31.3 [Haas 85]. While the administrative titles will vary from agency to agency, the three categories of users—senior administrative, program development, and technical—and their respective requirements are representative of most agencies. As well, the requirements in Figure 31.3 should be common to most agencies.

The stages of Minnesota's pavement management system implementation are shown in Figure 31.4 [Haas 85], together with the expected products or major outputs. Since the network level of pavement management in Minnesota was most urgent, the first three shorter term stages address that part of the implementation. Then, the two subsequent stages of the longer term implementation are directed to maintenance management and programming. Stage 6 addresses the project level and is concerned with implementing major improvements to an existing system. It should be noted that while the various stages are shown sequentially in Figure 31.4, the actual scheduling required considerable overlap and concurrency of stages over the total implementation time of several years. Also, the major outputs listed in Figure 31.4 were translated into detailed product expectations.

Work plans were generated for the various stages, showing specific activities and responsibilities, assignment of resources, and schedules. Much of the work was carried out by consultants [Maurer 87] in order to accelerate the implementation.

### 31.2.5 Selection Procedures

Step 5 has been included in Figure 31.1 to reflect the need for consistent strategy selections or decisions at two basic levels:

1. Network level (i.e., selection of programs)
2. Project level (i.e., selection of the best within-project design or maintenance alternative)

In some agencies both levels of decisions are made by the same senior administrators or engineers. In other agencies a group or committee is formed to make the selection or to make recommendations to senior management.

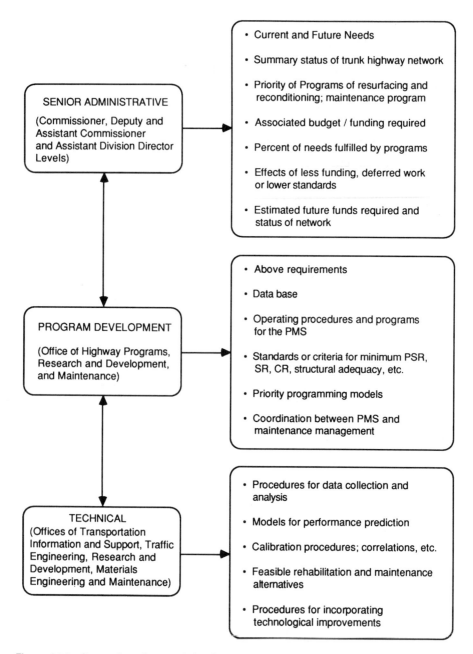

**Figure 31.3** Categories of network level pavement management users, and their requirements, within the Minnesota Department of Transportation [Haas 85].

# Implementing a Pavement Management System

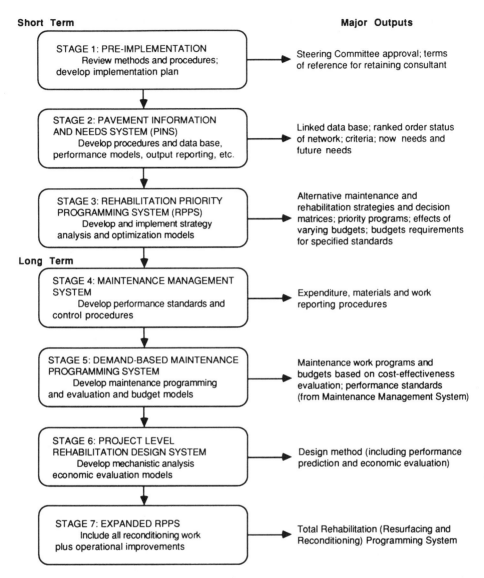

**Figure 31.4** Staged implementation plan for Minnesota's pavement management system [Haas 85].

The basic function of the network-level activities is to select a priority program, recognizing budget constraints, and this may involve a selection committee.

The basic function of the project-level activities is to make a final selection of the best strategy for each project. Normally, the design engineer makes project recommendations. However, some agencies have formed groups or committees to handle final approvals on an agencywide basis. The purpose is not to take responsibility away from the individual designer but rather to ensure consistency and continuity across the agency.

### 31.2.6 Actual Implementation Activities

Step 6 in Figure 31.1 represents the actual implementation work in terms of:

1. Carrying out the planned activities of step 4
2. Updating the implementation plan and schedule if necessary as the work progresses
3. Documenting computer programs, test methods, procedures, etc.
4. Training personnel in data collection and the use of the system, and installing software on the agency's computer facilities (smaller agencies, particularly local, sometimes choose to have the software operated and maintained externally by a consulting firm)
5. Doing progress checks on the work

### 31.2.7 Monitoring the System and Carrying Out Periodic Improvements

Responsiveness to user needs, operating problems, capability of handling situations or data not foreseen in the development of the system, reliability of the system, and various demands are all items that should be monitored in step 7 of Figure 31.1. As well, system maintenance, particularly of software and hardware, is facilitated by good monitoring.

Periodic improvements will be required in any system if it is to properly serve the agency. Such improvements are made necessary by new or better technology, changing scope and requirements of the system, and by new or better information. If the system has been designed to be modular (i.e., each stage and its components can be changed or replaced easily) then periodic improvements can be carried out efficiently.

### REVIEW QUESTIONS

1. Either from a PMS that is well documented in the literature, or information available on a locally implemented PMS, determine to what degree the implementation has in fact followed the steps of Figure 31.1.
2. Figure 31.1 has described these major levels of PMS users for the Minnesota Department of Transportation. Develop a similar description for your local or state agency PMS.

Chapter 32

# Construction

## 32.1 INTRODUCTION

A complete pavement management system must follow through from design of new pavements or rehabilitation to the implementation phases of construction and maintenance, plus data feedback. The transition from design to construction is one of the most important and difficult organizational boundaries in the pavement management system.

Construction converts a design recommendation into physical reality. Successful construction meets the planning and design objectives, within budget and time constraints. Contract tendering, contract award, construction schedule, materials supply and processing, and actual construction and quality control are usually conducted in a relatively routine manner. It is not sufficient, however, that design should function with little thought to construction or maintenance. Yet this is the case in many agencies, especially larger ones, where the traditional organization involves separate divisions for design and construction.

If conditions in the field are found to differ from the design inputs and assumptions, changes must be made. Such changes, at least major ones, should normally be reviewed and approved by the design group. Proper pavement management will ensure that sufficient communication occurs, that new or innovative design or construction solutions are not inhibited, and that as-constructed documentation is complete and understandable. This latter point is very important. Because of the usual hurry and day-to-day problems of construction, it is easy to delay or avoid such documentation of the pavement as it was actually built. There have been many experiences where searches through the files at some time after construction yield incomplete or erroneous information. Proper feedback and plans for rehabilitation depend on reliable and complete as-constructed documentation.

This chapter does not attempt to provide details on construction practices, construction control, or construction management per se. These are comprehensive subjects which are treated in a variety of construction manuals, operating procedures, and guides published by various agencies. Attention in this chapter is focused on (1) the interrelationships of construction with other phases of pavement management and (2) the documentation data that construction should produce for both construction and other uses. If these functions are carried out in a systematic manner, then the normal and expected variations in construction methods, equipment, and materials can be properly taken into account.

This chapter also summarizes the various levels of construction management and their relationship to pavement management. Finally, the chapter considers information acquired in construction and construction control as it relates to the pavement management system.

### 32.1.1 Construction Documents

In order to move the selected design to the construction phase, a set of definitive documents expressing the details of the selected design, or of the alternative offered for contract tenders, is needed. These documents not only convey details to construction but also serve as legal documents in procuring the services of a contractor or construction agency.

The traditional documents of design and construction are plans and specifications. These usually consist of the following:

1. A set of drawings that give detailed dimensions and other design aspects
2. A set of specifications that describe in detail the materials to be provided, their arrangement, required characteristics, etc.
3. A set of standards and specifications that have been previously approved by the agency and are in general use

The first purpose of these implementation documents is to describe the proposed pavement to the construction group. Because the pavement is usually built in conjunction with bridges, drainage facilities, and other items, the interrelationships involved are also described. In some agencies the construction group is equipped to construct the pavement as designed, and in such cases construction can immediately begin. Usually, however, the construction group has a supervisory or control function and the actual construction is independently done under a legal contract. In these cases, the documents become highly important because they serve a second purpose as the basis for bidding, pricing, and agreements regarding payment for the work.

# Construction

It has been common practice to provide "hard copy" plans and details from computer calculations to bidders. It is also possible to furnish this information in the form of computer disks, but the legal implications of any errors or omissions have to be considered.

## 32.1.2 Construction Management

Construction management involves the use of physical, financial, and personnel resources to convert designs to physical realities. This general concept intersects pavement management in the actual building of the pavement structure. The processes are successful when the stated planning and design objectives are met and the pavement is put into service.

The process of construction management is comprehensive and complex. It involves many considerations such as estimation, designation and scheduling of activities, organizational and personnel aspects, legal aspects, finances and cost control, records, and quality control. There are many published books and manuals available on the subject. Although it is beyond the scope of this book to treat this important subject in depth, the levels of pavement construction management as they apply to public agencies are subsequently discussed.

A major problem in construction over the years has been payment for the final product. Historically, the approach has been one of full payment for a pavement constructed according to (procedurally based) specifications and judged to be acceptable. Some agencies have now moved to what they term end product specifications, which cover items such as-built density, gradation, voids, and thickness. Also, some agencies reduce full payment through penalty clauses if the specifications are not met, and a few have bonus clauses if the specifications are exceeded.

It should be possible however to eventually develop true end product specifications based on long term pavement performance [Vlatas 89, Kuzyk 91]. The measures used could include as-built roughness, deflection, and surface friction and the specifications would place limits on these measures (i.e., the rate of deterioration) plus surface distress over the life-cycle of the pavement. Such an approach would transfer most of the risk to the contractor, but also provide for a more rational payment scheme, and a greater incentive for innovation.

## 32.2 CONSTRUCTION AS RELATED TO OTHER PHASES OF PAVEMENT MANAGEMENT

Effective pavement management depends on communication and coordination among all phases. Figure 32.1 illustrates the general information flows to construction from the other phases of a pavement management system. Details of these phases have been discussed in previous chapters. The diagram shows that construction receives vital information from all the phases.

On the other hand, construction also provides information to the other pavement management phases. The general nature of this set of information flows is given in Figure 32.2. The diagram again shows that information from construction is vital to efficient management in the other phases.

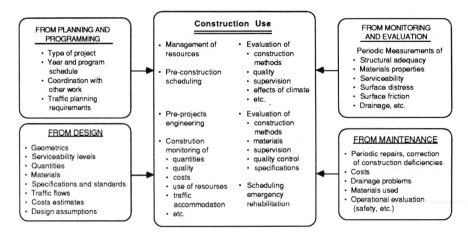

**Figure 32.1** Information flows to construction from other phases of pavement management.

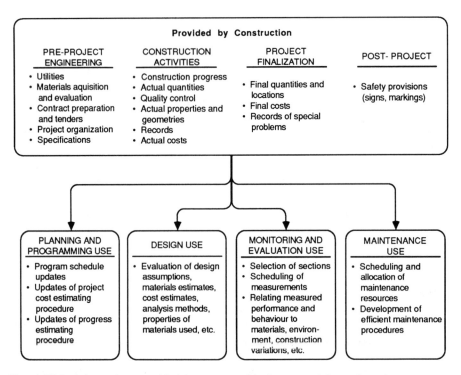

**Figure 32.2** Information provided by construction for potential use by other pavement management phases.

**Construction**

Five phases are considered in Figures 32.1 and 32.2: planning and programming, design, construction, monitoring and evaluation, and maintenance. It might seem that construction should interface only with design and maintenance, but this is an incomplete formulation. Construction is important to the entire pavement management system, and these relationships can be examined in terms of the two-way information flows illustrated in Figures 32.1 and 32.2.

### 32.2.1 Planning and Programming and Construction

The planning and programming phase provides the "what, when, and where" type of information to construction and the other phases. Information on type of work required and location of the work can provide direction for both short- and long-term construction planning for work force and other resources. Often, this preconstruction information, involving sufficient lead time, can provide better contract prices based on better availability of contractors and less crowding of the work. It is also possible to accomplish the preconstruction activities of surveys, right-of-way acquisition, utilities relocation, and so on, in a much more timely, economical, and efficient manner. The overall system or network economic planning and analysis will provide information for scheduling contract tenders, personnel, and resources and provide for consideration of time restrictions and related items.

In the reverse direction, construction relates to programming in many ways. For example, comparisons of final contract costs and materials quantities with the original programming estimates can help in updating future programming. Monitoring of construction in progress can point out the need for schedule adjustments. Money is not always available to cover cost overruns and early knowledge can make it possible to reprogram without penalty to other contracts. Finally, construction completion and road openings to traffic dates are directly important to the next round of programming.

### 32.2.2 Design and Construction

The plans and specifications from design provide direct input for construction. This is the most obvious and direct interaction between the two phases. In addition to the contract documents and details for construction, however, proper interaction in the early stages can assist the construction in the preliminary project field engineering phase and in setting up preliminary methods for locating materials and establishing quality and quantity control techniques.

Feedback from construction to design is equally vital. A seemingly economical design is not effective if it results in unusual or unmanageable construction problems. For example, a material may appear to be economical based on previous data, but a particular job or location may result in short supply and greater expense than expected. On the other hand, new construction techniques may lead to design changes. For example, most major highway departments at one time required thin lifts in asphalt concrete construction to gain adequate compaction. However, with the advent of heavier and variable-tire-pressure pneumatic rollers, it was shown that thick lifts may give better compaction under many situations. This type of feedback from construction led many agencies to change their specifications. A pavement management system can help in keeping new ideas from being discarded because of old methods.

### 32.2.3 Pavement Monitoring and Evaluation and Construction

In its broadest sense, pavement evaluation provides input to the construction phase. Periodic monitoring and evaluations of existing pavements can show that certain construction methods and technology, although acceptable under existing specifications, lead to premature failure or rapid deterioration. For example, aggregate gradation is controlled on a before-compaction basis. Subsequent testing in later evaluations may show that construction has actually degraded the material excessively, which may lead to structural weakness or drainage problems. A good data feedback system can assist in solving many such potential construction problems.

The input from construction to evaluation is direct and important. The as-built construction records provide the initial or zero-age evaluation of the pavement. Furthermore, if the construction data is good and effectively recorded, it will form the backbone of all future evaluation because the data will be more detailed than can ever be obtained again. In addition, these records can be used to assist in selecting the initial location of pavement sections to be periodically monitored. This is important, because monitoring is a sampling process.

During periodic monitoring, unusual problems observed in the pavement should be checked directly against existing construction records.

### 32.2.4 Maintenance and Construction

The last phase considered in relation to construction is maintenance. Both corrective repairs and preventive maintenance work intended to delay deterioration, such as a seal coat, are included in the definition of maintenance. Careful consideration of maintenance required and where it is done may show some patterns that feed back to construction and show weaknesses in specifications, construction methods, and materials and quality control methods.

Regarding the information flows from construction to maintenance, effective maintenance techniques and procedures are dependent on the actual materials and construction methods used. They are also related to the problems that were encountered during construction.

For example, poor pavement performance can often be shown to be the result of a period of unusually wet weather encountered during construction, perhaps interrupting compaction of a stabilized base or disturbing finishing operations. In such cases, the as-built construction records form the keystone of the maintenance data base. The information on final acceptance, claims, and adjustments can serve as the basis for preliminary scheduling of maintenance programs and resources. Likewise, knowledge of actual quantities of the materials used is essential to future maintenance planning.

## 32.3 PAVEMENT CONSTRUCTION MANAGEMENT

A significant part of the construction manager's activities is concerned with the documentation of the work and any factors that might affect the finished product. This documentation should also be useful to the other phases of pavement management and should be able to provide such information as the geometry of the pavement structure actually built, initial structural adequacy, initial roughness, the actual materials and their quality, and so on. Although much of this information is periodically acquired in pavement monitoring and evaluation, it is usually best known and provided by the construction manager for the as-constructed condition. This may well be done in cooperation with those responsible for periodic evaluation.

### 32.3.1 The Levels of Construction Management

Construction management can be considered in terms of a field level and a centralized level. For many agencies, these basic levels are also applicable to programming, design, construction, and maintenance.

The degree of activity and responsibility associated with each level of management usually depends on the size and complexity of the pavement network being managed. Large organizations may have more than one centralized level of management. In smaller jurisdictions, such as many of the smaller to medium-size municipalities, both levels may be combined under one individual manager who is responsible for the total road program.

Both levels of management have a number of common concerns. These include the following:

1. Scheduling and progress of the construction
2. Quality of construction
3. Costs incurred
4. User safety and convenience
5. Environmental protection
6. Industrial safety
7. Effective use of work force and resources
8. General operations efficiency

Any of these factors may affect other phases of pavement management, but the first three—construction progress, quality, and costs—are major concerns for all levels of construction management. These first three factors are also the source of most of the information documentation needed from the construction phase.

### 32.3.2 Centralized Levels of Management

The number of centralized levels of construction management depends on the size and complexity of the system being managed. Commonly there are one or two levels. With two levels, especially where highway agencies are involved, they are usually (1) regional and (2) central or head office. If only one level is used, both sets of duties are assumed jointly.

The regional manager's responsibility may include all construction operations taking place within a major geographical portion of the total pavement network. Alternatively, the regional manager may be concerned with the management of one specific construction type (bridges, earthwork, pavements, etc.) within a region or over the total network, depending on the organizational structure of the agency involved.

The regional construction manager's basic responsibilities include the monitoring of progress and expenditures, the control of quality and uniformity, and the assurance that design objectives, standards, and specifications are met. In addition to being accountable for the progress, quality, costs, and so on, of a set of projects, this manager also becomes concerned with the integration of construction with overall pavement management needs.

Some highway agencies have districts rather than regions, with district managers, whereas in others (i.e., larger departments), there may be districts within regions.

The central or head office level is involved in the management of construction for the total network. The primary concerns are progress (and completion), and expenditures (in comparison to budgets) for the total construction program. The central level of management also has a direct responsibility toward achieving an optimum set of construction operations for the entire roadway or transportation network. This requires explicit recognition of and close liaison with the other transportation management activities of the organization.

### 32.3.3 The Field Management Level

The field or project manager is in charge of the actual construction site. The project manager is normally expected to fulfill three basic functions in cooperation with the contractor's construction superintendent:

1. Successful completion of the project, as scheduled
2. Control of costs and quality
3. Documentation of costs, quality, and any other relevant construction factors

In order to fulfill these functions satisfactorily, the project manager must measure and assess the quantity and quality of the work and make day-to-day judgments on its acceptability. These functions are normally closely interrelated. For example, the documentation of construction quality and quantities for the acceptance and payment of the work is direct input to the pavement management system as a whole.

Because project managers actually experience, measure, observe, document, and report on the work "as built," on a project-by-project basis, they provide most of the basic construction information on the pavement network. Such field-level project managers are not ordinarily involved with the total concept of pavement management, as outlined in this book. Their concern is primarily with one phase, the successful construction of each individual project. Consequently, a project manager might well question the need to supply information that will be used only by higher levels or by other phases of management. It is therefore important that the project manager be well informed regarding the needs and uses of the information being generated. If at all possible, the information that project managers gather and report should be as directly useful to them and to their own project management activities as it is to any other management level or phases. It is also important that other pavement management levels or users not place unrealistic demands for information on field-level managers.

## 32.4 PAVEMENT CONSTRUCTION AND THE ENVIRONMENT

The construction of roads has had a major effect on the immediate environment at least since the time Roman roads were built, and perhaps even before that. The movement of people by foot, horseback, boat, or car has always changed the environment. In many ways this change has been good, or at least desirable, to the populace. Seldom has a group receiving its first all-weather road been unhappy because of environmental impact. In developing countries, where available funds must be stretched as far as possible to build new feeder roads, little attention has been given to environmental concerns. They are seen as extraneous for that type of road. However, in most areas of North America, Europe, parts of Asia, etc., environmental concerns are a major factor, often the major factor, involved in construction.

In developed areas, a substantial portion of pavement costs are now expended to address environmental concerns. The highway sector of transportation has in recent years received increasing criticism from those concerned with both the natural and social environment. Some of this criticism is warranted, and highway agencies are generally making serious efforts to protect the environment. At this time, however, they are faced with providing transportation links in response to public need.

The overall impact of a highway is basically its very presence, but its most visible impact on the natural environment is often during the construction phase. The construction manager should be knowledgeable about such impacts and should try to minimize them where possible, within his or her basic responsibility to manage construction of the work.

### 32.4.1 Construction and the Natural Environment

A number of major considerations apply to protection of the natural environment during and as a result of pavement construction. They include the following:

1. Dust-producing activities should be controlled to a reasonable degree. Most agencies have standards applying to dust-collection equipment for mix plant operations, and there are environmental regulations on dust and other emissions. In addition, watering, oiling, or other surfacing of haul roads may be required to control dust.
2. Gravel pits and rock quarries should be properly utilized so that the resources are conserved and so that these sites can be restored to some other useful purpose when they are exhausted. Most agencies have guidelines or controls for such land use.
3. Construction management should work closely with those agencies or groups charged directly with the protection, use, or development of natural resources. In addition, contracts and field work practice may include suggestions from environmental impact studies.
4. Natural water bodies and stream crossings should receive minimum disturbance through pavement construction. Major problems can arise where such construction is near a body of water or a stream and pollutants such as asphalts, lubricants, fuel oil, construction wastes, and so on, are allowed to enter the water, or where natural drainage patterns are disturbed. Also, haul roads for

aggregate supply to the site can cause problems where temporary stream crossings are involved. Such temporary crossings should minimize disturbance to the natural flow patterns, fish migration, or spawning beds, etc.
5. Construction operations should be carried out within the designated right of way. In addition, careful attention should be given to cleanup and trimming after construction.
6. Costs and particular problems associated with environmental protection should be documented. The costs of environmental protection should be justified in relation to the accrued benefits, both monetary and nonmonetary. Information in this regard will probably be expected from the construction manager at some postconstruction time.

### 32.4.2 Construction and the Social Environment

Various types of construction can sometimes have far-reaching and lasting effects on the social environment. As far as pavement construction is concerned, the major considerations involved include the following:

1. The effect of noise from quarry or gravel pit operations, mix plant operations, and haul trucks can cause considerable disturbance, especially in urban centers and during early or late hours. Postconstruction noise levels can be affected by the type and size of aggregates included in the surface, by joint spacing, by the pavement roughness, and so on.
2. Archaeological sites may be involved in certain situations. Although this can apply to construction of the road as a whole, pavement construction may be particularly involved through the use of gravel pits and quarries. Many agencies now initiate archaeological investigations prior to construction. Such preconstruction investigations, plus cooperation from construction management, can result in the desired protection while minimizing delays or extra costs to construction.
3. Construction work forces and operations can have both positive and negative effects on a community. The positive effects are primarily economic in nature, whereas the negative effects can involve attitudes and behavior.

## 32.5 CONSTRUCTION CONTROL

The control of construction is a vital part of construction management and of the pavement management system as a whole. Control can be considered in many ways and from several points of view. Administrative control of a job or of a construction site is one such aspect. Physical control of materials sources, properties, and grades, etc., of right-of-way access are other important aspects. The term *construction control* as usually applied, and as described in this chapter, identifies the process of controlling or assuring the quality of construction of the pavement. Other forms of control, such as administrative, are treated at appropriate places elsewhere in this book.

Construction quality can be controlled in many ways, all of which involve plans and specifications and certain measurement techniques for ensuring that the final

# Construction

pavement is constructed according to the design that has previously been selected and described in the plans and specifications. However, there is often some lost information between the design of the intended pavement and the plans and specifications that describe it.

The primary example of this is riding quality or roughness. Not many agencies (although the number is increasing) enforce a specification requiring a minimum serviceability index or maximum roughness for the completed pavement. This situation was perhaps best described many years ago by Jake Roberts, a district highway engineer in Texas, who had worked through all the ranks of design and construction. He said in 1964 "You know, we specify every damn thing about the road from the grain size of the sand to the producer of the asphalt but we don't specify the final quality [*the serviceability index*] of the pavement itself. That, I must judge myself by riding the road with the construction superintendent on the final inspection . . . " Roberts recognized that the PSI concept would someday give us the measuring tool for a more objective method of final acceptance of the total pavement as a package. This of course does not mean that accurate evaluation of strength, thickness, and so on, can be omitted.

Unfortunately, the difficulties of providing inexpensive, accurate, and reliable roughness measuring devices, whose results are acceptable in judging legal requirements of a contract, has slowed the process of using PSI or roughness per se in construction control. Adequate equipment and techniques are now available, but old specifications and methodologies are hard to change.

### 32.5.1 Types of Specifications

Basically, there are three types of specifications:

1. Detailed or procedural type of specifications
2. End-point or end product specifications
3. Final product or performance specifications

Detailed specifications are those that define the specific characteristics of the components, such as grain size, plasticity index of the fines, viscosity of the binder, and so on. These often also include specification of the details of the process, such as type and size of equipment, number of passes, operating temperatures, etc.

End-point or end product specifications would, for example, specify a minimum density (i.e., 95 percent of the laboratory standard) and water content range (i.e., 10 to 12 percent) for the base. But such specifications do not specify how to obtain these minimum values or what equipment to use.

Final product or performance specifications could take two forms: the most complete would be to specify the quality or performance of the whole pavement for a time period; the intermediate form would be to specify the final strengths or moduli and thicknesses of the pavement layers, instead of density, gradation, water content, and so on.

Consider a simple illustrative example of the difference in the three types of specifications. A company desires to connect two parts of a machine together, and the forces in each part are known. The detailed specifications might require a hex-head bolt made of annealed 60,000-psi steel, 2-in. long, 32 threads/in., and 1/2 in. in diameter. The end-point specifications might require a 2-in.-long bolt capable of handling a 12,000-lb force. The final product specification might require a fastener capable of connecting part $X$ to part $Y$ as shown in an accompanying diagram.

Unfortunately, the pavement problem is never as simple as a single fastener. Nevertheless, the analogy given provides a basic illustration for the concept of the three different types of specifications.

In road construction, the detailed type of specification has been most often used because of the complexity of the materials and procedures involved. Although the use of this type of specification might originally have been warranted, far too many detailed and confusing standards and specifications are currently in common use. A move toward end-point specifications is underway in many areas, and this is encouraging. Hopefully, the state of technology will continue to progress toward final product or performance specifications.

### 32.5.2 Types of Quality Control

Quality control or assurance is a most important aspect of construction and of the entire pavement management system. A considerable volume of information exists in the literature and this book does not treat the subject in detail.

Quality control is required to ensure that the finished pavement meets minimum standards consistent with the service desired for the planned life of the facility. This conformity is obtained by checking, through tests, the quality of the materials and the various phases of the work. There are administrative aspects of controls that deal with legal and qualitative matters, and technical aspects that refer to what we normally think of as quality control. In fact, these aspects are closely interrelated, and adequate quality control is essential to effective administrative and legal control.

Historically, the control of quality has been as confusing as the area of specifications. Basically, control has been based on a concept whereby an observed condition is assumed to define reality. A test is made; if the results equal or exceed the specified value, they pass. If the results are below the specified value, they fail. Little account is given to the fact that each test represents only a small sample of the material being judged. Individual inspectors often establish rules for retesting failed values and for judging how much material is to be judged by a single test.

During the 1950s, the aspects of statistical quality control began to enter the pavement field. Really important strides forward were made at the AASHO Road Test, where statistical methods were used to control the special construction sections and also to define the results. A great many government agencies began to adopt statistical quality control methods in the 1960s to the extent that now most agencies have such methods. The underlying concept is based on statistical theory and admits the existence of variability and the impracticality of obtaining 100 percent compliance with any reasonable specification because of the distribution of natural results. Of course, specifications and quality control must go hand in hand. It is not possible to take an old, absolute-type specification and control it statistically. Rather, it is necessary to determine the quality needed and then to derive the statistical control limits that represent these results effectively.

### 32.5.3 Testing for Quality Control

The mechanism for all quality control is testing each of the factors being controlled. They must be tested in some fashion for each unit of construction being controlled.

The type of tests to be used for each specification, and especially their number or frequency, is a subject of much controversy. Each organization has to decide on its own method of measurement according to its specifications and its resources. Figure 32.3 [RTAC 77], provides some sample guidelines for such selection, or for comparison with existing procedures, by showing the type of tests done and their frequency.

The number of tests for a certain volume of work done is a basic rate that must be adjusted using principles of statistics. First, the time or place for doing a test should be randomized. Then the rate can be increased or diminished depending on the variation of the first results obtained. If the deviation is very high, the frequency should be increased. If the deviation is relatively small, the frequency should be decreased. Each organization establishes its frequency formula for test according to the degree of control that it decides is necessary from experience, quality results, etc.

In order to set up a reasonable quality control program, an agency should employ the following basic steps:

1. Determine test methods and testing frequencies such that the given test properly describes the quality of the end product, while also permitting corrective action to be taken if deficiencies are noted.
2. Acquire test data and enter into the data file by location within the project so that individual points can be assessed. The information should be organized so that statistical analysis (mean, standard deviation, coefficient of variation, regression, and so on) can be readily performed.
3. Store in the data files only those test results that describe the finished product. Measurements or tests that result in further action should be reported, but because they do not represent the final pavement condition they need not be stored.

### 32.5.4 Quantities and Costs

Another major component of construction control involves quantities and costs of all construction actions and materials. A cost breakdown, in relation to the total cost of the project, will provide useful information for a variety of purposes, including that of relating individual items to the ultimate performance of the pavement.

The recorded costs are broken down or grouped to suit the particular policies or methods of the agency involved. The following minimum cost data should be recorded for each of the earthwork, subbase, base, and surfacing phases of construction [RTAC 77]:

1. Capital costs
2. Field supervision costs

| Tests or measurements | Frequency | Results to report |
|---|---|---|
| **a. Subgrade** | | |
| Reference specific weight (Proctor) and water content | Two for each type of material | a. Number of performed tests<br>b. Minimum and Maximum of specific weights (Proctor) and optimum water content for each material |
| Specific weight and water content of subgrade | One per 2000 m$^2$ with a minimum of three per section<br>For depth greater than 1 m: one per 4000 m$^2$ | a. Statistical distribution of specific weight and water content for each material<br>b. Average value<br>c. Standard deviation<br>d. Percent within the requirements |
| Classification of soil | One for each type of material<br>In the last top meter: One per 4000 m per layer | a. Number of performed tests |
| **b. Subbase** | | |
| Thickness | One per 2000 m$^2$ for each layer | a. Number of performed measures<br>b. Average value<br>c. Standard deviation<br>d. Percent within the requirements |
| Sieve analysis or sand equivalent | Two initially, then two per production day or per 2000 m$^2$ per layer | a. Number of performed tests<br>b. Any deviation from specifications<br>c. Range of results for maximum size, percent smaller than 4.8 mm, 1.2 mm, and 0.074 mm<br>d. Percent within the requirements |
| Reference specific weight (Proctor) and water content | Two for each type of material | a. Number of performed tests<br>b. Minimum and maximum of specific weights (Proctor) and optimum water content for each material |
| Specific weights and water content, on road | One per 2000 m$^2$ per layer with a minimum of three per section | a. Statistic distribution of specific weight and water content for each material<br>b. Average value<br>c. Standard deviation<br>d. Percent within the requirements |

**Figure 32.3** Guidelines for selecting quality control tests, frequency of testing, and reported results [RTAC 77].

**Construction**

| Tests or measurements | Frequency | Results to report |
|---|---|---|
| **c. Base: same tests as for the subbase plus:** | | |
| Abrasion<br>Soundness<br>Petrographic number | Two initially and one at each change of material | a. Number of performed tests<br>b. Any deviation from specifications<br>c. Range of results |
| **d. Bituminous pavement** | | |
| Sieve analysis | Two per shift per type of aggregate during production | a. Number of performed tests<br>b. Any deviation from specifications<br>c. Range of results for maximum size, percent smaller than 4.8 mm, 1.2 mm, and 0.074 mm<br>d. For c, average value, standard deviation, and percent within the requirements |
| Abrasion<br>Soundness<br>Petrographic number | Two initially and one at each change of material | a. Number of performed tests<br>b. Any deviation from specifications<br>c. Range of results |
| Quality of bitumen | One per delivery to the plant | a. Number of performed tests<br>b. Any deviation from specifications<br>c. Range of results<br>d. Origin of crude, refinery hauler |
| Marshall test:<br>• Resistance<br>• Void content<br>• Specific weight<br>• Bitumen content (by extraction) | One per day per mix | a. Range of results<br>b. Average values<br>c. Standard deviation<br>d. Percent within the requirements |
| Specific weight of layer mix | One per 2000 $m^2$ per layer with a minimum of three for a given pavement (mix) | a. Number of performed test<br>b. Statistical distribution of specific weight<br>c. Average value<br>d. Standard deviation<br>e. Percent within the requirements |

**Figure 32.3** *Continued*

3. Quality control costs
4. Overhead costs
5. Costs caused by special conditions

Where special circumstances or situations occur, such as flooding, strikes, expropriation, delays in moving utilities, and so on, which can seriously affect the costs of a project, they should be carefully documented so that an assessment can be made at the completion of the project. Also, the documentation may be required for claims or legal actions.

## 32.6 DOCUMENTATION OF CONSTRUCTION DATA

It has previously been emphasized in this chapter that the function of construction does not end with providing the actual physical facility but that it should also document the as-built properties or condition of the pavement. This represents important management information, and it should be acquired and processed in a manner that is easily understandable and usable.

A considerable amount of data can be generated in carrying out a pavement construction project. Records of this data have the following basic functions:

1. Provision of documented evidence that the project was constructed according to the contract specifications, for construction management purposes (payment, administration, etc.)
2. Provision of a data base to assess the adequacy of standards and practices used, effectiveness of the quality control methods, costs of the various operations, and the effects of construction practices on pavement performance
3. Provision of data for other phases of pavement management

Each highway agency has forms for documenting construction project data. It is beyond the scope of this book to present these forms, but a complete example set for pavement projects is available [RTAC 77]. The major functional classes of pavement construction project data that should be included are shown in Figure 32.4. This is not an absolute classification in that there is considerable overlap, but it follows the basic functions previously listed.

Some of the important types of as-built construction information can be classified as follows:

1. Type and thicknesses of pavement structure actually built, and over what lengths
2. Actual material quantities used, their properties, and the variations in properties
3. Actual costs (total and unit)
4. Actual construction dates or times
5. Pertinent records of weather, rainfall, drainage problems, traffic during construction, and so on

# Construction

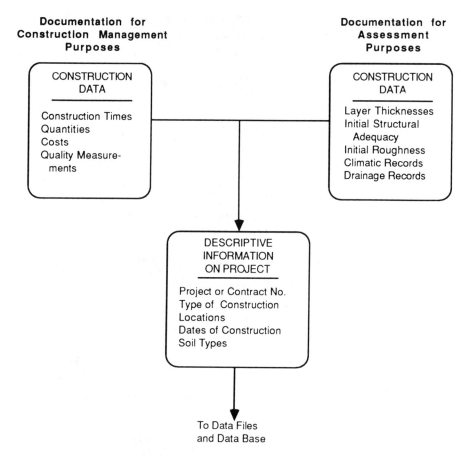

**Figure 32.4** Major functional classes of construction information documentation.

6. Initial structural adequacy of the pavement
7. Initial roughness of the pavement

The latter two types of information may be handled as a part of the pavement evaluation function. Nevertheless, they do represent important pieces of as-built construction information and should not be neglected.

The extent and quality of a pavement management system are dependent, to some degree, on the extent of the available data. For construction management, two basic levels of data collection can be used:

1. A minimum level aimed at providing minimal job history information
2. A higher level, depending on the capabilities and requirements of the particular agency, to serve as a more detailed basis for a pavement management system

When the level of data to be collected has been established, the information has to be coded and organized for use in a computerized processing system. The coding methods adopted should be consistent with those employed in other phases of the pavement management system.

The following sections consider the subject of construction documentation and information in somewhat more detail.

### 32.6.1 Descriptive Information on Project

General descriptive information on pavement projects should include the project or contract number, the dates of construction, types of construction and specific locations (i.e., beginning and end points), and subgrade soil types and locations.

In addition, information on such items as contractors, subcontractors, various external factors (project shutdowns, materials supply problems) may be acquired, depending on the agency and the nature of the project.

### 32.6.2 Data for Construction Management Purposes

Figure 32.4 identified one class of data related to construction management purposes. Although the component data types are part of the overall project data that can be used for subsequent assessment and other pavement management purposes, they have been separately identified to indicate their initial role for purposes of contract administration and payments.

The major types of data within the classification are construction times, actual quantities of subbase, base, surface and other materials processed or used, actual costs incurred, and quality measurement data (aggregate gradation, compaction, percent of binder, voids).

Quality control, as previously discussed, is one of the most important aspects of construction management on a pavement project. Quality control not only provides a basis for preventive or corrective action but also a means for observation or documentation for subsequent analysis. Although construction quality control should contribute to uniformity and better performance, it is not a substitute, for example, for good design and specifications. Rather, it should be consistent and compatible with other aspects of pavement management.

### 32.6.3 Construction Data for Assessment Purposes

The third major class of construction data documentation of Figure 32.4 relates to its use in assessment. It includes information on geometry (i.e., layer thicknesses), structural adequacy, and roughness immediately after construction and on such climatic and drainage factors as rainfall, temperature, water table, and scour problems.

Some of this data may be acquired during construction and some is obviously acquired on the new, as-built pavement. Again, there is considerable overlap of this class of data with the other two classes shown in Figure 32.4. However, they have been separately identified to emphasize the importance of construction data for not only assessing construction practices, quality control procedures, standards, and specifications, but also for assessing the effects of construction on pavement performance.

# Construction

## 32.6.4 Types of Data to Be Collected

Figure 32.4 illustrated the major classes of construction information that should be acquired. Figure 32.5 [RTAC 77] gives a summary set of guidelines related to typical construction data, data items and their collection, use, and frequency. The data in this figure is related to three levels of construction management: central, regional, and field. The data is also divided into a preconstruction or preparation stage, a during-construction stage, and an after-construction or final stage.

Certain types of climatic and drainage data during the construction period should also be acquired, because these two factors can affect pavement life and performance. Other environmental factors reflected mainly in design considerations also influence performance; data for these is usually gathered in the design or pavement evaluation phases. These other factors include the general geographical-climatic environment, the climatic environment expected and actually encountered subsequent to the construction, and so on.

Climatic conditions during the construction period can have a highly significant effect on the performance of a pavement and therefore provide an important input into a pavement management system.

Climatic data acquired during the construction period can be used for the following purposes:

1. Provision of vital background information on actual job conditions for any subsequent analysis, negotiation, or adjudication of claims with a contractor.
2. Adjustment of design and construction standards in a particular region or environment to minimize the influence of climatic conditions in subsequent construction.
3. Explanation of any unusual variations in pavement performance as a result of climatic variables. The information could also be used to include the effects of climate in serviceability-age (or traffic) models used in design.
4. Assessment of the effects of climate on materials and maintenance practices.
5. Development of realistic construction schedules and completion dates.

The minimum climatic data that should be acquired during the construction of a pavement includes general written summaries and diaries of climatic conditions and time lost due to climatic conditions. For a more complete evaluation, the climatic factors of temperature, rainfall, snowfall, and wind can be recorded in terms of daily high, low, and average temperatures, daily rainfall, daily snowfall, and daily maximum wind velocity. Such climatic data can be summarized in tabular or graphical format as required.

The drainage characteristics of a pavement and its surrounding area can also have a significant effect on performance.

Information on drainage can be used for the following purposes:

1. To provide advance warning of likely maintenance problems resulting from drainage deficiencies such as high water table, ponded water, and scour potential

| Management level | Data items |
| --- | --- |
| | **a. Preconstruction stage** |
| Central | 1. Credit allowance  6. Contract signing<br>2. Bid publication  7. Schedule of contract<br>3. Bid opening  8. Number of working days<br>4. Bid analysis  9. Number of contracts in active<br>5. Choice of contractor      status by the low bidder<br>                10. Pavement design data |
| Regional | 1. Expropriation  6. Order to start work<br>2. General project description  7. Contractor's list of<br>3. Quantities and unit price      personnel and equipment<br>4. Interference of other projects,  8. List of personnel for<br>   municipalities, companies, etc.    surveillance and quality<br>5. Mix designs       control |
| Field | 1. Survey  5. Terrain classification<br>2. Preparation of "right-of-way"  6. Public utilities: telephone,<br>   and "land access" plans     underground ducts, and<br>3. Determination of quantities    electricity<br>4. Climatic zone  7. Material sources and<br>        construction products |

| Management level | Daily | Monthly |
| --- | --- | --- |
| | **b. Construction stage** | |
| Central | 1. Approval of work overruns<br>2. Notice of claims | 1. Pay quantities for each type of work and percent complete<br>2. Payment |
| Regional | 1. Work overruns checked against design and credit allowance<br>2. Actions against claims<br>3. Minutes of field meetings | 1. Verification of paid quantities and estimation<br>2. Analysis of delays<br>3. Construction control test results summary |
| Field | 1. Work overruns<br>2. Contract daily statistics<br>3. Quantities and costs<br>4. Working days and downtime days<br>5. Material quality control test results; mix design test results; construction control test results<br>6. Diary of operations | 1. Résumé of decisions and actions taken at field meetings |

| Management level | Information items |
| --- | --- |
| | **c. Postconstruction stage** |
| Central | 1. Construction control summary  3. Owner's surveillance costs<br>2. Résumé of statistics on costs,  4. Quality variations from<br>   working days, downtime days     design<br>   on account of strikes, expropria-  5. Quality of finished product<br>   tion, public utilities, etc. |
| Regional | 1. Construction data  4. Résumé of quantities and<br>2. Claims        cost<br>3. Constructed pavement test  5. Last field inspection for<br>   results       final payment |
| Field | 1. Engineer's report  3. Drainage<br>2. Subgrade soils |

**Figure 32.5** Some general guidelines related to construction data collection [RTAC 77].

2. To improve subsequent construction cost estimates where drainage is likely to be a problem
3. To evaluate design methods and materials employed to control drainage

The minimum drainage data that should be acquired includes a general description of the drainage characteristics of the project, departures from design requirements or standards, and reports of deficiencies in subbase, base, and surface courses as a result of drainage factors during construction.

## 32.7 SUMMARY

This chapter identifies and briefly describes those aspects of construction that contribute to the operation and success of a pavement management system. Successful construction meets the planning and design objectives.

The chapter focuses on the interrelationships of construction with other phases of pavement management and the construction documentation that should be acquired. Details of construction practices, construction control, and construction management as a whole are not provided in the chapter. These are comprehensive subjects in their own right and are treated extensively in other publications. This chapter also considers levels of construction management and their relationship to the pavement management system.

Particular attention has been devoted to construction information. In addition to the use of such information for construction management purposes, it provides key input to planning, design, and maintenance. In the planning area, it is used for updating estimates, reprogramming projects, updating schedules, and so on. In the design area, construction information can be used to update and improve design models and to improve the accuracy of individual project economic evaluation. Maintenance management can benefit from construction information on materials used and special problems encountered.

## REVIEW QUESTIONS

1. List as many departments or divisions as possible in your local highway agency that interact with the construction process of a road. Show on a sketch how these units interact or overlap in their duties and responsibilities.
2. List and discuss three possible changes that could be made in the agency to improve the final product of the road or that could reduce costs by reducing duplication.
3. How many different levels of the organization become involved in the construction process? Show their relationships on a sketch. Can they be streamlined? How?
4. Outline on a large sheet of paper a modern highway departmental structure that you feel could improve the implementation of good pavement management systems. What new problems, however, are created by your proposed organization? List and discuss four such problems.

Chapter 33

# Maintenance

## 33.1 INTRODUCTION

It is recognized by many authorities that the type, frequency, and degree of maintenance on pavements can significantly influence performance. In addition, it can influence the time at which major rehabilitation, such as an overlay, is required.

A complete pavement management system must therefore include maintenance and rehabilitation. In fact, the original formulation of the system concept in NCHRP Project 1-10B [Hudson 73] explicitly included maintenance effects, but only over strenuous opposition from several leading highway engineers. The feeling was that maintenance, construction, and design should be kept separate. Several systems engineering authors attribute this to the historic structuring of organizations where the solutions to a problem are segmented according to the divisions in the organization [Hall 62]. Thus, a design division should deal with designing, a construction division with construction, a maintenance division with maintenance, and so on.

Although most highway agencies will undoubtedly continue with these types of organizational structures, the systems concept and modern organizational training have helped to overcome some artificial division barriers. There is no reason why maintenance, design, construction, and pavement monitoring and evaluation cannot be effectively coordinated, through pavement management, even though separated in administrative terms [RTAC 77]. Relationships between maintenance and the other areas of pavement management will be discussed further in this chapter.

The definition of *maintenance* varies among agencies. In a physical sense, maintenance consists of a set of preventive activities directed toward limiting the rate of deterioration of a structure, or corrective activities directed toward keeping the structure in a serviceable state. For pavements, this includes such preventive work as chip seals and such corrective work as patching. Service-type activities, such as paint striping, sweeping, and litter removal, would be included in a "maintenance management system," but because they have little effect on pavement performance, they are not usually considered directly in the pavement management system.

In an administrative sense, maintenance may be separated from rehabilitation by budgetary identification and by being performed in discontinuous sections of no more than, say, several hundred feet in length. For example, a repair followed by a 1 1/2-in. overlay of a 300-ft section of badly deteriorated pavement might be classed as maintenance, whereas the same thickness of overlay covering an entire project length might be classed as rehabilitation and become a capital budget item. Thus, the division between maintenance and rehabilitation is somewhat vague, but there is little need for a more detailed definition for the purposes of this chapter. It should be noted, however, that AASHTO has a series of procedures and reports relating to various areas of maintenance, including fairly specific definitions. NCHRP has defined a performance-based budgeting procedure appropriate to maintenance management, as shown in Figure 33.1 [Jorgensen 72].

This chapter deals primarily with the information associated with maintenance and rehabilitation and its use in pavement management, rather than with maintenance practices themselves. Although such practices are not considered in any comprehensive manner in this chapter, their importance to pavement performance should be emphasized, especially in relation to "quality" and frequency of work. There are various manuals available on the details of maintenance practices, such as those of The Asphalt Institute [TAI].

## 33.2 MAINTENANCE MANAGEMENT SYSTEMS

A maintenance management system is a technique or operational methodology for managing or directing and controlling maintenance resources for optimum benefits. It has been pointed out that this involves the following major components [Butler 75]:

1. An inventory of the physical elements of the system that can be maintained, plus the physical, operational, and environmental factors that can influence the amount of maintenance work generated
2. Performance standards that define maintenance procedures, resources in terms of labor, equipment, and materials, and the average accomplishment production rate expected from following the standards
3. Predictions of the workload generated in terms of maintenance accomplishments units, by a physical element of the highway, that is, a pavement of a given design subjected to specific traffic loadings in a given environment

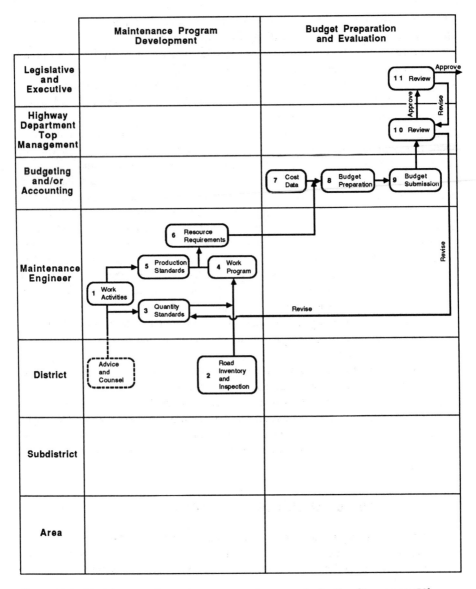

**Figure 33.1** Model system for maintenance performance budgeting [Jorgensen 72].

4. Allocations of available resources through objective budgeting mechanisms based on the specific requirements of the system and policy decisions related to the quality or level of maintenance desired
5. Feedback reports to monitor and update the system
6. Planning and scheduling procedures directed toward efficient use of resources

# Maintenance

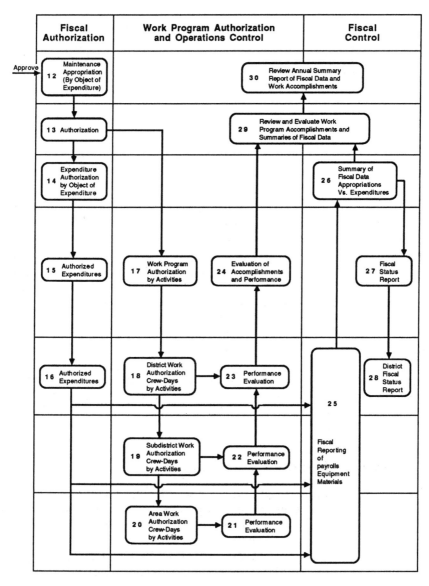

**Figure 33.1** *Continued*

A prime relationship of maintenance management to pavement management is to provide cost information associated with various levels of planning and programming, design, and construction. The type and degree of maintenance can also influence the rate of serviceability loss for a pavement.

The objectives of maintenance may be more specifically stated as follows:

1. To plan, direct, and control maintenance activities so that an acceptable level of service, consistent with the class of pavement, is achieved
2. To evaluate the methods and materials used in maintenance so that economical and efficient practices are developed
3. To acquire and report maintenance cost data so that unit costs for specific items for specific sections may be determined

It has been emphasized that maintenance management requires careful planning and implementation, efficient reporting methods, easy information retrieval, and accurate assessment of maintenance practices and problems [RTAC 77].

Thus, the maintenance management system as a whole involves managing highway maintenance, which includes the pavement. The pavement management system involves managing the pavement system, including its maintenance. The two concepts are complementary. They do not interfere with each other; rather, they reinforce and supplement each other. In some organizations, pavement maintenance and rehabilitation will be handled through a pavement management concept. In others, the maintenance section will carry the prime responsibility, with input from the pavement management group. The pavement management system is a necessary tool for analyzing and predicting the effect of various maintenance and rehabilitation policies.

A variety of literature is available on maintenance management systems. Those readers interested in additional information may refer to current publications of such organizations as the Transportation Research Board, AASHTO, and others.

The functional extent of maintenance management is illustrated in Figure 33.2 [RTAC 77]. The extent of the activities within each of the functions shown in the diagram depends largely on the agency involved. A larger highway agency would likely be well represented by Figure 33.2 while a smaller urban agency might have a less extensive system.

### 33.2.1 Levels of Maintenance Management

Most highway agencies in North America establish a relatively comprehensive organizational structure, inventory of equipment, work force, and other items, for maintenance of their network. This is in contrast to construction, for example, where the work is commonly done under contract.

The basic levels of maintenance management for such highway agencies are usually organized as field, regional, and central, as shown in Figure 33.3 [RTAC 77]. Relationships among the three levels, and the usual frequency of reporting, are also shown on the diagram. In the larger agencies, the organization of programs and budgets, operations, reports, procedures, and so on, will probably occur at the levels shown in Figure 33.3. For smaller agencies, such as rural municipalities, all these functions would probably be the responsibility of one "manager", such as the roads engineer.

The central or head office level of maintenance management has overall responsibility for determining the maintenance program according to the priorities established and within the available budget. Performance standards are taken into account in

# Maintenance

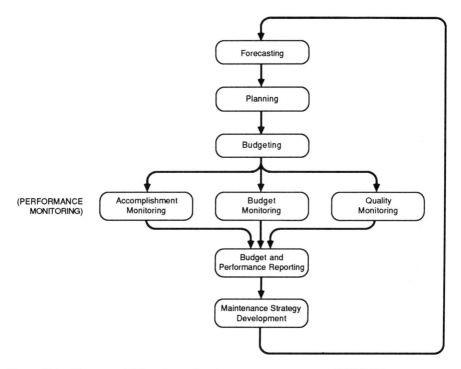

**Figure 33.2** The essential functions of maintenance management [RTAC 77].

establishing the program. Standards for the system are also established and controlled at this level, based on studies of the costs and benefits involved.

The regional level, or the district level for some agencies, has responsibility for determining maintenance priorities over an area covered by a number of field units or locations. These priorities are related to the specified system standards and are determined with the use of pavement evaluation information.

The field level is responsible for doing the actual work according to the standards. Because these standards should also reflect the efficiency of maintenance operations, it is the field-level work that provides the basic input for updating budgets and lists of priorities. Although the field level of maintenance management works within specified budgets and standards, it is also responsible for correcting localized deficiencies, and this requires on-site judgment and experience.

## 33.3 MAINTENANCE POLICIES

Policies for pavement rehabilitation and pavement maintenance vary widely from organization to organization, from place to place, and from time to time. Several factors usually govern these policies: (1) funds available, (2) historical precedent, and (3) political considerations. One or all of these items may be involved with a particular maintenance policy.

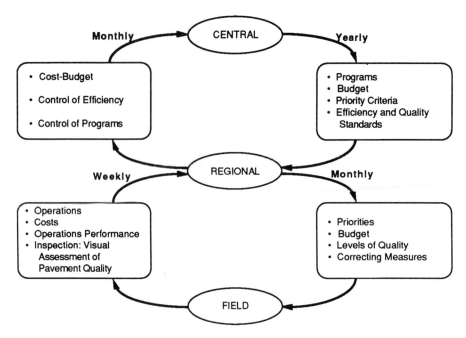

**Figure 33.3** Levels of maintenance management and frequency of reporting [RTAC 77].

### 33.3.1 Funds Available

Pavement maintenance is a continuing process that applies to a wide range and amount of physical area. Both the time and geographic extent of its application mean that even low-unit-cost maintenance will accumulate rapidly. If a fixed maintenance budget is low, then available unit expenditure funds will be low. Safety policies usually dictate that available funds be applied to fill the most extreme needs, and these require more than a unit share of the budget. The remaining budget then proves inadequate to serve the total area involved. Consequently, routine maintenance usually suffers or is omitted altogether. The subsequent budget period usually shows that the pavement has deteriorated more rapidly than expected, as a result of the lack of routine maintenance; thus, more of the small budget is required for heavy remedial work, and the downward cycle of deterioration continues.

Budgeting for maintenance is an important part of pavement management. The reader will recall that the decision-making process in the original design of a new pavement or of rehabilitation involves an estimated maintenance input. It is difficult, however, to predict such fund availability for 20 or more years, the typical analysis period for a pavement.

A properly developed pavement management system is adaptable to this problem by reanalysis or updating at any time. In fact, with a working pavement management system such reanalysis should be done periodically to evaluate the real performance of the pavement under the conditions that actually exist in service, rather than the originally predicted design inputs.

### 33.3.2 Historical Precedent

Until recently, maintenance has not been a widely hailed and researched area of endeavor. As a result, many maintenance policies or activities have been tried and adopted without adequate proof of their general applicability or cost-effectiveness. For example, a practice such as joint or crack sealing works well in wet, freeze zones and thus becomes standard policy in Ontario [Chong 87]. As another example, a precedent set by temporary budget restrictions such as no overlays shall exceed 2 in. may become accepted policy even when budget rules change and/or other, more critical situations come into existence.

Even though we often think of the highway system as being a homogeneous entity, it is certainly far from that and continually changing. For example, the interstate system in the United States was started in 1954, and a great deal of construction was completed in the early 1960s. At that time a very high quality maintenance policy was adopted. The pavements were all new and the maintenance required was light. By the mid 1970s quite a different situation prevailed, and maintenance at the same high quality level became extremely costly.

### 33.3.3 Organizational and Political Considerations

Over the years, road maintenance has suffered perhaps as much as any other part of the transportation field from political problems. In difficult financial times, maintenance is frequently one of the first programs sacrificed to resolve immediate budget needs. Organizationally, pavement maintenance is lumped with all right of way maintenance activities. Thus, if an unusual event occurs, such as a heavy snowfall winter, which requires a greater share of the maintenance funds than originally anticipated, the available budget for pavement maintenance may be substantially reduced. Adequate attention must be given to organizational considerations of maintenance, especially with respect to the overall problem of integrating the pavement management system, or, for that matter, the maintenance management system. The organizational problems are twofold in that they are related to both time and space.

Time considerations relate to the fact that much of the activity and emphasis of the pavement management concept is applied at the planning and programming, design, and construction stages whereas the maintenance portion of the problem becomes critical and active at some later time in the process. This time delay tends to separate the earlier activities from the maintenance functions. It is important in pavement management to ensure against this separation and to assist in involving maintenance personnel in the pavement management team at an early stage.

Space considerations relate to the management function often being centralized in a planning or design office. Maintenance functions, on the other hand, are field-related. The traditional maintenance organization is decentralized, with maintenance foremen in charge of relatively small areas. This field decentralization is very functional in ensuring that attention is given to all segments of the highway system. It also makes it considerably more difficult to ensure that there is central coordination of maintenance costs and practices.

### 33.3.4 Reanalysis at Any Time

The concept of an analysis cycle can be expanded to show that it is possible to run a reanalysis at any time the pavement engineer chooses. Thus, if an unexpected increase in traffic or change in other factors becomes evident, it should be possible to rerun the analysis, particularly the deterioration and economic evaluation models, with input data including the projected traffic and the existing condition of the pavement as obtained from the appropriate monitoring methods. In some cases, unexpectedly good or poor performance of the pavement materials will make it desirable to run a reanalysis cycle even though the pavement may not have reached its minimum serviceability level.

To generalize, we may think of the pavement as a physical system that is functioning all the time. We may examine that system at any time with available evaluation techniques and the system analysis computer program. To expand the concept, the pavement evaluation techniques of all types become the monitoring methods for the physical system.

## 33.4 EFFECTS OF POLICY VARIATIONS

It is one thing to establish rehabilitation and maintenance policies. It is quite another to ensure that these are followed in practice. The pavement management system has two concerns here. The first is to provide information that will assist, if possible, in encouraging consistent handling of pavement maintenance with minimum variation from or in policy. The second, and perhaps more important, is to provide a realistic method of considering and evaluating the effect of variations on the life and economy of the pavement sections on an individual basis.

The controlling factor in which is the better maintenance policy to follow is usually money. Policies are set over the long term; budgets are usually developed annually. Thus a policy may call for both cleaning culverts and fog sealing the pavement every year. If funds are not available for both, then the person in charge of the field level will likely make a personal preference choice in spite of policy.

Another important factor is the condition of the pavements within the maintenance subunit. In the best-devised budgeting techniques, "average" conditions will usually be considered. Often, local personnel will expend excessive funds on pavements in bad condition at the expense of possible preventive maintenance on other sections of pavement. Similarly, because a great deal of supervisor approval will be based on current inspections and how things look, there may be a tendency to overspend on visible maintenance such as mowing and trash pickup, at the expense of preventive pavement maintenance that has been programmed and considered in the life and performance of each pavement section.

# Maintenance

### 33.4.1 Overall Policy Shifts

Another major aspect of maintenance is the result of some overall shift in policy by the central agency involved. Such a situation occurred for many state highway departments in the mid-1970s when the oil crisis and poor economic conditions combined to drastically increase costs and reduce available funds. This situation occurred again in 1983 and in 1990–91.

It is even more pronounced in foreign countries. For example, in Brazil in 1975 the available maintenance budgets shrank to near zero because more than 70 percent of all oil products are imported, and increased oil prices created a major financial crisis. The decision was made that continued development of new roads was more desirable than maintenance, with the hope that there would be more benefit to the overall economy.

It is highly desirable to apply a pavement management system to assist in making these decisions rather than to make them ad hoc.

### 33.4.2 Analysis of Effects of Policy Changes

An important function of a pavement management system is to assist with the problem of analyzing policy shifts or required variations from policies or plans. In such cases, it is mandatory that each section of the pavement network be reanalyzed to see predicted effects on the expected life. This can be done on a regular basis and can provide invaluable information for use in showing budgeting agencies the true effects of stringent budgets. In many cases the investment in existing pavement must be protected with maintenance or rehabilitation, or the resulting long-term loss is immense.

Finally, it is possible and essential to reanalyze any given pavement in terms of original programming of rehabilitation and maintenance. If budget constraints are going to interfere with programmed work, then studies can and should be made to evaluate the effects and identify possible alternative actions.

## 33.5  COSTS, ECONOMICS, AND DECISION CRITERIA

The entire pavement design and management concept is related to costs and economics. In the pavement management process, the costs that must be considered include not only initial cost, but also routine maintenance costs and user costs. User costs are those costs that the pavement user pays, both directly and indirectly, in relation to the pavement facility or lack of it. User costs are related primarily to a pavement in poor condition that results in excessive roughness. A second major user cost is related to the detour and delay cost that the user suffers with relation to time required for maintenance and rehabilitation of a given facility. Considering these costs, along with the initial construction costs and the time value of money, makes it possible to evaluate true relative costs of various pavement maintenance and rehabilitation strategies and to select those that are optimal.

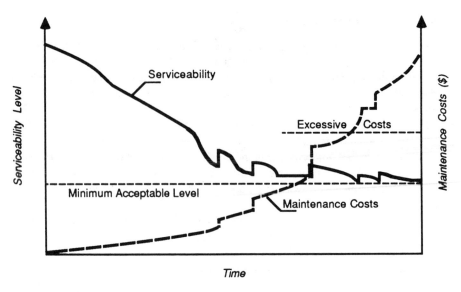

**Figure 33.4** Sample pavement record illustrating failure or unacceptability due to excessive maintenance costs.

### 33.5.1 Timing

The second major aspect of economics requires that there be good coordination between the various types of maintenance and rehabilitation. Thus, for example, if a pavement has inadequate skid resistance and a maintenance treatment, such as a seal coat, seems warranted, it should not be applied until a structural evaluation determines that a structural overlay is not required. Without this coordination, the seal coat might be applied one year, and then a major reconstruction or overlay required the next year would obviate the value of the prior seal coat.

### 33.5.2 Excessive Costs Criterion

A final important aspect of economy is the concept of excessive maintenance costs. It is possible to extend the life of a severely distressed pavement by providing extensive, heavy maintenance. This is sometimes done when a major highway is suffering damage but available funds are inadequate for rehabilitation. Figure 33.4 illustrates such a concept where a pavement is at or very near its unacceptable level but remains slightly above it because of the amount of maintenance expended. Keeping the deteriorated pavement above the minimum acceptable serviceability requires an acceleration in maintenance expenditures to the point where the costs become unacceptable.

Thus, a pavement might be considered to be "failed" when it reaches an unacceptable level of serviceability or when the cost of maintenance becomes excessively high. This might be controlled in an actual field situation by programming maintenance funds according to a predetermined formula. When it becomes impossible to maintain the road adequately for the formula funds, then a detailed evaluation should be performed to determine the problem.

### 33.5.3 Evaluation Decision Criteria

The pavement evaluation process includes decision criteria for threshold levels (minimum or maximum) of each of the items being monitored or evaluated. In some cases the threshold levels act directly as trigger values, providing information on which the required action can be based. A minimum acceptable level of serviceability is an example of a trigger value where an agency policy might establish a minimal level of, say, 2.5 for interstate highways. Pavement sections dropping below this level would then require a specific type of rehabilitation or maintenance.

Other types of decision criteria may be more flexible, either to serve as indicators to the engineer or to be varied with a particular pavement design as needed. Thus, in the first instance, the amount of cracking in a particular pavement section may have no absolute limit, but as cracking progresses it can be of significant concern to the engineer and indicate that rapid changes are taking place and that structural evaluation is needed. In the case of adjustable criteria, a particular pavement design method might predict an acceptable deflection of 0.015 in. for a particular section. Assume that subsequent evaluation then indicates a deflection significantly above 0.015 in., say, 0.020 in. This indicates to the pavement engineer that the pavement is behaving differently than expected, and thus a more complete evaluation and/or analysis seems justified. In each case, three possible paths may by followed: (1) continue maintenance for predicted life as programmed; (2) perform a more complete evaluation and analysis of the pavement section to provide a better basis for decision; or (3) take immediate appropriate rehabilitation or maintenance action as required based on the violated criteria.

Decision criteria for selecting a maintenance treatment can vary with highway class. Figure 33.5 [Jorgensen 72] describes example criteria for rutting and waves, sags, and humps (Agency C) and rutting (Agency F).

### 33.5.4 Combinational Decision

Certain highway agencies combine two or more of the measures previously outlined into a single index. Such combined indexes can prove very useful in overall analysis of a pavement network and in providing a simple method or ranking pavements into relative condition categories. But combining the information is not very useful for evaluating and analyzing a specific pavement section for maintenance. In that case, specific disaggregated information (i.e., distress, roughness) is needed.

## 33.6 MAINTENANCE INFORMATION NEEDS

Maintenance feedback information is vital in any pavement management system. It invariably finds its use in pavement rehabilitation and maintenance. The original analysis of a given pavement section incorporates a particular expected maintenance schedule.

If there is any variation from the mean or expected value, then the serviceability history of the pavement section may vary from the predicted value; that is, the performance curve may drop more rapidly or more slowly than predicted. When this

## 1.200 FLEXIBLE AND RIGID PAVEMENT MAINTENANCE

The purpose of this standard is to establish the guidelines by which the roadway surface of the interstate and other function classes of highways shall be maintained and to establish the degree each type of distress can be tolerated before remedial physical maintenance measures must be taken.

### 1.210 FLEXIBLE PAVEMENTS

**1 Rutting**

Rutting will be tolerated to the degree specified in the table below. Where rutting occurs in excess of that specified, the deficiency shall be corrected at the earliest opportunity.

|  | Interstate | Principal | Major | Collector | Other |
|---|---|---|---|---|---|
| Maximum allowable depth of rut on multilane or 70 mph highways | $\frac{1}{2}$ in. | $\frac{1}{2}$ in. | $\frac{1}{2}$ in. | $\frac{1}{2}$ in. | $\frac{1}{2}$ in. |
| Maximum allowable depth of rut on two-lane 60 mph or under highways | $\frac{1}{2}$ in. | $\frac{1}{2}$ in. | $\frac{1}{2}$ in. | $\frac{3}{4}$ in. | 1 in. |
| Maximum allowable depth of rut on bituminous treated roadway surfaces | $\frac{1}{2}$ in. | $\frac{3}{4}$ in. | $\frac{3}{4}$ in. | 1 in. | 1 in. |

**2 Waves, Sags, and Humps**

These types of flexible pavement distresses add to the discomfort of the road user and can become a hazard if allowed to severe.

On all functional classes of highways where 70 mph speed limits are allowed, repairs will be made wherever 50 percent or more of any given 100 ft of roadway is observed to exhibit characteristics of waves, sags, or humps in excess of 1 in. in height per 10-ft section.

All other roadways with posted speed limits of 60 mph or less will be considered for repair when observed to exhibit these characteristics in excess of 2 in. in height per 10-ft section except that waves, sags, or humps 4 in. in height or over will be corrected as soon as practicable after reported or observed.

(a) Agency C

**Figure 33.5** Sample codified maintenance levels: (a) Agency C (b) Agency F [Jorgensen 72].

happens, as shown in Figure 33.6, it is essential to have a reanalysis of the project needs for maintenance and/or rehabilitation.

The availability of feedback data from the pavement data base is important to maintenance for the following reasons:

1. It provides the maintenance administrator with important data and information on trends applicable to programming and planning maintenance expenditures.
2. It permits evaluation of the validity of existing maintenance models.

# Maintenance

---

**ROAD CONDITIONS**

+ Rutting at driveways, mailbox turnouts, rural intersections, etc.
+ Edge ruts at curves and other isolated locations.

**DESCRIPTIONS**

**Rutting at Driveways, Mailboxes, Rural Intersections, etc.**

These are specific places where traffic gose into the shoulder often and where the shoulder can't take it.

**Edge Ruts at Curves and Other Isolated Locations**

These are places where traffic goes off the actural surface and beats the shoulder material away leaving an edge rut.

**NEED FOR REPAIR**

This work can be done on an inclement weather schedule. It should be done whenever conditions show a problem. On edge ruts, let them get at least one inch deep before repair, and catch them before they get about two inches deep.

**HOW TO REPAIR**

Spot dump and hand spread material. If there is very much to be done, it may be wise to schedule a reshaping at the same time.
    Try to compact gravel material into place using the truck wheels or a portable roller.

**HOW MUCH TO DO**

Patch only the specific locations that show a need. Do not try to rebuild a shoulder by this method; you are usually working with places that repeatedly have trouble, so put out a little extra gravel when you do this job.

---

(b) Agency F

**Figure 33.5** *Continued*

Figure 33.6 illustrates the use of a maintenance data system to evaluate models by plotting deviations in predicted histories. Such information is extremely important to a system designer in determining how the formulated models are behaving.

Some discussion of maintenance data requirements is useful at this point. As a first step, pavement damage prior to performing maintenance should be recorded. This can be done on a sample basis or on a complete section basis depending on time and personnel available. It has been pointed out that the sampling technique is usually adequate, considerably cheaper, and more practical than a complete survey [Butler

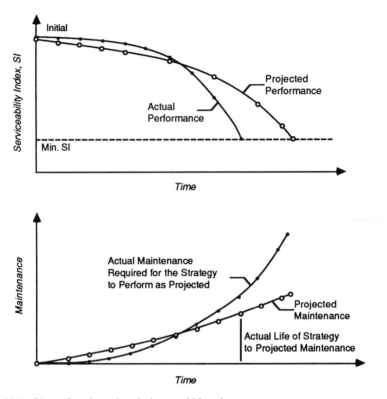

**Figure 33.6** Plots of projected and observed histories.

74, 75a, 75b]. This is a feasible approach where sections are homogeneous with respect to design, year of construction, traffic, subgrade soil, and general environment. Such designated homogeneous sections should be the same for all pavement management activities, and be given a control or reference number (actual locational referencing may be for example by milepost, or Geographic Information System, GIS).

The length allocation should be referenced and recorded in such a way that any point within the section can be located with reasonable accuracy.

The second important step involves actual units of the various type of maintenance performed, and the related costs, as subsequently discussed in more detail. The actual maintenance performed can be obtained from file records in the maintenance residencies if the maintenance foremen are encouraged to provide accurate records of the work accomplished.

Unfortunately, there are many pressures on maintenance personnel that may inadvertently encourage them to "generalize" records to average values rather than actually to record the work required on a bad section. Human nature does not, for some reason, like to report extreme conditions within a set of controls.

The costs of the performed maintenance are even more difficult to obtain. Actual expenditures may vary over quite wide ranges depending on material costs, unit prices,

# Maintenance

work schedules, efficiency, and so on. But it is not sufficient, for example, to record average patching costs over a 5-mile section when most of the total cost may be incurred over only a small portion of the section. Where it is extremely difficult or impossible to obtain true costs from existing records, it may be necessary to apply average unit prices to the work units performed and recorded in the field records.

In every case it is essential to work toward true records rather than to accept unrealistic averages because it is actually these variations from the averages that are important in the maintenance management and pavement management systems.

### 33.6.1 Work Activities and Accomplishment Units

This section began by considering maintenance information needs in general terms. In a more specific sense, the component work activities, and the accomplishment units in which they are measured, have to be identified and properly recorded in order to fulfill the information needs. It is, of course, essential that both the work activity and the accomplishment unit be clearly understood by the field-level people (i.e., those assigned to handle specific tasks).

Table 33.1 [Jorgensen 72] contains a typical listing of major work activities and associated accomplishment or work units. It covers the overall highway maintenance function, and therefore more than just pavement maintenance activities are included.

Each agency would have a set of forms for recording daily, weekly, biweekly, etc., the activities listed in Table 33.1. A typical set of such forms appropriate to the pavement part of maintenance is available [RTAC 77].

Guidelines for performance standards (i.e., accomplishment units expected) previously noted in this chapter should be established for each work activity. Analysis of performance achieved can be used to update existing standards.

### 33.6.2 Uses of Information by Other Areas of Pavement Management

It is perhaps self-evident that maintenance data may be used to varying degrees by other phases of pavement management. In order to do this effectively, the separation of maintenance and construction costs must be clearly identified, because the designation of various items and quantities as construction or maintenance can vary from agency to agency.

When a maintenance management system is functioning effectively, and proper records have been kept, analysis of the data can point out inadequate designs, inaccurate traffic estimates, inaccurate evaluation of materials, construction problems, and so on. A summary of how maintenance information can be used for such other pavement management activities is given in Figure 33.7 [RTAC 77]. For example, in addition to the forgoing analysis, maintenance data may be used in planning and programming and in design for future cost estimates.

## 33.7 SUMMARY

This chapter has considered the major components of maintenance practices and management that relate to the pavement management system. Type and degree of main-

## TABLE 33.1 ILLUSTRATIVE LIST OF MAJOR WORK ACTIVITIES [Jorgensen 72].

| | Maintenance activity | Work unit |
|---|---|---|
| Roadway surface | • Patch with premix | • Tons of premix |
| | Level with premix | Tons of premix |
| | Deep patch with premix or full depth replacement of surface course | Tons of premix |
| | Skin/spray/surface treatment patch | Cu. yd. of aggregate |
| | Fill or seal cracks with joints | Gal. of sealant |
| | Seal coat or surface treat full lane widths and in continuous form | Lane mile of surface |
| | Patch with P.C.C. | Cu. yd. of P.C.C. |
| | Blade or reshape unpaved surface | Road mile bladed |
| | Patch unpaved surface with gravel | Cu. yd. of material added |
| Shoulders and approaches | • Blade or reshape unpaved shoulders | • Shoulder mile bladed |
| | Patch with aggregate or other stabilized material | Cu. yd. of material added |
| Drainage | • Clean and reshape ditches | • 100 lineal feet of ditch cleaned |
| | Clean and repair drainage strructures | Each installation |
| | Replace minor structures | Lineal foot of structure |
| Roadside | • Mowing | • Acre mowed |
| | Chemical control of vegetation | Gallon of spray |
| | Cut brush | Acre mowed |
| | Cut/trim trees | Each tree |
| | Rest area maintenance | Each rest area |
| | Erosion Control | None |
| | Litter pickup | Mile of R/W |
| | Fence repair | Lineal foot of fence |
| Major structures | • Bridge inspection | • Each bridge |
| | Structure painting | Gallon of paint |
| | Seal or repair expansion joint | Lineal foot of joint |
| | Other structure repair | none |
| Snow and ice | • Snowplowing with /without sanding | • Lane mile plowed |
| | Snow removal loader/blower | Cu. yd. of snow removed |
| | Spot and/or continuous sanding | Cu. yd. of sand |
| | Spot and/or continuous salting | Cu. yd. of salt |
| | Stockpiling sand or salt | Cu. yd. handled |
| | Erect, remove, repair snow fence | Roll of fence (50') |
| | Clean up after storm | None |
| Traffic services | • Sign maintenance of any kind | • Each sign |
| | Lighting maintenance of any kind | Each installation |
| | Painting guideline | Pass mile striped |
| | Paint pavement messages | Each site |
| | Guardrail maintenance | Rail section worked on |
| | Ferry operation | Mile of operation |
| | Movable bridge operation | Each opening |
| | Road patrol | Odometer mile |
| Extraordinary maintenance | • Unusual or disaster maintenance | • None |
| Service functions and overheads | • Crushing/mixing | • Cu. yd. of material |
| | Stockpiling | Cu. yd. handled |
| | Equipment service and repair | Each vehicle |
| | Housekeeping | None |
| | Buildings and grounds maintenance | None |

# Maintenance

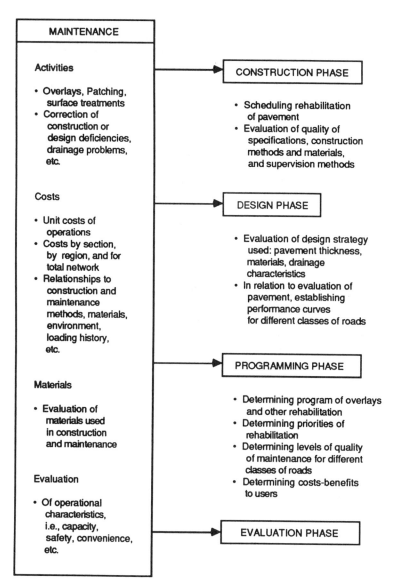

**Figure 33.7** Relationship of maintenance to other phases of pavement management. After Pavement Management Committee [RTAC 77].

tenance can significantly influence pavement performance and the time and type of any major rehabilitation required.

Maintenance management, if conducted properly, should be capable of determining if the maintenance activities are being carried out in the most economical and efficient manner. In addition, the records kept should be capable of providing information to the other phases of pavement management.

The basic components and functions of maintenance management systems have been identified and discussed in the chapter. Particular attention has been given to the levels of maintenance management, to maintenance policies, to the effects of policy variations, and to costs, economics, and decision criteria associated with maintenance management.

Information needs associated with maintenance have also been given particular attention. In addition to the use of the data generated for maintenance management purposes, it is necessary and, in fact, sometimes essential for other phases of pavement management. Although there are relatively well-developed methods available for collecting and reporting maintenance data, considerable work is still required for proper use in these other pavement management phases.

## REVIEW QUESTIONS

1. How do pavement management systems and maintenance management systems relate to each other? How do they differ?
2. Show in a sketch how maintenance and rehabilitation relate the initial design cycle and the total life-cycle of a pavement. You may want to use a PSI history (performance) curve.
3. What is meant by the concept of "reanalysis at any time" during the life of a pavement? Discuss how old-fashioned design concepts miss this important point, which is easily handled by a pavement management system.
4. What is the concept of "excessive cost criterion," and why is it useful and even necessary for realistic pavement management?

Chapter 34

# Research Management

## 34.1 INTRODUCTION

Most agencies are directly or indirectly interested in research, and a few spend significant sums for in-house and/or contracted research. Contracted research programs are exemplified by the ongoing National Cooperative Highway Research Program (NCHRP), and the Strategic Highway Research Program (SHRP) involving the United States, Canada, and many other countries, initiated in the latter part of the 1980s.

There is no clear or universally accepted definition of research. Many use the term *research and development* (R&D), particularly in the highway field, to signify the intent of the programs toward application. What falls under a research budget in one agency may fall within an operations budget in another agency. Thus, the definition of research is driven by budget, application intent, and interpretation considerations. In a general sense, research constitutes the tackling of problems to achieve new or better processes, materials, methods, procedures, decisions, or economy.

The framework for a research function within a PMS and the broad issues of technology improvement were introduced in Chapter 5, while the following sections consider levels of research management, the major elements of developing and carrying out research programs, and the effective utilization of research results. Concentration is on the management of research rather than the identification of research needs.

Subsequently, Chapter 45 will expand on the subject of strategic research and innovations for future pavement management systems.

## 34.2 LEVELS OF RESEARCH MANAGEMENT

The levels of management for construction and maintenance are relatively well defined, as described in the preceding chapters. Research management is, however, less clearly defined and can vary considerably in scope and organizational structure, depending on the agency concerned. It is usually a central function, particularly in state or provincial highway agencies, although a significant amount of the actual work may take place at the field level.

Some of the larger state/provincial agencies, and federal agencies such as the U.S. Federal Highway Administration, have distinct research divisions, with a designated budget, within their organization. In such cases, the research management is administrative and technical in nature. For example, the Ministry of Transportation of Ontario has the following major administrative/technical subdivisions:

- Pavements and roadway
- Materials
- Structures
- Highway innovations and strategic research
- Branch services

Smaller agencies, including most larger urban agencies, perform research within regular operations. For example, a materials supplier may convince the design engineer to try a new product, such as a polymer modified asphalt. A test section would then be built with regular construction funds and periodically observed. The problem with this informal approach to research is that the experiment may not be conducted in a way that will provide the maximum benefit or information to the agency. Also, such informal research will frequently be undocumented. When the engineer who conducted the experiment leaves the agency, the knowledge is lost.

Levels of research management are thus almost entirely a function of the agency involved, its size, and its perceived needs. Appreciable variation in the way research is organized and managed exists even among agencies of similar size. Nevertheless, any agency interested in initiating research, or improving, expanding, or assessing its own research management, might study the organization of research management in other, similar agencies.

## 34.3 MAJOR ELEMENTS IN GENERATING AND CARRYING OUT A RESEARCH PROGRAM

The major elements of a long-term pavement research framework for state transportation agencies have been defined [Hudson 92]. As shown in Figure 34.1, the framework addresses the development and management of short-, and long-range projects that fit a coherent life-cycle and pavement management concept. The broad focus in

# Research Management

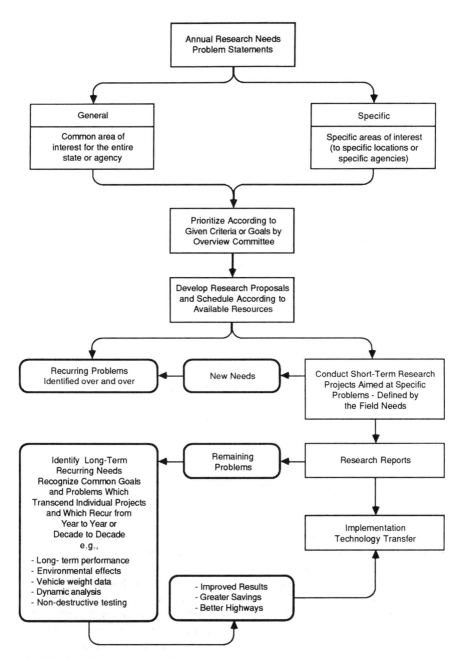

**Figure 34.1** Pavement research framework for state agencies [Hudson 92].

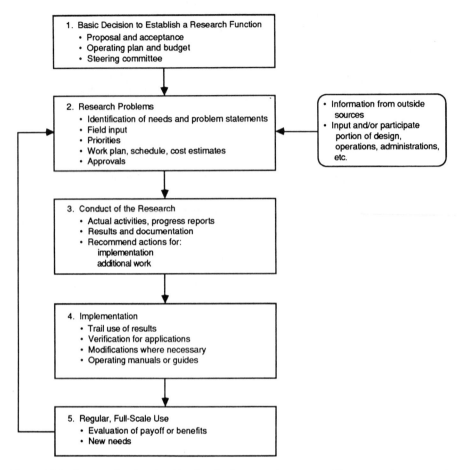

**Figure 34.2** Progression involved in developing, carrying out, and implementing research.

Figure 34.1 allows for efficient transfer of information and knowledge between research projects, it preserves the long-term integrity of pavement research, and it permits more efficient management of the overall program.

The development, initiation, execution, and implementation of the actual research, particularly for those agencies who are starting a research function, are summarized in Figure 34.2. It should be noted that an agency may not necessarily have its own designated research group or personnel, equipment, etc. This is particularly applicable to smaller agencies where the research function may consist mainly of evaluating the work of the others for its potential applicability to their needs. The progression in Figure 34.2 begins with a decision to actually develop a research function or program, and goes on to identify problems, conduct, and implementation of the research.

### 34.3.1 Decision to Establish a Research Function

The basic decision to establish a research function would likely begin with a proposal by someone within or outside the agency to senior administration. It would include an operating plan and budget, and likely the designation of a steering or coordinating committee. While a proposal to establish a research function would have to provide the rationale for doing so, it would only address research needs in a general way. Detailed technical proposals for individual research projects would follow, as shown in Figure 34.2.

The general research plan might well be updated periodically as the actual research is conducted. But it should continue to include a clear statement of objectives and an identification of the various agency resources and people that will be required.

The steering committee is perhaps most relevant to larger agencies and should include people from design, construction, maintenance, materials, planning, and traffic. Field input is important. It may be impractical in smaller agencies to have a distinct committee, and monitoring, coordination, etc., could be carried out through a senior technical administrator.

### 34.3.2 Research Problems

The development of problem statements should be preceded by an identification of the research needs of the agency, consistent with the objectives stated in the general research plan. Research problem statements sometimes reflect the interests of the researchers more than the actual needs of the agency. There should be strong input from operations people, including those in the field, for the identification of needs. In addition, information and research results from outside the agency may be used as a guide for the development of a research program.

Some agencies find it useful to classify their research needs, and this is reflected, for example, in their major administrative/technical subdivisions (see Section 34.2). One broad approach would be to use the following classifications:

1. Analytical research
2. Physical research (processes, materials, equipment, test methods)
3. Administrative, planning, economic, and traffic research (decision processes, models, data bases, management, heavy vehicle effects)

Since the results of much of the research will eventually find their way into field use, it is important that input from the field be obtained for identifying research needs and evaluating the results when they are implemented. However, because of the day-to-day nature of this work, field people may not always be aware of the long-term needs or implications of certain research efforts. Thus, research people should make an extra effort to communicate and justify their proposals. Also, it should be remembered that field people often solve problems on better methods of construction or maintenance, new equipment, and better use of materials. The research function can assist with the documentation of these solutions.

The priorities placed on research needs are usually subjectively determined, based on expected payoff, risk, time requirements, estimated costs, and resources available to do the work.

A well-developed work plan, schedule, and cost estimates are essential to the successful conduct of any research. Time and cost estimates should be realistic. Unfortunately, overruns often occur, largely because of the uncertainties associated with research. Nevertheless, justification for excessive time and cost overruns should be mandatory; otherwise curtailment may be necessary.

### 34.3.3  Conduct of the Research

The actual research activities would presumably follow the work plan and schedule previously established. For certain smaller agencies, these activities may consist entirely of evaluating the work done by others for its potential applicability to their needs. These can be valid and useful research activities.

The results of any research activities should be well documented, in the form of reports, guides, computer programs, and user manuals. Progress reports during the conduct of the research are necessary and useful for both administrative and technical purposes. The frequency and extent of such progress reports should be realistic and should not unduly detract from actual time spent on the research. Quarterly progress reports are often required for large, longer term projects.

In addition to serving administrative and technical management purposes, progress reports are quite useful to the researchers themselves in "forcing" internal discipline regarding schedule, allocation of effort, documentation, and communication of results.

The results of research should not only be well documented but be clear and understandable. Where applicable, sensitivity analyses of the results to changes in various factors are often useful. As well, recommendations for implementation, and for any additional work, should be part of the reporting of results of any research effort.

### 34.3.4  Implementation

The first step in implementing research usually consists of trial or pilot studies. These can serve not only to verify the applicability of the results, per se, but also to make any necessary modifications. Implementation should be a cooperative effort between research and operational people, as previously emphasized in Chapter 5.

Experience gained from implementation can be quite useful to subsequent, similar, or updated research efforts. As well, it can assist in preparation of the final operating manuals or guides.

An example of a comprehensive implementation study is contained in National Cooperative Highway Research Program Report 160, "Flexible Pavement Design and Management: Systems Approach Implementation" [Lytton 75].

### 34.3.5  Regular, Full-Scale Use

When research results find regular, full-scale use, an evaluation of the payoff or benefits can be made. Again, this should be a cooperative effort between research and operating people. Such evaluation should include a monetary value if possible; however, the payoff is often difficult to quantify and thus subjective assessments are usually also necessary. The evaluation of the benefits of research can be most useful as feedback to updating research needs.

## 34.4 ELEMENTS OF GOOD RESEARCH MANAGEMENT

The elements of successful research lie in developing answers to such questions as the following:

1. How much funding and resources should be committed to research?
2. Who does the work (i.e., contracted, in-house, or some combination) and how can innovation be encouraged?
3. Why not just "borrow" research results from others?
4. What degree of cooperation should exist between public agencies such as state transportation departments, private industry, and universities or research agencies?
5. How necessary is it to disseminate the results of research through publications, seminars, conferences, courses, etc.?
6. How can the chances of future, continuing research success be increased?

The amount of resources committed by an agency, including funding, indicates the degree of importance placed on this area of pavement management. Although universally applicable guidelines on the amount of funds that an agency should spend on pavement research do not exist, larger agencies should be spending at least 0.5 percent of their capital budget on research in order for the research function to be meaningful.

Who does the work depends largely on the size, resources, and needs of the agency. In the case of smaller agencies, contracted research would likely be for specific problems. For larger agencies, some balance of in-house and contracted research is usually desirable. The advantages of contract research include being able to draw on special expertise, accelerate projects, handle overload, provide a degree of objectivity that may not always exist in the agency, and avoid buildup of an oversized establishment within the agency. On the other hand, in-house research can provide a valuable training function, make implementation easier, facilitate administration, and be more responsive to the special needs of the agency.

Innovation in pavement management has clearly involved universities and private industry to a large degree, and future success will continue to involve the interaction of these two groups with public agencies. It must be remembered that a large number of the key people in pavement management, within both private industry and public agencies, have come from universities. Moreover, the supply of this basic human resource product has been made possible largely through funded research at universities.

The question of borrowing the results of research rather than spending money on conducting it has been partially answered in the preceding paragraphs. For smaller agencies with limited resources and research capabilities, this is probably a necessity rather than a question. Even though borrowed research is a definite alternative for larger agencies, there are disadvantages. These include implementation delay, the need for setting up some screening mechanism to judge what is useful, and the time required for familiarization and understanding of the borrowed research. Consequently, it is not desirable for larger agencies to rely exclusively on borrowing research. Moreover, larger agencies have a degree of responsibility for conducting research that can be used by the smaller agencies within their region.

The question of cooperation between government, private industry, and research institutions or universities is easily answered. In general, unless proprietary interests are involved, such coordination is desirable. It can bring a great degree of combined expertise to bear on problems and accelerated solutions can be found. Private industry and consultants are in business, of course, to make a profit; this certainly need not preclude their cooperation with government and other nonprofit organizations in tackling research problems of joint concern.

The question of the extent to which an agency should go in disseminating its research results cannot be answered on a universal basis. Certainly, all research should be properly documented, at least within the agency. From this point, the question of whether to publish depends largely on the motives of the agency and the individual research people. A major advantage to the presentation and publication of research results is the discipline it "forces" in conducting the research because when it is exposed to others it is also "tested." There is also an inherent benefit in terms of morale to the research people and the agency through the recognition by other people. As well, the training offered through seminars, conferences, and short courses on the results of research can be most valuable.

## REVIEW QUESTIONS

1. Describe how Figure 34.1 should be modified to address the structure and needs of a local highway agency.
2. Develop a one page research problem statement for the evaluation of a Type IV slurry mix by a local highway agency.
3. Identify issues that should be considered by a state highway agency before implementing a pavement overlay design method that was developed by a national research agency such as the National Cooperative Highway Research Program.

# References to Part Five

[Butler 74]     Butler, B.C., and L. G. Byrd, "Maintenance Management Concepts," Public Works Magazine, August 1974.

[Butler 75a]    Butler, B.C., "Highway Maintenance Research Needs," Paper Presented to Transportation Research Board, Washington, DC, January 1975.

[Butler 75b]    Butler, B.C., and L. G. Byrd, "Maintenance Management," Section 25 of the *Handbook of Highway Engineering*, Van Nostrand Reinhold, 1975.

[Chong 87]      Chong, G. J., and W. A. Phang, "Improved Preventive Maintenance: Sealing Cracks in Flexible Pavements in Cold Regions," Ontario Ministry of Transportation Rept. PAV-87-01, December 1987.

[Clark 87]      Clark, G. N., "Pavement Management in Kansas: Problems and Issues in Implementation," Proc., Vol. 3, Second North American Conf. on Managing Pavements, Toronto, November 1987.

[Haas 85]       Haas, R., "Minnesota's Pavement Management System: Implementation Recommendations," Report prepared for Minnesota Dept. of Transportation, June 6, 1985.

[Hall 62]       Hall, A.D., *A Methodology for System Engineering*, Van Nostrand, 1962.

[Hudson 92]   Hudson, W. R., and R. Haas, "Research and Innovation Required to Standardize Pavement Management," ASTM STP1121, 1992.

[Jackson 87]  Jackson, L. B., and R. G. Grauberger, "Implementation of a PMS: Institutional Hurdles, Pitfalls and the Role of Good Communications," Proc., Vol. 3, Second North American Conf. on Managing Pavements, Toronto, November 1987.

[Jorgensen 72]  Roy Jorgensen Associates, "Performance Budgeting System For Highway Maintenance Management," NCHRP Report 131, 1972.

[Kuzyk 91]    Kuzyk, P., R. Haas, and R. W. Cockfield, "Performance Based Specifications for Pavements", Canadian Journal of Civil Engineering, December 1991.

[Lytton 75]   Lytton, R. L., W. F. McFarland, and D. L. Schafer, "Flexible Pavment Design and Management: Systems Approach Implementation," NCHRP Report 160, 1975.

[Maurer 87]   Maurer, F. V., and E. E. Ofstead, "Minnesota's Pavement Management System: How it Came About and the Steps Taken," Proc., Vol. 3, Second North American Conf. on Managing Pavements, Toronto, November 1987.

[RTAC 77]     Roads and Transportation Association of Canada, "Pavement Management Guide," 1977.

[TAI]         The Asphalt Institute, "Asphalt in Pavement Maintenance," Manual Series MS-16.

[Vlatas 89]   Vlatas, D. A., and R. E. Smith, "Implications of Life Cycle Performance Specifications," Transportation Research Record 1215, Transportation Research Board, 1989.

Part Six

# Examples of Working Systems

Chapter 35

# Basic Features of Working Systems

## 35.1 INTRODUCTION

Much has been written and spoken about systems concepts and pavement management systems during the past 25 years, and surely more will be said in the future. The key issue, however, is that to achieve the major benefits available in a systems approach, the user must move past the conceptual stage and develop an actual working system. This working system will not be perfect at any stage of development, but it is the necessary starting point for subsequent improvements. Essential elements of the working system include a basic set of models that take appropriate specific inputs such as those discussed in earlier chapters and transform them into useful predicted outputs in some reasonable format. To qualify, the working system must calculate some type of appropriate cost function and, with a well-defined algorithm, order the potential alternative strategies in an informative way for use by the appropriate designer or administrator in making a final decision. As pointed out earlier, this decision may be relative to design, planning, programming, rehabilitation, or maintenance.

A working system is not a computer program per se, but the detail, complexity, and repetition of the calculations required make it almost essential that a computer program be developed as a useful mechanism for applying the working system. Thus, all working systems covered in this part of the book do have associated computer programs.

In this chapter we outline a number of working pavement management systems developed in the period 1968–1990. No attempt is made to include all such systems or to defend the omission or inclusion of any particular system in those listed here. This is not a survey text, and we leave summarization of all methods to others. Obviously, new methods and greatly upgraded methods will continue to appear. This is expected and indeed desired. Nevertheless, the format and details of most available methods will remain current for many years.

## 35.2 STRUCTURE OF AVAILABLE PROJECT LEVEL SYSTEMS

Two of the first working project level systems actually available for practical use were FPS (Flexible Pavement Design System) and SAMP (Systems Analysis Method for Pavements). FPS is a continuing series with sequential numbers such as FPS-1, FPS-2, ... , where each number indicates an improvement or change in the method from previous versions. SAMP employs the same sequential numbering system in order to provide for upgrading. FPS was developed jointly by a pavement research team at the Texas Highway Department, the Texas Transportation Institute, and the University of Texas in 1969 [Hudson 70]. More details on these methods will be given in subsequent sections.

SAMP was developed by a research team of Fred N. Finn, B. Frank McCullough, and W. R. Hudson under contract to the National Cooperative Highway Research Program through Materials Research and Development, Inc., in Oakland, California [Hudson 68]. It also has a sequential numbering system, with the programs SAMP-5 and SAMP-6 being the versions that are best known and most often used at this time.

In the following sections we outline the main structure and approach of several existing working systems. In Chapter 37 we present the details of SAMP-6 as a specific example working system. We compare SAMP-6 against the conceptual system and point out in detail the type of changes needed to continue improvement of the system. We then look at SAMP-5 and the changes that were made to obtain SAMP-6, as an example of how upgrading can take place.

### 35.2.1 Flexible Pavement Design System (FPS)

FPS was the first major working pavement management system. The basic working models were developed at the Texas Transportation Institute [Scrivner 68] as part of a major cooperative research project between the Texas Highway Department and the Texas Transportation Institute to "adapt the findings of the AASHO Road Test to Texas conditions." It is interesting to note that after seven years research, Scrivner found that major changes required in Texas design methods included not so much the adoption of the structural findings of the AASHO Road Test as the addition to the design approach the consideration of climate, maintenance costs, user costs, etc. These findings were independent of the work done for NCHRP [Hudson 68], but when put in the same general format as the conceptual pavement system developed there, the

## Basic Features of Working Systems 449

similarity is striking (see Figure 35.1). The framework of the original design method project was further developed in a subsequent project, and with some modifications the computer program FPS-1 was developed. Subsequent changes were made, and the first published version was presented as FPS-3 [Hudson 70].

**Structure of FPS as a Pavement Management System**—As a pavement management process, the FPS computer program is part of a larger, more complete pavement management system. In fact, it was the development and use of FPS that truly helped clarify the broader management concepts. This may be confusing for the reader, but it should be studied carefully. The pavement management system involves:

1. The application of FPS to develop a list of alternative choices
2. Selection of one of the alternatives by the designer or administrator
3. Preparation of implementation documents including plans and specifications
4. Construction of the pavement as planned
5. Maintenance of the facility as required
6. Collection and analysis of feedback data for determination of subsequent actions required in the system

Thus, we see the steps required for pavement management are more comprehensive than the analysis performed with a computer program. In order to obtain the details of all these steps, it will be necessary for the reader to study carefully several project reports [Hudson 70, Butler 73, Hudson 73b]. In the summary presented here, we concentrate on the details of the computer program and generalize the related steps, such as construction and preparation of plans and contract documents.

**FPS Computer Program**—The FPS computer program is basically outlined in Figure 35.1. The top half of the diagram outlines the structural inputs and subsystems to the program, followed by the accumulation, summarization, or transformation of the outputs into a system output function, which can be used for comparing various designs.

The bottom half of the diagram outlines the decision criteria used to array and select final pavement strategies. The program uses these criteria to organize and list the potential choices based on minimum total cost considering the time value of money.

**Input Variables**—The number of input variables in the FPS program is a function of the maximum number of layers being considered in the design. Ten categories of variables are considered as follows:

1. *Program controls* are required to control the operation of the program.
2. *Unit costs* are the economic inputs required for the computation of the costs of each pavement design.
3. *Material properties* define the characteristics of each material.
4. *Environmental factor* is a district temperature constant based on the mean temperature of the area where the pavement is to be constructed and used in the prediction of the behavior of each pavement design.
5. *Serviceability index values* are used to predict the life of an initial design or an overlay by determining the serviceability level of the pavement after initial

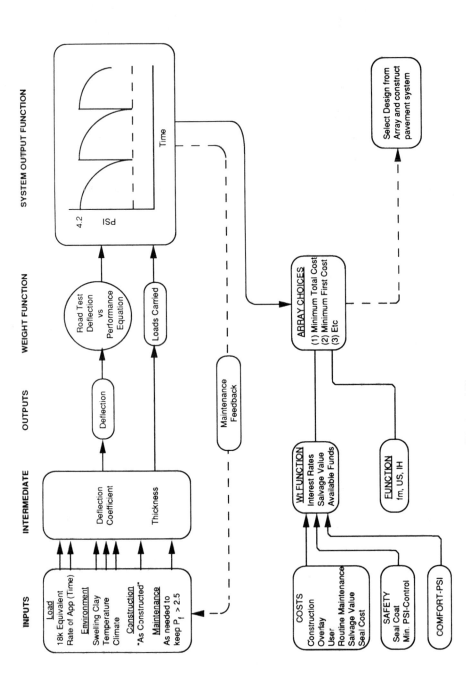

**Figure 35.1** Working pavement system [Hudson 68].

## Basic Features of Working Systems

**Table 35.1 Summary of Program Inputs Required for Use in FPS-3**

| Type of input | Number of inputs[1] |
|---|---|
| Program control | $n + 3$ |
| Unit cost | $2n + 4$ |
| Material property | $n + 4$ |
| Environment | 1 |
| Serviceability index | 3 |
| Seal coat schedule | 2 |
| Constraint | $2n + 6$ |
| Traffic demand | 6 |
| Traffic control | 13 |
| Miscellaneous | 2 |
| Total | $6n + 44$ |

[1] $n$ is the number of materials considered for use above the subgrade. For example, one surface material, two bases and two subbases has $n = 5$ and therefore, the number of program controls is eight, and the total number of inputs is 74.

construction and after overlay construction and the minimum value of serviceability that will be allowed during the analysis period.

6. *Seal coat schedule* variables describe the restraints imposed on seal coats by the designer and are used in the determination of a seal coat schedule for each pavement design.
7. *Constraints* are variables that are often implicit in a design problem but that must be explicitly stated in FPS. They are important in controlling the design and management scheme produced by the program, and they are also vital controlling factors in keeping computer run time for a given problem within reasonable limits.
8. *Traffic demand inputs* describe the expected traffic that the pavement must serve during its lifetime.
9. *Traffic control inputs* are used in the computation of user costs by determining how traffic will be handled during overlay construction. The traffic models are too complex to repeat here, but are given in Appendix B of [Butler 73].
10. *Miscellaneous* parameters are variables that do not fit in any other group.

The number of variables in each category is related to the number of materials being tried, and the distribution of these is summarized in Table 35.1 by category for FPS-3. Subsequent versions of the program have similar input requirements, although many changes have also been introduced as each version has been developed.

Many sensitivity analyses have been run on the program, showing that some of the more detailed input variables have little or no effect on the resulting system outputs [Hudson 71]. We are not attempting here to justify the inclusion of extra variables, merely to advise the reader of the input and output structure of the program.

**General Description of the Program and Outputs, with Mathematical Models**—In order to understand FPS, it is necessary to know generally how the input data is handled once entered. Figure 35.2 shows the mechanics of the program in flow chart form. Figure 35.3 shows the concepts of FPS in pictorial form.

First, the computer reads all inputs and prints them out for verification. Then all possible initial design combinations are calculated for individual analysis. The initial cost of a particular design is computed and compared against available funds. If costs are exceeded, the design is discarded and the computer program goes to the next design. For feasible cost designs, the design thickness is checked against the total thickness constraint and more designs are discarded. If a design meets both of these first two constraints, its expected design life is calculated using serviceability-performance concepts, swelling clay parameters, and anticipated traffic.

If the first calculated design life is less than the minimum time to first overlay, the design is discarded and the program passes on to the next design. For each of those designs that have passed all feasibility checks, an optional overlay policy must be selected that will last for the entire analysis period. Total costs are computed for each feasible design.

The program then considers the next design and continues until all possible designs are considered and either selected as feasible or discarded. The feasible designs are sorted by total cost and a set of optimal designs are printed in order of increasing total cost as shown in the sample output (Figure 35.4). Four types of mathematical models are used to estimate design feasibility.

*Physical Models*—Physical mathematical models are used to estimate what will happen to a real-world pavement under traffic throughout the analysis period. In order to do this in FPS, three kinds of mathematical models are used.

1. Traffic models predict the amount, type, and distribution of traffic and consist of a traffic equation that predicts the amount of traffic that will have passed at any time, an equivalency equation that relates traffic volume and weight distribution to equivalent 18-kip (18,000-lb) single axles, and a set of traffic handling models for use in overlay construction.
2. Environmental models predict the effect of environmental conditions on pavement behavior including (a) an ambient temperature model, (b) an *in situ* stiffness coefficient model, (c) a regional factor model, and (d) a swelling clay model. These models will require improvement in subsequent versions.
3. Performance models predict the behavior of the pavement based on the Present Serviceability Index (PSI) concept developed at the AASHO Road Test and include a pavement strength model that can be based on either structural number and soil support models such as those developed for the AASHO Interim Pavement Design Guides or on surface curvature index [Scrivner 68].

*Economic Models*—Economic models are used to estimate the component and total cost of a design and a breakdown of that cost. All costs are converted to present value at discount rates supplied by the program user. There are seven types of economic models in FPS:

1. An initial construction cost model, which determines the cost of the initial construction based on the cost per compacted cubic yard of each material used

## Basic Features of Working Systems

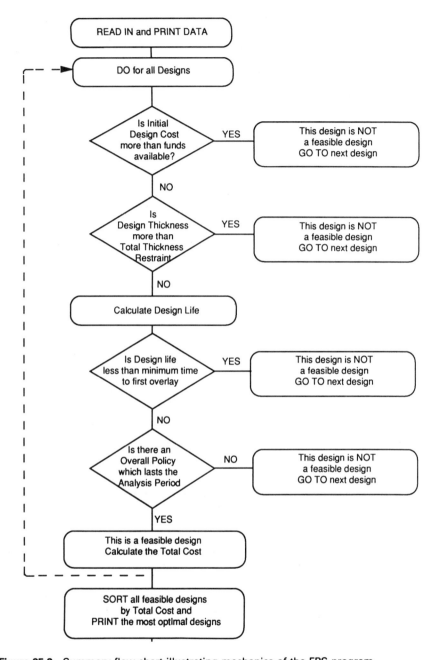

**Figure 35.2** Summary flow chart illustrating mechanics of the FPS program.

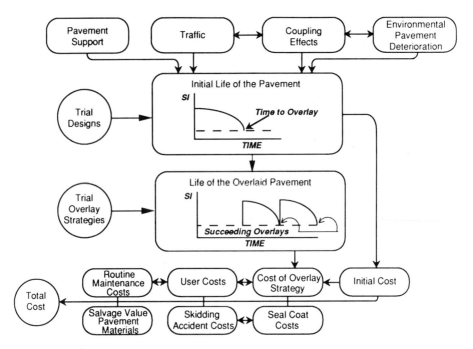

**Figure 35.3** Schematic computation diagram of the FPS pavement design system [Lytton 75].

2. A seal coat cost model, which determines when a seal coat is needed and how much it will cost
3. An overlay construction cost model, which together with a physical model determines time, thickness, and cost of overlay
4. A routine maintenance model, which predicts the annual cost to maintain the pavement
5. A users' cost model, which determines the cost to users resulting from traffic delays because of overlay construction
6. A salvage value model, which determines the value of the pavement remaining at the end of the analysis period
7. A total overall cost model, which combines all costs during the analysis period to their present value at the beginning of the period

*Optimization Models*—Two optimization models are used in FPS to determine a set of optimal designs, based on present value of overall cost. The two methods used in FPS-3 are as follows:

1. A modified branch and bound technique which systematically determines which initial construction designs will lead to a set of optimal designs. The criterion used in this technique is a relationship between strength and cost. Stated in simple terms, if a design is more expensive and at the same time

PROB 1      FPS EXAMPLE WITH TYPICAL DESIGN VALUES
SUMMARY OF THE BEST DESIGN STRATEGIES,
IN ORDER OF INCREASING TOTAL COST

|  | 1 | 2 | 3 |
|---|---|---|---|
| MATERIAL ARRANGEMENT | ABC | ABC | ABC |
| INIT. CONST. COST | 22.000 | 22.278 | 24.056 |
| OVERLAY CONST. COST | 4.051 | 4.051 | 3.784 |
| SEAL COAT COST | 3.600 | 4.112 | 3.111 |
| USER COST | 0.235 | 0.237 | 0.232 |
| ROUTINE MAINT. COST | 1.302 | 1.311 | 1.372 |
| SALVAGE COST | -4.533 | -4.586 | -4.921 |
| TOTAL COST | 26.655 | 27.403 | 27.634 |
| NUMBER OF LAYERS | 3 | 3 | 3 |
| LAYER DEPTH (INCHES) | | | |
| D(1) | 7.00A | 8.00A | 8.00A |
| D(2) | 7.00B | 5.00B | 7.00B |
| D(3) | 5.00C | 5.00C | 5.00C |
| NO. OF PERF. PERIODS | 3 | 3 | 3 |
| PERF. TIME (YEARS) | | | |
| I(1) | 5.039 | 5.234 | 5.664 |
| I(2) | 12.578 | 13.281 | 14.883 |
| I(3) | 22.617 | 24.297 | 27.930 |
| OVERLAY POLICY (INCH) (INCLUDING LEVEL UP) | | | |
| O(1) | 1.5 | 1.5 | 1.5 |
| O(2) | 1.5 | 1.5 | 1.5 |
| SEAL COST SCHEDULE | | | |
| SC(1) | 9.01 | 9.10 | 9.21 |
| SC(2) | 17.51 | 18.00 | 19.01 |

**Figure 35.4** Summary of the three optimal designs for the FPS sample problem.

has less strength than some other design, it cannot produce a better design and is therefore discarded.

2. A minimum total present worth model where the total present worth of cost for each design is evaluated and sorted in order of least value. The design engineer can then evaluate this list to select the optimal design. The optimal design may not be the lowest total cost design, as no economic analysis can capture all of the factors which need to be considered in the final selection of the design strategy. In the determination of the optimal overlay policy for each initial design, it was found that the best technique is to look at all possible policies within the bound limits.

*Interaction Models*—An interaction model is an algorithm that defines the interactions among two or more other models. For example, in finding the life of initial and overlay construction designs, a time must be determined that will satisfy both the performance and traffic models. Because of the complexity of these models, it is necessary to use an interactive technique.

*Other Subsystems*—In addition to computer programs or software subsystems, there are three other major subsystems: (1) the hardware subsystem, (2) the organization subsystem, and (3) the information subsystem. The hardware subsystem involves (a) the Dynaflect for measuring pavement deflection and estimating material properties, and (b) a computer for use with the software subsystem.

The organization subsystem relates the functional requirements within the operating organization of the using agency for setting up and operating the pavement management system. The information subsystem involves: (1) the research information, (2) the design input data, and (3) the feedback information that must be developed and used if the FPS pavement management system is to function. Hudson presents a detailed description of the subsystems [Hudson 70].

### 35.2.2 Ontario Pavement Analysis of Costs

The Ontario Pavement Analysis of Costs (OPAC) is an advanced pavement design and management system developed for use in the Canadian province of Ontario [Jung 70, Kher 74]. It is a computerized system that compares the performance and cost of hundreds of design alternatives for flexible pavements within a short time. Using the system, pavement design engineers can select the most effective pavement design at the least cost.

**OPAC Features and Background**—OPAC provides a fast, economical, and comprehensive method of flexible pavement design. Before the system was available, a thorough analysis of only a few designs could take months to complete. Within just a short time, OPAC gives an analysis of the factors that influence pavement design: traffic, environment, subgrade, pavement configuration, construction materials, available funds, and limits of pavement roughness.

OPAC predicts pavement life as deterioration in Riding Comfort Index (RCI) as a function of traffic loading and annual cyclic environmental changes. The deterioration caused by the two factors have been respectively modeled using data from the AASHO Road Test and the Brampton (Ontario) Road Test. Subgrade surface deflection under a standard wheel load is used as a predictor of future behavior of a pavement.

Successful thickness designs in the province were analyzed using elastic layer theory, and several interactions of such analyses resulted in a set of modulus values which were thereafter assigned to various paving materials and six different categories of subgrades in the province. These modulus values, called modulus coefficients, when assigned to AASHO Road Test sections, gave a good correlation among subgrade deflection, traffic, and pavement performance. For a pavement section, this correlation submodel allows calculation of the component of pavement deterioration that is caused by traffic. The other component, caused by environment, was modeled from the Brampton Road Test data. It is based on the difference in deterioration predicted due to traffic, as previously indicated, and actual measured deterioration of Brampton

## Basic Features of Working Systems

Road Test sections over a period of eight years. The two submodels of pavement deterioration are combined to predict the performance of any pavement section.

**Cost Analysis**—OPAC predicts pavement costs throughout the life of a pavement. Various cost components include initial capital expenditure, subsequent resurfacing, maintenance expenditures, and salvage. Models are incorporated to calculate the initial capital expenditure for different types of construction. These include conventional, deep strength, and full-depth pavements, full-width granular construction, treated bases and subbases, and earth, granular, or paved shoulder construction. Varying lane widths, shoulder widths, shoulder roundings, side slopes, and two different types of facilities (divided or undivided) can be handled by changing system inputs. Future resurfacing cost includes the cost of isolated patching, leveling course(s), lift(s) of surface and binder course, and shoulder upgrading. The model for maintenance cost, which includes the cost of pavement and shoulder maintenance, uses the Ministry's past maintenance data. The model for salvage return is based on the remaining utility of the pavement structure at the end of the analysis period.

OPAC also includes road user cost in its economic analysis of design alternatives. There are two components of road user cost. The first component consists of costs incurred by motorists when they are inconvenienced by detours or delays that cause idling, speed changes, and slow driving. The other component is incurred when the pavement surface becomes rougher, riding comfort declines, and road user cost increases. Road user costs are calculated using models for gasoline and oil consumption, tire wear, vehicle maintenance and depreciation, and driver time costs caused by reduced speeds on rougher roads.

OPAC provides an evaluation of the various cost components of a pavement on the one hand and the various consequent costs to the users on the other hand and makes it possible to examine design trade-offs. For example, pavement design engineers may weigh the advantages of selecting a pavement with higher initial costs (fewer future overlays, therefore lower future costs) versus the selection of a pavement with a lower initial cost (more future overlays, therefore higher future costs).

OPAC extends the capabilities of pavement design engineers. It combines their experience with economic analysis and performance prediction. The engineer specifies the range of design alternatives to be analyzed. OPAC then generates many possible alternatives, provides a detailed cost analysis, and presents a set of optimum designs. The optimization can be based on agency cost alone, on agency plus total user cost, or agency plus any specified fraction of user cost. The final decision regarding the selection of a design remains with the pavement design engineer, who must also consider such information as construction problems, aggregate depletion, and traffic safety.

**Interactive Computer Operation**—Communication between the pavement design engineer and OPAC is achieved through a PC or terminal. A question-and-answer dialogue is established between the computer and the engineer using the keyboard of the terminal.

In response to questions posed by the computer, the pavement design engineer enters basic design specifications such as subgrade condition, performance requirements, traffic projections, available materials, and their costs.

Within seconds the computer returns its analysis to the terminal, printing out various design alternatives (maximum 24) to meet the engineer's specifications. For each design the computer printout lists the estimated life of the initial pavement and its total life history. It also lists the initial construction cost and discounted future costs of subsequent overlays, annual maintenance, salvage value, and user costs.

If the pavement design engineer decides to modify the basic design criteria, this can be done immediately and another set of design alternatives printed within minutes.

**Sample Problem**—In many ways OPAC is similar to FPS, the Texas-based system, as shown by the similarity of the output sheet, Figure 35.5. On each output sheet near optimal designs are listed in order of increasing total cost (net present value). As many such sheets as desired can be called for, although usually the first 8 or 16 designs will be adequate.

OPAC is an excellent operating pavement management system, and a good example of pavement systems technology.

### 35.2.3 Rigid Pavement Design System

A working pavement management system has been developed for rigid or portland cement concrete pavements, Rigid Pavement System (RPS) for the Texas Highway Department [Kher 71]. In general, models used in rigid pavement design are better developed and better defined than those used in flexible pavement design. On the other hand, the understanding of overlay relationships and designs for cracked concrete pavements is incomplete and difficult to model. In general, RPS follows the conceptual system laid out in NCHRP Project 1-10 [Hudson 68]. However, several factors differ for rigid pavements. A conceptual Rigid Pavement Design System was laid out by Kher as shown in Figure 35.6. The central feature in the concept is still a system output function in the form of a pavement serviceability versus traffic history relationship.

The concept goes one step further toward the overall management concept than the original Hudson diagram, because it illustrates the use of feedback data and a research program to continually improve the system computer models.

**Systems Formulation of RPS-1**—Comprehensive formulation of a rigid pavement design system is an ultimate goal that may be achieved through stages of implementation and feedback as well as research. In the systems framework the development of RPS-1 can be described by the following terms:

1. Objectives
2. Inputs
3. Constraints
4. Decision criteria
5. System analysis
6. Output

*Objectives*—A large amount of research has been done in the past on various individual models or groups of models defining various parts of the comprehensive RPS system.

The RPS computer program was developed with the following main objectives:

MINISTRY OF TRANSPORTATION AND COMMUNICATIONS - FLEXIBLE PAVEMENT DESIGN SYSTEM, OPAC

| PROB | DIST | WORK | HIGHWAY | DATE | PAGE 10 |
|---|---|---|---|---|---|
| 1 | 3 | PROJECT 118-68-01 | 6N | 2/20/73 | |

SUMMARY OF THE BEST DESIGN STRATEGIES IN ORDER OF INCREASING TOTAL COST

| | 1 | 2 | 3 | 4 | 5 | 6 | 7 | 8 |
|---|---|---|---|---|---|---|---|---|
| MATERIAL ARRANGEMENT | ABCD | ABCD | ABCD | ABCD | ABC | ABC | ABC | ABC |
| INIT. CONST. COST | 144057 | 138476 | 147716 | 152567 | 156496 | 150956 | 159466 | 153981 |
| OVERLAY CONST. COST | 42774 | 48874 | 41188 | 41188 | 43659 | 52359 | 43659 | 52395 |
| USER COST | 445 | 512 | 433 | 432 | 455 | 567 | 454 | 567 |
| ROUTINE MAINT. COST | 17771 | 17572 | 18066 | 18066 | 17572 | 17572 | 17572 | 17572 |
| SALVAGE VALUE | -3067 | -3312 | -3094 | -3215 | -3330 | -3939 | -3432 | -4041 |
| TOTAL COST | 201981 | 202122 | 204310 | 209039 | 214853 | 217552 | 217720 | 220475 |
| LAYER DEPTH (INCHES) | | | | | | | | |
| D(1) | 1.50 | 1.50 | 1.50 | 1.50 | 1.50 | 1.50 | 1.50 | 1.50 |
| D(2) | 4.00 | 4.00 | 4.00 | 5.00 | 6.00 | 6.00 | 7.00 | 7.00 |
| D(3) | 7.00 | 6.00 | 6.00 | 6.00 | 9.00 | 8.00 | 7.00 | 6.00 |
| D(4) | 9.00 | 9.00 | 12.00 | 9.0 | | | | |
| PERF. TIME (YEARS) | | | | | | | | |
| T(1) | 10.4 | 10.0 | 10.8 | 10.7 | 10.4 | 10.0 | 10.4 | 10.0 |
| T(2) | 20.5 | 19.9 | 21.1 | 21.0 | 20.4 | 19.7 | 20.3 | 19.7 |
| T(3) | 30.7 | 30.1 | 31.5 | 31.4 | 30.5 | 29.6 | 30.4 | 29.6 |
| T(4) | | | | | | 39.5 | | 39.5 |
| OVERLAY + LEVEL-UP | | | | | | | | |
| O(1) | 3.0 | 3.0 | 3.0 | 3.0 | 3.0 | 3.0 | 3.0 | 3.0 |
| O(2) | 3.0 | 4.0 | 3.0 | 3.0 | 3.0 | 3.0 | 3.0 | 3.0 |
| O(3) | | | | | | 3.0 | | 3.0 |

A = SURFACE; B = BINDER; C = GRAN. BASE; D = GRAN SUBBASE

**Figure 35.5** Typical summary output of OPAC [Kher 72].

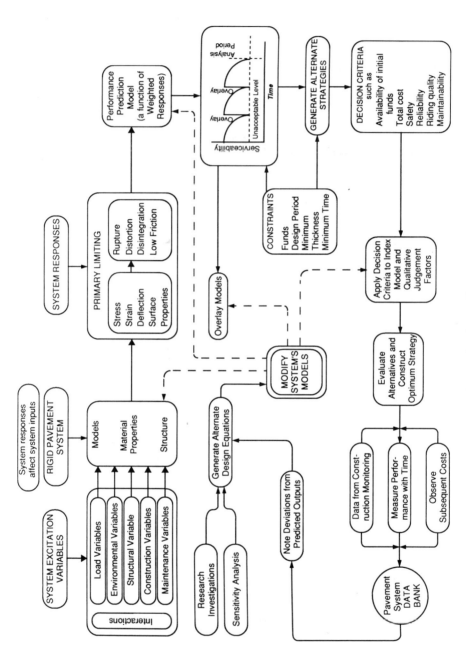

**Figure 35.6** Conceptual Rigid Pavement Design System [Hudson 70].

## Basic Features of Working Systems

1. To evolve an efficient solution process for rigid pavements
2. To serve as a first block in continuing systems research
3. To provide a generalized procedure so that future modifications can be incorporated with a minimum effort

*Inputs*—Inputs to the system consist of about 115 parameters. These inputs are dictated by the models used in the system. Enough inputs are provided so that, in general:

1. All traffic loads can be accounted for effectively.
2. Existing performance models can be evaluated with the help of the required material properties.
3. Serviceability-age histories can be estimated.
4. Different concretes, subbases, and reinforcements can be tried.
5. Subbases can be effectively designed and evaluated.
6. Joints in initial construction can be designed.
7. Seal coats can be provided where required.
8. Sufficient maintenance can be provided.

*Constraints*—RPS generates alternative designs based on the range of allowable layer thicknesses and concrete properties. This can generate a large number of alternative designs for analysis. Constraints serve to restrict the number of alternatives that will be considered by any one run of the program. RPS uses constraints on the maximum pavement thickness, maximum initial cost, and minimum initial life. Once an alternative survives these constraints, an overlay analysis is performed if the computed pavement life is less than the design period.

*Decision Criteria*—Total present worth of costs is the prime decision criterion available from the output of RPS. However, in developing the input for the program the designer must input decisions on the maximum funds available for initial construction, the minimum allowable serviceability level, and the minimum initial pavement life. These constraints would be decision criteria in a traditional pavement design procedure.

*System Analysis*—The concepts of stage construction are used for designs that reach the minimum specified serviceability levels at times less than the analysis period. Reinforcement and joints are designed for each initial design. Subbase, concrete, and overlay thicknesses are computed for each strategy designed.

All costs of initial and future construction are calculated. Future costs include those for overlays, maintenance, seal coats, and traffic delays during overlay operation. Initial costs consist of subgrade preparation, subbase, concrete, reinforcement, and joints. The procedure is outlined in Figure 35.7.

*Outputs*—The quantitative decision criteria included in the RPS system are not comprehensive enough to yield final judgments solely on total present worth of cost, so the designer is presented with a set of alternative design strategies and other pertinent information in a summary table. The most economical design for each pavement-overlay combination and a complete analysis of the number of initial designs, strategies, and relative constraining effects of various restraints are also printed out.

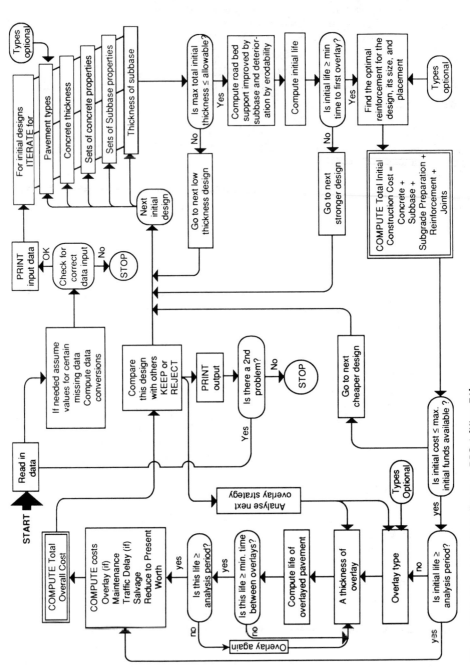

**Figure 35.7** Summary flow diagram, RPS-1 [Kher 71].

**Basic Features of Working Systems**

For each strategy in the output, a complete description of thickness, materials used, overlays, serviceability lives, joint and reinforcement detailing, and each cost involved is printed out.

**System Mathematical Models**—The working system RPS-1 was developed using various mathematical relationships called systems models. Some of these relationships exist in the literature as a result of test roads, other experiments, or theoretical analyses. Certain other relationships necessary for developing a rational working system were derived by Kher. The system models are subdivided into the following major categories:

1. Performance models
2. Models for traffic analysis
3. Subgrade-effect performance models
4. Foundation strength models
5. Stochastic variations in the material properties
6. Models for overlay design
7. Models for reinforcement design
8. Economic models
9. Miscellaneous

The models are too comprehensive to be discussed here in detail. However, a number of the models are sufficiently different from previous models that the reader may want to review the original program documentation [Hudson 70].

### 35.2.4 System Analysis Method for Pavements

SAMP has the broadest application of any existing working pavement system. This does not mean that it is necessarily the best, but rather that it was developed under support from the National Cooperative Highway research program and the American Association of State Highway and Transportation Officials, which is a nationwide organization. The original SAMP program was developed in 1966 and the first published version was SAMP-5 [Hudson 68, Hudson 73b].

A followup contract was conducted at the Texas Transportation Institute. In this contract, an improved version of SAMP was developed (SAMP-6) [Lytton 75]. This is the best illustration available to date of the feasibility of upgrading an existing working system as new information becomes available, or, in this case, of how new ideas can be infused by others if adequate support funds can be made available.

In Chapter 37, the SAMP working system will be presented in detail as a sample system for illustration and analysis. Thus, no additional details of this system will be presented here.

### 35.2.5 The Highway Design Model

The Highway Design and Maintenance Standards Model (HDM) has been developed by the World Bank [Watanatada 87]. It is a versatile and useful tool which permits the user to test and compare the economic viability of various alternative road improvement and maintenance strategies. HDM can be used at both the network and the project level, and is one of the few new working PMS tools developed since 1970. It is well suited to the study of economic benefits of pavement strengthening and relatively minor geometric improvements, as well as of providing a smoother riding surface. The basic concept and structure of the model are described in Chapter 38.

The broad concept of the HDM model is simple; three interacting sets of cost relationships are added together over time. Costs are determined by predicting physical quantities of resource consumption which are then multiplied by unit costs or prices. The HDM model is used to make comparative cost estimates and economic evaluations of different policy options, including different time staging strategies, either for a given road project on a specific alignment or for groups of links on an entire network. It can quickly estimate the total costs for large numbers of alternative project designs and policies, year by year, for up to 30 years, discounting the future costs if desired at different discount rates.

## 35.3 NETWORK LEVEL SYSTEMS EXAMPLES

Much of the progress in pavement management since the early 1980s has been focused on the development and implementation of pavement management systems for network level decisions. The Federal Highway Administration has required the development of systems by all state highway agencies. Many local government agencies have recognized the benefit of pavement management and have also implemented systems. While both state and local government systems share a generic framework for PMS, the specifics of the implementation vary to meet the specific needs of each agency.

### 35.3.1 Arizona Department of Transportation

In the early 1980s, the Arizona Department of Transportation contracted for the development of a pavement management system. The resulting system is one of the few PMS which uses a true optimization procedure for budget allocation [Kulkarni 80].

The inventory for the pavement management system is based on route and mileposts. Pavement conditions are monitored for ride quality with a Maysmeter, a distress survey and skid measurements with a Mumeter. The items included in the distress survey are percent of area cracked, patching, flushing index, and rutting (interstate pavements only). The distress data is collected at each milepost on two-lane roads and at each milepost in each direction for routes with four or more lanes. The distress survey sample size is 1,000 $ft^2$ per mile. The condition data base also contains fields for the date of construction, most recent project, regional factor, traffic level, and functional classification. The construction history data base contains the route number and starting and ending milepost, in 0.01 of a mile, for each construction project constructed by the state. The pavement structure is identified by the thickness and material type for each layer in the structure.

Based on traffic level, roughness, cracking, and regional factors, each milepost in the network is assigned to a condition state. The percent of pavement sections in each condition state is determined. Transition probability matrices are used as performance models in the system. These matrices define the probability of transitioning

### Basic Features of Working Systems

from one condition state to another, given the current condition state of the pavement and a rehabilitation strategy. These strategies range from routine maintenance to reconstruction. The optimization process, using a Simplex solution method, determines the percent of pavements which should receive each rehabilitation strategy to minimize the rehabilitation budget subject to constraints. The constraints on the optimization are a minimum percent of pavement in good condition and a maximum percent in poor condition for each functional class.

The pavement management staff then works with district engineers to select the actual sections of highway that will be rehabilitated. A five-year plan is established and updated annually.

#### 35.3.2 Minnesota DOT

The Minnesota Department of Transportation initiated development of a network pavement management system in the mid-1980s [Maurer 87, PMS Ltd. 89, Hill 90]. The system was developed on a DOS personal computer which interfaces with the department's mainframe computers. Data files are maintained on the mainframe and accessed as needed by the PMS microcomputer. The data base consists of separate files for inventory and pavement condition data. There are 14 FORTRAN programs for the analysis of the data and report generation. Project selection recommendations are generated with a "near optimization" technique based on cost-effectiveness. Further information about the Minnesota DOT PMS is presented in Chapter 36.

#### 35.3.3 Local Agency Network Systems

Numerous systems have been implemented by local agencies. Historically, agencies had to individually develop systems to meet their specific needs. However, there are now a range of systems available, both through consultants and in the public domain, that can be used by the local agencies. Some of the most widely used public domain systems are:

1. PAVER, developed by the Corps of Engineers and supported by the American Public Works Association
2. MTC, developed for the San Francisco Bay Area Metropolitan Transportation Commission [MTC 86, Smith 88]
3. CALTRANS systems for microcomputers are available for both rigid and flexible pavements
4. ITRE, developed by the North Carolina Institute for Transportation Research and Education for use on local road networks

Many of the local agency systems are less sophisticated than the state systems. Generally, pavement condition is evaluated by human observation rather than with devices such as roughness meters. The pavement performance models frequently are based on extrapolations of limited condition data. Finally, many local systems use a priority method for selecting projects rather than optimization.

The MTC system is a second generation of the PAVER system. It is readily available, implemented on a microcomputer and supported and updated by the commission [Smith 88]. This system is treated in more detail in Chapter 36.

## REVIEW QUESTIONS

1. Summarize in your own words the details of the FPS system. Where was it developed? Is it a total pavement management system? Does it operate at the project or design level?
2. Summarize in your own words the details of the OPAC (Ontario Pavement System). Where was it developed? Is it a total pavement management system? Does it operate at the project design level?
3. Summarize in your own words the details of the RPS. Where was it developed? Is it a total pavement management system? Does it operate at the project design level?
4. Summarize in your own words the details for SAMP. Where was it developed? Is it a total pavement management system? Does it operate at the project design level?
5. Summarize in your own words the details of HDM3. Where was it developed? Is it a total pavement management system? Does it operate at the project or network level?

Chapter 36

# Network Level Examples of PMS

This book has previously dealt with general concepts of pavement systems and basic features of working systems including several examples. This chapter presents a more detailed description of two working network level systems. The first is a state level system which features optimization, the Minnesota DOT system. The second is the MTC system developed for the Metropolitan Transportation Commission in San Francisco, California, for use by municipalities.

## 36.1  MINNESOTA DOT NETWORK LEVEL PMS

MNDOT desired to develop and implement a pavement management system (PMS) specifically geared to the state's requirements, resources, and conditions. One resulting action was to establish a pavement management section within the Office of Research and Development. An initial part of this section's responsibilities was to review the available information on pavement management and the implementation experience of several other states and provinces [Maurer 87].

The requirements of several different types of PMS users in MNDOT were defined in a study [Haas 85] and used as a basis for identifying the building blocks and key elements of the actual PMS.

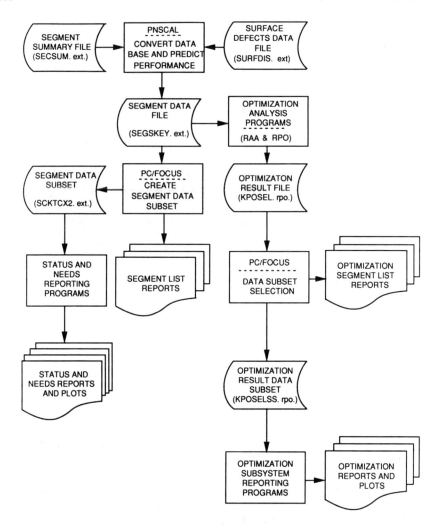

**Figure 36.1** Overview of Minnesota DOT PMS [PMS Ltd. 89].

The overall logic of MNDOT's PMS is shown in Figure 36.1. It operates on two primary data files (segment data, SEGSUM, and segment surface defect data by milepost, SURFDIS) which are transferred from the Transportation Information System (TIS) on the mainframe to a PMS microcomputer [PMS Ltd. 89].

The major outputs of the PMS, as shown in Figure 36.1, are Status and Needs Reports and Plots, and Optimization Reports and Plots. The MNDOT PMS operates on a DOS personal computer and uses PC/FOCUS for data base management functions. The system requires 640K bytes memory (RAM) and hard disk storage of approximately 20 Mbytes (not including space required for PC/FOCUS).

PC/FOCUS also provides the menu based user interface, segment list reporting,

# Network Level Examples of PMS

**Table 36.1 Major FORTRAN Programs in the Minnesota PMS**

| PROGRAM NAME | MENU | PURPOSE |
|---|---|---|
| PNSCAL | SU | Performance Predictions |
| SEGSUMR | SNR | Segment Summary Report |
| NEEDS | SNR | Segment Needs Report |
| PERF | SNR | Segment Performance Graphs |
| INDHIST | SNR | Index Distributions |
| HIST3D | SNR | Index Distributions (3-D) |
| BARCHRT | SNR | Needs Year Distributions |
| RAA | ROS | Rehab. Alternatives Analysis |
| RPO | ROS | Rehab. Priority Optimization |
| RORRPS | ROR | Performance Summary Report |
| RORAPH | ROR | Annual Performance Distributions |
| RORH3D | ROR | Performance Distributions (3-D) |
| RORPRF | ROR | Segment/Alternative Performance |
| RORRLR | ROR | Network Remaining Life Report |

Menu:  SU = SYSTEM UTILITIES
SNR = STATUS AND NEEDS REPORTS (PDR in PMS.FMU)
ROS = REHABILITATION OPTIMIZATION
ROR = OPTIMIZATION SUBSYSTEM REPORTS

and segment data subset screening. The analysis programs and all reporting programs, other than segment lists, are FORTRAN programs. SPF/PC is used to edit user parameter files for program execution.

### 36.1.1 Major Programs In the PMS, and Their Outputs

The MNDOT PMS incorporates 14 (FORTRAN) programs, listed in Table 36.1, together with their key purpose or outputs. Illustration of these outputs, with summary descriptions of the analysis procedures (i.e., equations, models), are provided in the following sections. These illustrations are subdivided into Status and Needs, which is represented by the first 7 programs of Table 36.1, and Optimization (see Figure 36.1), represented by the latter 7 programs of Table 36.1.

It is emphasized that the example outputs on the following pages should be viewed as illustrative only, since they are based on preliminary runs of a newly installed system, and in some cases with incomplete data. Consequently, the illustrative numbers should not be considered to be absolute.

### 36.1.2 Status and Needs

Of the seven total FORTRAN programs in this subsystem, one is used for performance prediction, two produce printed reports, and four produce graphical outputs. Performance predictions in the MNDOT PMS involve the following for each segment. Details of the models used and the input factors are contained in a report by the consultant [PMS Ltd. 89], along with typical curves.

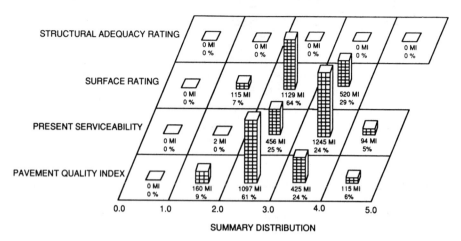

**Figure 36.2** Summary distribution of present status for the Interstate network.

1. Input the data (SEGSUM) record.
2. Predict surface distresses (for 20 years) and calculate Surface Rating (SR); calculate average SR's.
3. Predict Present Serviceability Rating (PSR), Structural Adequacy Rating (SAR), and the composite Pavement Quality Index (PQI).

**Status Reports**—A (present status) SEGMENT SUMMARY REPORT can be presented in tabular form for any particular route. Also shown at the bottom of the table are the totals for the network. The distributions for these totals can also be displayed in graphical form, as shown in Figure 36.2.

**Needs Reports**—The SEGMENT NEED REPORT indicates the needs or "trigger" years, based on application of the performance models. The trigger year occurs when the predicted performance for any particular index or parameter reaches a minimum acceptable level, as specified by the user. An example application of the performance prediction models for a particular segment is shown in Figure 36.3. The needs year distribution summary can be tabulated or shown graphically based on Structural Rating (SR).

### 36.1.3 Rehabilitation Optimization

The Rehabilitation Optimization Subsystem incorporates seven FORTRAN programs (see Table 36.1), the first two of which carry out the analysis and optimization while the five remaining are used to produce reports.

**Rehabilitation Alternatives Analysis**—Minnesota used 34 rehabilitation alternatives in 1989. An override option allows the user to specify a subset of these alternatives for the analysis of any particular section. There are three basic decision trees, one for concrete (CON and CRC), one for bituminous (BIT and BOB), and one for bituminous over concrete (BOC). The decision tree is applied to each possible implementation year (i.e., projects are advanced and delayed from their needs or trigger

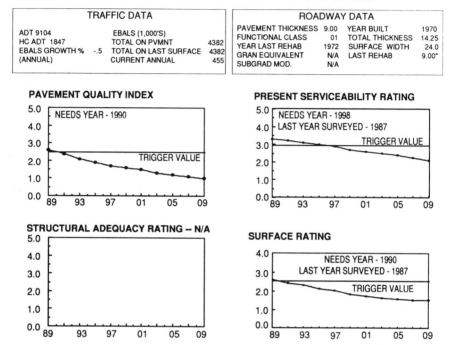

**Figure 36.3** Example application of the performance prediction models to a segment.

year) because the conditions used in the decision tree can change from year to year. Performance models for the rehabilitation alternatives, as well as the improvement in the performance parameter when the rehabilitation is applied, are used to estimate the pavement condition over time [PMS Ltd. 89]. For some alternatives, such as minor joint reseal, the improvement is 0 for some of the parameters (i.e., PSR).

The present worth of costs is calculated for rehabilitation for each implementation year, ongoing maintenance, and user costs. As well, cost-effectiveness for each alternative–implementation year combination is calculated [PMS Ltd. 89, Hill 90]. Of particular interest is the comprehensive user-delay cost model, which may or may not be included in the analysis as an option. It considers such factors as traffic handling method, traffic volumes, length of the rehabilitation or maintenance zone, number of days, speeds, capacity, type of facility (multilane, two-lane, shoulders, etc.), and direction of travel.

Detailed reports of the alternatives analyzed for each segment for each year of the program period can be generated in the MNDOT PMS. However, only network summary report examples are provided here.

**Optimization Analysis**—The optimization used in MNDOT's PMS is based on marginal cost-effectiveness calculations [Hill 90]. It is therefore a "near optimization" rather than a true optimization method but the results are the same for practical purposes.

The analysis can be performed to accomplish either of the following objectives:

1. Effectiveness maximization, where the primary constraints are specified budget limits for each year of the program period, or
2. Cost minimization, where the constraints are either:
    a) minimum average network performance (in terms of PSR, SR, SAR, or PQI), or
    b) maximum percent of mileage below the minimum acceptable or trigger level

An example summary report for the effectiveness maximization mode is shown in Figure 36.4 for an annual budget limit of $15,000,000 (actual predicted costs, which are slightly lower, are listed in the third row of the diagram). This example is based on effectiveness maximization for PSR.

Figure 36.4 shows, on the left part of the diagram, how average PSR will change for the expected budget (solid line) from about 3.4 in 1989 to about 3.2 in 1988, and for 0 budget (dotted line) to a low of about 2.7 in 1998. The right part of the diagram shows the associated accumulation of deficient mileage, in terms of percent below the minimum or trigger PSR level. For the expected budget (solid line), this would rise from about 10 to 32 percent over the program period, while for 0 budget (dotted line), it would rise to about 65 percent.

### 36.1.4 Summary

In conclusion, Minnesota has a comprehensive, flexible operational PMS in place, tailored to its requirements. Among the key reasons for the successful development and implementation were careful, pre-implementation planning, strong support throughout the department including senior management, a sound technical basis for the system in terms of the database, models, programs and reporting functions, and a firm commitment by those responsible for its operation and use.

## 36.2 METROPOLITAN TRANSPORTATION COMMISSION PMS

### 36.2.1 Introduction

The Metropolitan Transportation Commission (MTC) Pavement Management System is a simple and efficient system designed for network level pavement management. It is a tool well suited to quickly and efficiently assess the condition of a road network, define and adjust maintenance treatments and costs, and examine the effects on road condition of specific maintenance budget allocation strategies. The reports generated from the use of this system aid as a powerful tool to obtain funding from decision-makers, to allocate funds for the most cost-effective repairs, and ultimately, to maintain the network in a satisfactory condition. The system is designed primarily for use by

| Mn/DOT : REHABILITATED PERFORMANCE SUMMARY REPORT | | | | | | | | | | |
|---|---|---|---|---|---|---|---|---|---|---|
| YEAR | 1989 | 1990 | 1991 | 1992 | 1993 | 1994 | 1995 | 1996 | 1997 | 1998 |
| BUDGET (MILLIONS) | 14.9 | 14.9 | 14.9 | 15.0 | 14.9 | 15.0 | 15.0 | 14.8 | 15.0 | 15.0 |
| NON-REHAB AVG. PSR | 3.3 | 3.2 | 3.2 | 3.1 | 3.0 | 3.0 | 2.8 | 2.8 | 2.8 | 2.7 |
| REHAB AVG. PSR | 3.3 | 3.3 | 3.2 | 3.2 | 3.2 | 3.2 | 3.2 | 3.2 | 3.1 | 3.1 |
| NON-REHAB % PSR < MIN | 9 | 12 | 15 | 19 | 26 | 33 | 43 | 52 | 60 | 64 |
| REHAB % PSR < MIN | 8 | 10 | 11 | 14 | 14 | 20 | 22 | 27 | 31 | 31 |

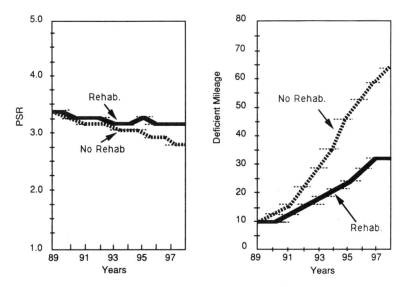

**Figure 36.4** Example performance summary report for network optimization (effectiveness maximization mode).

city and county jurisdictions. The basic concept and structure of the system summary follow [MTC 86].

### 36.2.2 Background

MTC is the transportation planning agency for 103 cities and counties in the San Francisco Bay Area. The origins of the PMS efforts by MTC are traced to a study of local road and street maintenance needs and revenue and short falls, conducted by MTC in 1982 [Smith 88]. The results of the study indicated that local jurisdictions in the Bay Area were spending only 60 percent of the funds required to maintain roads in a condition considered adequate.

This demonstrated the need to improve pavement management techniques and practices in the Bay Area. A committee of engineering and maintenance personnel from local public works departments was formed to evaluate the pavement management efforts in the Bay Area, while six public works directors reviewed a proposal to develop a prototype PMS. The two groups emphasized that the PMS has to be simple to use and recommended incorporation of only proven techniques and practices that their personnel could understand and use. Also, it was indicated that the system must match the needs and resource capabilities of the jurisdictions.

In 1983, a three-way partnership was formed to develop the pilot PMS. ERES consultants, Inc., with R. E. Smith as the principal investigator, was retained to develop a PMS. MTC provided most of the funding, programming expertise, and staff time to assist the participating agencies. Six Bay Area agencies, including three counties and three cities, participated in the study by providing experienced personnel for user-guidance in the development of the system, ensuring continual feedback to developers and programmers.

### 36.2.3 Description of the System

The basic concept of the system is illustrated in Figure 36.5. The implementation of the system begins with a network inventory. Necessary data from the pavements must be collected in terms of construction, traffic, date of last maintenance, etc., and entered in the PMS data base. The entire pavement network is then divided into uniform manageable sections/units for record and analysis purposes. Condition data is collected for each management section.

The Pavement Condition Index (PCI) is computed based on the observed distress types, severity, and extent. The PCI of each management section is then projected and aggregated to estimate the average network condition over time. These projections are made from a family of performance curves which are based on the current condition of the existing pavements in terms of PCI versus age or, if adequate data is available, PCI versus age times the natural logarithm of average daily traffic.

A decision tree was developed which assigns the most cost-effective maintenance or rehabilitation strategy from a specified list of maintenance and rehabilitation (M&R) strategies, depending on the PCI information, functional class, and cost information. The present condition and projected condition of the management units are analyzed for each year of a five year analysis period to identify when they should be scheduled for maintenance or rehabilitation and when the funds will be needed.

When the funds needed exceed those available, the management sections which should be funded for each year of the analysis period are selected using a cost-effectiveness analysis. An analysis procedure that allows the manager to consider different budget levels and different splits in the budget between preventive maintenance and rehabilitation options is developed. This is done to ensure the best overall network condition for the funds expended. The data base is updated after the maintenance and rehabilitation work has been completed and is ready for the next analysis cycle or period. Figure 36.6 identifies the inputs to the system and the types of output reports generated.

The MTC PMS, also called as BAPMS (Bay Area Pavement Management System), runs on DOS personal computers. At least 640K RAM and a 20 megabyte hard

# Network Level Examples of PMS

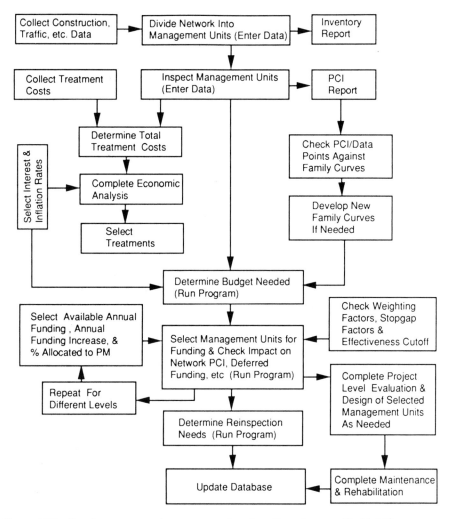

**Figure 36.5** Bay Area pavement management system flow diagram [Smith 88].

drive are required to run the system. The program uses a commercial data base software package, Rbase System V, a product of Microrim.

The Bay Area PMS was initially designed to handle only four distinct surface types that are most common in the Bay Area. They are:

1. Asphaltic concrete pavements never overlaid (AC)
2. Asphaltic concrete pavements overlaid with asphaltic concrete (AC/AC)
3. Portland cement concrete pavements overlaid with a layer of asphaltic concrete (AC/PCC)

| INPUTS | |
|---|---|
| SECTION DATA<br><br>　Section identification*<br>　Street identification*<br>　Street name*<br>　Location Begins (from)*<br>　Location ends*<br>　Functional class*<br>　Length*<br>　Width*<br>　Area*<br>　No. of lanes*<br>　Surface type*<br>　Date of present surface constructed*<br><br>　General code<br>　Traffic index<br>　ADT<br>　Area ID<br>　Parking width<br>　Funding source<br>　Comments | INSPECTION DATA<br><br>　Distress code<br>　Severity<br>　Quantity |
| | MAINTENANCE AND REHABILITATION DATA<br><br>　Month/ year treatment applied*<br>　Treatment description ID no.*<br>　Thickness<br>　Cost of treatments |
| | DEFAULT VALUES IN THE SYSTEM - USER MODIFIABLE<br><br>　PCI condition category definitions<br>　PCI breakpoints by functional class and surface type<br>　Treatment descriptions<br>　Preventive maintenance treatments and costs<br>　Rehabilitation treatments and costs by functional class<br>　Weighting factors |
| | OTHER - Inflation EUAC%, Interest %, Budget |

| OUTPUTS / REPORTS | |
|---|---|
| PAVEMENT CONDITION INDEX<br>　PCI calculation report<br>　PCI history report<br>　Section PCI listings by ascending PCI order<br>　Section PCI listings by alphabetical street order with sect. ID<br>　Section PCI listings by alphabetical street order with street ID | MAINTENANCE AND REHABILITATION<br><br>　Reinspection schedule<br>　M & R history report<br>　Inspection inventory report<br>　M & R change report |
| BUDGET NEEDS<br>　Projected PCI values and costs summary report<br>　Projected PCI values and costs detailed report<br>　Preventive maintenance costs<br>　Rehabilitation costs<br>　Treatments and costs by section and year of application<br>　Costs by functional classification and condition category | LONG TERM PROJECTION REPORTS<br><br>　PCI projection reports - years 1-5<br>　PCI projection reports - years 6-10<br>　PCI projection reports - years 11-15<br>　PCI projection reports - years 16-20<br>　Projection reports : yearly budget<br>　　allocation by condition category<br>　　and maintenance type |
| BUDGET SCENARIO REPORTS<br>　Summary budget scenario report<br>　Detailed - Five year PCI projections<br>　　Costs of stopgap and deferred maintenance<br>　　Cost effectiveness and treatment by section<br>　　Sections not selected for maintenance/ rehab<br>　　Sections with errors | OTHER<br><br>　Network replacement value report |

**Figure 36.6** Inputs and outputs of the Bay Area pavement management system [Smith 88].

**Network Level Examples of PMS** 477

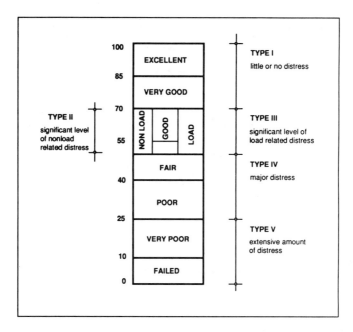

**Figure 36.7** Pavement condition categories used in Bay Area pavement management system [Smith 88].

4. Multiple surface treatments never overlaid with a subsequent layer of asphaltic concrete

Version 5.09 of the system, however, includes PCC pavement surfaces never overlaid.

Seven major types of surface distress that occur predominantly in the Bay Area are considered by the condition assessment. They are alligator cracking, distortions, patching and utility cut patch, rutting/depression, longitudinal transverse cracking, block cracking, and weathering and ravelling. The first four are grouped as primarily structural related distresses and the next three are grouped as primarily environmental or material related distresses.

Deduction curves are used to calculate the Pavement Condition Index (PCI) based on the type, severity, and amount of the various distresses. Five pavement condition categories are defined based on a PCI scale with values ranging from 100 to 0 (excellent to failed) and depending on the load or non-load related distress. Distress is measured in square feet, except for longitudinal and transverse cracking which are measured in linear feet. The pavement condition categories along with the PCI scale boundaries are shown in Figure 36.7.

Future PCI condition is predicted as

$$\text{PCI} = 100 - Ae^{-(R/\text{AGE})^B} \tag{36.1}$$

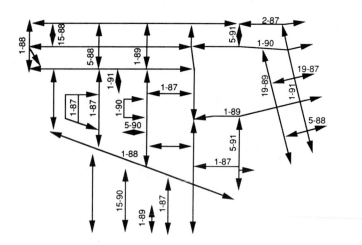

**Figure 36.8** Example network divided into management sections with treatment number and year of application (1-86) from Bay Area PMS budget scenario output [Smith 88].

where

    AGE = the age of the current pavement surface, or function of age and traffic
$A, R, B$ = regression coefficients.

The projected PCI is adjusted by considering the PCI based on latest inspection data and most recent maintenance on the pavement.

The five category condition definition provides for a realistic maintenance and rehabilitation assignment. A total of 17 types of maintenance treatments along with four major rehabilitation strategies are considered in the maintenance and rehabilitation actions. The four major rehabilitation actions considered are:

1. Reconstruction, surface and base
2. Reconstruction surface
3. Thick asphaltic concrete overlay (2.5″)
4. Thin asphaltic concrete overlay (1.5″)

A decision tree identifies potential actions and an economic analysis is applied to select the maintenance and repair actions. An example network with treatment number and year of application from PMS budget scenario output is given in Figure 36.8.

Considerable modification in M&R treatment for the management section selections needs to be done in developing "contract or maintenance packages". For example, no agency would want to provide a treatment such as overlay for two blocks, heater scarify overlay for one block, skip two blocks, then apply chip seal to next two blocks, and so on. They would apply the same treatment for an area under one contract package. Similarly if two streets are connected by a third one, it is only logical to do the treatments under one contract package for all three streets rather than repair two streets and do the treatment for the connecting street six months later.

A Weighted Effectiveness Ratio (WER) is used to rank and prioritize projects:

$$WER = [(AREA/YR) \ WF]/(EUAC/SY) \qquad (36.2)$$

where

- WER = weighted effectiveness ratio
- AREA = area under PCI curve
- YR = years affected
- WF = weighting factor for usage (modifiable default values based on functional class)
- EUAC = equivalent uniform annual cost
- SY = square yards in the management units

The user can specify a percentage of the budget to allocate to the rehabilitation. The computer software selects projects identified for rehabilitation from highest-weighted effectiveness ratio to lowest until the funds allocated for rehabilitation are assigned. Funds not assigned to rehabilitation are allocated to preventive maintenance. The management units identified as needing rehabilitation but not selected for funding are assigned stopgap maintenance funds. The total budget allocated to rehabilitation, stopgap maintenance, and preventive maintenance is calculated. The budgeting process is repeated for each of the five years in the analysis period.

## 36.3 CONCLUSION

The MTC PMS has been widely applied in California and has also been used by local agencies in other states. It is a complete package in the sense that each of the components required for a pavement management system is included in the MTC PMS. One of the greatest advantages of this system is the central support that MTC provides for the users. This includes training in the use of the system and periodic upgrading of the program.

## REVIEW QUESTIONS

1. Summarize the strengths and weaknesses of the Minnesota DOT network level PMS.
   a. Would you adapt this system for use in your own highway department?
   b. What changes would be necessary for you to use the system in your state or nation?
   c. What changes would you propose?
2. Is the MTC pavement management system primarily a network or a project level system? Please explain your answer.
3. Is the MTC pavement management system primarily applicable to state level or city level pavement management requirements? Explain your answer.

Chapter 37

# The SAMP System

## 37.1 INTRODUCTION

The SAMP (Systems Analysis Method for Pavements) programs, first developed for the NCHRP Program [Hudson 73a] are widely applicable, and relatively simple to use. These two aspects make SAMP most useful for illustration even though it is not the latest or most comprehensive system available.

The basic structural analysis models in SAMP are based on the AASHO Road Test results as modeled by the AASHO Interim Pavement Design Guides [AASHO 61]. These design equations are used in some form by 35 state highway departments [Van Til 72], and thus the basic models find wide application.

The purpose of SAMP is to identify from available input data a pavement that can be maintained above the specified minimum serviceability over the specified design period at a minimum overall cost. The SAMP computer program provides the decision-maker with a set of feasible pavement designs arranged in priority order of increasing total present worth of costs. Other pertinent information necessary for use in making rational design decisions is also provided. A variety of input data is necessary for the solution of each problem. Figure 37.1 is a typical input data sheet from a SAMP computer printout, and illustrates a sample set of input for the program. An example output of the program is shown in Figure 37.2. The information provided

**The SAMP System**   481

for each alternative design includes: (1) initial construction, (2) overlay schedule, and (3) a cost breakdown. The cost breakdown includes initial construction, overlays, routine maintenance, salvage value, and user cost during overlay construction. Present worth of costs is computed using a discount rate selected by the user.

## 37.2 INPUT FOR SAMP-5

The particular version of the Systems Analysis Model for Pavements presented is SAMP-5, the fifth in the series. There are seven classes of input variables required by the program. These are: (1) material properties, (2) environmental and serviceability, (3) load and traffic, (4) maintenance, (5) program control and miscellaneous, (6) constraints, and (7) traffic delay. Each class of variables is important in the solution of a problem by the computer. Figure 37.1 is an example input for a typical analysis problem.

**Material Properties**—For each material type, the user must define the type of material, in place cost per cubic yard, strength coefficient, minimum and maximum allowable thickness, salvage value, and soil support value (for unbound materials only). The salvage value is expressed as the percent of the new material cost.

**Environmental and Serviceability Parameters**—The environmental and serviceability program considers two types of environmental variables. The most important defines the effect of swelling clay on the performance of the pavement. This effect is calculated using two parameters. The first is a function of the material properties and controls the rate of change of serviceability with time. The other is the level of serviceability that the pavement would reach as a result of swelling clay alone, in the absence of traffic. The second category of environmental variables involves the general climatic and geologic effects of the region, as specified by the AASHO Interim Pavement Design Guide [AASHO 61]. This effect has not yet been quantified, although some states use it in their design methods [Van Til 72].

Closely related to the environmental variables are the serviceability parameters. These include a serviceability index at initial construction, PSI, the serviceability index after overlay construction, and the lower limit of the serviceability index or the level when overlay construction or failure occurs.

**Load and Traffic Variables**—The load variables used in this program include the total accumulation of the equivalent 18-kip axle loads expected to be carried by the pavement during the analysis period, and the average daily traffic (ADT) at the beginning and at the end of the analysis. These variables are combined in the program to give a load-time function that distributes the traffic loads carried throughout the analysis period. The program uses this information to calculate user costs resulting from traffic delay and uses the loads to be carried in the model to predict pavement life. In addition to the three important load variables, two other traffic load variables are important: (1) the percent of ADT passing through the overlay zone per hour and (2) the type of road under construction (rural or urban).

**Maintenance Variables**—Four maintenance variables are required to implement the NCHRP maintenance study, which is used as the basis for calculating projected maintenance costs in this program. They are: (1) the number of days per year that the

```
PROGRAM SAMP5 (SYSTEMS ANALYSIS MODEL FOR PAVEMENTS_ REVISED 26MAR80
RUN NO. 1 EXAMPLE FOR HAAS, HUDSON, AND ZANIEWSKI

    PROB PRO1    SAMP5 EXAMPLE WITH TYPICAL DESIGN VALUES

    THE CONSTRUCTION MATERIALS UNDER CONSIDERATION ARE
              MATERIALS           COST      STR.     MIN.     MAX.    SALVAGE    SOIL
    LAYER CODE    NAME          PER CY    COEFF.   DEPTH    DEPTH     PCT.     SUPPORT
      1    A    ASPHALT CONCRETE  74.00     0.44     6.00     8.00    50.00     0.00
      2    B    CRUSHED STONE     32.00     0.14     5.00     8.00    50.00     9.20
      3    C    GRAVEL            10.00     0.11     5.00    10.00    50.00     6.90
               SUBGRADE            0.00     0.00     0.00     0.00     0.00     4.25

    ENVIRONMENTAL AND SERVICEABILITY VARIABLES
        R-REGIONAL FACTOR.                                                       1.7
        PSI-THE SERVICEABILITY INDEX IN THE INITIAL STRUCTURE.                   4.2
        P1-THE SERVICEABILITY INDEX IN AN OVERLAY.                               4.2
        P2-THE MINIMUM ALLOWED VALUE OF THE SERVICEABILITY INDEX. (POINT AT
            WHICH AN OVERLAY MUST BE APPLIED).                                   2.5
        P2P-THE LOWER BOUND ON THE SERVICEABILITY INDEX WHICH WOULD BE
            ACHIEVED IN INFINITE TIME WITH NO TRAFFIC.                           1.5
        BONE-THE RATE AT WHICH NON TRAFFIC FACTORS REDUCE THE SERVICEABILITY
            INDEX.                                                             0.120

    LOAD AND TRAFFIC VARIABLES
        RO-THE ONE-DIRECTION AVERAGE DAILY TRAFFIC AT THE BEGINNING OF THE
            ANALYSIS PERIOD.                                                  10000.
        RC-THE ONE-DIRECTION AVERAGE DAILY TRAFFIC AT THE END OF ANALYSIS
            PERIOD.                                                           20000.
        XNC-THE ONE-DIRECTION ACCUMULATED NUMBER OF EQUIVALENT 18-KIP
            AXLES DURING THE ANALYSIS PERIOD.                               5000000.
        PROP-THE PERCENT OF ADT WHICH WILL PASS THROUGH THE OVERLAY ZONE
            DURING EACH HOUR WHILE OVERLAYING IS TAKING PLACE.                   6.0
        ITYPE-THE TYPE OF ROAD UNDER CONSTRUCTION. (1-RURAL, 2-URBAN).           1

    MAINTENANCE VARIABLES
        X2-THE NUMBER OF DAYS PER YEAR THAT THE TEMPERATURE REMAINS BELOW 32F   60.
        CLW-THE COMPOSITE LABOR WAGE.                                            4.25
        CERR-THE COMPOSITE EQUIPMENT RENTAL RATE.                                5.00
        CMAT-THE RELATIVE MATERIAL COST (1.00 IS AVERAGE).                       1.00
```

**Figure 37.1** Typical computer listing of SAMP-5 input data (1 of 2).

temperature remains below 32°F, (2) the composite labor index, (3) the composite equipment rental rate, and (4) the relative material costs for the locality. The details covering the development of this model can best be obtained in the original NCHRP report [Bertram 67, Nair 73].

**Program Control and Miscellaneous Parameters**—Five input variables are shown in the program control and miscellaneous category. These include the number of output pages, the total number of construction materials being considered, the width of each lane in feet, and the discount rate or time value of money.

**Constraint Variables**—The pavement designer (program user) must input values for six variables which constrain the number of alternatives considered by the program. These variables are probably new to the reader, because they are used implicitly in other design methods or are ignored completely. They include: (1) minimum time to the first overlay, (2) minimum time between overlays, (3) maximum funds available for construction, (4) maximum total thickness of construction, (5) minimum thickness of a single overlay, and (6) maximum accumulated thickness of all overlays.

The minimum and maximum thicknesses of the construction materials listed under material properties are also constraints on the system and limit the range of feasible designs.

## The SAMP System

```
PROGRAM CONTROL AND MISCELLANEOUS VARIABLES
    NMB-THE NUMBER OF OUTPUT PAGES FOR THE SUMMARY TABLE (10 DESIGNS/PAGE).         2
    NM-THE TOTAL NUMBER OF MATERIALS AVAILABLE, EXCLUDING SUBGRADE.                 3
    CL-THE LENGTH OF THE ANALYSIS PERIOD (YEARS).                                  20.
    XLW-THE WIDTH OF EACH LANE (FEET).                                             12.
    RATE-THE INTEREST RATE OR TIME VALUE OF MONEY (PERCENT).                        5.0

CONSTRAINT VARIABLES
    XTTO-THE MINIMUM ALLOWED TIME TO THE FIRST OVERLAY.                             5.0
    XTBO-THE MINIMUM ALLOWED TIME BETWEEN OVERLAYS.                                 5.0
    CMAX-THE MAXIMUM FUNDS AVAILABLE FOR INITIAL CONSTRUCTION.                     30.00
    TCKMAX-THE MAXIMUM ALLOWABLE TOTAL THICKNESS OF INITIAL CONSTRUCTION.          32.0
    OVMIN-THE MINIMUM THICKNESS OF AN INDIVIDUAL OVERLAY.                           0.50
    OVMAX-THE ACCUMULATED MAXIMUM THICKNESS OF ALL OVERLAYS.                        2.5

TRAFFIC DELAY VARIABLES ASSOCIATED WITH OVERLAY AND ROAD GEOMETRICS
    ACPR-ASPHALTIC CONCRETE PRODUCTION RATE (TONS/HOUR).                           75.0
    ACCD-ASPHALTIC CONCRETE COMPACTED DENSITY (TONS/COMPACTED CY).                  1.80
    XLSO-THE DISTANCE OVER WHICH TRAFFIC IS SLOWED IN THE OVERLAY DIRECTION.        0.60
    XLSN-THE DISTANCE OVER WHICH TRAFFIC IS SLOWED IN THE NON-OVERLAY DIRECTION.    0.60
    XLSD-THE DISTANCE AROUND THE OVERLAY ZONE (MILES).                              0.00
    HPD-THE NUMBER OF HOURS/DAY OVERLAY CONSTRUCTION TAKES PLACE.                   8.0

TRAFFIC DELAY VARIABLES ASSOCIATED WITH TRAFFIC SPEEDS AND DELAYS
    THE PERCENT OF VEHICLES THAT WILL BE STOPPED BECAUSE OF THE MOVEMENT OF
    PERSONNEL OR EQUIPMENT.
    PPO2-IN THE OVERLAY DIRECTION.                                                  5.00
    PPN2-IN THE NON-OVERLAY DIRECTION.                                              5.00
    THE PERCENT OF VEHICLES THAT WILL BE STOPPED BECAUSE OF THE MOVEMENT OF
    PERSONNEL OR EQUIPMENT.
    DDO2-IN THE OVERLAY DIRECTION (HOURS).                                          0.150
    DDN2-IN THE NON-OVERLAY DIRECTION (HOURS).                                      0.150
    AAS- THE AVERAGE APPROACH SPEED TO THE OVERLAY AREA.                           50.
    THE AVERAGE SPEED THROUGH THE OVERLAY AREA
    ASO-IN THE OVERLAY DIRECTION (MPH).                                            30.
    ASN-IN THE NON-OVERLAY DIRECTION (MPH).                                        50.
    MODEL-THE TRAFFIC HANDLING MODEL USED.                                          0
```

**Figure 37.1** Typical computer listing of SAMP-5 input data (2 of 2).

**Traffic Delay Variables**—Traffic delay variables are divided into two parts. The first contains six variables associated with the overlay procedure and the road geometrics. They are: (1) asphalt concrete production rate, (2) asphalt concrete density, (3) distance over which traffic is slowed in the overlay direction, (4) distance over which traffic is slowed in the nonoverlay direction, (5) distance around the overlay zone, and (6) the number of hours per day that overlay construction actually is expected to take place. The second category contains four variables associated with the traffic speed and delays. They are: (1) percent of vehicles stopped because of construction traffic, (2) average delay per vehicle stopped, (3) average approach speed through the overlay area, and (4) the particular model used to handle the traffic. These variables are used to calculate vehicle user costs during the overlay of the various proposed pavement designs.

## 37.3 SAMP PROGRAM OPERATION

The SAMP-5 computer program uses an algorithm similar to the FPS system discussed in Chapter 35. The main differences involve the model used for calculating routine maintenance cost, the performance models, and the use of seal coats.

PROB PRO1    SAMP5 EXAMPLE WITH TYPICAL DESIGN VALUES

SUMMARY OF THE BEST DESIGN STRATEGIES,
IN ORDER OF INCREASING TOTAL COST

| | 1 | 2 | 3 | 4 | 5 | 6 | 7 | 8 | 9 | 10 |
|---|---|---|---|---|---|---|---|---|---|---|
| MATERIAL ARRANGEMENT | ABC | ABC | ABC | ABC | ABC | ABC | ABC | ABC | ABC | ABC |
| INIT. CONST. COST | 20.681 | 21.056 | 20.986 | 21.125 | 21.194 | 21.222 | 21.153 | 21.361 | 21.431 | 21.500 |
| OVERLAY CONST. COST | 4.133 | 4.051 | 4.133 | 4.051 | 4.051 | 4.051 | 4.133 | 4.051 | 4.051 | 4.051 |
| USER COST | 0.138 | 0.136 | 0.138 | 0.136 | 0.137 | 1.015 | 0.138 | 0.135 | 0.136 | 0.137 |
| ROUTINE MAINT. COST | 1.301 | 1.303 | 1.301 | 1.304 | 1.307 | 1.303 | 1.301 | 1.302 | 1.304 | 1.306 |
| SALVAGE VALUE | -4.285 | -4.355 | -4.342 | -4.368 | -4.381 | -4.387 | -4.374 | -4.413 | -4.426 | -4.439 |
| TOTAL COST | 21.968 | 22.189 | 22.216 | 22.248 | 22.308 | 22.325 | 22.351 | 22.437 | 22.496 | 22.555 |
| NUMBER OF LAYERS | 3 | 3 | 3 | 3 | 3 | 3 | 3 | 3 | 3 | 3 |
| LAYER DEPTH (INCHES) | | | | | | | | | | |
| D(1) | 6.75A | 6.50A | 6.25A | 6.75A | 7.00A | 7.25A | 7.00A | 6.00A | 6.25A | 6.50A |
| D(2) | 5.00B | 6.00B | 6.50B | 5.50B | 5.00B | 5.00B | 5.50B | 7.50B | 7.00B | 6.50B |
| D(3) | 8.50C | 8.50C | 8.50C | 8.50C | 8.50C | 6.75C | 6.75C | 8.50C | 8.50C | 8.50C |
| NO. OF PERF. PERIODS | 3 | 3 | 3 | 3 | 3 | 3 | 3 | 3 | 3 | 3 |
| PERF. TIME (YEARS) | | | | | | | | | | |
| I(1) | 5.039 | 5.078 | 5.039 | 5.078 | 5.195 | 5.039 | 5.039 | 5.039 | 5.078 | 5.195 |
| I(2) | 12.461 | 12.656 | 12.461 | 12.773 | 13.086 | 12.617 | 12.461 | 12.578 | 12.773 | 13.047 |
| I(3) | 22.383 | 22.852 | 22.383 | 23.164 | 23.789 | 22.813 | 22.383 | 22.656 | 23.125 | 23.711 |
| OVERLAYING POLICY (INCH) (INCLUDING LEVEL-UP) | | | | | | | | | | |
| O(1) | 1.5 | 1.5 | 1.5 | 1.5 | 1.5 | 1.5 | 1.5 | 1.5 | 1.5 | 1.5 |
| O(2) | 1.5 | 1.5 | 1.5 | 1.5 | 1.5 | 1.5 | 1.5 | 1.5 | 1.5 | 1.5 |

**Figure 37.2** Typical SAMP-5 output (1 of 2).

PROB PRO1   SAMP5 EXAMPLE WITH TYPICAL DESIGN VALUES

SUMMARY OF THE BEST DESIGN STRATEGIES,
IN ORDER OF INCREASING TOTAL COST

|  | 11 | 12 | 13 | 14 | 15 | 16 | 17 | 18 | 19 | 20 |
|---|---|---|---|---|---|---|---|---|---|---|
| MATERIAL ARRANGEMENT | ABC | ABC | ABC | ABC | ABC | ABC | ABC | ABC | ABC | ABC |
| INIT. CONST. COST | 21.528 | 21.569 | 21.597 | 21.708 | 21.639 | 21.667 | 21.694 | 21.736 | 21.764 | 21.806 |
| OVERLAY CONST. COST | 4.051 | 4.051 | 4.051 | 3.973 | 4.051 | 4.051 | 4.051 | 4.051 | 4.051 | 4.051 |
| USER COST | 0.135 | 1.137 | 0.136 | 1.135 | 0.138 | 0.137 | 0.135 | 0.137 | 0.136 | 0.137 |
| ROUTINE MAINT. COST | 1.302 | 1.310 | 1.304 | 1.323 | 1.316 | 1.306 | 1.302 | 1.310 | 1.304 | 1.306 |
| SALVAGE VALUE | -4.444 | -4.452 | -4.457 | -4.478 | -4.465 | -4.470 | -4.476 | -4.483 | -4.489 | -4.497 |
| TOTAL COST | 22.572 | 22.616 | 22.631 | 22.661 | 22.679 | 22.690 | 22.708 | 22.751 | 22.766 | 22.803 |
| NUMBER OF LAYERS | 3 | 3 | 3 | 3 | 3 | 3 | 3 | 3 | 3 | 3 |
| LAYER DEPTH (INCHES) |  |  |  |  |  |  |  |  |  |  |
| D(1) | 6.75A | 6.75A | 7.00A | 7.25A | 7.00A | 7.25A | 7.50A | 7.50A | 7.75A | 6.00A |
| D(2) | 6.50B | 6.00B | 6.00B | 5.00B | 5.50B | 5.50B | 5.50B | 5.00B | 5.00B | 8.00B |
| D(3) | 6.75C | 8.50C | 6.75C | 8.50C | 8.50C | 6.75C | 5.00C | 6.75C | 5.00C | 8.50 |
| NO. OF PERF. PERIODS | 3 | 3 | 3 | 3 | 3 | 3 | 3 | 3 | 3 | 3 |
| PERF. TIME (YEARS) |  |  |  |  |  |  |  |  |  |  |
| I(1) | 5.039 | 5.234 | 5.078 | 5.352 | 5.313 | 5.195 | 5.039 | 5.234 | 5.078 | 5.195 |
| I(2) | 12.578 | 13.242 | 12.773 | 13.672 | 13.477 | 13.047 | 12.578 | 13.242 | 12.773 | 13.047 |
| I(3) | 22.656 | 24.102 | 23.125 | 25.117 | 24.648 | 23.711 | 22.656 | 24.102 | 23.125 | 23.594 |
| OVERLAYING POLICY (INCH) (INCLUDING LEVEL-UP) |  |  |  |  |  |  |  |  |  |  |
| O(1) | 1.5 | 1.5 | 1.5 | 1.5 | 1.5 | 1.5 | 1.5 | 1.5 | 1.5 | 1.5 |
| O(2) | 1.5 | 1.5 | 1.5 | 1.5 | 1.5 | 1.5 | 1.5 | 1.5 | 1.5 | 1.5 |

**Figure 37.2** Typical SAMP-5 output (2 of 2).

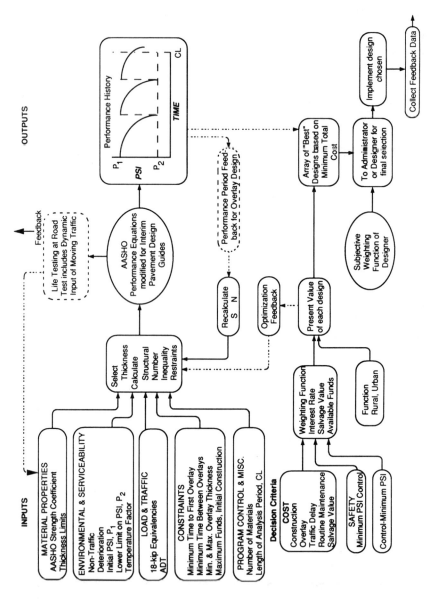

**Figure 37.3** Block diagram of SAMP pavement system.

# The SAMP System

The operation of the program, Figure 37.3, can be broken down into four sections: (1) initial construction, (2) overlay construction, (3) cost computation, and (4) optimization.

**Initial Construction**—The variables used in the initial construction phase of the program are material strengths, including the AASHTO strength coefficients as well as soil support values as described in the report NCHRP Project 1-11 [Van Til 72]. The system takes the thickness ranges of the materials and increments through all possible layer and thickness combinations, calculating the structural number for each layer selection. The asphalt concrete is incremented by 1/4 in. steps and the increment of other layers is in steps equivalent in cost to 1/4 in. of asphaltic concrete rounded off to the nearest 1/4 in. That is, if the cost of asphaltic concrete is $40 per cubic yard and the cost of gravel is $10 per cubic yard, then the increment for gravel is (1/4) · (40/10) = 1 in. Each structural number is then combined with the loading and nontraffic deterioration parameters in the performance model to determine the first period of the performance history.

Each initial construction design must first satisfy three constraints. It must cost no more than the maximum funds available for initial construction, be no thicker than the total thickness constraint, and have an initial life at least as long as the minimum time to the first overlay. For each design that meets these requirements the program continues to the overlay phase.

**Overlay Design**—The variables used in the overlay calculations are: (1) the lower bound on the serviceability index, (2) the length of the analysis period, and (3) minimum and maximum overlay thickness constraints. If the serviceability of a pavement reaches the minimum allowed, as determined by the performance model, then an overlay is required. It is assumed that overlays will be constructed of asphaltic concrete in multiples of 1/2 in. starting with the minimum overlay thickness constraint. All possible combinations of thicknesses and numbers of overlays are then tried to determine which overlay policy is most economical for each initial design. This interaction with the cost computation is shown as optimization feedback in Figure 37.3. The number of possible overlay policies is controlled by the maximum overlay thickness constraint and by the length of the analysis period.

**Cost Computation**—The costs computed in the SAMP programs are: (1) initial construction cost, (2) overlay cost, (3) maintenance cost, (4) traffic delay cost, and (5) salvage value [Hudson 73a]. Salvage value in SAMP-5 is based on a percentage of the structure existing at the end of the analysis period. A different percentage may be used for each material. Maintenance costs are calculated by the road and shoulders model of the NCHRP Report 42 [Bertram 67]. Each cost is then discounted to its present value using the input discount rate. The sum of these discounted costs is called the total present worth of cost.

**Optimization**—The main part of the optimization of the SAMP-5 program is in the overlay procedure as explained above. The remainder of the optimization is to sort each initial construction design together with its optimal overlay policy, in order of increasing total cost. Then the 10 best designs are printed out in a summary table, Figure 37.2. The number of pages in the summary table (10 designs per page) is specified by the program user.

## 37.4 PERFORMANCE MODEL

The performance model used in the SAMP programs consists of several equations and a set of inequality restraints. The equations are: (1) a traffic equation, (2) a structural number equation, and (3) performance equations.

**Traffic Equation**—A traffic equation is used to predict the equivalent number of 18-kip axles during the life of the pavement. The equation was originally developed by the Texas Highway Department and is as follows:

$$N_k = \frac{N_c}{C(r_o + r_c)} 2r_o t_k + \frac{r_c - r_o}{C} t_k^2 \qquad (37.1)$$

where

$t$ = time in years since initial construction
$N$ = total number of equivalent applications of 18-kip axle that will have been applied in one direction during time $t$, millions
$C$ = length of the analysis period, yrs.
$N_c$ = $N$ when $t = C$
$N_k$ = $N$ when $t = t_k$
$r_o$ = ADT (one direction) when $t = 0$
$r_c$ = ADT (one direction) when $t = C$
$t_k$ = value of $t$ at the end of the $k$th performance period of the beginning of the next period; $t_o = 0$

**Structural Number Equation**—The structural number equation was taken from the AASHO Road Test [AASHO 61] and is shown below:

$$SN^* = A_1 D_1 + A_2 D_2 + \ldots + A_n D_n \qquad (37.2)$$

where

$SN^*$ = structural number of the pavement
$A_i$ = AASHTO strength coefficient of the $i$th layer
$D_i$ = thickness of the $i$th layer
$n$ = number of layers above subgrade

The load-associated performance equation is a modification of that contained in the 1972 AASHO Interim Guide and may be expressed as follows:

$$P = P_1 - (P_1 - 1.5) \left[ R(N - N_k) \left( \frac{1.051}{(SN + 1)X_j} \right)^{9.3633} \right]^\beta \qquad (37.3)$$

where

$P$ = Present Serviceability Index (PSI at time $t$)
$P_1$ = initial PSI, either after construction, or after an overlay
$N_k$ = total 18-kip equivalent axle loads to the end of the $k$th performance period (where the first performance period is to the time when the first overlay is placed)
$SN$ = structural number
$R$ = regional factor
$X_j$ = $10^{0.03973(S-3)}$
$S$ = soil support value for the subgrade material
$\beta$ = $0.4 + 0.081(19)^{3.23}/(SN + 1)^{5.19}$

The environmental performance equation is based on an approximation of a diffusion model for the swelling clay process and may be expressed as follows:

$$P = P_1 + 0.335 C_1 C_2 (e^{-\theta t} - e^{-\theta t_k}) \tag{37.4}$$

where

$P, P_1$ = as previously defined
$C_1$ = probability of surface activity (that part of the project length, in decimal form, over which environmentally caused roughness is expected to occur)
$C_2$ = maximum differential movement expected on the pavement over a long period of time if the pavement is not overlaid
$\theta$ = surface rise factor, which is roughly proportional to the diffusion constant
$t$ = as previously defined
$t_k$ = time to the end of the $k$th performance period

There is a users guide [Lytton 75] that contains methods for estimating the quantities in equation 37.4.

The total performance equation is a result of adding the serviceability loss due to traffic to the serviceability loss due to environmental factors and may be expressed as follows:

$$P = P_1 - (P_1 - 1.5) \left\{ R(N - N_k) \left[ \frac{1.051}{(SN+1)X_j} \right]^{9.3633} \right\}^{\beta} + 0.335 C_1 C_2 (e^{-\theta t} - e^{-\theta t_k}) \tag{37.5}$$

It is much too cumbersome to develop a numerical example of performance prediction using these equations. They have been included as subroutines in the total SAMP package. It should be noted that the performance prediction in this method covers the entire performance curve, including overlays, over the design period or life of the pavement, as contrasted to the AASHO Interim Guide method itself.

**Inequality Restraints**—The inequality restraints are also from the Final Report, NCHRP Project 11-1 [Van Til 72]. They are:

$$D_1 \geq \frac{SN_2}{A_1} \tag{37.6}$$

$$SN_2^* = A_1 D_1 \geq SN_2 \tag{37.7}$$

$$D_1 \geq \frac{SN_{i+1} - SN^*i}{A_i} \quad i = 2, \ldots, n \tag{37.8}$$

$$SN_{i+1}^* = SN_i^* + A_i D_i \geq SN_i + 1 \quad i = 2, \ldots, n \tag{37.9}$$

where all symbols are previously defined.

## 37.5 INTERPRETATION AND USE OF RESULTS

The results of SAMP-5 have many uses. These vary from the obvious direct design studies and comparisons that can be made to the use of the program to study or investigate the effect on pavement life or performance of various types of materials for the various layers. Program users can doubtlessly find many additional ways to use the program to assist them.

In order to interpret the results, let us look at Figure 37.2. Across the top of the page are the numbers 1 through 10. These are numeric designations for the 10 best possible pavement designs that the program has calculated. These are listed in increasing order of total cost. On the left side of the page are arranged a series of titles of items given in the summary table.

The first item, material arrangement, identifies the exact materials used in each trial and the order in which they are placed in the pavement, first number on top. The next six items represent the various costs and salvage values calculated by the program. The number and depths of layers, shown next, are self-explanatory, with the letter designation of material type printed beside the respective thicknesses.

The number of performance periods reflects the number of times the pavement is renewed by new construction, overlay, or other rehabilitation to a high level of serviceability and the history of serviceability loss to the subsequent overlay. The performance time is the total time from original construction in years to the end of the performance period to renew the pavement and initiate the new performance period.

## 37.6 IMPROVEMENT OF SAMP SYSTEM

As pointed out earlier, continued updating and improvement are not only possible but desirable in pavement management systems. SAMP is a good illustration of this, because an improved version, SAMP-6, is now available. This program modification has grown out of NCHRP Project 1-10A to implement and evaluate SAMP-5. Although

## The SAMP System

the SAMP-5 program uses up to 100 input variables that were thought to cover the range of variables normally considered in the pavement design process, it still needed to be implemented, that is, to be applied to actual pavement design problems. If discrepancies were noted between the program and practice, then the program needed to be changed to reflect as closely as possible the real decision-making process. Full implementation of the computer program required detailed descriptions of how it was to be used, how data was to be input, and how data was to be obtained from the field using data feedback storage systems.

**Objectives**—The primary objectives of this project were the further development of the SAMP-5 program to the field application stage and its pilot testing in one or more state highway departments. It was anticipated that meeting these objectives would involve:

1. Pilot testing SAMP-5, including a sensitivity analysis on one or more state highway departments using the current pavement structural design procedure of the test state as the structural subsystem. It was recognized that considerable development work in pilot testing a pavement design system similar to SAMP-5 was in progress in the Texas Highway Department. However, it was desired that the pilot testing activities of this project would be undertaken in state highway departments other than Texas in order to develop a broader base of experience.
2. Revising the working system as necessary in accordance with the experience gained during pilot testing.
3. Finalizing the SAMP-5 working system as a pavement design and management tool, including the preparation of detailed descriptions for the user's guide, input forms, and data feedback storage systems.
4. Determining research needs in each of the subsystems of SAMP-5, using sensitivity analysis as needed.

The main aim of the project was to pilot test an overall system. The essential element in the pilot test was an in-depth evaluation by the states of the systems approach. The test states ran the SAMP-5 program on their own computers using input variables that were well understood and simple to collect.

**Major Program Revisions**—As a result of the first pilot study cycle, extensive revisions were recommended by each of the states. While these revisions were being made, the states were asked to assemble data for additional runs they might wish to make on the improved computer program. In this phase of the research the objectives were to:

1. Develop SAMP-5 into its final form.
2. Rerun the typical problems and ascertain whether results were as desired.
3. Make selective sensitivity analyses.
4. Develop a user's guide as well as a program documentation check and a dictionary computer code.
5. Develop information on data feedback systems.

```
SAMP6 RUN   = 'SAMP6 EXAMPLE PROBLEM  4 LANE INTERSTATE HIGHWAY (NO SWELLING CLAY)
PROB        = ' 1 MINIMUM TIME TO OVERLAY OF TWO YEARS.

DESIGN TYPE  4, A 4 LAYER DESIGN
MATERIAL ARRANGEMENT A3LM

        EXCLUDING TACK, PRIME, BITUMEN, AND THE SHOULDERS,
        THE MATERIAL LAYER COSTS/(SQ.YD.) ARE . . .

LAYER NO.        ----------MATERIALS----------      -DOLLARS-PER-SQUARE-YARD-
                 CODE  DESCRIPTION                 MINIMUM  MAXIMUM  INCREMENT

    1         A        ASPH.CONC.TYPE 3             0.750    0.750
    2         3        ASPH.CONC.TYPE 3             1.000    6.000
    3         L        LIME STAB.S-C-G              0.778    2.778
    4         M        SELECT MATERIAL              0.222    0.556

       4              THE OPTIMAL DESIGN FOR THE MATERIALS UNDER CONSIDERATION--
                      FOR INITIAL CONSTRUCTION THE DEPTH SHOULD BE
                A     ASPH.CONC.TYPE 3          1.50     INCHES
                3     *ASPH.CONC.TYPE 3         8.00     INCHES
                L     LIME STAB.S-C-G           4.00     INCHES
                M     SELECT MATERIAL           4.00     INCHES

       THE LIFE OF THE INITIAL STRUCTURE = 8.6 YEARS  STRUCTURAL NUMBER 4.46
       THE OVERLAY SCHEDULE IS
              1.00 INCH(ES)  (EXCLUSIVE OF LEVEL-UP AND WEAR-COURSE
                                                      AFTER 8.6 YEARS.
       THE TOTAL LIFE = 22.6 YEARS

       THE TOTAL COSTS PER SQ.YD. FOR THESE CONSIDERATIONS ARE
              INITIAL CONSTRUCTION COST            7.653
              TOTAL ROUTINE MAINTENANCE COST       0.453
              TOTAL OVERLAY CONSTRUCTION COST      0.781
              TOTAL USER COST DURING OVERLAY
                  CONSTRUCTION                     0.028
              SALVAGE VALUE                       -1.135
              TOTAL OVERALL COST                   7.780

SAMP6 PROGRAM ACTIVITY REPORT, DESIGN TYPE A3LM
  INITIAL DESIGNS
       72      WITHIN COST AND THICKNESS CONSTRAINTS
       45      FEASIBLE TO FIRST OVERLAY
  OVERLAYS
      163      CONSIDERED
       98      FEASIBLE
       72      FEASIBLE OVERLAY POLICIES
  COMPLETE DESIGNS
       43      FEASIBLE
```

**Figure 37.4** Sample output summary of an optimum design strategy for a four-layer system. (Note: These costs are for 1975. The user is required to input current costs.) [Lytton 75].

## 37.7  IMPLEMENTATION STUDY

As a result of the program revisions outlined above, SAMP-6 was created. The project team then assisted Kansas, Florida, and Louisiana with running the program for typical pavement design problems using data collected by their own state personnel. This was assumed to be a realistic test of implementation of a pavement management system.

The researchers concluded the implementation was successful [Lytton 75]:

> The major finding of this project is that the SAMP-6 computer program is a working implementable systems analysis model for pavements . . . The states in which the com-

SAMP6 RUN = 'SAMP6 EXAMPLE PROBLEM 4 LANE INTERSTATE HIGHWAY (NO SWELLING CLAY)
PROB = '1 MINIMUM TIME TO OVERLAY OF TWO YEARS.
PROBLEM SUMMARY OF THE BETTER FEASIBLE DESIGNS
IN ORDER OF INCREASING TOTAL COST

| | 1 | 2 | 3 | 4 | 5 | 6 | 7 | 8 | 9 | 10 |
|---|---|---|---|---|---|---|---|---|---|---|
| MATERIAL ARRANGEMENT | A3 | A3 | A3LM | A3LM | A3L | A3L | A3LM | A3L | A3LM | A3LM |
| INIT. CONST. COST | 6.049 | 7.215 | 7.653 | 7.349 | 7.296 | 7.808 | 7.831 | 8.074 | 7.527 | 8.009 |
| OVERLAY CONST. COST | 2.302 | 0.680 | 0.781 | 1.533 | 1.338 | 0.818 | 0.742 | 0.781 | 1.457 | 0.701 |
| USER COST | 0.082 | 0.026 | 0.028 | 0.055 | 0.050 | 0.029 | 0.027 | 0.028 | 0.053 | 0.026 |
| ROUTINE MAINT. COST | 0.115 | 0.515 | 0.453 | 0.231 | 0.400 | 0.480 | 0.452 | 0.453 | 0.276 | 0.484 |
| SALVAGE VALUE | -1.101 | -0.965 | -1.135 | -1.300 | -1.202 | -1.252 | -1.169 | -1.390 | -1.334 | -1.202 |
| TOTAL COST | 7.446 | 7.471 | 7.780 | 7.867 | 7.882 | 7.883 | 7.883 | 7.946 | 7.979 | 8.019 |
| NUMBER OF LAYERS | 2 | 2 | 4 | 4 | 3 | 3 | 4 | 3 | 4 | 4 |
| LAYER DEPTH (INCHES) | | | | | | | | | | |
| D(1) | 1.50 | 1.50 | 1.50 | 1.50 | 1.50 | 1.50 | 1.50 | 1.50 | 1.50 | 1.50 |
| D(2) | 8.00* | 10.00* | 8.00* | 6.00* | 8.00* | 6.00* | 8.00* | 4.00* | 6.00* | 8.00* |
| D(3) | | | 4.00 | 8.00 | 4.00 | 12.00 | 4.00 | 20.00 | 8.00 | 4.00 |
| D(4) | | | 4.00 | 4.00 | | | 6.00 | | 6.00 | 8.00 |
| STRUCTURAL NUMBER | 3.86 | 4.66 | 4.46 | 4.10 | 4.30 | 4.38 | 4.54 | 4.46 | 4.18 | 4.62 |
| NO. OF PERF. PERIODS | 4 | 2 | 2 | 3 | 3 | 2 | 2 | 2 | 3 | 2 |
| PERF. TIME (YEARS) | | | | | | | | | | |
| T(1) | 3.3 | 11.4 | 8.6 | 4.9 | 6.8 | 7.6 | 9.6 | 8.6 | 5.6 | 10.8 |
| T(2) | 9.8 | 28.6 | 22.6 | 14.0 | 18.4 | 20.4 | 25.0 | 22.6 | 15.6 | 27.5 |
| T(3) | 15.5 | | | 21.7 | 21.6 | | | | 24.1 | |
| T(4) | 20.7 | | | | | | | | | |
| OVERLAY POLICY (INCH) EXCLUSIVE OF LEVEL-UP & WEAR-COURSE | | | | | | | | | | |
| O(1) | 1.00 | 1.00 | 1.00 | 1.00 | 1.00 | 1.00 | 1.00 | 1.00 | 1.00 | 1.00 |
| O(2) | 1.00 | | | 1.00 | 1.00 | | | | 1.00 | |
| O(3) | 1.00 | | | | | | | | | |

**Figure 37.5** Sample output summary of the better overall design. (Note: These costs are for 1975. The user is required to input current costs.) [Lytton 75].

puter program was pilot tested expect to use it in their design system: Louisiana for design; Florida, for design studies and as a building block for a future, more mechanistically oriented design system; and Kansas, as a research tool and supplement to their current design system. There were no problems encountered in interfacing computer programs between states.

In the NCHRP project SAMP-6 was modularized into distinct subsystems, which can be reprogrammed with a minimum of effort. In addition, a detailed user's guide, a program documentation deck, and flow charts of the program were prepared for subsequent use. All of these documents point to the practicality of the pavement management system approach, and the reader interested in more detail is urged to read NCHRP Report 160 and its supplement, which can be obtained from the Transportation Research Board in Washington, D.C.

**The SAMP-6 Computer Program Input**—The SAMP-6 program requires 12 classes of input variables:

1. Program control and miscellaneous input
2. Environment and serviceability variables
3. Traffic and reliability variables
4. Constraint variables
5. Traffic delay variables
6. Maintenance variables
7. Cross-section, cost model, and shoulder variables
8. Tack coat, prime coat, and bituminous materials variables
9. Wearing surface variables
10. Overlay variables
11. Pavement material variables
12. Shoulder layer material variables

Figure 37.4 illustrates a sample output of an optimum design strategy and Figure 37.5, a sample output summary of the best overall designs from a sample SAMP-6 run.

**Continuing Changes**—A major finding of NCHRP 1-10A was the need to improve the structural subsystems of the SAMP programs. As a result, Project 1-10B was initiated and sounder mechanistically based structural subsystems developed for inclusion into the SAMP programs [Finn 77]. There is an ongoing need to implement research findings into the SAMP programs. For example, the programs should be updated to use the models in the current AASHTO Pavement Design Guide [AASHTO 86].

## REVIEW QUESTIONS

1. Outline the basic concepts and details of SAMP-5 in a simple and understandable way.
2. What does the "5" mean in the title SAMP-5? How does SAMP-5 differ from SAMP-4, for example?
3. What do you see as the basic strengths of SAMP-5?
4. What do you see as the basic weaknesses of SAMP-5?

Chapter 38

# The Highway Design and Maintenance Standards Model

## 38.1 BACKGROUND

The Highway Design and Maintenance Standards Model (HDM), developed by the World Bank, is a versatile tool for studying the economic viability of alternative road improvement and maintenance strategies. HDM uses empirical relationships for estimating pavement performance and vehicle operating costs. These relationships were developed from data collected by the Transportation and Road Research laboratory in Kenya and by GEIPOT and Texas Research and Development Foundation in Brazil. Other studies which provide data for the development and calibration of HDM were conducted in India and the Caribbean. The World Bank developed this current version of HDM.

HDM can be used at both the network and the project level. It is well suited to the study of economic benefits of strengthening pavement, making relatively minor geometric improvements, and providing a smoother riding surface. The basic concept and structure of the model is summarized as follows [Watanatada 87].

## 38.2 DESCRIPTION OF THE MODEL

The broad concept of the HDM model, as illustrated in Figure 38.1, is quite simple. Three interacting sets of costs are added together over time and discounted to present

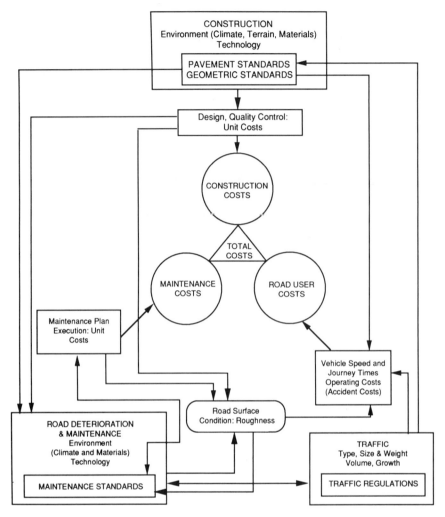

**Figure 38.1** The HDM model: Interaction of costs of road construction maintenance and use [Watanatada 87].

values. Costs are determined by first predicting physical quantities of resource consumption which are then multiplied by unit costs or prices. The cost factors considered in the model are highway construction and maintenance costs and road user costs.

Vehicle speed is a major determinant of vehicle operating costs. Speed is estimated through complex probabilistic functions relating to road geometric design, surface condition, vehicle type, and driver behavior.

The HDM model is used to make comparative cost estimates and economic evaluations of various policy options, including different time-staging strategies, either for a given road project on a specific alignment or for groups of links on an entire

# The Highway Design and Maintenance Standards Model 497

network. It can quickly estimate the total costs for large numbers of alternative project designs and policies, year by year, for up to 30 years. It can discount the future costs if desired at different postulated interest rates, so that the user can search for the alternative with the lowest discounted total cost. Alternatively, the user can request comparisons in terms of rate of return, net present value, or first year benefit. Another capability, using the Expenditure Budgeting Model, is finding the set of design and maintenance options that would minimize total discounted transport costs or maximize net present value of an entire highway system under multiyear budget constraints.

In addition to comparing alternatives, the model can be used to analyze the sensitivity of the results to changes in assumptions about key variables such as unit cost, traffic growth rates, the discount rate, and the value of passengers' time. Table 38.1 is a general summary of the scope of the model in terms of input requirements, limits, and the outputs.

As seen from Table 38.1 in a single computer run, the model can evaluate up to 20 different road links, each having up to 10 sections with different design standards and environmental conditions. Each link can have a different traffic volume. Further, different maintenance standards can be implemented on different sections. At any time, any section can be upgraded (e.g., from earth to gravel or from gravel to paved) and the road can be realigned or widened. Altogether, up to 50 alternatives can be compared in one run.

In order to make these comparisons, of course, the model must be given detailed specifications of the various alternative sets of construction programs, design standards, and maintenance and other policies to be analyzed, together with unit costs, projected traffic volumes, and environmental conditions. Since there is always the possibility of error in coding these inputs, the model includes an extensive checking program which examines the inputs for format errors and internal inconsistencies. Warning messages are automatically produced when such errors or inconsistencies are found, or when the program is requested to extrapolate relationships beyond their empirically validated range.

Once any apparent input errors have been corrected, the model estimates speeds and resource consumption of the vehicles as well as road deterioration and resources for maintenance for all the combinations. The resource requirements for road construction for each design option may be endogenously estimated in the model or may be directly specified by the user if more specific information or local data exists. After physical quantities involved in construction, maintenance, and vehicle operation are estimated, user-specified prices and unit costs are applied to determine financial and economic costs. Comparisons in terms of relative benefits, present value, and rate-of-return calculations then follow. The user has a wide range of options in specifying what results are included in the printed report. Modeling of a link alternative is illustrated in Figure 38.2.

Because some of the model relationships have highly complex nonlinear forms, simulation, rather than any formal optimization technique, is employed in the HDM model itself, and the "optimization" which takes place in that model is merely the selection of the group of alternatives with the highest discounted net benefits among those specified by the user. There is the possibility that an untried policy combination

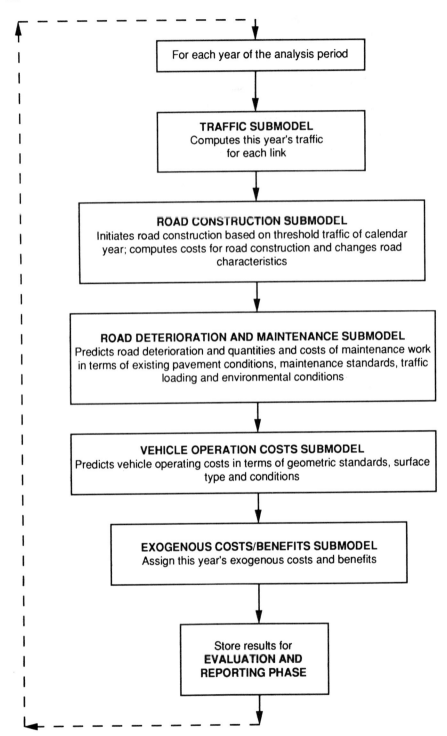

**Figure 38.2** Simulation of link alternative [Watanatada 87].

could exist which would provide superior results to any of those specified for analysis. However, the ease of specifying and analyzing large numbers of alternatives reduces the practical importance of this limitation.

### 38.2.1 Background of the Models

In both Kenya and Brazil, an initial step was to measure the permanent characteristics of each test section. These included the geometric characteristics such as width, rise, fall, and curvature. More importantly, the properties of the pavement layers were determined. These included measurement of strength, layer thickness, particle size distribution, density, moisture content, and plasticity. The basic pavement strength indicator in both studies is the structural number, $SN$, as conceptually derived during the AASHO Road Test. There were slight differences in the evaluation of $SN$ in the Kenya and the Brazil study. In the latter, structural coefficients for bound materials were based on compressive strength for cement-stabilized materials and resilient modulus for asphalt concrete and asphaltic bound materials. In both studies, however, layer coefficients for natural soils and gravel, as well as subgrade contribution to $SN$, are based on the CBR test.

The condition of each test section was regularly monitored during the respective studies. Again, both studies assessed the same type of parameters, namely cracking, potholes, rut depth, roughness, and deflection. But in this respect, there was a notable divergence of instrumentation and of modeling concepts, particularly in regard to the definition and analysis of cracking and potholes, and in the definition of road roughness and instrumentation for measuring it.

This eventually led to considerable differences in the formulation of performance prediction equations. The Kenya equations are characteristically of a continuous function type, each distress function being independent of other distress types; in other words, a parallel development in all distress modes. This makes for relatively straightforward functional forms, but it misses some of the causalities involved in road deterioration. The Brazil study builds on the causality of events, but in so doing introduces formulating discontinuities, which increase the computational effort.

### 38.2.2 Results on Road Deterioration and Maintenance

The Kenya and Brazil studies have advanced our ability to predict road deterioration under normal maintenance regimes, and in particular have provided quantitative relationships that give reasonable results when extrapolating over the life of a road. The effects of alternative maintenance policies have been well quantified for unpaved roads, and reasonably quantified for paved roads, except that the longer term effects of repeated maintenance on subsequent deterioration need further research.

For paved roads, strong relationships have been developed from the Brazil study for original pavements under normal maintenance. Important features of the relationships are their incremental form and inclusion of both traffic and time variables. This permits evaluation of the marginal effects of a vehicle transit, and of time and climate which are important issues in user taxation. From the Brazil study, both roughness and cracking predictions have developed beyond simple correlation models, and incorporate most major mechanistic effects. In the cracking and ravelling relationships,

**Table 38.1 HDM-III Model Inputs and Outputs [Watanatada 87]**

| Inputs | Input limits |
|---|---|
| Link characteristics (existing road and environmental factors) | 20 links |
| Construction projects and costs (widening or new construction standards for assigning to links) | 50 projects with maximum duration of 5 years for any one project. |
| Maintenance standards and unit costs (intervention criteria, properties and costs for assigning to links) | 30 standards |
| Vehicle fleet characteristics and unit costs (common to all link-groups) | 9 vehicle types |
| Traffic volumes, distribution and growth (sets for assigning to links) | 20 traffic sets |
| Exogenous costs and benefits (sets for assigning by link) | 20 sets |
| Link alternatives (assign to links the above construction and maintenance standards, traffic and exogenous C-B sets) | 100 |
| Group alternatives (assign link-alternatives to link-groups) | 100 group alternatives involving not more than 20 groups or 100 link-alternatives |
| Number of studies, economic comparisons and sensitivity analysis (defines groups to be compared and type of analysis) | Up to 5 studies with maximum group comparisons of 50 with the number of alternatives compared not to exceed 200; 5 discount rates (in addition to zero) per study. |
| Report requests (uniform per run) | Maximum of 500 reports |
| Analysis period (uniform per run) | Up to 30 years with the product of link alternatives and number of analysis years not to exceed 800. |

**Table 38.1 Continued**

| Outputs |
| --- |
| Road maintenance summary (by link or group) |
| Annual road maintenance costs and physical quantities (by link or group) |
| Annual traffic (link only) |
| Annual road conditions (link only) |
| Annual road user costs and physical resources consumption (link only) |
| Financial costs of alternative (link or group) |
| Economic and foreign exchange costs of alternative (link and group) |
| Comparison of costs of alternatives (link and group) |
| Summary of comparison of alternatives by discount rate (link and group) |
| Summary of costs and comparisons by discount rate (link alternatives only) |

however, construction practices and material properties, not easily quantified for network analysis, were found to have strong effects, and local linear calibration of these models is recommended.

The effects of maintenance on the rate of paved road deterioration, and its initial effect on condition, are as yet only moderately well quantified. In the Brazil study, the major differences in behavior before and after maintenance were in cracking but, apart from initial effects quantified in the models, any significant differences in roughness progression were well quantified through the change in strength parameters. Recent follow-on studies in Kenya, however, have shown very strong reduction in the progression of roughness following multiple reseal applications under apparently negligible changes of strength. Further study is thus required on long-term effects over several successive maintenance phases.

The effects of the limited range of observed rainfall in both the Kenya and Brazil studies were negligible on paved roads. On unpaved roads the Kenya study showed small effects on gravel loss, but the Brazil study showed small effects on both roughness and gravel loss. The relatively small rainfall effect in the paved road relationships may not extend to high rainfall, or low-intensity rainfall climates, and requires further study. An important related effect was that the amount of the cracking was found to affect rut depth and roughness progression, but the effects may be understated for situations where the pavement layers become saturated.

### 38.2.3 Calibration of the Models

As pointed out in the previous subsection, certain general relationships, such as those regarding behavior of materials, operation of vehicles, and the effect of surface roughness on vehicle operating speeds and costs, can be applied in different countries of similar climates. However, a degree of calibration will almost always be required for some parameters of the model in order for the model to simulate reasonably well the conditions in a particular country.

It is advisable that effort should be dedicated to make adjustment at least to the pavement deterioration factors in the HDM III program which increase or decrease the different distress manifestations to simulate the local conditions.

## 38.3 CONCLUSION

In addressing such a broad and complex set of phenomena over the worldwide diversity of conditions it must be recognized that some factors are better understood, and some relationships are better determined, than others. Life-cycle cost modeling is still in its infancy, and further research will be needed: (1) to refine and strengthen the validation of the various prediction models, and to encompass other important phenomena, particularly traffic congestion, and (2) to evaluate the transferability of general model forms and further broaden their empirical validation for diverse physical and economic environments.

## REVIEW QUESTIONS

1. Describe the background and development concepts of HDM III. Where did the basic data for development derive from?
2. Is HDM III a network or a project level system?
3. Is HDM III a traditional pavement management system? Does it have more capabilities or fewer capabilities than the traditional system?

Chapter 39

# Municipal Pavement Design System

## 39.1 INTRODUCTION

Most project level pavement management systems were developed for state or provincial applications. In 1985, a prototype project level pavement management system was developed for the City of Austin, Texas. This procedure combined the systematic concepts of FPS, RPS, and SAMP with the design practices used by the Texas State Department of Highways and Public Transportation. During the development of this program, emphasis was placed on providing input and output formats that were suitable for use by the city engineers [Seeds 86]. During 1989 and 1990 ARE Inc. engineers, in an internally funded project, modified the system to use the structural design models from the AASHTO Guide for the Design of Pavement Structures. The resulting system, named the Municipal Pavement Design System, MPDS-1, provides engineers in local agencies with a project level pavement management system that uses a state-of-the-art pavement design method. Since the basic structure of the model is similar to the systems previously described, the following discussion focuses on applications of the system rather than describing the details of the model.

## 39.2 CHARACTERISTICS OF MPDS-1

While the structure of MPDS-1 is similar to FPS and RPS, it is also more advanced than these systems in that it combines the analysis and design of both flexible and rigid pavements in one design system. Use of the model is similar to the other models in that the user is required to generate inputs for the model, perform the analysis and review the resulting designs, and select the best one from a series of options that were analyzed.

Input to the model is accomplished through a series of interactive screens. As the data is input, the program performs logic and limit checks to assist the engineer with the proper entry of the data. This is an improvement over FPS and RPS which were developed for batch mode operations. The types of input data required are similar to either RPS or FPS depending on the type of pavement being designed.

During the analysis, MPDS-1 generates a series of alternative designs based on the range of allowable thicknesses for each layer. The program starts with the minimum thickness of pavement and incrementally increases the surface thickness by an amount specified by the user. Each design is checked against constraints on the maximum initial cost and thickness. The initial service life of each alternative is analyzed and checked against a constraint. An overlay analysis is performed for each pavement which passes the initial life constraint. In addition to overlays, MPDS-1 will consider the application of a seal coat as a preventive maintenance measure prior to the initial pavement or overlay reaching the minimum acceptable level. Figure 39.1 demonstrates the performance of three strategies that would automatically be generated during a MPDS-1 analysis.

MPDS-1 has an internal constraint on the designs that are analyzed by the program. Iterations of increasing the thickness of a layer are terminated once a design reaches the required service life, e.g., if a pavement with a 3 inch surface layer will survive the required design life, then the program will not consider a 3.5 inch surface layer. The additional thickness increases costs without increasing the life of the pavement considered in the analysis.

As the program is analyzing each alternative, the costs of the alternative are computed. The program determines costs of initial construction, routine maintenance, seal coats, overlays, and user delay during seal coats and overlay construction. Figure 39.2 is an example of the costs associated with a particular design strategy. All future costs are discounted to present value and the total present worth of costs is calculated.

The output of the program is a table with several design strategies, sorted in order of least present worth of costs. The design engineer can then select the best strategy. Engineering judgment can be used in the selection of the strategy that best meets the design requirements. This may not be the design that has the lowest total present worth of costs, but at least the engineer is making an informed decision.

## 39.3 MPDS-1 EXAMPLE

The following example is for the design of a "typical" city flexible pavement. The seal coat policy applies the treatment when the serviceability is 3.2 and extends the

# Examples of Working Systems

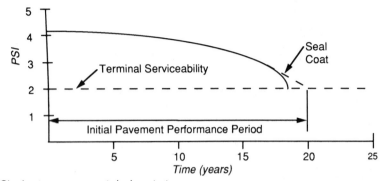

a) Single stage pavement design strategy

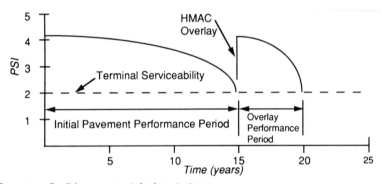

b) Two-stage flexible pavement design strategy.

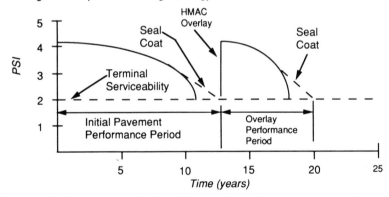

c) Two-stage flexible pavement design strategy with effect of seal coats considered.

**Figure 39.1** Illustration of alternative pavement design strategies.

# Municipal Pavement Design System

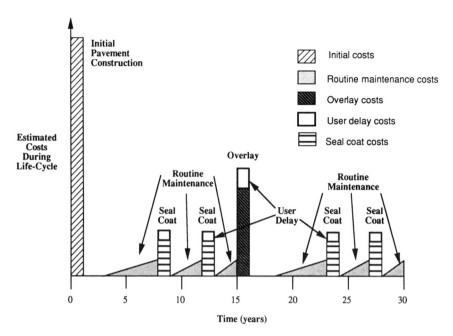

**Figure 39.2** Illustration of life-cycle costs associated with a given pavement design strategy [Seeds 86].

life of the pavement or overlay for two years. All overlays are assumed to be 2 inches thick and are applied when the pavement reaches a serviceability of 2.0.

The ranges of layer thicknesses considered are:

| Thickness | Asphalt Concrete | Flexible Base | Flexible Subbase |
|---|---|---|---|
| Minimum | 1.5 in. | 4.0 in. | 4.0 in. |
| Maximum | 4.0 in. | 6.0 in. | 8.0 in. |
| Increments | 0.5 in. | 1.0 in. | 2.0 in. |

This produces a total of 54 combinations of initial pavement designs that are considered by MPDS-1. Table 39.1 is a portion of the feasible strategies that would be generated for the allowable combinations of layers. Of the 10 strategies presented, 5 do not meet the initial condition constraints, and 3 meet the initial condition constraints and are analyzed for an overlay. The remaining 2 strategies are not analyzed since a thinner, lower cost pavement meets the performance requirements.

Table 39.2 presents the cost analysis for the three feasible designs from Table 39.1. This analysis shows the pavement with the lowest initial cost also has the lowest total present worth of costs even though it requires an overlay after 13 years.

Table 39.1  Example of Strategy Generation Process of Pavement Structural Design

| Strategy No. | Initial Pavement Layer Thickness (in) | | | Initial Pavement Performance Period (years) | | 2-in AC Overlay Performance Period (years) | | Life of Strategy (years) | Remarks |
|---|---|---|---|---|---|---|---|---|---|
| | Asphalt Concrete | Flexible Base | Flexible Subbase | Before Seal Coat(s) | After Seal Coat(s) | Before Seal Coat(s) | After Seal Coat(s) | | |
| 1  | 1.5 | 4 | 4 | 3  | —  | —  | — | 3  | Discarded—1, 2, 4 |
| 2  | 1.5 | 4 | 6 | 4  | —  | —  | — | 4  | Discarded—1, 2, 4 |
| 3  | 1.5 | 4 | 8 | 6  | 8  | —  | — | 8  | Discarded—2, 4 |
| 4  | 1.5 | 5 | 8 | 7  | 9  | —  | — | 16 | Discarded—3, 4 |
| 5  | 1.5 | 6 | 8 | 8  | 10 | —  | — | 17 | Discarded—4 |
| 6  | 2.0 | 6 | 8 | 11 | 13 | 5  | 7 | 20 | Feasible Strategy |
| 7  | 2.5 | 6 | 8 | 15 | —  | 5  | 7 | 20 | Feasible Strategy |
| 8  | 3.0 | 6 | 8 | 18 | 20 | 5  | — | 20 | Feasible Strategy |
| 9  | 3.5 | 6 | 8 | —  | —  | —  | — | —  | Discarded—5 |
| 10 | 4.0 | 6 | 8 | —  | —  | —  | — | —  | Discarded—5 |

Notes:
1. Initial pavement did not reach seal coat year.
2. Initial pavement did not survive until first overlay.
3. Overlay did not reach seal coat year.
4. Strategy did not last analysis period.
5. Overdesign of initial pavement.

Table 39.2 Comparison of Three Flexible Pavement Design Strategies

| | | Design Strategy | | |
|---|---|---|---|---|
| Design/Cost Factor | | 5 | 6 | 7 |
| Initial Pavement | Performance Period (years) | 13 | 15 | 20 |
| | HMAC Surface Thickness (in) | 2.0 | 2.5 | 3.0 |
| | Flexible Base Thickness (in) | 6 | 6 | 6 |
| | Subbase Thickness (in) | 8 | 8 | 8 |
| | Seal Coat Year(s) | 11 | – | 18 |
| HMAC Overlay | Performance Period (years) | 7 | 5 | – |
| | Thickness (in) | 2 | 2 | – |
| | Seal Coat Year(s) | 18 | – | – |
| Present Value of Costs ($/SY) | Initial Construction | 30.84 | 33.20 | 35.55 |
| | Initial Maintenance | 0.40 | 0.47 | 0.61 |
| | Initial Seal Coat(s) | 0.47 | – | 0.33 |
| | Initial Pavement Salvage Value | –2.91 | –3.13 | –3.55 |
| | Overlay Construction | 3.45 | 3.13 | – |
| | Overlay Maintenance | 0.19 | 0.13 | – |
| | Overlay Seal Coat(s) | 0.33 | – | – |
| | Overlay Salvage Value | –2.45 | –2.45 | – |
| | Total Net Present Value | 30.32 | 31.34 | 33.14 |

## 39.4 SUMMARY

This chapter demonstrates that project level pavement management is available to local agencies. Initial use of this method requires more data collection and input than is required for traditional design methods. However, this is a one time investment to start the use of the systems method. Based on careful selection of inputs and thorough review of the output for different pavement design projects, users of the procedure should become confident about the cost-effectiveness of the use of available funds. If this practice is implemented every time a pavement is constructed, reconstructed, or even overlayed, the local agency would ultimately get to the point where it would be getting the most efficient use of its available funds over the entire network.

## REVIEW QUESTIONS

1. How does municipal pavement management differ from state or national level pavement? Please list at least three differences.
2. At the municipal level, which is more important: network level or project level?

# References to Part Six

[AASHO 61]     American Association of State Highway Officials, "AASHO Interim Guide for the Design of Pavement Structures," 1961.

[AASHTO 86]    American Association of State Highway and Transportation Officials, "AASHTO Guide for Design of Pavement Structures," Volumes 1 and 2, Washington, D.C., 1986.

[Bertram 67]   Bertram D. Tallamy Associates, "Interstate Highway Maintenance Requirements and Unit Maintenance Expenditure Index," NCHRP Report 42, 1967.

[Butler 73]    Butler, L., and H. Orellano, "Implementation of a Complex Research Development of Flexible Pavement Design System into Texas Highway Department Design Operations," Research Report 123-20, Texas Highway Department, 1973.

[Finn 77]      Finn, F. N., C. Saraf, R. Kulkarni, K. Nair, W. Smith, and A. Abdullah, "The Use of Distress Prediction Subsystem for the Design of Pavement Structures," Proc., Fourth Intl. Conf. on Structural Design of Asphalt Pavements, Univ. of Michigan, 1977.

[Haas 85]      Haas, R. C. G., "Minnesota's Pavement Management System: Implementation Recommendations," Report prepared for Minnesota Department of Transportation, 1985.

# References to Part Six

[Hill 90] Hill, L., and R. C. G. Haas, "Module E: Multi-Year Prioritization", Developed for FHWA Advanced Course on Pavement Management Systems, June 1990.

[Hudson 68] Hudson, W. R., F. N. Finn, B. F. McCullough, K. Nair, and B. A. Vallerga, "Systems Approach to Pavement Design System Formulation, Performance Definition, and Material Characterization," Final Report, NCHRP Project 1-10, 1968.

[Hudson 70] Hudson, W. R., B. F. McCullough, F. H. Scrivner, and J. L. Brown, "A Systems Approach Applied to Pavement Design and Research," Research Report 123-1. Center for Highway Research, University of Texas at Austin, 1970.

[Hudson 71] Hudson, W. R., B. F. McCullough, and R. K. Kher, "A Sensitivity Analysis of Flexible Pavement System FPS2," Research Report 123-8, Center for Highway Research, University of Texas at Austin, 1971.

[Hudson 73a] Hudson, W. R., R. K. Kher, and B. F. McCullough, "A Systems Analysis of Rigid Pavement Design," Research Report 123-5, Center for Highway Research, University of Texas at Austin, 1973.

[Hudson 73b] Hudson, W. R., and B. F. McCullough, "Flexible Pavement Design and Management Systems Formulation," NCHRP Report 139, 1973.

[Jung 70] Jung, F. W., R. K. Kher, and W. A. Phang, "OPAC, A Performance Prediction Subsystem, Flexible Pavement," Research Report 200, Ontario Ministry of Transportation and Communications, 1970.

[Kher 71] Kher, R. K., "A Systems Analysis of Rigid Pavement Design," Ph.D. dissertation, University of Texas at Austin, 1971.

[Kher 74] Kher, R. K., and W. A. Phang, "OPAC, Economic Analysis Elements," Research Report 201, Ontario Ministry of Transportation and Communications, 1974.

[Kulkarni 80] Kulkarni, R. G., K. Golabi, F. Finn, and E. Alviti, "Development of a Network Optimization System," Woodward-Clyde Consultants, San Francisco, 1980.

[Lytton 75] Lytton, R. L., W. F. McFarland, and D. L. Schafer, "Flexible Pavement Design and Management Systems Approach Implementation," NCHRP Report 160, 1975.

[Maurer 87] Maurer, F. V., and E. E. Ofstead, "Minnesota's Pavement Management System: How it Came About and the Steps Taken," Proceedings, Second North American Conference on Managing Pavements, Vol. 3, Toronto, 1987.

[MTC 86] *MTC Pavement Management System User's Guide*, Metropolitan Transportation Commission, 1986.

[Nair 73] Nair, K., and C. Y. Chang, "Flexible Pavement Design and Management Materials Characterization," NCHRP Report 140, 1973.

[PMS Ltd. 89] Pavement Management Systems Limited, "Minnesota's Department of Transportation Pavement Management System: System Documentation," Report prepared for MNDOT, 1989.

[Scrivner 68]   Scrivner, F. H., W. F. McFarland, and G. R. Carey, "A Systems Approach to the Flexible Pavement Design Problem," Research Report 32-11, Texas Transportation Institute, Texas A&M University, 1968.

[Seeds 86]   Seeds, S. B., R. F. Carmichael, III, and B. F. McCullough, "Municipal Pavement Structural Design and Life-Cycle Cost Analysis System," User's Manual for MFPS-1 and MRPS-1, Developed for the Transportation and Public Services Department, City of Austin, Texas, 1986.

[Smith 88]   Smith, Roger E., et al., "A Comprehensive Ranking System for Local Agency Pavement Management," Transportation Research Record 1123, Transportation Research Board, 1988.

[Van Til 72]   Van Til, C. J., B. F. McCullough, B. A. Vallerga, and R. G. Hicks, "Evaluation of AASHO Interim Guides for Design of Pavement Structures," NCHRP Report 128, 1972.

[Watanatada 87]   Watanatada, T., C. G. Harral, W. D. O. Paterson, A. M. Dhareshwar, A. Bhandari, and K. Tsunokawa, "The Highway Design and Maintenance Model: Description of the HDM-III Model," The Highway Design and Maintenance Standards Series, Volumes I and II, World Bank, Transportation Department, Washington, D.C., 1987.

Part Seven

# Looking Ahead

Chapter 40

# Analyzing Special Problems

## 40.1 INTRODUCTION

The pavement management process provides engineers and administrators with the means for dealing with special problems that occur from time to time. A wide variety of problems can be handled by having a structured pavement management system which includes engineering, economics, environmental effects, and other factors. This chapter provides examples of some of these special problems, and their treatment within a modern, broadly based pavement management system.

## 40.2 HISTORICAL PROBLEMS

One of the problems that has long plagued highway pavement engineers has been how to adapt empirical design methods to a changing situation. In the 1930s crude pavement design methods were used based on soil index numbers of various kinds. Those were useful design approaches, but when abnormal situations occurred, not covered by the average conditions represented, failures resulted from the designs.

During and following World War II the CBR pavement design method was developed and used extensively for highway pavements and for heavy military airfields. The method, after several adjustments through the early years, worked fairly well for the conditions that governed its development. In general, the CBR method was based on preventing subgrade failure under average traffic conditions, which were usually low to medium volumes. For heavy traffic conditions, failures developed and traffic adjustment factors had to be developed.

More dramatic, however, was the inadequacy of the method to account for new materials. The CBR method does not give credit for greater strength of overlying materials. It requires only a "minimum thickness of cover" regardless of whether the overlying material was crushed stone or asphalt-stabilized material. This inadequacy of the method is symptomatic of the problems of applying seemingly "simple" methods to complex problems and is not intended to downgrade the CBR method, which was and is serving well in many areas.

Of course, no single design method is adequate to account for all the factors involved in pavements. That is one of the basic reasons for developing pavement design systems in the first place. For example, the design method developed within the Texas Highway Department in the 1950s using a simplified triaxial test was based on a more rational evaluation of materials strength than CBR. The method became known as the Texas Triaxial Design Method and was used to design many miles of pavement from 1954 to 1965. It was not originally capable of considering load repetitions or stabilized materials and thus had to be modified as traffic volumes increased. In trying to evaluate the method and improve it in light of AASHO Road Test conditions it was deduced that a broader concept was needed and a simple pavement design system was born [Scrivner 68a, 68b].

## 40.3 ENERGY ISSUES

Since 1974 the United States and many of the major countries of the world have come face to face with several major energy shortages. The prospect of continuing cheap energy is not good, particularly petroleum. Moreover, there is a decreasing acceptability of the environmental implications of excessive energy use. As a corollary, the availability of low cost asphalt is over, because refinery technology can now fractionate asphalt to obtain more gasoline and other lighter oil products. In addition to the basic cost of the asphalt, a great deal of energy is used in producing highway construction materials. Aggregates require crushing, drying, and hauling, all of which consume energy. Portland cement requires high temperature burning and then grinding and transportation. Steel requires heat for production and processing. The preparation of asphalt concrete requires heating for drying, mixing, or distribution. Cutback asphaltic materials all use some type of petroleum distillate as a carrier and thus are seldom used. Emulsions are less energy intensive but they have other undesirable trade-offs which must be carefully evaluated. In order to assess the implications of these energy issues, it is necessary to make a thorough and intensive economic analysis of all factors for the many options open to the designer or manager, rather than rule-of-thumb decisions.

A pavement management system provides the designer with the information and analysis needed to make correct decisions and to evaluate changes required in existing pavement maintenance and rehabilitation strategies. For example, in some cases the consumption of fuel or energy itself is more important than cost alone. It is possible to evaluate energy utilization in the management system by considering energy consumed in lieu of cost (e.g., energy cost instead of dollar cost). In some countries, Brazil for example, the consumption of energy is the overriding concern in highway investment decisions rather than the present value of money cost.

Agencies managing pavements should carefully evaluate their old rules of thumb in the face of major shifts like those we have seen with regard to energy. Policies should be reevaluated with respect to: (1) when to use rigid pavements, (2) when to reseal pavement surfaces, and (3) when to regravel unpaved roads.

## 40.4 ALTERNATE SOURCES OF MATERIALS

How does one evaluate a new source of materials? For the past few decades most agencies have had quite rigid specifications which required the use of only good or high quality materials. This was probably considered essential to providing high quality pavements. However, in many places high quality materials are difficult to find, and the cost of transportation and energy consumption is excessive. Techniques are required to evaluate alternative sources of materials and alternative design and construction methods with these materials.

It is no longer feasible to label a material only as "acceptable" or "unacceptable." Rather, its performance potential and cost must be evaluated. In this way, optimum cost-benefit relationships can be achieved. A pavement management system provides all the tools necessary to carry out the evaluation. Furthermore, as the feedback portion of the management system is implemented, the true performance capability of alternative materials will be readily obtainable with statistical confidence.

## 40.5 NEW TYPES OF PAVEMENTS AND NEW MATERIALS

The past few decades have been a period of rapid technological change. New materials and pavement types have emerged during this period. Existing pavement design formulas have not been adequate to handle these new conditions. The CBR test, for example, does not give proper credit to stabilized materials even though the AASHO Road Test has shown clearly that such materials performed considerably better than even the best granular materials.

Likewise, various agencies have tried to deal with the design of continuously reinforced concrete pavements by attempting some type of uniform thickness reduction (e.g., 2-in. reduction) from a "standard" jointed pavement design. In some cases this approach is adequate, but in many others the resulting designs have failed prematurely and thus caused significant problems. The design problem, therefore, is not just one of selecting a thickness.

A pavement management system provides the means for making design comparisons between alternative materials or alternative pavement types. Cost and performance comparisons should be able to determine the relative benefits of using an old material or a new one. Similarly, new pavement types can be compared for cost and benefits.

It should be noted, however, that the management system is no panacea. Good models are required and the user needs to understand the relationships of new materials and new designs before accepting them as potential new solutions. This, of course, does not preclude the use of new processes on an experimental basis to obtain information for the systems models and future design.

## 40.6 CHANGES IN LOAD CONDITIONS

Since the mid-1950s a serious problem of load limits and their effects on the highway system has arisen. What is the true economy of raising load limits, or of holding them constant? Since the completion of the AASHO Road Test, significant information has been available on the relative damage effect of various axle loads. This damage is a result of both axle-load magnitude and number of load repetitions (in addition to the environment in which the pavement must serve). Significant pressure will always exist from trucking companies and industry for larger allowable loads, because heavier payloads for a fixed vehicle and crew produce correspondingly lower rates for the shipper and thus higher profits.

What is the corollary cost to society or taxpayers as a whole, in terms of increased pavement maintenance costs and decreased performance? This issue has remained a political football for years. Many studies have been carried out [Duzan 62; Langsner 62; AASHO 74; Smith 89; McNerney 91] but most if not all stop short of a thorough analysis for various reasons.

Prior to development of pavement systems methodology, it was impossible to look effectively at all the factors involved, or at least enough of them, to provide administrators and legislators with rational information. Pertinent questions related to any proposed change in axle loads are:

1. What true saving will actually result in the cost of transporting goods (e.g., what benefits will accrue to the consumers)?
2. What will be the effect on the performance life of existing pavements designed under previously existing standards?
3. What will be the effect on maintenance costs for existing highways?
4. What changes, if any, should be made to pavement rehabilitation policy and scheduling?
5. What changes will result in the costs of rehabilitating existing pavement to provide required performance under new loads?
6. What changes should be made in the design of future pavements to accommodate the proposed change in load limits?
7. What total costs (agency plus user costs, minus trucker savings) will result from the change in design standards?
8. What cost/benefit relationship exists for the load limit change for: (a) individual roads under consideration and (b) the highway system as a whole?

A pavement management system provides a tool for answering each of the foregoing questions. In fact, studies may well be made of existing legal load limits to determine if they are economical. Such studies of the existing and new load limits

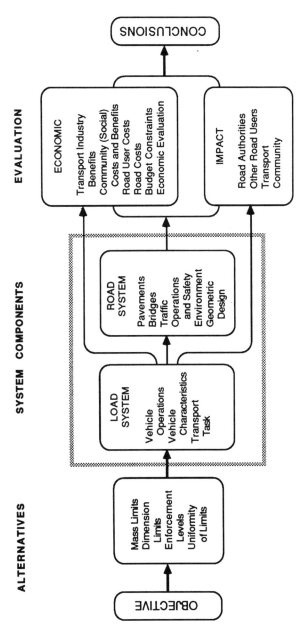

**Figure 40.1** Network flow diagram for NAASRA systems analysis of pavement [Fry 75].

can provide highway administrators with facts to present to commissioners and legislatures as well as user groups.

An excellent example of the use of a systems approach to this problem has been made by the National Association of Australian State Road Authorities (NAASRA) in its Study of the Economics of Road Vehicle Limits [Fry 75]. An illustration of this approach is given in Figure 40.1. It shows a structure of two basic subsystems of interrelated models:

- Load system (vehicle operations and vehicle characteristics)
- Road system (pavements, bridges, traffic operations, and safety)

Each subsystem included a number of computer models to predict the response of various system components to alternative sets of vehicle mass and dimension limits. The load system models predicted the response of vehicle characteristics and their operations to possible changes in the limits. The changes in loading and vehicle operations were then applied to the various road system models to examine the effects of these changes. The form and basis of all the models used are reported in [NAASRA 74]. The road models used relate directly to the trend of pavement serviceability with age or increase in roughness with age for various types of roads and environmental conditions. The change in these road life history curves is predicted as a function of the legal load limit.

## 40.7 SUMMARY

This chapter contains only a sample of the special problems that face highway administrators and that can be analyzed with a modern pavement management system. It would be useful to provide a comprehensive example solution to one or more of these problems, but such an example would of necessity be oversimplified because of space limitations. From the material presented, however, it is hoped that the reader can visualize the power of the methods available in finding solutions.

## REVIEW QUESTIONS

1. Name the key historical problems associated with pavements, particularly since the World War II.
2. Is the pavement management applicable in any way to energy consumption problems for transportation?
3. What applications can be made of pavement management in other new technologies?

Chapter 41

# Applications of Expert Systems Technology

## 41.1 INTRODUCTION

Knowledge-based expert systems (KBES) technology is in the field of artificial intelligence (AI). An expert system can be defined as an intelligent computer program that uses knowledge and inference procedures to solve problems difficult enough to require significant human expertise for their solutions [Harmon 85]. Artificial intelligence applications include not only expert systems but also robotics, computer vision, and natural language understanding, all of which can benefit pavement engineering.

Expert systems emulate human expertise and can solve problems which do not have algorithmic solutions. They provide an explanation of reasoning processes and conclusions and may provide useful solutions even with incomplete data. In particular, expert systems can produce inference engines and can act as a tutor.

## 41.2 USEFULNESS OF EXPERT SYSTEMS TECHNOLOGY

Because of the nature of the data used for design, construction, maintenance, and pavement evaluation processes (i.e., qualitative and quantitative, heuristic rules), knowledge-based expert systems (KBES) technology seems well suited to handle or integrate one or more of these processes. KBES do not depend on a single compu-

**Table 41.1  Expert Systems Versus Conventional Software [Schwartz 88]**

| Expert Systems | Conventional Software |
|---|---|
| • Symbolic programming languages (e.g., Prolog, Lisp). | • Conventional programming languages (e.g., Fortran, C, Pascal). |
| • Usually based on rules of thumb (or other knowledge representation schemes) that are generally reliable, but not always correct. These are concepts and cannot be reduced to formulas or numbers. | • Stated equations which can be proven. If correct numerical data is provided, a correct answer will result. |
| • Can explain its logic and reasoning. | • Provides answers only. |
| • Can function with incomplete data. | • Needs all data called for to operate. |
| • Development team includes domain experts by definition. | • Often programmed in isolation from domain experts and users. |
| • Provides capability to easily examine knowledge base. | • May be extremely difficult to examine embedded knowledge. |

tational path. Their ability to solve problems does not come principally from knowledge manipulation but from the richness, pertinence, and redundancy of the knowledge itself [Harmon 85]. These systems can combine the analysis of quantitative results, expert knowledge and engineering judgment, and knowledge of local conditions into an integrated procedure.

In the field of pavement engineering a great deal depends on the judgment and experience of technologists and engineers. Many experienced people, however, are nearing retirement. Replacement with younger people, generally with little experience in pavement technology, will be necessary.

Consequently, it is important to capture as much as possible of the knowledge and experience currently existing, not only for the benefit of younger people entering the area but also for the benefit of contractors, road authorities, and consultants.

## 41.3 EXPERT SYSTEMS VERSUS CONVENTIONAL SOFTWARES

A significant feature of expert systems when compared with conventional softwares is their ability to use knowledge when it is needed. In an expert system, the knowledge is declarational in nature and remains untouched until a situation occurs where it is required to draw conclusions. The order in which the chunks of knowledge are used is not determined in advance, as in conventional programs. Relationships between knowledge elements are given symbolic expressions whose final interpretations are fixed at run time. Table 41.1 summarizes the differences between expert systems and conventional software.

A typical expert system is able to explain how it reached its conclusions and why it asked the questions that the user received. These explanations can be requested whenever desired, while the problem-solving process is proceeding.

Even with incomplete information about the problem situation, expert systems are able to make conclusions that are possible with the incomplete data. Also the expert systems themselves can be tested and demonstrated while parts of their knowledge are still missing. They are able to use efficiently all the knowledge that they contain. On the other hand it should be remembered that expert systems are not able to give warnings in situations where they in fact do not have enough knowledge to solve the given problem.

The symbolic and declarative way to express the knowledge, as used in expert systems, creates one more advantage. In most cases, the experts themselves can quite easily understand the contents of the knowledge in the form in which it is expressed in the system. This reduces the possibility of misunderstandings between the experts and the programmer.

## 41.4 EXPERT SYSTEMS ARCHITECTURE

Figure 41.1 shows the basic elements that exist in an expert system. These include a knowledge base, a working memory, and an inference engine, as subsequently described.

### 41.4.1 Knowledge Base

The knowledge base contains permanent rules, facts, and other types of knowledge (that are usually used by an expert to derive a conclusion) about the specific problem. To create a "detective knowledge base," it is certainly required that the objects or elements of the problem be included as well as a list of characteristics of each of these objects. Also, some way is needed to link things together to be able to establish the problem space which consists mainly of the following:

1. Patterns of elements or symbols, each representing a state or way the task situation may occur
2. Links between elements corresponding to operations that can change one state to another [Harmon 85]

There are different methods, however, for representing knowledge. Each has advantages and disadvantages in terms of the encoding process and the nature of the problem itself. The most popular approach to representing the domain knowledge is by producing rules referred to as *IF* (premise) *THEN* (action) rules.

### 41.4.2 Working Memory

The working memory resembles a data base of conventional programs. It keeps track of the program status, such as values of incorporated variables and rules to be employed, and also contains input data for the given problem [Harmon 85].

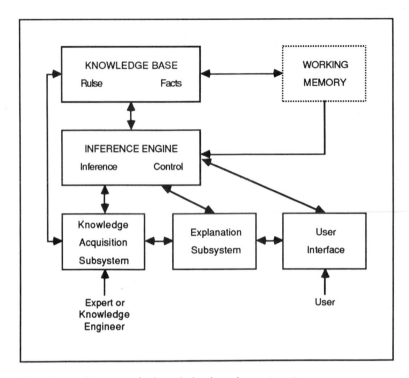

**Figure 41.1** The architecture of a knowledge-based expert system.

### 41.4.3 Inference Engine

The inference engine, also called rule interpreter or reasoning structure, controls the execution of the program. It receives data about a problem to be solved from the user, from another information system, or from some sensors. Then it uses the content of the knowledge base to reach conclusions while aiming to solve the given problem. The reasoning process continues by using the conclusions as new data and by acquiring further data from the user.

Figure 41.1 shows that the user interface provides access to the system for expressing queries and entering data. Expert systems should achieve friendly and efficient communication.

## 41.5 EXPERT SYSTEM DEVELOPMENT TOOLS

Expert system tools provide the specific means for handling knowledge representation, inference, and control to model particular classes of problems. In terms of software complexity, the available tools can be divided into: General Purpose Programming Language (e.g., C and Lisp), General Purpose Representation Language (e.g., Prolog and OPS5), and ES Building Framework (e.g., Insight). Hardware sophistication ranges from special Lisp machines to mainframes to microcomputers. Most expert

**Table 41.2  Criteria for Selecting an Expert System Development Tool**

1. Type of application
2. The ability to interface with other programs
3. Availability of complex mathematical routines
4. Type of control strategy and inference mechanism
5. Response time (in solving problems or asking questions)
6. Programming aids (e.g., editors, debuggers, and help facility)
7. Portability
8. User support
9. Cost

systems have been developed to run on DOS microcomputers. However, a mainframe-based expert system for highway maintenance strategies has also been developed [Haas, C. 87].

Selection procedures for development tools are not yet well established. However, criteria for selecting a tool can be found in a variety of references [Harmon 85, Maher 87], as summarized in Table 41.2. Basically, the type of application and the ability to interface with other external programs are very important.

Table 41.3 provides a summary of some of the available expert system development tools (as of 1991). An up-to-date, comprehensive review is almost impossible because of the continuous advances that result in new versions of these tools being marketed. Table 41.3 shows that EXSYS and INSIGHT2 were used more for pavement-related problems than any of the other tools. If the software will be used for several years, commitment of the software developer to continuous support and enhancement of the product can also be an important selection factor.

## 41.6 EXAMPLES OF EXISTING EXPERT SYSTEMS FOR PAVEMENT ENGINEERING

The application of expert systems technology to highway pavement engineering started in the early 1980s. However, few commercial expert systems have been described in the literature.

Examples of expert systems applications that are operational or under development are described in the following sections. These include ROSE, SCEPTRE, PRESERVER, and DMMD expert systems (see Table 41.3).

### 41.6.1  Routing and Sealing Expert System (ROSE)

ROSE is an expert system developed for the Ministry of Transportation of Ontario (MTO) to advise on a particular highway preventive maintenance activity, namely routing and sealing of asphalt concrete pavements in cold areas [Hajek 89]. System recommendations are given as a probability of routing and sealing being an appropriate, cost-effective maintenance treatment. The objective of routing and sealing is to prevent surface water from entering and damaging the pavement structure. ROSE is based on the MTO pavement monitoring and evaluation system, and interacts with an existing

Table 41.3  Some Expert System Development Tools

| Tool | Developed Expert System | Programming Lang. | Knowledge Represent. | Reasoning Strategy | Certainty Factors | Access to External Files | Explanation Facility | Editing Facility | Manufacturer and Product | Remarks | Tool Ref. No. |
|---|---|---|---|---|---|---|---|---|---|---|---|
| EXSYS | 1. Life-Cycle Pavement Behavior Modeling (ROSE) [Hajek 89]<br>2. Managing Pavements (SCEPTRE) [Ritchie 87]<br>3. Pavement Overlay Design (OVERDRIVE) [Ritchie 89]<br>4. Network Level Pavement Management (GRAPEE) [Haas 90] | C | IF-THEN Production Rules | Backward or Forward Chaining | Yes | .COM and .EXE | Yes | Yes | Exsys Inc. Contr. St. 14 Albuquerque NM | No Telephone Support or Consulting | [Haas 90] |
| INSIGHT2 | 1. Diagnosis of Marshall Mix Design Using Expert Systems Technology (DMMD) [Hozayen 90]<br>2. Diagnosis Hot Mix Asphalt Segregation (SEG) [Elton 89]<br>3. Concrete Pavement Evaluation and Rehabilitation [Hall 87]<br>4. Forest Road Design [Fricker 91]<br>5. Using ES to Select Traffic Analysis Software [Chang 87] | PASCAL | IF-THEN ELSE Production Rules | Backward or Forward Chaining | Yes | Any File Type | Yes | Yes | Level Five Research Inc., Indialantic, FL | Telephone Support and Consulting Available | [Hozayen 90] |
| EXPERTEASE | . . . . . . | PASCAL | IF-THEN Production Rules | Backward Chaining | No | No | No | Yes | Ex. Softw. Int. Ltd., Scotland | . . . . . . | [Elton 89] |

| Shell | Example Applications | Language | Knowledge Representation | Inference | | | Wordstar | Vendor | Notes | Reference |
|---|---|---|---|---|---|---|---|---|---|---|
| M.1 | 1. Designing Optimal Transportation Network [Tung 87]<br>2. Diagnosis of Hazardous Waste Incineration [Huarng 86] | PROLOG | IF-THEN Rules and O-A-V or O-A Pairs | Backward Chaining | Yes | Yes | Yes | Wordstar | Teknowledge, Palo Alto, CA | Sometimes Can Deal with Inheritance | [Teknowledge 84] |
| 1ST-CLASS | ... | PASCAL | Decision Tree | Backward or Forward Chaining | Yes | Any | No | Yes | Prog. In Motion Wayland, MA | No Tracing Facility | [Thomas 86] |
| VP-EXPERT | ... | PASCAL | IF-THEN Rules | Backward Chaining | Yes | .COM and .EXE | Yes | Yes | Paperback Software, CA | Interacts with Lotus | [Moose 87] |
| PERSONAL CONSULT. | 1. Field Inspection of Concrete Dams [Franck 89] | LISP | IF-THEN Rules and O-A-V | Backward Chaining | Yes | .COM and .EXE | Yes | Yes | Texas Instr. Inc., Austin, TX | Run Time May Be Long | [Elsevier 89] |
| CLIPS | ... | C | IF-THEN Rules | Backward or Forward Chaining | No | Fortran Files | Yes | Yes | COSMIC, Univ. of Georgia, CA | Telephone Support | [NASA 87] |
| OPS5 (Mainframe) | 1. PRESERVER: A Knowledge-Based Pavement Maintenance Consulting Program [Haas 87] | OPS5 | IF-THEN Rules | Backward Chaining | Rating Expressions | Fortran Files | Yes | Yes | Official Production Systems | ... | [Forgy 81] |

pavement management data base. A knowledge base containing decision logic about when to route and seal was developed using the MTO research results. The system incorporates 26 numerical variables, such as pavement surface distress, and contains about 360 rules. It was developed using the microcomputer-based EXSYS knowledge engineering environment. A FORTRAN version of the system was applied to about 900 pavement sections to illustrate the consequence of different routing and sealing policies. The system was also tested on a number of different pavement maintenance problems. ROSE is operational; however, the developers indicate that its direct application by other jurisdictions may be difficult or even inadvisable. The methodology and general approach may have general applications but the specific data base and rules are reflective of the conditions of encountered by the MTO.

### 41.6.2 A Surface Condition and Pavement Rehabilitation Expert System

SCEPTRE is an acronym for Surface Condition Expert System for Pavement Rehabilitation. It is an integrated set of expert systems under development for analysis and design of pavement rehabilitation strategies [Ritchie 87]. SCEPTRE is intended to be an expert pavement engineering advisor, particularly in local agencies at the city and county level. It basically evaluates project level pavement surface distress and other user inputs, to recommend feasible rehabilitation strategies for subsequent detailed analysis and design. SCEPTRE has been developed using the knowledge engineering shell EXSYS on a Compaq portable microcomputer, and runs on any MS-DOS compatible PC. The system is rule-based and uses a backward-chaining inference method. SCEPTRE 1.4 addresses state-maintained flexible pavement in the state of Washington. Ongoing research is expected to refine and adapt the knowledge base for local agencies.

### 41.6.3 Maintenance Strategies Expert System

PRESERVER is an expert system developed to advise field engineers and maintenance foremen on road maintenance strategies [Haas, C. 87]. The system is conceptually similar to SCEPTRE but focuses on routine maintenance activities, whereas SCEPTRE currently emphasizes major rehabilitation strategies. Both systems now address state or provincially maintained highways. PRESERVER incorporates maintenance treatment actions designed for Ontario road conditions. The system includes rules for a subset of distress types defined in a Pavement Maintenance Guidelines Manual of the Ministry of Transportation of Ontario. This manual was the principal knowledge source. Based on observed distress information provided to PRESERVER by the user, sets of feasible treatments for each distress condition are generated.

The existing system has been used to illustrate and test concepts. It was developed in OPS5 for a VAX mainframe computer and is a rule-based system. The main sections of the program are sequential, with subsections using both forward and backward chaining inference methods.

### 41.6.4 Marshall Mix Design Expert System

DMMD (Diagnosis of Marshall Mix Design) is an expert system developed at the University of Waterloo [Hozayen 90]. DMMD is a consultative tool that can advise the pavement engineer on adjustments to be made if the mix properties do not meet the specifications (i.e., if mix stability, flow, air voids, and/or voids in mineral aggregates are not satisfactory).

The DMMD expert system consists of 401 rules that contain the knowledge of human experts. A combination of both forward and backward chaining are used by the inference mechanism (control structure) of the INSIGHT 2 expert system development tool.

## 41.7 EXPERT SYSTEMS TECHNOLOGY BENEFITS AND LIMITATIONS

Some valuable benefits can be gained by applying expert systems technology to pavement engineering. For example, they can be used as a consultant, as a training tool, or in an advisory role. However, like any other technology, expert systems have limitations that make the encoding of certain human expert knowledge difficult.

In addition to the current limitations in developing an expert system, there are some barriers in obtaining the cooperation of the experts in certain fields regarding the encoding their knowledge. These include:

- user/expert communication gaps, and possible misinterpretations
- fear and uncertainty regarding legal/ethical responsibilities
- lack of time on the part of the expert

There is some concern that the experts would be expendable after their knowledge is encoded. However, this concern is unfounded, because experts are continually advancing their knowledge. Nevertheless, an effective interaction with the user is still needed to further the ability of both the user and the expert. Experts will still be needed for problems beyond the capability of the expert system.

### 41.7.1 Future Expert Systems Technology

There are areas of pavement management that are more amenable than others to expert systems application in both the short and long term. In the short term, applications will continue to be at the project level, particularly in performance evaluation, diagnosis, feasible solution testing, and project cost estimation. In the long term, expert systems can be applied at the project level to defining homogeneous sections, performance predictions, and life-cycle cost analysis. At the network level, expert systems can be applied to long-term performance prediction and constrained optimization.

Future prospects are numerous and certainly include the following:

1. Better self-teaching capability as expert systems mature and as we gain experience in their development
2. Feedback on the performance of expert systems themselves and updating capabilities
3. Better ability to resolve differences among experts
4. Enhanced human participation in decision making

5. Better integration of network and project levels of pavement management
6. Enhancement of the understanding of pavement fundamentals
7. Improvements in the underlying logic base
8. The development of more advanced tools and generic systems to improve transferability

### 41.7.2 Selecting Promising Areas for Applying Expert Systems Technology

Several areas in pavement management hold promise for future expert systems development. In each case a decision must be made on whether the advantages of applying the technology to the given problem are dominant. A set of evaluation criteria should be used to arrive at such a decision. These criteria involve the following questions:

1. Is engineering judgment and experience required to solve the problem?
2. Is the problem repetitive?
3. Is the problem bounded?
4. Is the problem well understood?
5. Is the problem symbolic in nature (i.e., not a pure mathematical problem)?
6. Are the human experts available? Do they have time?
7. Is the problem worth solving?

If the answers to most of these questions are yes, expert system technology can be applied to the problem. However, it is difficult to guarantee success because of unforeseen problems in encoding human knowledge. In other words, human experts have skills that the knowledge engineer (developer) may not be able to simulate using the available expert system development tool.

In summary, the main role of expert systems in pavement management will likely be an advisory or consultant one. Expert systems may take more of a decision-making role in smaller agencies. In addition, expert systems have the potential to be used as a training tool.

### REVIEW QUESTIONS

1. Does expert systems technology offer a perfect solution to pavement management?
2. What are the problems with developing expert systems to a useful level?
3. What are the most promising areas for expert systems in the decade of the 1990s?

Chapter 42

# New and Emerging Technologies

## 42.1 INTRODUCTION

New technologies applicable to pavement management are continually evolving. Some of these have direct and immediate benefits while others provide indirect or longer term benefits. It is not possible in this chapter to cover all new technologies. Thus people active in the pavement management field at any level should continually survey and study areas of new technology for possible application.

This chapter briefly considers, as sample, geographic information systems (GIS), new data base methodologies, new computer hardware and software, new measurement technologies, applications of PMS to other areas, and interfaces with other management systems. Each of these areas deserves a lot of time and energy, but space limitations preclude such detailed treatment.

## 42.2 GEOGRAPHIC INFORMATION SYSTEMS

A geographic information system can be defined in several ways. The U.S. Geological Survey (USGS) defines it as "a computer hardware and software system designed to collect, manage, manipulate, analyze, and display spatially referenced digital data" [PennDOT 90]. The USGS then defines spatially referenced digital data as "computer-readable, geographically referenced features that are described by both geographic position and attributes" [PennDOT 90].

PennDOT defines GIS as "an automated system designed to allow users to more easily filter, manage, analyze, display and share location-oriented data and associated explanatory information," [PennDOT 90]. From these and other sources the following working definition for GIS is considered appropriate: *A methodology involving a computerized data management system designed to capture, store, retrieve, analyze, and display spatially referenced data.*

Considerable attention has been given to the application of GIS to the pavement management process. Several states, including Texas and Wisconsin, have studied the possible applicability of a GIS [WisDOT 90b]. One of the most urgent needs related to PMS, defined by Texas field personnel, is the need for district headquarters to produce maps highlighting the location and condition of substandard pavement sections. This is an ideal application for GIS. The districts also outlined the need for accessing, manipulating, analyzing, displaying, and reporting information on the road network graphically. Again, GIS is directly applicable. Program people to field people also require the means to present the benefits of their priority listings for pavement needs in graphical form, in addition to lists. A graphic display shows the distribution and location of various sections and has a stronger visual impact.

Wisconsin, which monitors and manages a system of 8,000 miles of highways, after thorough review of their needs, mandated the application of a GIS. Their pavement management steering committee particularly highlighted the spatial data concepts needed for use in design of the large decision support data base and spatial analysis routines for evaluating pavement performance [WisDOT 90a, 90b].

In many states, the department of transportation acts as the state cartographers as well as cartographers for the highway system itself. For this reason, the DOTs are usually a prime choice for implementing a GIS that may have statewide applicability.

Automated mapping is not a GIS, but a GIS can perform automated mapping functions. If automated mapping were the only application needed in a pavement management system, it would be better to purchase a dedicated, automated mapping system. However, a GIS is a good way to integrate into graphical files, large amounts of data and other information necessary for the pavement management process. The ability of a GIS to integrate data and then to display it graphically, as well as to integrate an analytical method based on geographic location, is a primary advantage.

There are many other possible applications of GIS to the pavement management process. For example, a GIS can be used in performing traffic flow modeling and interactive traffic simulation. This type of modeling could permit examination of proposed changes in the physical characteristics of the highway, and in turn evaluate expected performance changes in the pavement without having to reanalyze the data base every few years and develop and rewrite a new model.

It may be possible in the future to collect pavement inventory and condition data digitally at highway speeds using global positioning system (GPS) coordinates. A demonstration project at Ohio State University, sponsored by 38 states, has shown the feasibility of this concept [Ohio DOT 90]. This trial successfully integrated a van equipped with GPS, a color video camera for highway information, and a digital stereo camera system capable of making metric measurements of the highway. Improvements are still needed in the quality of the photography and the individual measurements of condition elements before they can be directly used, but the feasibility was demonstrated.

## 42.3 NEW SOFTWARE, HARDWARE, DATA BASES, AND PERSONAL COMPUTERS

Computer technology continues to advance at a remarkable rate. Many engineers now have personal computers with data processing capabilities exceeding the computer availability for an entire highway department in the early 1960s. Equally important, the technology for data storage has increased by several orders of magnitude in recent years. Currently, relational data bases are available on many personal computers and 100 megabytes or more of data storage is relatively inexpensive. Other data storage concepts and new integrated microchips will increase both the size and the speed of data processing capabilities incredibly in the coming decades. All of these new technologies should substantially aid pavement management, but at the same time they will also require intelligent application. Data processing and computational power should never replace good engineering thought and judgment. Rather, they should supplement and enhance engineering applications.

The next decade will likely see voice input computers, portable battery powered computers capable of accessing central data base storage via satellite transmission, and numerous other innovations not imagined at the present time. These can enhance the pavement management process if intelligently applied, but the knowledge and skills of pavement management engineers and administrators who must apply them correctly will still be essential.

Along with increasingly powerful hardware will come more sophisticated software. GIS and expert systems are two examples of generic classes of software which provide powerful tools to the pavement managers. In addition to these generic classes of software, pavement managers will have the freedom to develop increasingly sophisticated analysis tools such as finite element methods for mechanistic analysis of pavement structures and true optimization algorithms for determining network programs. For example, the Arizona DOT network optimization system now can be performed on a 486 microcomputer operating under OS/2 [Wang 93].

## 42.4 NEW MEASUREMENT TECHNOLOGY

New measurement technology continues to be developed and yet some problems remain intractable as far as pavement measurements are concerned. This is due primarily to the relative cost of the technology. For example, the AASHO Road Test staff in 1958 sought a gyroscopic horizontal reference for the Road Test [Hudson 58]. Such a gyroscope was available in NASA, but the cost of the device would, in fact, have consumed nearly all of the total budget of $30 million. The availability of technology in the space and military industries can have direct application to the pavement field, but the relative costs are often orders of magnitude higher than typically considered in the pavement field. This relative cost/benefit aspect of pavement management measurement technology must be sorted out for the full benefits of new technology to impact on our transportation system.

A survey of the full range of needed or available measurement technology is not possible herein, but some example possibilities are briefly described in the following sections.

### 42.4.1 Integrated Survey Vehicle

There remains a need for the development of a truly integrated survey vehicle which can travel over a highway system at ordinary road speeds and record and process data related to the highway and pavement including ride quality, condition, structural adequacy, safety, and other features. Various attempts have been made toward developing such vehicles as outlined in Part Two of this book. However, all have various limitations, including cost, reliability, operational ruggedness, and maintenance requirements. As the value of good pavement management data, and its application, becomes more evident, and the benefits can be quantified, (which should run into billions of dollars annually), we will be able to move effectively into development and operation of automated survey vehicles.

### 42.4.2 High Speed Structural Analysis

High speed structural measurements may not need to become a part of the total integrated survey vehicle because there are times when it is important to do structural analysis independently. Thus, a separate need exists for the development of high speed structural measurement equipment. Attempts to produce such equipment have been made over a number of years and certain advances have been made. Among these are the automated Benklemen Beam developed by California and the Deflectograph developed in France which travels at speeds of 5–7 miles per hour. These are certainly not high speed, but they do exceed the speed of other deflection measuring equipment which requires a stop each time readings are to be made. New technology, such as surface wave analysis, can perhaps also be applied as well as radar and laser ranging techniques. Ideally, the instrumentation will not only take data at travel speeds, but process that data in real time or at least in the field to permit rapid evaluation of existing pavements as well as newly constructed pavements.

### 42.4.3 Direct Imaging and Analysis Techniques

New techniques are rapidly developing which will permit the direct imaging of pavements including surface distress, texture, and many other elements. This methodology is also capable of being applied to quality control of aggregate gradation. A quantitative measurement of aggregate texture can undoubtedly go a long way toward solving the problem of aggregate shape and other factors which have a strong effect on the stability of asphaltic and concrete mixtures.

### 42.4.4 Automated Testing Procedures

Materials testing for mix design and quality control now consumes considerable agency resources. At the same time the Strategic Highway Research Program is recommending increasingly sophisticated test methods. Clearly there is a need to develop automated test equipment which will reduce the potential demands while improving the quality of the measurements. This will probably be more expensive than highway agencies are accustomed to spending on equipment. However, if the result is better performing pavements, the payoff for the investment can rapidly amortize their costs.

## 42.5 INTERFACE WITH OTHER SYSTEMS

Pavement management has shown real leadership in integrating various disciplines into a useful process. Beginning in the 1960s, pavement management provided a framework for integrating all aspects of pavements ranging from planning through design and construction to maintenance and rehabilitation. The success and benefits have carried over to bridge management where significant work was done during the decade of the 1980s and is currently being applied to congestion, traffic flow, and safety aspects of the highway network [Hudson 87]. There is no better proof of the success of pavement management than that it has been widely emulated in other highway and transportation areas. Now it is important to successfully apply these techniques and to coordinate their application. Tremendous synergism can be gained from interaction of pavement management systems and bridge management systems, and clearly the programming of funds for pavement rehabilitation must be carefully and completely coordinated with the needs for: (1) bridge maintenance and rehabilitation, (2) capacity expansion (congestion management), and (3) safety corrections and improvements. While these areas each have their own specific requirements, they are all truly subsystems of an overall highway management system, or an executive management system as it is sometimes called in the transportation field. Clearly, there are many common elements which must be considered at the beginning. These include traffic data, vehicle loadings, and geographic location, just to mention a few.

## 42.6 SUMMARY

It is essential that the application of new methodologies and improvements to pavement management over the next decade is carefully and thoroughly coordinated with the improvements in and applications of other management systems in the transportation field. There are many applications where pavement management can be effectively integrated with the total engineering infrastructure. This is particularly true in cities and counties where utilities (water, sewer, wastewater, telephone, gas, electricity, etc.) often occupy the same or contiguous right-of-way with the streets or pavement. Clearly, a real opportunity exists to carefully and effectively integrate individual infrastructure elements into a total infrastructure management system over the next decade.

## REVIEW QUESTIONS

1. Name and discuss at least two new methodologies that may be applied to pavement management in the decade of the 1990s.
2. Are there any problems in trying to develop pavement management as a stand-alone system?

Chapter 43

# Institutional Issues and Barriers Related to Pavement Management Implementation

## 43.1  INTRODUCTION

Successful implementation of a pavement management system in any organization requires that the institutional and organizational factors be considered along with the technical components. Broadly speaking, institutional issues are those tactical aspects of a PMS which affect its acceptance and success of the implementation. These institutional issues must be understood and addressed by the agency prior to, during, and after the system implementation, and should begin in the initial stages along with technical issues. The first step in PMS implementation, gain management support, is also the first institutional issue [Kher 90, Smith 91].

For about the first decade of their existence (1966–1975), pavement management systems emphasized the project level. From 1976 to at least the early 1990s, great interest developed in the application of pavement management concepts at the network level, and this focus may well continue. It should be remembered, however, that the true effectiveness of a PMS requires a balanced application at both levels.

Good pavement management is not business as usual. It is not proclaiming that pavement management is being done, when the reality is that existing practices have simply been relabeled. Implementation of a PMS requires a change in the way some existing agency people think, react, and do their day-to-day work. No one is initially comfortable with change, yet change is necessary in life and in successful pavement management. For example, information may be acquired in great quantities, but it is essential to process and evaluate it so that it is meaningful.

In essence, intelligence must be applied to the coordination, proper processing, and interpretation of information. Wisdom is then the combination and application of intelligence and experience. There are many more intelligent people in the world than there are wise people. There are many more systems using raw data than are sorting, sifting, and applying it wisely in combination with codified experience.

Models, methodologies, and procedures for pavement management are aimed at providing the essential service of transforming raw data into information and intelligence to administrators and decision-makers. They must then combine that with experience to make wise decisions. This requires several prerequisites to be fulfilled.

1. Decision-makers must appreciate the importance of this service (be wise); they should also be capable of understanding the usefulness and limitations of the process (be educated and well trained).
2. Pavement management systems must be available, usable, and credible. This has not always been the case because inadequate resources have been applied to the process.
3. The ongoing nature of pavement management must be recognized and supported. Most if not all the activities (data collection, analysis, budgeting, etc.) are annual. An agency that relies on a pavement management report produced 5 years ago does not have a pavement management system.

Public policies involving risks are more likely to be acceptable when based on credible technical and engineering information, and where sound trade-off and impact analyses have been performed.

All public and private systems, including pavements, are planned, designed, constructed, operated, and modified or rehabilitated under conditions of uncertainty and risk. Thus, the assessment, quantification, evaluation, and analysis of risk (reliability) in the context of systematic pavement management are essential to the process of good decision making.

## 43.2 SKEPTICISM OF PAVEMENT MANAGEMENT

While pavement management has been around for a number of years, there is still skepticism at middle and upper management levels in some highway agencies. The sources of this skepticism, which are largely related to modeling and to systems analysis in general, include:

- Misuse and incorrect applications of models; insufficient research basis; proliferation of models and methods without systematic inventory; improper calibration, validation and verification of models
- Too much use of the PMS delegated to people who do not understand the process
- Lack of incentives to properly document the PMS

**Institutional Issues and Barriers Related to Pavement Management Implementation** 539

- Overemphasis on use of computers as black boxes
- Lack of communication among developers, users, and beneficiaries of PMS
- Lack of recognition of the PMS as a means rather than an end; lack of recognition by decision-makers on strengths, weaknesses, and limiting assumptions of the PMS
- Lack of an interdisciplinary team; for example, lack of statistical expertise
- Insufficient data available to properly use the methods in early phases of implementation
- Inadequate understanding of the true costs and benefits of PMS

How is the use of pavement management "sold" in an agency? The process of communicating ideas about pavement management may be as important as the content. Determining what change should occur (what to do) is easier than determining how to bring about the change (how to do it). Implementation of a change such as the application of PMS requires a carefully coordinated and sequenced set of changes in training, policy, and personnel practices executed over a period of time.

A number of organizational barriers always exist which work against change in general and in particular against the adaptation of modern systems technology. Major changes such as developing and implementing a PMS tend to provoke complex and compensating reactions. Unless the interrelationships between different divisions or parts of an organization are accurately identified, the organization may react to subvert the intended change rather than to welcome it. Consequently, it is necessary to:

- Clarify the issues and fully inform the decision-makers.
- Achieve substantive, effective, organizational agreement on a course of action that is feasible, equitable, and desirable.
- Train the related personnel (or retrain as necessary).
- Carry out the actual implementation, including the ongoing followups.

Good, well-documented case study examples of pavement management can be useful in communicating the benefits. On the other hand, any case study is necessarily an oversimplified example and cannot cover all aspects or details. Thus, it may mislead if taken literally.

Any approach to these issues requires consideration of the costs and the benefits of implementing a PMS. These are further considered in Chapter 44.

## 43.3 MANAGERIAL AND ORGANIZATIONAL ISSUES

### 43.3.1 Personalities and Personnel Factors

One of the most complicated issues involved in pavement management activities are the personnel factors. Not many people relish change, and fewer still are risk-takers, particularly in government agencies. The risk-takers usually enter the private sector, such as business, law, or consulting. Of course circumstances can put risk-takers in government positions and security-seekers in private jobs, but not as a general rule.

Consider five factors of personnel and personality: (1) Overall personal attitudes toward change plus four levels of action or involvement, (2) ownership or political level, (3) top management level, (4) middle management level, and (5) working level.

### 43.3.1.1 Personal Attitudes

Inherent in most human beings is the general tendency to be comfortable with the status quo. Cliches are abundant, such as: "I don't need a computer to help me do my job, I have made these decisions for the last twenty years," or "I don't have time to attend training sessions," or "if it ain't broke, don't fix it."

In general people tend to feel that the way they are currently doing their job is the right way. Furthermore, they are in a comfortable niche which often does not require great thought and this feels good.

On the other hand, many people have busy jobs where they are saturated with the pressure of keeping up with everyday activities. They feel they do not have the time to learn how to use a computer so that they could make their data summary on a spreadsheet. They don't know that they could reduce an eight hour job into a two hour job.

The typical pavement management agency is a public agency with no profit motive. There is no way to judge the quality of the work in terms of getting the best product for the money; thus, there is little pressure to change. Even people who have a strong personal incentive to change, are often resisted either consciously or unconsciously by their co-workers. The possibility exists that a senior level person may actually see a threat in implementing pavement management. For example, an aggressive young person in a lower level job wants to implement new computer-related activities or new methods of measurement or data processing, but the superior is comfortable with the old way, thoroughly understands the old forms, and resents the intrusion.

Finally, if a major change or a new management style occurs, such as the method of programming expenditures, this may tend to suggest that people have not been doing a good job. The implication is that if this new method is so good, why wasn't it thought of sooner.

All of these personal attitudes can be barriers to the development of good, new management styles including pavement management.

### 43.3.1.2 Political or Ownership Level

This level can have great effect on the acceptance of pavement management. There are a number of examples of successful implementation where the political leadership demanded that the agency adopt a better method of programming funds. One good example is the Kansas legislature which passed a funding bill that required the state highway agency to develop a pavement management system in order to determine the best expenditure of funds. The Arizona Department of Transportation had a similar situation, where the assistant state highway engineer, who subsequently became the state highway engineer, wanted a new methodology for programming funds and followed through until pavement management was adopted.

There are better examples, of course, in private industry. Perhaps one of the best is that of Gibson's discount store chains who started out with good management and

### Institutional Issues and Barriers Related to Pavement Management Implementation

merchandise but did not change and subsequently went bankrupt. By contrast, Walmart stores, developed by Sam Walton, who amassed one of the largest fortunes in the world, has innovative checkout with automatic inventory control, automated credit card processing, and many other features which allow them to keep the margin of profit low on each individual item, but to produce large absolute profits from the entire organization.

#### 43.3.1.3 Top Level Management Concerns

It may be hard for top level managers to admit that they can do better. If they change their method of decision making, it may imply that they could have been making better decisions all along. Fortunately, fewer and fewer managers feel threatened by computer use, and regard it as a friendly tool, which aids their decision making. The good top level managers realize they have to exercise their own experience and wisdom in interpreting the computer output to get good results.

#### 43.3.1.4 Mid-level Management Concerns

Many people have worked very hard over the years to reach their current positions in mid-level management in the agency. These people have established a routine that works for them. They are currently in the design division or the maintenance division and in some cases have been there for 20 years or more.

The introduction of a new system, such as a pavement management system, implies training and the need to change the status quo. It may require reorganization and moving from a comfortable location.

There are examples in highway agencies where permanent records were kept by people trained in the art of uniformity and quality. But they had no desire for a transition to the impersonal level of computer data entry. Certainly changes in training become necessary and in some cases the individual personalities do not mesh with the new realities.

Everyone knows of new ideas put forth in an organization that did not take root and last. Thus, a tendency may exist to feel that pavement management is just the new and latest fad style of management and it will eventually pass away.

#### 43.3.1.5 Lower Work Levels Concern

At the operating levels of highway agencies, pavement management may involve retraining for different skill requirements. The transition is the difficult part. People with certain existing skills may not be amenable to the new skills required. Finding jobs for the old staff and finding and training new staff can be difficult.

For example, knowledge of statistics and its applications in the broadest sense and knowledge of operations research and benefit cost-analysis are essential in pavement management. These are not always available in the current beginning level professionals within an agency. Computer skills are also required, but these are much more prevalent in the average highway agency now than before, although in some agencies such as counties or small cities, this is still a problem.

Modern pavement management also requires high speed measurement and data collection, which means instrumentation technology is a requirement. The ability to measure, collect, process, and analyze data is essential to good pavement management in both the lab and the field.

As another example, personnel who are currently in the materials division may be required to collect materials information from pavements all over the state. This may mean several days of travel each week, and may not be acceptable to the existing staff.

Finally, some staff at the lower levels are accustomed to operating equipment routinely that is not thought of as requiring adjustment or calibration. It either works or does not work. Yet, most if not all the measurements for pavement management require careful calibration and careful observation. Making people aware of the need for good calibration and good data collection is a critical part of pavement management and one that has yet to be totally solved.

### 43.3.2 Organizational and Management Realities

Each institution has a formal organizational structure usually shown with an organogram. Most are pyramidal, large at the bottom and small at the top. In such organizations it is necessary for the chief executive to oversee a wide range of activities, but the effective span of control of an executive is generally considered to be 7 to 10 people. That is, a good executive can effectively deal directly with no more than about 10 people as direct subordinates. These people then must deal with others multiplying their effectiveness and building the pyramid.

For example, say that a chief engineer of a small highway agency has four major divisions and four districts directly reporting to him or her, as shown in case A in Figure 43.1. With more divisions and districts this could be a difficult, if not impossible task. Consequently, it may be necessary to change to an organizational structure such as case B in Figure 43.1, where intermediate or deputy chiefs are introduced to coordinate and/or direct activities of several units. If there are many districts, they may be organized into regions, where six or more report to one regional engineer. This is the case of the Federal Highway Administration.

Where does the pavement management group (PMG) fit into either of these or any other structure? In case A, it involves all of the districts and several of the divisions. In case B, it involves the deputy chief engineers and the regional engineers.

There are many examples where the PMG or pavement management office was placed in the design or materials group division originally. In Washington State for example, pavement management began in the materials group because it was primarily developed and operated at the project level. But this subsequently became somewhat awkward for network level pavement management to function in the process of planning, programming, and funding.

In the Texas Department of Transportation, the development of pavement management also began in the late 1960s, at the project level, and as a result was located in the pavement design section. In the 1980s, network level pavement management received major attention and the pavement management group (PMG) was placed in the maintenance division. This PMG was charged with pavement evaluation and rehabilitation programming. But the project level activity remained in the design di-

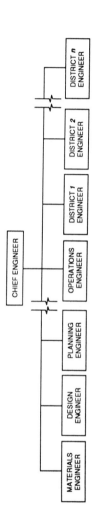

**A) Potential Structure for Small Highway Agencies**

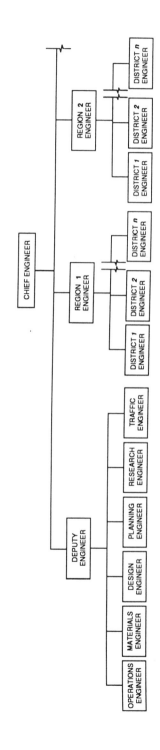

**B) Pyramid Structure for Large Highway Agencies**

**Figure 43.1** Example organizational structures.

vision, with the result that some problems were caused in developing an overall, integrated pavement management system.

In the state of Arizona a special group was formed in the 1970s, reporting directly to the assistant state highway engineer. Their PMS was a network level system intended to assist with the programming of pavement maintenance and rehabilitation on a statewide basis. When the pavement management system was fully developed for initial implementation, it was given a permanent home in the organization.

The PMS activity should function at as high a level as possible in the organization, whether it be city, county, state, or national, where it can be given full-time attention and direct supervision. There are cases of a PMG reporting at a high level, but through a committee to a top deputy or assistant chief engineer. In these situations, if the top level executive considers the responsibility is only to chair a committee, where the actual reporting is through an administrative assistant, the PMG will not function effectively. There must be a broad overview otherwise the PMG will not succeed. The job must be a live responsibility.

The development and implementation may be one of separate divisions who control network level and project level pavement management. Frequently, the computer division will exert strong influence on the system because it controls the programming resources. In such cases an outside consultant can be extremely valuable in rationally utilizing the resources of the entire agency without being bound by the existing structure.

In summary all elements of an organization, formal and informal, must be considered if pavement management is to be successfully implemented. It is important that the chief engineer or the equivalent strongly supports the pavement management system, wants it to function effectively, and instills this desire into the lower levels.

### 43.3.3 Ability to Withstand Organizational Changes

Highway agencies, at all levels, operate under varying management styles and under various organizational structures. These styles and structures influence and control the environment under which the pavement management system will be expected to operate. But this operating environment rarely remains static for long. In some cases it may change with every political administration. Certainly it changes with major promotions and changes of personnel within the agency.

With the large number of personnel changes and retirements anticipated in highway agencies in the next few years (the average age of highway personnel exceeds 50 years in most state highway departments in the United States, for example), major changes will undoubtedly occur in the way they operate and in their organizational structure. It has also proven to be true that the aggressive, forward-thinking young engineers who are willing to take on jobs in the pavement management section are also those who succeed in getting promotions. Of the over 100 engineers from state and federal agencies who attended a six-weeks intensive pavement management course at The University of Texas from 1983 to 1986, fully 75 percent have been promoted to new jobs within a 5 year period of attending this course. This is clear evidence that any robust pavement management system must be able to withstand organizational changes.

In some cases the success of a PMS can only be assured through organizational change. In older organizations there may be no particular unit or division within the agency broad based enough to provide a home for pavement management, and it may be necessary to form a new division to handle the PMS. In other cases it may be necessary to modernize the data collection unit or transfer personnel from one division to another to form the required PMS team. In a small agency, such as a city or a county, it may be necessary to upgrade computing capabilities to carry out the PMS activities.

### 43.3.4 Involvement of the Computer Staff or Section

Any medium to large size agency undertaking PMS development must carefully consider how the data processing is to be carried out. More than one pavement management system has failed due to lack of support and cooperation from the computer or data processing division.

One reason was the tendency in the 1970s and 1980s toward centralization of computing facilities to reduce duplication, proliferation of equipment, incompatibilities between files, etc. But this all too frequently hindered access by engineering staff. In many cases program development or modifications were out of control of the user.

This problem should be eliminated with the advent of more powerful and inexpensive personal computers, but it has not been. Many pavement management systems, even at the state level, can function effectively on a personal computer. However there are benefits to be gained by coordinating data bases across a number of systems. For example, there must be good coordination between pavement, maintenance, and bridge management. This can be handled by using the main frame for central data storage and using the personal computers for data processing, analysis, reporting, etc.

## 43.4 LEGAL AND REGULATORY REQUIREMENTS

In recent years the time between project inception and fulfillment has grown dramatically. For new projects it can be up to 10 years with an average of 8 to 9 years in most U.S. state highway agencies. Rehabilitation projects can be shorter but still average 4 to 5 years in some agencies. This scenario will likely continue for some years.

By contrast maintenance activities may require immediate action to fulfill legal or regulatory requirements for safety. Thus, pavement management activities must cover time spans from immediate to 10 years of lead time. These times are required for environmental impact analyses, route locations, right-of-way acquisition, and preliminary and final designs as well as routine advertising for bids and contract letting. Public hearings are also required and can be very time consuming.

Consequently, a PMS must be capable of flexible, multiyear programming if it is to fulfill the needs of modern, complex highway or public works agencies. Provision must also exist to shift projects from year to year based on possible delays, changes in costs and/or funding, and various external factors.

For example, when a project is delayed several years, and the pavement is allowed to deteriorate to a worse condition, this may mean a shift from minor rehabilitation to a major reconstruction effort. Not only will the program list have to be revised to accommodate the project delay, but additional funds may have to be obtained to address the further deterioration.

## 43.5 AGENCYWIDE PMS

While a PMS will likely involve several major divisions within an agency directly, it must also interface with literally all of the divisions or departments in one way or another. The following discussion considers some of these interfacing requirements.

### 43.5.1 Routine Maintenance—Maintenance Management

Maintenance is typically an independent division within most highway agencies. Its usual role in the programming process is essentially that of a specified maintenance budget which is divided among districts on an allotment basis, year after year, in association with a maintenance management system.

Because of this independence, little attention may be focused on the trade-offs which exist between increased maintenance expenditures and their effect on future rehabilitation needs. In many cases, there is little interaction with the maintenance staff once a program has been developed. The result is minimal coordination of efforts between maintenance needs and upcoming rehabilitation projects.

It is important, therefore, to interface PMS with maintenance management system (MMS) activities and to include MMS personnel in the PMS information loop. Likewise, a maintenance module must be included in the PMS for maximum efficiency and economy.

### 43.5.2 Coordination and Integration With Other Management Systems

In any state highway agency, the management of the highway network involves more than simply building, repairing, and maintaining the pavement surface of the roadways. Pavement management systems were initiated in the mid-1960s and early 1970s, but now there are management systems for bridges, congestion, geometrics, and safety. It is important that the PMS interface with these other management systems, or at least with the expenditures and projects for such activities to get the best overall expenditure of public funds.

Today, most projects include many work items other than those that are pavement related. Because of user costs associated with lane closures and often complex contracting procedures and traffic control costs, it is imperative to coordinate all necessary repairs and upgrading needed on a particular section of highway within the same construction project.

In some cases, the need for the highway project is triggered by congestion, bridge repairs, safety requirements, geometric inadequacies, etc., rather than the need for pavement rehabilitation. The incorporation of all such elements into the development of a comprehensive multiyear program requires an executive management system. A pavement management system must be capable of working in harmony and providing the necessary information to the executive management, and to bridge management,

# Institutional Issues and Barriers Related to Pavement Management Implementation 547

congestion management, safety management, etc. However, the other programs should not override the PMS and thus compromise the quality of the largest public investment which is the pavement network.

### 43.5.3 Including a Benefit Reporting Mechanism in the PMS

The identification and documentation of tangible and intangible benefits to the organization from implementing a PMS are very important. Has the system saved money or freed funds for other purposes? Has the system helped in explaining the decisions made? Has it helped senior management in understanding and explaining to the public and politicians that the best use of tax dollars is being made? Has the system facilitated decision making and time savings to organizational personnel managing the pavement network. These and other questions must become part of the output of the system.

### 43.5.4 Accommodating Changing Technology in the PMS

Highway agencies are continually faced with keeping abreast of changes in technology. Many if not all find it difficult to integrate new technology at an accelerated pace. The ability and flexibility of a PMS to accommodate changes will play a large part in successful implementation.

Changing technology affects several aspects of pavement management. The most obvious changes occur in data collection. For example, many agencies initially developed methods of collecting pavement distress data using visual inspections. Today, a number of agencies are using automated survey vehicles in their data collection efforts, and this trend will undoubtedly continue.

The crucial issue before adopting any new technology is a plan for correlating previous data with the new procedures. To continue with the previous example, in most cases, visual distress survey procedures do not directly correlate with new, automated methodologies and a large historical data inventory could be voided if these issues are not addressed.

The PMS must also be able to accommodate new rehabilitation procedures, including treatments and equipment used, as they develop. As an agency becomes comfortable with the new techniques, they should be added to the list of feasible or approved alternatives and performance trends should be documented for future needs assessments.

Perhaps the most difficult aspect of the rapid emergence of new technologies is finding the time to really evaluate them for applicability to the agency and thus staying ahead of the game.

## REVIEW QUESTIONS

1. What is meant by institutional issues in discussing pavement management?
2. Name several sources of skepticism on pavement management in highway agencies. Is this skepticism valid?
3. What personal concerns arise in personnel applying pavement management technology?

Chapter 44

# Cost and Benefits of Pavement Management

## 44.1 INTRODUCTION

Previous chapters have covered various institutional issues related to the pavement management process. However, an issue not adequately treated by many agencies is that of comparing the costs and benefits of implementing a PMS. This will undoubtedly have to be addressed more fully in the future.

Any such analysis must be done on an individual agency basis. Being convinced that the benefits of implementing a PMS in an agency outweigh the costs is not sufficient. It must be demonstrated. This chapter provides some aspects of the benefit-cost analysis that can be carried out. As well, it suggests some methodologies to be considered for application in any agency.

## 44.2 PAVEMENT MANAGEMENT COSTS

There are several types of costs associated with pavement management. The two primary ones are:

1. The costs of developing and operating a PMS, along with the necessary and appropriate data for using the PMS and keeping it current; that is, the cost of the PMS itself

Cost and Benefits of Pavement Management 549

2. The actual expenditures on the pavements or the highway system itself, which are associated with evaluating a PMS

In reality, it is the ultimate savings in real highway expenditures that measure the effectiveness of a PMS. The initial pavement investment and related costs must be considered along with savings and benefits that can be realized from the effective implementation of a PMS.

Many problems occur in assessing either of the costs outlined above. Apparent costs can vary greatly depending on accounting procedures and methodologies within a given organization. Some highway agencies do not account for overhead or indirect costs when they do work with their own staff, which is misleading. The same activities done by contract or the use of outside experts, however, clearly have to include indirect costs in the final contract price for the work.

True cost information for evaluating the costs and benefits or cost-effectiveness of a PMS is difficult to obtain. This is partially because few agencies have a fully implemented PMS, and fewer still have kept effective comparisons and records of costs. In the case of the highway facility or pavement itself, the costs associated with construction can be documented, but they have usually occurred over many years and a common basis of comparison is often difficult to obtain. More substantially, the costs of maintaining pavements are extremely difficult to define and very few highway agencies have truly good maintenance cost information defined by specific pavement location.

## 44.3 BENEFITS AND COSTS ASSOCIATED WITH A PAVEMENT MANAGEMENT SYSTEM

Tables 44.1 and 44.2 present a variety of benefits and costs associated with pavement management. They are categorized in terms of general, elected representatives, senior management, and technical level people in Table 44.1. Costs and benefits particular to the state level are listed in Table 44.2.

Some benefits and some costs are quantifiable but others are subjective and general in nature. As subsequently described, benefits must be brought to a common basis of measure to be compared analytically with costs. However, benefits are generally excluded from pavement management decision making, largely because of the philosophy that it is adequate to provide a safe and comfortable highway to serve the public. The idea of improved benefits accruing to the user based on better ride quality and lower vehicle operating costs has not been widely exploited within North America, although it is widely used in World Bank evaluations for developing countries. One of the very few quantitative assessments of the benefits and costs associated with a PMS has been provided by Falls as summarized in the next section [Falls 92].

### 44.3.1 Quantitative Assessment of Management Benefits and Costs

It has been suggested that the true indications of the cost-effectiveness of a pavement management system involve the ultimate savings in real highway expenditures plus

Table 44.1 Notes on Benefits and Costs of a Pavement Management System

|  | A. General | B. Elected Representatives | C. Senior Management | D. Technical Level People |
|---|---|---|---|---|
| Benefits | • Realize magnitude of the pavement investment<br>• Better chance of correct decisions<br>• Improved intra-agency coordination<br>• Improved technology use<br>• Improved communication | • Justify maintenance and rehabilitation programs<br>• Assurance of best expenditure of tax funds<br>• Less pressure for arbitrary program modifications<br>• Objective answers to effects of lower funds or lower standards | • Comparative view of network status (current and future)<br>• Objective answers to: funding level effects on status, implications of deferred work and/or lower standards<br>• Justifying programs to elected representatives<br>• Assurance of best use of available budget<br>• Defining the "management fee" (percent of budget) | • Improved recognition of various agency elements<br>• Increased awareness of available technology<br>• Improved communication between design, construction, maintenance, planning, and research<br>• Satisfaction of providing best value for available funds |
| Costs | • Software development<br>• Data collection, processing, storage and analysis<br>• Actual operation of the system; computer hardware, staff<br>• Indirect costs | • Some general costs<br>• Reporting<br>• Processing special requests | • Developing, installing and operating costs of the PMS<br>• Data collection, processing and analysis<br>• Staffing and organizational changes | • Making changes in procedures<br>• Time and effort to upgrade skills; training costs |

## Table 44.2 Additional Costs and Benefits of State Level Pavement Management

| Costs | Benefits |
|---|---|
| • Establishment of a department task force/steering committee<br>• Consulting services<br>• Data collection<br>  — Agency personnel (engineers, technicians, equipment operators), and travel costs<br>  — Training<br>  — Equipment (vehicles, data loggers, distress survey devices, non-destructive structural test devices, surface friction measuring equipment, drilling and coring equipment, roughness measuring equipment)<br>  — Traffic control<br>  — Traffic data acquisition<br>• Data processing (personnel, equipments, supplies, etc.)<br>• Data analysis and reports (personnel, computers and peripherals, supplies, etc.)<br>• System maintenance (personnel, equipment, etc.)<br>• Training agency personnel<br>• Administration | • Maintenance and rehabilitation needs and budgets; priority programming<br>• Justification for funding requests<br>• Effectiveness for expenditures through timely and appropriate action<br>• User cost control through level of service; savings in user costs<br>• More efficient usage of maintenance resources<br>• Improved planning, design, construction, research, performance models, safety, etc.<br>• Improve knowledge of statewide pavement conditions and needs<br>• Improved network serviceability |

user costs savings. Because the former presents difficulties in documenting the costs and benefits associated with highway investments, it has been suggested that if the user cost savings alone indicate a substantial degree of cost-effectiveness of a PMS, then the basis exists for quantitative justification of a PMS [Falls 92].

The Alberta PMS, initiated in 1980 and fully implemented by 1985, provided an excellent case application for testing the concept. It involves a network of over 10,000 km of primary highways; a well-documented history of roughness, surface distress, and structural adequacy; and a rehabilitation budget which remained fixed at $40 million annually over 10 years. The costs of the PMS development and operation were also well documented and it was considered that the vehicle numbers and annual mileage on the relevant network could be estimated within a reasonable degree of error. As well, it was considered that the value of the network could be reasonably well estimated and that vehicle operating cost relationships from the World Bank were applicable.

Based on the foregoing, user cost savings were calculated for an increase in average network serviceability, which actually occurred even though the budget remained constant (in real terms it decreased; thus the analysis was conservative). The benefits-cost ratio (B/C) for savings compared to costs, depending on the assumptions used, were generally in the order of 100:1 [Falls 92]. While this does not represent an exhaustive economic analysis, it seems to be quite a valid way to assess the value of a PMS. Moreover, it can be a very effective means for justifying the implementation of a PMS.

### 44.3.2 Additional Indirect Benefits

A significant indirect benefit of a PMS is improved awareness of the process of pavement management. This is illustrated by the teaching of such pavement management graduate level courses as the United States Federal Highway Administration's, "An Advanced Course on Pavement Management" during 1990 and 1991. The people taking the course, for a variety of reasons, became real advocates of PMS and returned to their organizations with renewed enthusiasm for providing good pavements.

A final set of indirect benefits of PMS that are hard to quantify is the spin-off technology to other infrastructure or facilities such as bridge management. Considerable attention began to be directed to the development of bridge management systems (BMS) in the mid-1980s. A major impetus was provided by the knowledge and improvement in pavements arising from the application of PMS [Hudson 87; Hudson, W. 87a, 87b; Hudson 89a, 89b].

## 44.4 EVALUATION METHODOLOGIES

A number of methodologies exist for comparing costs and benefits of a pavement management system. In the many references available, the alternative methods range from discriminant analysis to general decision theories. Among the most promising candidate methods [Morris 60; Dyckman 61; Edwards 67; Manheim 69, 70; Lee 85] are the ones briefly discussed in the following sections.

### 44.4.1 Benefit-Cost Criterion

Perhaps the best known method for measuring the efficiency of an activity is the benefit cost analysis or, more specifically, the benefit-cost ratio. Efficiency is generally described in this term because other variations, such as rate of return, are sufficiently similar to the benefit-cost analysis to have the same strengths and weaknesses. It has a sound foundation and provides a conceptually sound basis for effective comparisons. In practice, however, there are difficulties which tend to reduce its usefulness. The biggest drawback arises from the difficulty in breaking the factors into either the cost or benefit category; more specifically, the difficulty exists of actually measuring the true cost and the true benefit. There are many intangible factors in benefit-cost analyses, which can be treated as follows:

1. They may be rated subjectively and included in the analysis.
2. When subjective scaling is not possible, verbal descriptions of intangible benefits can be provided in addition to the measured costs and benefits and used as balancing aids by the decision-maker.
3. The intangibles can be simply ignored. Unfortunately, this is common.

## Cost and Benefits of Pavement Management

In other words, if the analyst becomes preoccupied with the mathematics of the benefit-cost analysis and the measurable impacts, the tendency is often to omit intangible benefits. The result can be an inflexible narrowness which leads to a less optimal decision.

Another problem associated with benefit-cost analyses is the question of who pays the cost? and who receives the benefits? For example, improved programming of maintenance funds may be a benefit of the PMS which accrues to the agency and to the public. The cost, on the other hand, may be seen by the maintenance director as a budget imposition. The cost may also involve a change in working assignments or the requirement to undergo additional training for some of the agency personnel.

### 44.4.2 Excess Benefits

One of the many variations of cost-benefit analyses involves the calculation of excess of benefits over costs. A simple case study comparison of this methodology involved a 373 mile arterial network where the 10 year program list from a PMS optimization and from a subjectively based needs study, for an annual budget of $10 million, produced a total of $11 million vehicle operating cost savings or net benefits for the optimized program [Smeaton 81].

### 44.4.3 Goal Achievement

As you attempt to use the various methodologies for analyzing costs and benefits and you review criticisms on benefit-cost and similar evaluation procedures, it is easy to be dismayed with the seemingly overwhelming complexity facing the decision-maker. You also gain some appreciation for the position of the politician or the manager who must react to and give solutions for complex problems every day. One technique for broadening the evaluation and decision-making process is known as goal achievement [Dickey 83]. It involves the assessment of potential alternatives in terms of impacts compared to objectives. Quantifiable measures, which can be probabilistic, are employed in this technique, although some subjective measures may also be used. In general, the procedure is to establish various criteria or goals for alternative methodologies. Next, quantitative measures or subjective estimates are given to each criterion for each of the alternatives. These are then standardized and compared on the basis of a total score of 100 to see which alternatives most completely fulfill or achieve the goals of the decision-maker(s).

### 44.4.4 Cost-Effectiveness Technique

The cost-effectiveness technique (C/E) is an alternative approach to the goal achievement procedure. C/E is actually a relatively simple procedure and works on the basic premise that better decisions will arise when clearer and more relevant data is supplied to the decision-maker. No specific attempt is made to put all benefits and costs in common units, such as the dollars. The following quote is relevant to this approach [Dickey 83]:

> Because many of the consequences and outputs from a transportation system are intangible and otherwise difficult to evaluate in some common metric, the decisions regarding the

conversion to a single dimension and hence the selection decisions are necessarily subjective in nature, at least at the present time.

What might be more useful at this time is a technique for providing the kind of informational support for the selection among plans which recognizes the complex nature of these decisions. Such a decision supporting framework does not attempt to make decisions, but instead structures the information required for making a subjective, but systematically enlightened choice.

Three criteria should be satisfied by any such framework [Dickey 83].

1. Capability of assimilating benefit-cost and similar methodological results in addition to other informational requirements
2. Strong orientation toward a system of values, goals, and objectives
3. Allowance for the clear comparison of trade-offs or compromises between objectives or making explicit the relative gains and losses from various alternatives

Effectiveness is defined as the degree to which an alternative achieves its objective, which may be, for example, the area under the performance curve weighted by traffic volume and section length (see Part Three). The definition, by itself, helps to overcome one of the major objections to the benefit-cost approach in that goals are specified explicitly and are not covered by an all-encompassing benefit term.

The value of the cost-effectiveness approach includes:

1. Simulation, to some extent, of the process by which actual decisions are made
2. Allowance for clearer delegation of responsibility between analysts and decision-makers
3. Easier provision of relative information, in an understandable format, so that the choice process is simplified

### 44.4.5 Search and Choice

In the field of transportation systems analysis, a technique termed "Search and Choice in Transport Systems Analysis" and "Problem Solving Process" (PSP) has been described for use in dynamic modeling of decision making [Manheim 69]. A schematic outline of the process is shown in Figure 44.1.

The focus of the PSP is on actions. Because search and selection procedures concern the basic processes of generation and selection of actions, these procedures are at the heart of the PSP. However, there are a variety of other activities that must occur to allow search and selection to operate, and to revise the context in which they operate. Goal formulation and revision procedures are particularly important.

While the PSP seems to be valid for a decision-maker dealing with various transportation systems, it has not yet apparently been applied to pavement management systems.

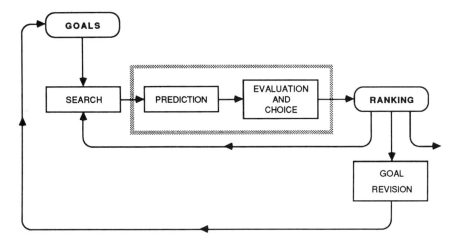

**Figure 44.1** Basic cycle of the PSP (Problem Solving Process) [Manheim 69].

### 44.4.6 Statistical Decision Theory

We live in a very uncertain world, but it is easy to become fascinated by quantitative complex models which produce deterministic results. In truth, there is uncertainty in any such analysis. Uncertainty in transportation, for example, lies in demand estimation (such as traffic), technology, and goals. A key requirement of the future PMS methodology is to more adequately incorporate uncertainty or risk analysis in various predictions, analyses and decisions. It is possible that such technology as [Manheim 70] will eventually become applicable to the PMS process.

### 44.4.7 Discriminant Analysis

Discriminant analysis and classification are multivariate techniques for separating distinct sets of objects and for allocating new objects to previously defined groups. Discriminants are sought whose numerical values are such that the classifications of variables are separated as distinctly as possible. The goal of classification is to sort objects into two or more labeled classes. The emphasis is on deriving rules that can be used to optimally assign a new object to the labeled classes.

A function that separates may serve as an allocator and conversely an allocatory rule may suggest a discriminatory procedure. In practice, the distinction between discrimination (or separation) and classification (or allocation) is not so clear.

The methodology of discriminant analysis while useful in dealing with a large number of objects does not appear to be appropriate for evaluating the costs and benefits of the pavement management process [Lee 85].

### 44.4.8 Other Methodologies

Many other methodologies have been used for decision making. Among these are found terms such as benefit/risk analysis and preference and value trade-offs. It is not considered useful to further assess the applicability of these methodologies to pavement management at this time.

Any approach used for assessing the benefits and costs of a PMS must be meaningful to the decision-maker. After all, the purpose of any such assessment is to present information which will be useful in convincing decision-makers to implement improved PMS methods.

## 44.5 CONCLUDING REMARKS

At least two of the methodologies outlined in the foregoing sections may be useful in assessing the value of a PMS. Certainly benefit-cost analyses are useful in comparing costs and benefits for a given network and informing decision-makers of the value of the PMS. Cost-effectiveness techniques are similarly applicable to such assessments.

Finally, the general concept of goal achievement methods is also worth consideration. It is not yet clear exactly how the method might be applied since it involves an examination of the goals of decision-makers on an individual basis. This might be handled with hypothetical examples if interviews with two or more decision-makers could be arranged for information-gathering purposes.

## REVIEW QUESTIONS

1. What make up the costs associated with pavement management?
2. Do the benefits of a pavement management process outweigh the costs?
3. Name some of the costs associated with the pavement management process.
4. What benefits are associated with pavement management process?
5. How can we measure pavement management cost-benefits effects?

Chapter 45

# Future Direction and Need for Innovation in Pavement Management

## 45.1 INTRODUCTION

Research costs money! Innovation saves money! Good research produces innovations! That is the focus of this chapter. Emphasis in highway research and particularly pavement management research for the past two decades or more has been largely on short-term needs and implementation. Lack of support for intermediate and long-term efforts unfortunately leaves many of the same problems faced in the early years of pavement management, such as lack of good, long-term performance prediction models.

Pavement management has progressed from a concept in the 1960s to a working process in the 1970s to a significant degree of implementation in the 1980s and 1990s. The principles have been formulated and much has been learned from implementation experience at the federal, state/provincial, and local levels in various countries. But this has not been matched by improvements in the component technology of pavement management, and by progress toward standardization.

A substantial amount of innovation will be necessary if we are to realize a standardized pavement management process with widespread or universal applicability. Such a PMS must be technically sound yet have sufficient flexibility for tailoring to individual agency needs and resources. The required innovation and research should range

Table 45.1 Changing Nature for Pavement Research and Associated Issues/Needs [Haas, R. 87]

| | Pre-1950 | 1950's | 1960's | 1970's | 1980's |
|---|---|---|---|---|---|
| Research Emphasis | • Empirical observation of what works<br>• Development of load spreading concept (Westergaard and Burmister theory)<br>• Development of empirical test methods (CBR, Marshall, Hveem)<br>• Specifications (materials, construction) | • Tying down basic properties of materials (asphalt, PCC); standardizing test equipment and procedures; developing improved specifications<br>• Better materials processing and construction technology<br>• Designing and carrying out AASHO Road Test | • Relating materials properties to observed pavement problems (i.e., cracking, durability)<br>• Analysis of AASHO results (load equivalency factors, serviceability-performance concept, design equations) and adoption to state practice<br>• Initiation of satellite tests and long-term pavement performance observations<br>• Initiation of computer-based layer methods of structural analysis<br>• Initiation of pavement management research | • Developing improved structural design technology<br>• Equipment and procedures for in-service measurements and evaluations<br>• Develop pavement design and management methods (FPS, RPS, SAMP, OPAC, etc.), including life-cycle economic analysis and priority programming<br>• Developing of recycling technology<br>• New materials (sulfur-asphalt, polymers) | • Network-level PMS application<br>• Automation of in-service pavement evaluation<br>• User cost relationships<br>• Micro-computer based PMS methods, models and procedures<br>• Maintenance and rehabilitation (treatments, performance predictions, economic evaluation)<br>• Initiation of SHRP studies<br>• Reliability concept in AASHTO guide<br>• Performance based specifications |

| Issues/Needs | | | | |
|---|---|---|---|---|
| Provide pavement structures for increasing loads and traffic; all-weather surfaces for rural roads; wartime "lessons" on needs for structural design procedures, specifications, test methods, improved construction, etc. | Providing materials and technology for initiation of the post-war road-building boom, including start of the interstate system; lack of knowledge on relative damage effects of heavy loads | Need for solving major distress and performance problems appearing in pavements; demands for better quality materials; adopting AASHO Road Test results (i.e. new guides); developing more fundamentally based methods of structural design; need for performance and distress models | Need for better, more comprehensive data bases, performance estimates for design alternatives, identifying most economic alternatives, better traffic and load inputs; need for improved priority programming procedures and more comprehensive pavement management in general | Energy conservation; user costs; effects of loads, environment and their interactions on pavement deterioration; cost allocation; premium or new, improved materials; implementation of pavement management at state and local levels; improved airport PMS. |

from short-term problem solving to strategic efforts for technology and application improvements.

The issues presented in this chapter are intended to complement those addressed in Chapters 5 and 34. In these chapters a historical perspective was presented to demonstrate how pavement management evolved from various research programs, and subsequently how application of a comprehensive pavement management system can focus a research program. The perspective of this chapter is to address future evolution of pavement research to complement management systems technology.

This chapter addresses the research and innovations needed to improve pavement management. More specifically, it has the following objectives:

1. Review the changing nature of pavement research and the associated issues.
2. Describe a standardized or generic structure for pavement management within which the component activities, and research toward their improvement, can be incorporated.
3. Identify the major types of research which must be carried out for a successful program of improvements in pavement management technology and application.
4. Define the major elements of successful pavement research.
5. Identity some of the opportunities for innovation and major advances in pavement technology and application of the process.

## 45.2 CHANGING NATURE OF PAVEMENT RESEARCH AND THE ISSUES AND NEEDS

In order to develop a program for innovation, it is useful to review in historical context the changing nature of pavement research and the issues it has addressed. Table 45.1 illustrates the emphasis of research and needs over the past 30 years. It is not meant to be exhaustive; rather, it shows that while the issues and the research emphasis have changed to a considerable degree, many of the problems remain. For example, the needs for better performance predictions, materials, construction and maintenance technology, good data bases, energy conservation, and traffic and load input data are as important as ever.

Table 45.2 shows some of the key issues for the 1990s. The breakdown into general system-related issues versus more specific technology-related issues is intended more for broad identification than for sharp classification. How pavement research will respond, and what the emphasis will be, still remains to be determined. Certainly, the Strategic Highway Research Program (SHRP), formulated in the mid 1980s and initiated in 1987 [SHRP 86], addresses a number of the issues from Table 45.2. However, it should be noted that SHRP is technology-related and does not address pavement management per se. Moreover, SHRP cannot address or solve all the pavement problems that exist. Adoption or incorporation of the results into the technology base for pavement management still largely remains to be carried out over the next decade or two.

Table 45.2  Key Pavement Management Issues in the 1990s

| A. General System-related issues | B. Specific Technology-related Issues |
|---|---|
| • Efficient, reliable tools for determining and interpreting physical conditions of pavements<br>• Rehabilitation of an agency highway network<br>• Automation of construction and maintenance<br>• Relevancy between specification and quality and performance of the end-product<br>• Toward true end-product specifications (based on long-term pavement performance)<br>• Reliable measurements and forecasts of traffic and loads<br>• Long-term monitoring of performance and behavior (in-service sections, test roads, relationships between controlled, laboratory measurements and in-service observations)<br>• Broader application and implementation of pavement management and equitable allocations of funds; assessments of long-term implications of funding decisions<br>• Fair, user-tax assessments from different classes of vehicles<br>• Better evaluation of variability and formal incorporation of risk management procedures and decisions<br>• Attraction of qualified people, improved quality of training and education (including continuing education)<br>• Improved productivity and better utilization of technology from other fields<br>• Energy conservation in construction, production and processing of materials, vehicle operation<br>• Integration of pavement management with other facilities management systems | • Solving specific asphalt distress problems (stripping, thermal cracking, rutting, binder aging, reflection cracking through overlays) and specific Portland cement concrete pavement problems (faulting at joints, spalling, cracking)<br>• Alternatives and timing for preventive and corrective maintenance treatments to maximize cost-effectiveness<br>• Recycling of waste and reclaimed materials (asphalt, concrete, bricks, tires, plastics, spent foundry sands, fly ash, roofing materials, etc.) into pavements<br>• Modified, premium-quality asphalts (through polymer or other related processes)<br>• Fundamentally based test methods and compositional analyses for binders, and relationships to in-service behavior<br>• Upgrading of marginal materials and/or selective use in pavement type and structure<br>• Performance prediction models which identify load, environment and interaction-related losses<br>• High-speed, automated, reliable methods of deflection and surface distress measurement |

## 45.3 STANDARDIZED (GENERIC) STRUCTURE FOR PAVEMENT MANAGEMENT

In order to realize maximum benefits from the pavement research, a structure or framework for incorporating the results into pavement management practice should exist. Such a structure facilitates the mechanisms for adapting and implementing innovative results.

One of the first definitions of a standardized pavement management structure for both the network and project-level sets of activities was set out in an NCHRP report and in two national workshops [Hudson 79, 80] and later updated in ASTM [Haas, R. 87]. Table 45.3 provides a summary outline of the structure and a listing of some of the key component activities or decisions. This is a framework only. A specific operating system for a particular agency would have a linked set of the specific models, methods, and procedures which comprise these activities. However, a framework combined with an agency's specific system can enhance the identification of issues and needs, research priorities, and the implementation of the results.

The question of whether the future evolution of pavement management will require a different structure has also been considered [Haas 88]. It was concluded that this should not be necessary for at least the next decade because the structure is quite amenable to progress, it allows for agencies to exercise flexibility, and it provides a consistent philosophy for addressing issues and needs. Moreover, because of its generic basis, it is in fact applicable to the management of other facilities, with of course some modifications of particular terminology.

## 45.4 MAJOR TYPES OF RESEARCH AND BENEFITS OF A COORDINATED PLAN

Many state and federal agencies have prepared statements of pavement research needs, research plans, and programs of technology transfer. These are necessary and a large amount of useful research has been carried out. However, what is often lacking is an overview of what is required for a successful program of research, and the associated long-term benefits or payoffs.

To achieve such success, the following four major types of research should be incorporated in the overall approach:

1. Developing solutions to short-term problems and applications
2. Intermediate-term research and development
3. Strategic or long-term research
4. Implementation, including technology transfer and the development of research capabilities

Emphasis in pavement research for the past several decades has been on short-term research and implementation. There has been a lack of support for intermediate- and long-term efforts. On the positive side, SHRP, which began in 1987, has provided a focal point for reevaluation of some overall pavement technology research needs. Of particular importance to pavement management are the Long-Term Pavement Performance (LTPP) Study, the asphalt studies, and the maintenance studies of SHRP.

In addition, much of the knowledge gained from past highway experience is being lost as staff retire. Literally, the experience gained in the 1950s, 1960s, and 1970s is

## Table 45.3  An Activity/Decision-Based Generic Structure for Pavement Management [Haas, R. 87]

| Basic Blocks of Activities | Network Level (Administrative and Technical Decisions) | Project Level (Technical Decisions) |
|---|---|---|
| Data | • Sectioning and data acquisition (field data on roughness, surface distress, deflection, etc., plus traffic, cost and environmental data);<br>• Portrayal of present status<br>• Data processing and evaluation | • Subsectioning and detailed data acquisition (materials, traffic, unit costs, etc.)<br>• Data processing and evaluation |
| Criteria | • Minimum or maximum acceptable levels (serviceability, surface distress, structural adequacy, etc.)<br>• Maximum program costs<br>• Maximum levels of traffic interruption<br>• Selection basis (i.e., cost-effectiveness) | • Minimum or maximum as-built conditions (roughness, structural adequacy, surface friction, etc.)<br>• Maximum project costs<br>• Selection basis (i.e., minimum net present worth of costs) |
| Analyses | • Present needs sections, deterioration predictions and future needs sections<br>• Maintenance and rehabilitation alternatives for needs sections, deterioration predictions, life-cycle costs and benefits<br>• Priority analysis for different budget levels or for specified performance standard(s) | • Within project rehabilitation or maintenance alternatives, detailed field and laboratory tests<br>• Deterioration predictions (serviceability and distress) for alternatives<br>• Economic evaluation of alternatives |
| Selection | • Determination of final programs of rehabilitation and maintenance<br>• Program recommendations, administrative and elected body approvals | • Best within project or section maintenance and/or rehabilitation alternatives |
| Implementation | • Establishment of work schedules, sequences, contract tenders and awards<br>• Program monitoring<br>• Budget and financial planning updates<br>• Inventory and data base updates | • Construction activities, work control and quality assurance, as-built records<br>• Maintenance activities and management records<br>• Data base updates |

rapidly disappearing from the pavement engineering field with the continued retirement of senior staff.

It is important to have an overall, coordinated plan to guide future funding and to address future needs. Benefits that can derive from such an overall plan include:

1. Providing the means for seeking and organizing results of research that is performed both nationally and internationally
2. Providing direction for future research funding and enabling personnel to tailor research to future national needs
3. Providing a coordinated avenue to implement innovation more readily
4. Identifying more rapidly the limitations and shortcomings of existing and historical methods
5. Integrating current knowledge, data, and research results into a coherent strategy that is consistent with a standardized PMS

## 45.5 ELEMENTS OF SUCCESSFUL RESEARCH

Management of a research function within a PMS was addressed in Part Five, while the planning aspects of a research function implementation were considered in Part One. This section expands on development of a strategic research program. Among the elements of a successful program of research are:

1. Having an overall plan for short-, intermediate-, and long-term research
2. Top-level commitment and support plus sufficient funds
3. Continuity of funding, not stop and start
4. Providing the flexibility and freedom for innovation
5. Developing research capability (people, facilities, etc.)
6. Cooperation between practitioners and researchers
7. Disseminating the results of the research through publications, conferences, workshops, seminars, short courses, etc.

### 45.5.1 An Overall Plan

An integrated, overall plan covering short-, intermediate-, and long-term research is particularly important for state and federal agencies. The issues of current concern might carry the primary focus but a macro approach will allow for better interaction between projects, allow better identification of priorities, preserve the long-term integrity of the research, and permit more efficient, overall program management.

### 45.5.2 Commitment and Funding Support

Successful pavement management systems at both the state and local levels have had, without any known exceptions, strong, top-level commitment and support in the organization. Similarly, pavement research programs must have such commitment and support, in addition to the expected commitment of the researchers themselves. Sufficient and consistent funding with a reasonable degree of flexibility is also necessary. This is not to say that justification for funding and identification of expected payoffs

**Future Direction and Need for Innovation in Pavement Management**  565

are not necessary. If the payoffs are to be realized and the opportunity for innovation is to exist, such funding support and flexibility are essential components. Organizational support—in terms of facilities, staff, opportunities to interact with practitioners and researchers both within and outside the agency, and very importantly, encouragement—is also important to successful research.

### 45.5.3 Continuity of Funding

To be successful, research funding must have reasonable continuity. This does not mean a blank check but rather the opportunity to meet real breakthroughs with adequate support and funding. Innovation does not occur on a schedule, it happens in unique and unexpected ways and should be nonrestricted.

### 45.5.4 Flexibility and Freedom for Innovation

A common thread of successful, innovative research has been the degree of flexibility and freedom provided to the researchers. Innovative results cannot be mandated. They come from hardworking, innovative people who are not placed in a bureaucratic straightjacket of administrative control. Particularly constraining is a detailed, procedural environment where more time is spent in progress reporting than in actually doing research. A research management team should select researchers in whom they have confidence. A multilayer mixture of administration, not control, is the key to good results. The AASHO Road Test is the prototypical example where the director, William Carey, had the authority and the freedom to fulfill the project mandate.

It must also be recognized that research may carry a considerable degree of risk, and that the payoff in terms of implementation may be some distance in the future. Thomas Edison tried more than 100 material combinations before he succeeded in producing the first electric light bulb. He "failed" his way to success.

### 45.5.5 Developing Research Capability

Research capability resides in universities, institutes, consulting organizations, and state and federal research groups. While much of this capability has been acquired on the job, the basic source lies in universities. Many persons who are active in pavement research have postgraduate degrees where they learned the basic concepts required for research success.

Development of research capability requires dedicated, competent students, research support, course work, and direction from professors. If one looks at the highly regarded pavement researchers in the United States, Canada, and abroad, in the public agencies, and in the private sector, a substantial number of them come from universities having an extensive track record of educational excellence and research accomplishments. It is essential that continued regeneration of research capability occurs, with universities playing an integral part, and that there be a strong interaction between the public and private sectors and the universities.

### 45.5.6 Cooperation Between Practitioners and Researchers

As previously emphasized in Chapters 5 and 34, successful innovation can best be implemented if the practicing engineer is involved from the beginning. A PMS makes this possible because the feedback loop for new innovation is hinged on the results of field use and upgrading of the PMS. It is important for practitioners to recognize that there is such a thing as appropriate research methodology which must be used to produce the best results.

### 45.5.7 Dissemination of Research Results

Research results need to be disseminated within organizations and externally for peer review. Of course much of the internal success is in terms of implementation and improved efficiency or cost-effectiveness, but external judgments are also important to followup work and its long-term success. There are many new techniques for dissemination of results which include, for example, videotapes, multimedia presentations, and user friendly computer software programs.

The forums for dissemination of research results include journal publications, conferences, workshops, and seminars. The latter two forums are also often applicable to internal dissemination. Another important forum is the example represented by the "Advanced Course in Pavement Management Systems" of the FHWA, which incorporates both up-to-date practice and recent research results [FHWA 91].

## 45.6 OPPORTUNITIES FOR INNOVATION

### 45.6.1 Evolution of Pavement Management

The evolution of the pavement management process provides a good context for identifying opportunities for innovation and major advances in technology, as well as the application of the process itself. The assumption is that the process per se will not likely change substantially for at least the next decade, for two basic reasons: (1) implementation experience has shown that PMS is acceptable in its present form to most agencies; moreover, a period of time is needed for consolidation and for the large remaining number of agencies to install their systems; and (2) the major current thrust is improvement of the technology within the process, rather than the process itself. Research is needed on the process of pavement management.

The evolution of public sector network-level pavement management can be summarized in terms of two streams, state/provincial/federal and local. Pavement management is expected to exist as a distinct and stand-alone process for the first stream during the 1990s. The reason is largely related to the size of the networks, the organizational structure, and the methods of budget preparation and administration for the state/provincial/federal situation. These authorities deal with a number of quite large management systems even within their transportation departments (i.e., airports, highways, pavements, safety, bridges) and it is extremely difficult to combine the benefits of each into a single stream. Consequently, the interfacing has to be done on a broad policy level.

For local agencies, it is quite likely that pavement management will evolve into a larger, integrated Total Facilities Management (TFM) type of system. This is indeed desirable where one office or individual, such as a commissioner of public works, is responsible for underground services, traffic, pavements, bridges, parks and recrea-

# Future Direction and Need for Innovation in Pavement Management

tion, etc. Pavement management systems can be valuable here because they represent the most advanced and comprehensive system development of all the facilities involved and can thereby provide the keystone or guidance for the development of TFM systems.

The opportunities for innovation thus lie in the areas of new and improved technology, process integration, and standardization of pavement management. At the project level, pavement management will likely continue in the same generic form for at least the next decade. The major opportunities for innovation will therefore lie in the areas of:

1. Technology improvements, materials, processes, automation, characterization, analyses, equipment, etc.
2. Long-term performance prediction models for pavements

The first area of opportunities is relatively self-explanatory but the second area represents something more long-term in nature, as subsequently described.

## 45.6.2 Areas of Opportunities

Literally, hundreds of specific opportunities exist for innovation in PMS technology and application of the PMS process. While comprehensive national and regional efforts to identify and prioritize these opportunities are very important, it is most useful within the scope of this section to identify several broad areas of opportunity as shown in Table 45.4. The table is not meant to be exhaustive, and the assignment of network versus project level applicability, degree of risk and short-to-long-term payoff is largely subjective. However, the areas listed represent a considerable range of opportunities and the context for many specific opportunities with regard to the following five areas:

1. Incremental improvements in technology
2. Utilization of experience from more widespread and longer-term implementation of PMS
3. Development of new equipment and methods, and their automation
4. Application of new technologies (i.e., expert systems)
5. Design, construction, and maintenance of long-term performance guaranteed pavements

The latter item depends on the development of long-term performance-based specifications (see Table 45.4, item 2). It represents a significant, albeit high-risk, opportunity which will require considerable standardization, particularly regarding the specification elements.

Figure 45.1 illustrates the concept of long-term performance-based specifications, using the generic structure described earlier in Table 45.3. Ultimately, this concept could function with the agency required to set the life-cycle, performance-based specifications and to carry out periodic monitoring. The contractor would be responsible for all project-level activities. This true end-product approach transfers the risk from the agency to the contractor. Its appeal lies in the innovation that can be explored between a premium pavement and a low initial cost pavement with more extensive

Table 45.4 Areas of Substantial Opportunities for PMS Advances and Innovation

| Opportunity | 1. Development of a widely accepted, standardized structure or framework for pavement management which: a) allows flexibility for alternative models and tailoring to individual agency situations, b) includes staged implementation guidelines, and c) identifies or specifies deliverables of each stage for various types of agencies | 2. Development of long-term performance based specifications for pavements | 3. Development of an equitable and efficient method of: a) determining pavement damage due to loads, environment and their interactions, b) assessing the component damage costs, and c) assigning them to vehicle classes. | 4. Quantification of the behavior, performance, rehabilitation strategy, and user-cost effects of various preventive and corrective maintenance treatments under various conditions | 5. Development of incentive programs for contractors, researchers, and public agency specifiers to realize full benefits of PMS improvements (i.e., encouragement of innovation by contractors, follow-through by researchers and incorporation of new ideas or research results by specifiers) |
|---|---|---|---|---|---|
| Network (N) and/or Project (P) level applicability | N & P | P | N & P | P | N & P |
| Degree of risk: High (H), Medium (M), Low (L) | L | H | M | M | M |
| Short (S), Intermed. (I) or Long-term (L) payoff | S | L | L | I | S |

| Opportunity | 6. Comprehensive identification and quantification of payoffs for technology improvements (truck suspensions which minimize damage, effective drainage systems, etc.) and solution of specific technical problems (rutting, reflection, cracking, etc.) | 7. Development of comprehensive programs for improving the technical capabilities of contractors, public agency specifiers and researchers to realize the full benefits of PMS research results | 8. Resolution of the inconsistencies between sophisticated analytical methodologies and the relative lower quality or approximation of input data (traffic, environment, material) | 9. Development of effective "interfacing" between network and project-level PMS so that decisions are consistent | 10. Planning and executing a major (funded) program to codify the next generation of pavement management (including the standardized framework, incorporation of SHRP and other results from practitioners and researchers, etc.) |
|---|---|---|---|---|---|
| Network (N) and/or Project (P) level applicability | N & P | N & P | N & P | N & P | N & P |
| Degree of risk: High (H), Medium (M), Low (L) | L | L | M | L | L |
| Short (S), Intermed. (I) or Long-term (L) payoff | S, I and L | S, I and L | L | M | M |

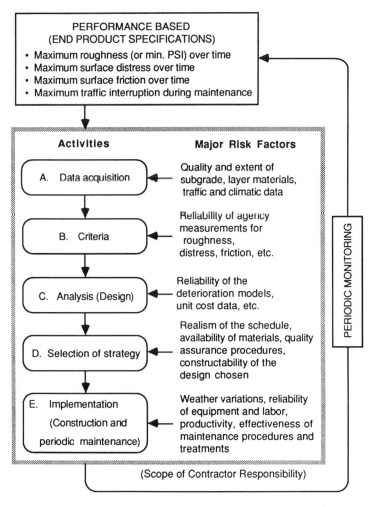

**Figure 45.1** Basic elements and long-term performance-based specifications.

periodic maintenance. A specification on maximum traffic interruption would limit the extreme of a low-cost initial pavement with subsequent extensive maintenance and/or rehabilitation over the life-cycle.

The long-term performance-based specifications will of course necessitate certain practices related to performance bonds, financing, and the like which are different than generally used today. However, the basic concept of performance-guaranteed pavements can be advantageous to both the consumer and the contractor.

The last item of Table 45.4 would involve an effort similar to that required for developing the 1986 AASHTO Pavement Design Guide. It is important that the next generation of pavement management be codified or comprehensively described, and that it incorporate the latest technology and research program results (including those

# Future Direction and Need for Innovation in Pavement Management

**Table 45.5  What Has Been Learned from 25 Years of Pavement Management Experience**

| From the PMS process itself |
|---|
| 1. The framework and component activities for the PMS process can be described on a generic basis. |
| 2. Existing technology and new developments can be effectively organized within this framework. |
| 3. The process allows complete flexibility for different models, methods, and procedures. |
| 4. PM operates at two basic levels: network, and project. |
| 5. A sound technological base is critical to the process and its effective application. |
| **From using PMS** |
| 6. Development and implementation of a PMS must be staged. |
| 7. Staging allows for understanding and acceptance by various users. |
| 8. Options almost always exist; they should be evaluated on a life-cycle basis; this means we need models for predicting deterioration of existing pavements and rehabilitation or maintenance alternatives. |
| 9. PM can make efficient use of available funds but it will not "save" a network if funding is below some threshold level. |
| 10. Good information is essential to the effective application of a PMS. |

from SHRP) if the potential benefits of the vast amount of knowledge and experience available are to be realized. In addition, it is important that the team chosen to do this includes practitioners and researchers, as was included for the AASHTO Guide effort, and that they have sufficient freedom to innovate.

## 45.7  THE FUTURE OF PAVEMENT MANAGEMENT

### 45.7.1  Learning from the Past

A lot has been learned since the pavement management system concept was initially developed in the 1960s. The use of a sound technological base plus good data; a staging requirement for implementation; and the fact that alternatives, deterioration models, and life-cycle economic evaluation are essential elements, are examples. Table 45.5 identifies some of the key lessons from 25 years of pavement management experience [Better Roads 89].

Equally interesting is the generic structure of pavement management that has evolved. This generic form can be applied to other infrastructure components, such as water, sewer, and bridges, and should remain reasonably stable as future technology is developed. Rather than hamper progress such a stable structure in fact provides a consistent philosophy for identifying technology improvement needs and realizing the benefits of such improvements.

### 45.7.2  Looking to the Future

Several issues and needs must be resolved for PMS to progress:

- Resolve the effects of different organizational structures.
- Identify the requirements and direction of local versus state or federal systems.
- Establish benefits in quantitative terms.
- Integrate pavement management with maintenance management and other areas or levels of transport system management.

There are pavement management process related issues to be resolved, including the need to:

- Establish relationships between pavement and other facilities or infrastructure management systems, with methods for comparing results.
- Effectively use automation in data acquisition and processing, decision making, construction and maintenance operations, and so on.
- Develop better interfacing of network and project levels of pavement management.
- Develop better methods of estimating existing pavement deterioration, as well as maintenance and rehabilitation treatments.
- Develop better ways to evaluate the impacts of different vehicle weights, types, and dimensions.

Clearly, good implementation and application must continue, but new ideas and new innovation are also badly needed. No innovative open-ended PMS research is currently underway anywhere in the world. It is badly needed. What are the essential elements of renewing innovation in pavement management?

1. A source of funds in reasonably sized programs; say, $500,000 over 3 years
2. Dedicated, zealous researchers; a small interdisciplinary team of three or four people including expertise in pavements, economics, and statistics
3. Reasonable flexibility and freedom for the research team to be innovative
4. A sponsor willing to accept failure as an outcome, since innovative research is risky
5. An open-minded, small group of advisors to interact and advise the direction of the team, but not to dictate or control
6. Trust placed in the team of researchers, with no manipulation of them or the results

### 45.7.3 Expected Changes

Advances in PMS will come from incremental improvement in current technology and changes in the process, greater use of PMS, new equipment and methods, and application of new technologies. New equipment and methods, along with automation, offer some very promising opportunities. Pavement management technology will undergo significant improvements, particularly in construction, maintenance, and evaluation. Feedback of this data can improve pavement design models and practices.

Technology exists for using laser, optical, and acoustical methods to measure profile or roughness. Instruments for measuring surface distress are entering the mar-

# Future Direction and Need for Innovation in Pavement Management

ket. High-speed deflection-measuring methods are under development and should be available for use during the 1990s.

Promising technologies for construction and maintenance will mainly involve robotics for equipment, plus microelectronic-based automated control procedures. An example of a very promising new technology in construction is a different method for asphalt compaction [Abdelhalim 92]. The strategic Highway Research Program is sponsoring the development of robotic equipment for pothole repair and crack sealing.

## 45.8 CLOSURE

It seems important to include a closure because details and time-related aspects of technology change, sometimes drastically, sometimes slowly.

Many specific opportunities exist for innovation and major advances in pavement management technology and application. Comprehensive national and regional efforts are required to identify and prioritize these opportunities, but their scope is illustrated by several broad areas of opportunity. These range from the development of long-term performance-based specifications to the planning and execution of a major program for codifying the next generation of pavement management.

This book is written in the 1990s, an update of the original 1977 version. While changes have occurred in perhaps one-third of the technology, the concepts and applications are timeless. Fully two-thirds of the contents of this book, including the basic principles, will be valid for some years.

## REVIEW QUESTIONS

1. Name some of the future directions expected to develop in pavement management as we approach the 21st century.
2. Name one or two of the major elements of future research needs for pavement management.

# References to Part Seven

[AASHO 74] "AASHO Interim Guide for Design of Pavement Structures," Washington, DC, 1974.

[Abdelhalim 92] Abdelhalim, O. A., R. Haas, and W. Phang, "Improving the Properties of Asphalt Pavements Through the Use of AMIR Compactor: Laboratory and Field Verification," Proc., Seventh International Conference on Asphalt Pavements, Vol. 4, Nottingham, Aug. 1992.

[Better Roads 89] "The Future of Pavement Management," Better Roads, July 1989, p. 44.

[Chang 87] Chang, Edmond Chin-Ping, "Using Expert Systems to Select Traffic Analysis Software," Transportation Research Board, Washington, DC, 1987.

[Derfler 85] Derfler, F. J., "Expert-Ease Makes Its Own Rules," PC Magazine, pp. 119–124, April 16, 1985.

[Dickey 83] Dickey, J. W., et al., *Metropolitan Transportation Planning*, Second Edition, Hemisphere Publishing Corporation, McGraw-Hill Book Company, New York, 1983.

[Duzan 62] Duzan, H. D., J. C. Oehmann, and J. M. Davis, "Use of Road Test Equations in Pavement Cost Allocation," Special Report 73, Highway Research Board, 1962.

# References to Part Seven

| | |
|---|---|
| [Dyckman 61] | Dyckman, J., "Planning and Decision Theory," Journal of American Institute of Planners, Vol. 27, 1961. |
| [Edwards 67] | Edwards, W., and A. Tversky, *Decision Making*, Penguin Books, Baltimore, 1967. |
| [Elsevier 89] | "Personal Consultant Easy, Release 2.0" Microcomputers in Civil Engineering 4, Elsevier Science Publishing Co. Inc., 655 Avenue of the Americas, New York, pp. 163–165, 1989. |
| [Elton 89] | Elton, D. J., "Expert System for Diagnosis of Hot Mix Asphalt Segregation", Transportation Research Board, Washington, DC, 1989. |
| [EXSYS 85] | "EXSYS—Expert System Development Package", EXSYS, Inc., Albuquerque, NM 87194, 1985. |
| [Falls 92] | Falls, L. C., S. Khalid, and R. Haas, "A Cost-Benefit Analysis of Network Level Pavement Management," Proc., Transportation Association of Canada Annual Conference, Quebec, Sept. 1992. |
| [FHWA 91] | "An Advanced Course in Pavement Management Systems," Course notes, Federal Highway Administration–U.S. Department of Transportation and Transportation Research Board, 1991. |
| [Forgy 87] | Forgy, C. L., The OPS5 Reference Manual, Department of Computer Science, Carnegie-Mellon University, 1987. |
| [Franck 89] | Franck, B. M., and T. Krauthammer, "An Expert System for Field Inspection of Concrete Dams: Part 1 Engineering Knowledge," Engineering With Computers, An International Journal for Computer-Aided Mechanical and Structural Engineering, 1989. |
| [Fricker 91] | Fricker, J., "Expert System for Forest Road Design," Under development, Department of Civil Engineering, Purdue University, 1991. |
| [Fry 75] | Fry, A. T., "A Study of the Economics of Road Vehicle Limits—Summary and Recommendations," NAASRA Study Team Report R3, October 1975. |
| [Haas, C. 87] | Haas, C., and H. Shen, "PRESERVER: A Knowledge-Based Pavement Maintenance Consulting Program," Proc., Second North American Conference on Managing Pavements, Toronto, 1987. |
| [Haas, R. 87] | Haas, R., "The Role of Standards in Pavement Management," ASTM Standardization News, April 1987. |
| [Haas 88] | Haas, R., "Future Direction of Pavement Management," Proc., Vol. VI, International Road Federation, Seoul, Sept. 1988. |
| [Haas 90] | Haas, R., T. Triffo, and M. Karan, "The Use of Expert Systems in Network Level Pavement Management," Proc., OECD Workshop on Knowledge-Based Expert Systems in Transportation, ESPOO, Finland, June 1990. |
| [Hajek 89] | Hajek, J., G. Chong, R. C. G. Haas, and W. Phang, "Knowledge-Based Expert System Technology Can Benefit Pavement Maintenance," Transportation Research Board, Washington, DC, 1989. |
| [Hall 87] | Hall, K. T., J. M. Connor, M. I. Darter, and S. H. Carpenter, "Development of an Expert System for Concrete Pavement Evaluation and Rehabilitation," Proc., Second American Conference on Managing Pavements, Toronto, 1987. |

| | |
|---|---|
| [Harmon 85] | Harmon, P., and D. King, *Artificial Intelligence in Business—Expert Systems*, A Wiley Press Book, J. Wiley & Sons Publishing Company Inc., 1985. |
| [Hozayen 90] | Hozayen, H., W. Schenk, and R. Haas, "Diagnosis of Marshall Mix Design Using Expert Systems Technology," Proc., Canadian Technical Asphalt Association, Winnipeg, Manitoba, 1990. |
| [Huarng 86] | Huarng, Y. W., S. Shenio, A. Mathews, F. Lai, and L. Fan, "Fault Diagnosis of Hazardous Waste Incineration Facilities Using a Fuzzy Expert System," American Society of Civil Engineers, 1986. |
| [Hudson 58] | Personal communication with W. R. Hudson, resident staff AASHO Road Test, 1958. |
| [Hudson 79] | Hudson, W. R., R. Haas, and R. D. Pedigo, "Pavement Management Systems Development," NCHRP Report 215, November 1979. |
| [Hudson 80] | Hudson, R., and R. Haas, "The Development, Issues and Process of Pavement Management," Paper prepared as background for National Workshops on Pavement Management, Phoenix, May 1980, Charlotte, Sept. 1980. |
| [Hudson 87] | Hudson, S. W., R. F. Carmichael III, L. O. Moser, W. R. Hudson, and W. J. Wilkes, "Bridge Management Systems," NCHRP Report 300, Transportation Research Board, December 1987. |
| [Hudson, W. 87a] | Hudson, W. R., "Approaches for Comparing the Costs and Benefits of Improved Levels of Pavement Management," ARE Report prepared for the FHWA, Contract No. DTFH61-87-P-00554, 1987. |
| [Hudson, W. 87b] | Hudson, W. R., C. Boyce, and N. H. Burns, "Improvements in On-System Bridge Project Prioritization," Research Report 439-1, Center for Transportation Research, The University of Texas at Austin, January 1987. |
| [Hudson 89a] | Hudson, W. R., N. Burns, R. Harrison, and J. Weissmann, "A Bridge Management System Module for the Selection of Rehabilitation and Replacement Projects," Preliminary Report 439-4F, Center for Transportation Research, The University of Texas at Austin, November 1989. |
| [Hudson 89b] | Hudson, W. R., "Development of Managing Systems Around the World," Engineering Foundation Conference Proceedings on Managing America's Aging Bridge Systems: Issues and Directions, New York, 1989. |
| [INSIGHT 86] | "INSIGHT 2 Reference Manual, Version 1.0," Level Research Inc., 503 Fifth Avenue, Indialantic, FL 32903, 1986. |
| [Kher 90] | Kher, R., "Institutional Issues," *An Advanced Course in Pavement Management Systems*, U.S. Department of Transportation, 1990. |
| [Langsner 62] | Langsner, G., T. S. Huff, and W. J. Liddle, "Use of Road Test Findings by AASHO Design Committee," Highway Research Board, Special Report 73, 1962. |
| [Lee 85] | Lee, H., W. R. Hudson, and C. L. Saraf, "Development of Unified Ranking Systems for Rigid and Flexible Pavements in Texas," Research Report No. 307-4F, Center for Transportation Research, The University of Texas at Austin, 1985. |

# References to Part Seven

[Luce 57]  Luce, R. D., and H. Raiffa, *Games and Decisions*, John Wiley and Sons, New York, 1957, Ch. 13.

[Maher 87]  Maher, M. L., "Expert Systems for Civil Engineers: Technology and Application," American Society of Civil Engineers, 1987.

[Manheim 69]  Manheim, M. L., "Search and Choice in Transport Systems Analysis," Paper sponsored by Committee on Transportation System Evaluation and presented at the 48th Annual Meeting of TRB, 1969.

[Manheim 70]  Manheim, M. L., "Decision Theories in Transportation Planning," Highway Research Board Special Report No. 108, 1970.

[McNerney 91]  McNerney, M., W. R. Hudson, and T. Dossey, "A Comparison and Re-Analysis of the AASHO Road Test Rigid Pavement Data," Presented at 70th Annual Transportation Research Board Meeting, Washington, DC, January 1991.

[Moose 87]  Moose, A., D. Shafer, C. Gossard, and D. V. Winkle, "VP-EXPERT Reference Manual," Paperback Software International, 2830 Ninth Street, Berkeley, CA 94710, 1987.

[Morris 60]  Morris, W. T., *The Analysis of Management Decisions*, Richard E. Irwin, Inc., Homewood, IL, 1960.

[NAASRA 74]  "A Study of the Economics of Road Vehicle Limits, Concepts and Procedures," Study Team Report R1, National Association of Australian State Road Authorities, Melbourne, Australia, July 1, 1974.

[NASA 87]  "CLIPS Basic Programming Guide Version 4.1," NASA, COSMIC, 382 E. Broud St., Athens, GA 30602, Sept. 1987.

[Ohio DOT 90]  "Application of the Global Positioning (GPS) for Transportation Planning," Proc., State Steering Committee, Ohio DOT, Columbus, OH, January 1990.

[PennDOT 90]  "The Development of a Geographic Information System in Pennsylvania," PennDOT, Harrisburg, PA, March 1990.

[Ritchie 87]  Ritchie, G. S., "Application of Expert Systems for Managing Pavements," Proc., Second North American Conference on Managing Pavements, Toronto, 1987.

[Ritchie 89]  Ritchie, G. S., "Expert System for Pavement Overlay Design," Transportation Research Board, Washington, DC, 1989.

[Schwartz 88]  Schwartz, C. W., Lecture on expert systems given at the Workshop on Expert Systems for Maintenance and Rehabilitation of Flexible Pavements, McLean, VA, March 1988.

[Scrivner 68a]  Scrivner, F. H., W. M. Moore, and W. F. McFarland, "A Systems Approach to the Flexible Pavement Design Problem," Research Report 32-11, Texas Transportation Institute, Texas A&M University, 1968.

[Scrivner 68b]  Scrivner, F. H., and W. M. Moore, "An Empirical Equation for Predicting Pavement Deflections," Research Report 32-12, Texas Transportation Institute, Texas A&M University, 1968.

[SHRP 86]  "Strategic Highway Research Program Research Plans," Final Report, Sponsored by American Association of State Highway and Transportation Officials (AASHTO), Federal Highway Administration–U.S. Department of Transportation (FHWA), and Transportation Research Board (TRB), May 1986.

[Smeaton 81]  Smeaton, W. K., M. A. Karan, C. Bauman, and G. A. Thompson, "Case Illustration of Long-Term Pavement Management and Update," Proc., Canadian Technical Asphalt Association, Montreal, Nov. 1981.

[Smith 89]  Smith, W., and Evans, *Road Work, a New Pricing and Investment Policy*, The Brookings Institution, 1989.

[Smith 91]  Smith, R. E., "Addressing Institutional Barriers to Implementing a PMS," Pavement Management Implementation, F. B. Holt, W. L. Gramling, eds., American Society for Testing and Materials, STP1201, Philadelphia, 1991.

[Teknowledge 87]  Teknowledge Inc., "M. I. Reference Manual," 525 University Avenue, Palo Alto, CA 94301, August 1987.

[Thomas 86]  Thomas, W., "1st Class Instruction Manual, Programs in Motion," Wayland, MA, 1986.

[Tung 87]  Tung, R. Shieng-I, and J. B. Schneider, "Designing Optimal Transportation Networks: An Expert Systems Approach," Transportation Research Board, Washington, DC, 1987.

[Wang 93]  Wang, C. P., J. P. Zaniewski, G. Way, and J. P. Delton, "Revisions to the Arizona Department of Transportation Pavement Management System," Presented at the Annual Transportation Research Board Meeting, Washington DC, 1993.

[WisDOT 90a]  "Pavement Management Decision Support Using a Geographic Information System," Wisconsin DOT, Madison, WI, May 1990.

[WisDOT 90b]  "A Network Data Base Model for GIS Application: Implementation at the Wisconsin Department of Transportation," Wisconsin DOT, Madison, WI, March 1990.

# Index

AASHTO Guide for the Design of Pavement Structures,
    data collection, 339
    layer strength coefficients, 488
    load equivalencies, 316, 317
    overlay design, 343
    performance models, 324–326, 332–334
    rehabilitation design, 337
    reinforcing steel design, 334
    reliability, 296
    variance, 261, 265, 295, 298, 299
Analysis period, typical, 350
APL profilometer, 86
Appian Way, 5
Automatic Road Analyzer, ARAN, 144
Axle loads,
    actual loads, 276
    axle types, 275
    effect of changing legal limits, 518
    equivalencies, 74, 518

Axle loads (*Cont.*):
    legal limit, 275
    variability, 295

Benefit maximization, 236
Benkleman beam, 113, 117
BISAR, 282, 283
Bureau of Public Roads roughometer, 89, 90, 92, 94, 102

California Bearing Ratio, 269, 294, 318, 319, 499, 515, 516, 517
Candidate projects, 35
CHEV5L, 282, 283
CHLOE profilometer, 82, 83, 96, 101
Construction,
    annual budget, U.S.A., 8
    component of pavement management, 33, 34
    control, 404–405
    costs, 407, 410

Construction (*Cont.*):
  data, 63, 168, 171
  documents, 396, 410
  environmental considerations, 403
  historical aspects, 4–6
  historical data needs, 63–66, 73
  influence of design, 399
  management, 397, 400, 401
  needed data, 75, 412
  planning and programming, 399
  planning tools, 22
  policies and specifications, 304, 305
  project analysis subsystem, 42, 43
  quality control, 30, 34, 73, 106, 270, 395, 397, 400, 406, 407, 408, 410, 412, 534, 535
  roughness measurement during construction, 87
  social considerations, 404
  specifications, 405–406
  stage, 461
  strategy, 11
  variability during construction, 294, 298
Cost and benefits, 265, 306, 346, 357
Cost-effectiveness, 206, 217, 342
Cost minimization, 236, 240
Creep compliance, 317–320

Data,
  analysis, 98, 111, 169, 394, 411, 414, 509, 538, 541, 542, 545, 547
  collection, 63, 70, 71, 73, 75, 86, 98, 111, 129, 134, 135, 168, 169, 295, 337, 339, 342
  management, 168, 169, 171, 532
  retrieval, 420
  structure, 71
Deflection,
  criteria, 110
  as decision criteria, 427
  factors affecting, 111
  maximum tolerable, 126
  measuring devices, 112–124
  need in network monitoring, 109
  sample rate, 342
  sampling, 111
  structural adequacy index, 127
  structural capacity index, 125
  uses in pavement management, 64
Deflectograph, 116
Delphi technique, 163
Design,
  alternatives, 11, 263, 311, 323, 369

Design (*Cont.*):
  components, 259
  constraints, 266, 267
  framework, 255, 258
  objectives, 255, 264, 265, 267
  strategy, 11, 262, 265, 266, 300, 304, 305, 306, 308, 310, 368, 369, 371, 372, 455, 492, 494, 505
Discount rate, 351, 352, 358, 452
Distress,
  applications of data, 150
  components of a distress survey, 132, 464
  composite index, 149
  definitions, 24, 26, 67
  development of a survey procedure, 132
  engineering evaluation, 47, 104
  equipment for distress,
    ARAN, 144
    GERPHO, 144
    PASCO, 144
    RST, 144
    surveys, 135
  evaluation of distress, 47
  image analysis, 534
  major distress modes, 263
  models for fatigue, 313
  models for permanent deformation, 318
  models for thermal cracks, 321
  pavement condition index, 474
  sample rate, 464
  survey methods, 131, 133
  types of, 132, 134, 223
  uses in pavement management, 64
Dynaflect, 117, 118, 127, 171
Dynatest equipment, 91, 121, 122, 172

Economic analysis and evaluation,
  basic principles, 345–346
  consideration for method selection, 353
  cost and benefit factors, 346
  cost-effectiveness analysis, 219
  design constraints, 266
  discount rate influence, 351
  evaluation of design alternatives, 263
  example of life-cycle costs, 241
  example problem, 360
  inflation, 351
  methods,
    benefit-cost ratio, 359
    cost-effectiveness, 359
    equivalent uniform annual cost, 353
    present worth, 354
    rate-of-return, 358

# Index

Economic analysis and evaluation (*Cont.*):
  models in,
    FPS, 452
    HDM, 495–497
    OPAC, 457
  network analysis, 40
  objective of, 352
  project analysis, 48, 65, 66
  project timing, 202
  purpose of, 351
Elastic modulus, 281, 292
ELSYM5, 282, 283, 330
Environment, 14, 17, 27, 31, 36, 37, 51, 61, 65, 70, 106, 134, 200, 261, 263, 272, 278, 279, 330, 337
Equivalent uniform annual cost method, 353, 479
Expert systems, 53, 207, 209, 521, 522, 523, 524, 525, 528, 529, 530, 533, 567

Falling Weight Deflectometer, FWD, 119, 120, 121, 122, 123, 124
Financing, 346, 388, 570
Finite element method, 281, 285, 317
Flexible pavement, 8, 10, 33, 36, 136, 163, 195, 256, 257, 266, 288, 302, 326, 328, 440, 448, 456, 458, 505, 528
Friction, 30, 39, 42, 44, 46, 64, 66, 67, 110, 154, 155, 157, 158, 188, 189, 281, 290, 291, 292, 334, 397

Geographic information systems, GIS, 71, 175, 430, 531, 532, 533
GERPHO, 144, 146

Highway Design and Maintenance standards model, HDM, 463, 464, 495, 496, 497, 500, 502, 503
Hveem, 79, 271

Implementation,
  activities, 394
  construction documents, 396
  current status, 7
  decisions, 388
  defined, 385
  institutional barriers, 245
  institutional issues, 537–547
  project level, 43
  requirements, 17
  research, 440
  staged, 391

Implementation (*Cont.*):
  steering committee, 388
  steps required, 386
  subsystem for, 40
  time requirements, 391
Indirect tensile test, 335
Inflation, 275, 351, 352, 362
Information needs, 13, 258, 261
Information systems, 71, 175, 531
Interest rate, 351, 354
International roughness index, IRI, 94, 98, 188

KUAB, 121, 123

Law profilometer, 86, 97, 98, 99, 106
LAYER, 281, 282, 283
Layer coefficients, 499

Maintenance, 416–442
  alternatives, 206, 305
  construction policies impact, 304–305, 400
  cost-effectiveness, 220
  costs, 74, 171
  decision criteria, 427
  definition, 417
  economic analysis factors, 346
  expert systems, 209
  funds, 422
  information needs, 427
  interaction with construction and design, 416
  level of influence, 32
  management, 43
  management systems, 417
  network level, 4, 28
  optimization, 230
  performance standard, 221
  policies, 421
  pre-rehabilitation strategies, 241
  priority programming, 224–238
  project level, 4, 42
  strategies, 206–223
  subsystem data, 43, 64, 66
  timing, 426
  trade-off with rehabilitation, 240
  work activities, 431
Materials characterization, 261
Mays ride meter, Maysmeter, 89, 90, 95, 96
Metropolitan Transportation Commission, MTC, 149, 162, 465, 467, 472, 473, 474, 479
Municipal Pavement Design System, MPDS, 504, 505, 507

Net present value, 458, 497
Net present worth, 352
Nondestructive tests and evaluation, 108, 110, 112, 114, 130, 339, 342

Ontario pavement management practices, 36, 53, 135, 143, 206, 207, 209, 215, 350
OPAC, 36, 327, 369

PASCO, 144, 145
Pavement evaluation,
  alternatives, 308
  combining measures, 68
  construction evaluation, 411
  construction influenced by, 400
  data uses, 46
  decision criteria, 427
  distress, 134–144
  expert systems, 521
  function, 3, 44
  influence on maintenance treatment selection, 302
  objectives, 161
  objectivity, 67
  performance related, 66
  structural, 113–124, 342
Pavement performance, 76–130
  compatibility, 105
  concept development, 11
  conceptual model, 295
  distress interaction, 263
  evaluation, 76
  Long Term Pavement Performance study, 37, 121, 144, 331, 562
  maintenance effect, 417, 433
  output function, 24
  related to roughness, 47
  research need, 263
  structural response interaction, 258
  uses, 64
Pavement quality index, 68, 162, 163, 165, 189, 470
PAVER, 71, 134, 149, 150, 152, 153, 162, 465
Permanent deformation, 132, 263, 271, 279, 283, 287, 314, 318, 319, 320
Phonix FWD, 121
Plate load test, 113
Poisson's ratio, 270, 281, 283, 290, 292
Power spectral density, 92, 94, 95, 97
Present serviceability index, PSI, 79, 96, 101, 102, 126, 187, 188, 189, 191, 192, 217, 225, 229, 276, 290, 360, 364, 365, 366, 405, 434, 452, 481

Present serviceability rating, PSR, 79, 80, 82, 99, 102, 104, 470, 471, 472
Present worth method, 352, 354, 355
Priority program, 28, 29, 32, 226, 234, 238, 393
Probabilistic concepts, 20
Profilometers, 82, 83, 84, 96, 97, 101, 107
Psychophysical scaling, 104

Ranking methods, 228, 236
Rate-of-return, 352, 353, 357, 358
Regional factor, 74, 176, 325, 452, 464, 489
Rehabilitation, 206–223
  alternatives, 206, 262, 470
  combined programs with maintenance, 240–247
  cost data, 74, 215
  decision process, 207
  design data, 338
  design procedures, 336–344
  economic analysis factors, 346
  expert systems, 207, 209, 528
  feasible alternatives, 204
  implementation, 43
  level of influence, 32
  needs years, 202
  non-overlay, 337
  overlays, 306
  performance models, 342
  policies, 424
  policy effects, 307
  priority programming, 224–239
  program level, 36
  project analysis, 42
  reconstruction alternative, 310
  selection decision tree, 211
  service life, 44
  timing, 426, 545
  types, 336
  work program, 39, 40
Reliability, 11, 21, 30, 111, 203, 294, 295, 296, 298, 299, 326, 333, 335, 346, 369, 371, 394, 494, 534, 538
Remaining life, 112, 126, 213, 343
Research management, 43, 54, 385, 435, 436, 441, 565
Research needs, 50, 51, 55, 435, 439, 440, 441, 491, 562, 573
Resilient modulus, 111, 285, 326, 499
Response type road roughness measuring system, RTRRMS, 82, 83, 84, 88, 89, 90, 91, 92, 93, 94, 96, 98, 99, 100, 101, 102, 103, 105, 106, 107

# Index

Ride quality, 39, 40, 76, 77, 79, 83, 92, 101, 103, 105, 107, 135, 155, 209, 263, 336, 349
Riding comfort index, 79, 80, 82, 102, 103, 126, 163, 188, 197, 211, 231, 349, 456
Rigid pavement, 8, 11, 33, 36, 138, 284, 302, 329, 331, 332, 333, 334, 343, 458, 460, 461, 505, 517
Rigid Pavement System, RPS, 36, 308, 331, 333, 335, 336, 458, 461, 463, 466, 504, 505
Road Rater, 117, 118–119
Road Surface Tester, RST, 144, 148
Roughness, 78, 79, 81, 82, 91, 92, 96, 98, 103, 105, 107, 176, 305, 362, 397, 401, 404, 405, 411, 412, 425, 427, 456, 464, 499, 502, 520, 551, 572
    compatibility of measurements, 104, 105
Roughness index, 81, 94, 98, 102, 188
Roughness measurement, 106
Rutting, 67, 101, 125, 134, 149, 150, 213, 263, 283, 314, 318, 341, 342, 427, 464, 477

Safety,
    components, 154
    data use, 65
    evaluation, 154–160
    measure, 67
    policies, 422
    requirement for pavement, 264
    response variable, 46
    skid resistance, 155
Salvage value, 352
Serviceability,
    compatible measures, 104–105
    concept evolution, 101
    construction quality, 405
    functional quality, 67
    fundamental assumptions, 78
    index, 92
    maintenance effect, 306, 419, 426
    model, 322
    performance measure, 27, 47, 76
    regional considerations, 105
    roughness, 47, 79, 82, 101, 103, 104
    scale, 78
    service life, 324
    traffic speed effect, 349
    trigger values, 110

Serviceability (*Cont.*):
    user-related evaluation, 47
    vehicle operating costs affected, 216, 551
Skid number, 157, 158, 198
Skid resistance, 24, 39, 46, 67, 110, 132, 154, 155, 156, 157, 158, 159, 160, 188, 189, 261, 263, 264, 339, 426
Slope variance, 83, 91, 96, 101, 102
Soil support, 452, 481, 487, 489
Stiffness modulus, 290, 291
Strategic Highway Research Program, SHRP, 6, 37, 53, 54, 121, 125, 134, 144, 258, 332, 435, 560, 562, 571
Structural,
    analysis, 44, 255, 256, 258, 261, 280, 386, 480, 534
    capacity, 39, 108, 111, 112, 125, 126, 129, 162, 163, 261, 262, 296, 310, 342, 343
    number, 325, 343, 452, 487, 488, 489, 499
    response, 258, 269, 280, 293
Surface dynamics profilometer, 86, 107
Surface wave evaluation, 81, 534
Systematic Analysis Method for Pavements, SAMP, 36, 327, 336, 448, 463, 466, 480–494, 504
Systems method, 16, 17, 18, 19, 20, 22, 337, 518

Traffic analysis, 463
Traffic handling, 65, 217, 262, 308, 452, 471
Traffic variables, 481
Traveling deflectometer, 116
TRRL profilometer, 84
Types of pavements, 8, 69, 328, 517

User costs, 34, 36, 44, 48, 66, 67, 70, 74, 217, 262, 263, 264, 266, 307, 321, 346, 347, 348, 349, 354, 355, 357, 360, 362, 371, 425, 448, 451, 457, 458, 471, 481, 483, 496, 518, 546, 551
User savings, 356

Vehicle operating costs, 30, 189, 191, 215, 216, 225, 235, 263, 264, 347, 350, 354, 356, 362, 365, 366, 495, 496, 549
VESYS, 283, 319

Wheel loads, 27, 113, 275, 282